Organosulfur and Organoselenium Chemistry

Organosulfur and Organoselenium Chemistry

Guest Editor

Ming Wang

MDPI

Basel • Beijing • Wuhan • Barcelona • Belgrade • Novi Sad • Cluj • Manchester

Guest Editor
Ming Wang
School of Chemistry and
Molecular Engineering
East China Normal University
Shanghai
China

Editorial Office
MDPI AG
Grosspeteranlage 5
4052 Basel, Switzerland

This is a reprint of the Special Issue, published open access by the journal *Molecules* (ISSN 1420-3049), freely accessible at: www.mdpi.com/journal/molecules/special_issues/T6GKHGO88U.

For citation purposes, cite each article independently as indicated on the article page online and using the guide below:

Lastname, A.A.; Lastname, B.B. Article Title. *Journal Name* **Year**, *Volume Number*, Page Range.

ISBN 978-3-7258-2762-6 (Hbk)
ISBN 978-3-7258-2761-9 (PDF)
https://doi.org/10.3390/books978-3-7258-2761-9

Contents

About the Editor

Ming Wang

Ming Wang, a professor at East China Normal University, received his Ph.D. degree in 2011 from the East China University of Science and Technology. From 2011 to 2014, he was a postdoctoral researcher under the guidance of Professor Yonggui Robin Chi at Nanyang Technological University, Singapore. In 2014, he joined East China Normal University as a lecturer and is currently a professor at East China Normal University. His research interest has focused on sulfur chemistry.

Preface

Organosulfur- and organoselenium-containing compounds play a crucial role in organic synthesis. These sulfur-containing molecules are widely applied in pharmaceuticals, materials, natural products, and food. Additionally, sulfur and selenium show divergent functions and potencies in different oxidative states, leading to the rich chemistry of their transformations. This *Molecules* Special Issue introduces organosulfur and organoselenium chemistry in fields such as organic synthesis, materials, chemical biology, and computational chemistry, and contains eleven articles and four reviews. We collected excellent contributions from scientists across these different research areas, demonstrating the global reach of organosulfur and organoselenium chemistry; the authors contributing to this Special Issue hail from China, Canada, France, Germany, America, Poland, Japan, the Czech Republic, and Croatia.

This Special Issue will be useful to scientists, researchers, and graduate students in the fields of organic chemistry, materials chemistry, chemical biology, and computational chemistry. We thank our many prominent colleagues for their contributions.

Ming Wang
Guest Editor

molecules

MDPI

Article

Modulation of Properties in [1]Benzothieno[3,2-*b*][1]benzothiophene Derivatives through Sulfur Oxidation

Aneta Rzewnicka [1], Rafał Dolot [2], Maciej Mikina [1], Jerzy Krysiak [1] and Remigiusz Żurawiński [1,*]

1 Division of Organic Chemistry, Centre of Molecular and Macromolecular Studies, Polish Academy of Sciences, Sienkiewicza 112, 90-363 Lodz, Poland; aneta.rzewnicka@cbmm.lodz.pl (A.R.); maciej.mikina@cbmm.lodz.pl (M.M.); jerzy.krysiak@cbmm.lodz.pl (J.K.)

2 Division of Bioorganic Chemistry, Centre of Molecular and Macromolecular Studies, Polish Academy of Sciences, Sienkiewicza 112, 90-363 Lodz, Poland; rafal.dolot@cbmm.lodz.pl

* Correspondence: remigiusz.zurawinski@cbmm.lodz.pl

Abstract: This study explores the impact of sulfur oxidation on the structural, optical, and electronic properties of [1]benzothieno[3,2-*b*][1]benzothiophene (**BTBT**) derivatives, specifically focusing on 2,7-dibromo BTBT (**2,7-diBr-BTBT**) and its oxidized forms, 5,5-dioxide (**2,7-diBr-BTBTDO**) and 5,5,10,10–tetraoxide (**2,7-diBr-BTBTTO**). The bromination of **BTBT** followed by sequential oxidation with *m*-chloroperoxybenzoic acid yielded the target compounds in good yields. They were characterized using a wide array of analytical techniques including different spectroscopic methods, X-ray analysis, thermal analysis, and quantum chemical calculations. The results revealed that sulfur oxidation significantly alters the crystal packing, thermal stability, and optoelectronic properties of **BTBT** derivatives. Notably, the oxidized forms exhibited increased thermal stability and enhanced emission properties, with quantum yields exceeding 99%. These findings provide valuable insights for designing advanced organic semiconductors and fluorescent materials with tunable properties, based on the **BTBT** core.

Keywords: benzothienobenzothiophene; sulfones; organic electronics; DFT calculations; reorganization energy; fluorescence; semiconductors

check for updates

Citation: Rzewnicka, A.; Dolot, R.; Mikina, M.; Krysiak, J.; Żurawiński, R. Modulation of Properties in [1]Benzothieno[3,2-*b*][1]benzothiophene Derivatives through Sulfur Oxidation. *Molecules* **2024**, *29*, 3575. https://doi.org/10.3390/molecules29153575

Academic Editor: Ming Wang

Received: 18 July 2024
Revised: 26 July 2024
Accepted: 27 July 2024
Published: 29 July 2024

1. Introduction

Organic semiconductors (OSCs) have been continuously attracting attention due to their distinctive properties, such as mechanical flexibility and chemical versatility [1]. Their π-conjugated systems can be readily adjusted to optimize molecular arrangement with specific optical and electronic properties [2]. Among these semiconductors, [1]benzothieno[3,2-*b*][1]benzothiophene (**BTBT**) derivatives have been extensively studied and applied in organic electronics due to their favorable characteristics, including high charge carrier mobility, tunable electronic properties, thermal stability, and solution-processability [3–5]. Due to these features, **BTBT** derivatives are highly promising materials for different applications, including organic light-emitting diodes (OLEDs) [6–10], organic field-effect transistors (OFET) [11–27], and organic sensors [28–32]. The structural modifications of the **BTBT** core have a profound influence on its electronic properties, intermolecular interactions, and crystal packing, which play a crucial role in the performance of electronic devices [33]. Considering that the presence of thiophene rings in the **BTBT** structure allows for the modulation of its optoelectronic properties through the oxidation of sulfur atoms, this opens up opportunities to develop new materials for organic electronics. Previous studies have demonstrated that the chemical oxidation of the thienyl sulfur atom in thiophene-based poly- and oligomers is a useful strategy to modulate the electronic structure and has an influence on the fluorescence and redox properties [34–36]. It was found that oligothiophene *S,S*-dioxides are recognized as outstanding candidates for use in light-emitting diodes [37–40]. Moreover, the oxidation of the sulfur atom in thiophene transforms its electron-donating character into the electron-accepting sulfonyl group. This transformation indicates that oligothiophene sulfones can be interesting from the point of view of their application as bipolar or n-type semiconductors [41,42]. Compared to **BTBT** derivatives, research concerning their oxidized counterparts is very limited, as evidenced by the negligible number of publications in this field. There have been only three publications dedicated to this class of compounds. Two of these focused on the synthesis of **BTBT** 5,5,10,10-tetraoxide (**BTBTTO**) and its reaction with amines [43,44], while one study highlighted the use of diphenylamino derivatives of **BTBT** dioxide and tetraoxide for the fluorescent imaging of lipid droplets [45].

In this paper, we present comparative studies of unoxidized and oxidized **BTBT** derivatives, focusing on the influence of sulfur oxidation on structural, physical, optical, and electronic properties. The investigations were conducted on three compounds: 2,7-dibromo[1]benzothieno[3,*b*][1]benzothiophene (**2,7-diBr-BTBT**), its 5,5-dioxide (**2,7-diBr-BTBTDO**), and its 5,5,10,10-tetraoxide (**2,7-diBr-BTBTTO**). The presence of bromine atoms in **2,7-diBr-BTBT** *S*-oxides provides an opportunity for easy further tuning of their desired properties through straightforward functionalization via Heck or Suzuki cross-coupling reactions, paving the way to new compounds with potential optoelectronic applications. To characterize the synthesized compounds, various techniques and methods were employed. Structural verification and analysis of oxidation's impact on crystal packing were conducted using single-crystal X-ray analysis. Optical properties were investigated through emission and absorption spectroscopy. Thermal stability was assessed using thermogravimetric analysis (TGA) and differential scanning calorimetry (DSC). Additionally, electronic properties, such as frontier orbital energies, ionization potential, electron affinity, and internal reorganization energy, were predicted using DFT and TDDFT methods.

2. Results and Discussion

2.1. Synthesis of 2,7-diBr-BTBT S-oxides

The synthesis of **2,7-diBr-BTBT** *S*-oxides, as depicted in Scheme 1, commenced with **BTBT** as the starting material, which was obtained using a one-pot procedure from commercially available *o*-chlorobenzaldehyde [46]. Next, **BTBT** was subjected to a bromination reaction to produce **2,7-diBr-BTBT [47]**. The resulting dibromo **BTBT** derivative was oxidized with *m*-chloroperoxybenzoic acid (*m*-CPBA) at room temperature, affording the dioxide **2,7-diBr-BTBTDO** in 76% yield. This compound was subsequently further oxidized to the corresponding tetraoxide **2,7-diBr-BTBTTO**. To avoid issues with separating the product from the unreacted substrate (both compounds have low solubility and similar retardation factors), the reaction was driven to 100% conversion. To ensure this, the reaction was conducted at an elevated temperature with a large excess of *m*-CPBA. The reaction proceeded cleanly, yielding **2,7-diBr-BTBTTO** as the sole product in 83% yield.

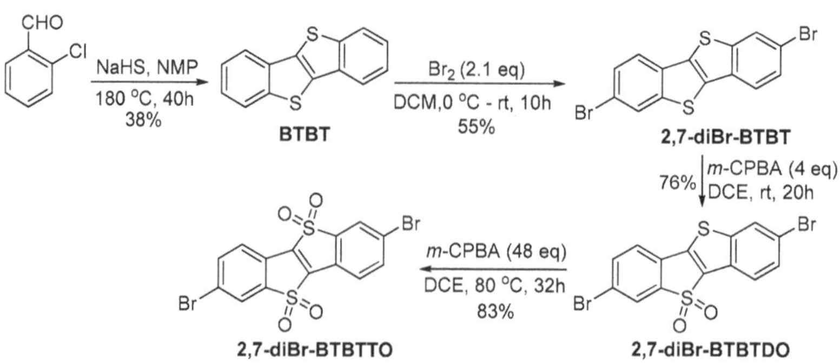

Scheme 1. Synthesis of 2,7-diBr-BTBT *S*-oxides.

2.2. Structural Analysis

X-ray diffraction analysis was performed to confirm the structures of the synthesized **2,7-diBr-BTBT** *S*-oxides. Single crystals of **2,7-diBr-BTBTDO** and **2,7-diBr-BTBTTO** for SC-XRD analysis (CCDC: 2368892 and 2368896) were obtained by heating the dissolved compounds in the corresponding solvents followed by slow cooling of the samples. All crystallographic data are listed in Tables S1–S3.5 in the Supplementary Materials. The analyzed molecules crystallized in different space groups, indicating that the degree of sulfur oxidation has a significant effect on their crystal packing (Figure 1). Furthermore, the crystal packing of the **2,7-diBr-BTBT** *S*-oxides differed from the well-known herringbone arrangement observed for the unoxidized parent compound [47]. **2,7-diBr-BTBT** forms a crystal structure where molecules are π-stacked, maintaining an interplanar spacing typical for flat halogenated OSCs of approximately ~3.46 Å [48]. **2,7-diBr-BTBTDO** and **2,7-diBr-BTBTTO** also crystallize in an π-stacked molecular arrangement, but the oxidation of one or two sulfur atoms has a significant effect on the arrangement of the crystals obtained. **2,7-diBr-BTBTDO** forms π-stacked molecular columns with an alternating arrangement of the oxidized sulfur atoms and an interplanar distance of ~3.55 Å between adjacent molecules. In contrast, **2,7-diBr-BTBTTO** forms oppositely oriented planes with a distance of ~3.46 Å, and each molecule is π-stacked with four neighboring molecules.

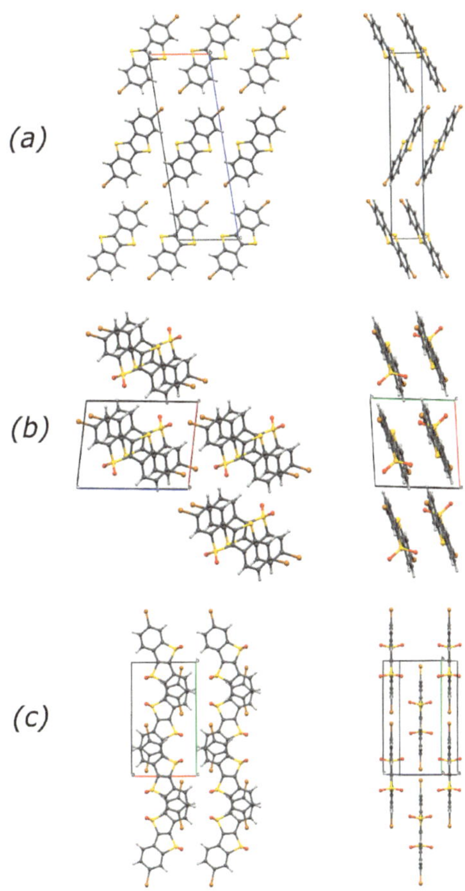

Figure 1. Perspective views of molecular packing in single crystals of (**a**) **2,7-diBr-BTBT**, (**b**) **2,7-diBr-BTBTDO**, and (**c**) **2,7-diBr-BTBTTO**.

The oxidation of the sulfur atoms in **2,7-diBr-BTBT** not only changes the molecular packing but also influences the number of close contacts between the atoms of neighboring molecules. For **2,7-diBr-BTBT**, ten contacts were observed, including two S···S and four Br···Br contacts per molecule (Table S4.1 in the Supplementary Materials). Oxidation prevents the formation of contacts by sulfur atoms, while oxygen atoms form close contacts, including O···H (in both *S*-oxides), O···Br, O···C(sp2), and O···S (only in **2,7-diBr-BTBTDO**). Br···Br contacts are also present in **2,7-diBr-BTBTDO**, whereas they are absent in **2,7-diBr-BTBTTO**. The number of contacts varies depending on the oxidation state of the molecules (Tables S4.2 and S4.3 in the Supplementary Materials). Fifteen contacts were listed for **2,7-diBr-BTBTDO**, while their number increased to twenty–four per molecule for **2,7-diBr-BTBTTO**.

2.3. *Physical and Optical Properties*

The obtained compounds were white (**2,7-diBr-BTBT**), yellowish (**2,7-diBr-BTBTDO**), and yellow (**2,7-diBr-BTBTTO**) solids. To evaluate the thermal stability of **2,7-diBr-BTBT** and its oxidized derivatives, thermogravimetric analysis (TGA) and differential scanning calorimetry (DSC) were carried out. The results show that the thermal stability increases with the number of oxygen atoms in the molecule (Table 1 and Figure S1 in the Supplementary Materials). The fully oxidized **2,7-diBr-BTBTTO** undergoes decomposition approximately 18 °C and 38 °C above the decomposition temperatures of its partially oxidized counterpart **2,7-diBr-BTBTDO** and its unoxidized parent compound **2,7-diBr-BTBT**, respectively. In addition, the compounds do not exhibit liquid crystal properties. DSC measurements showed that they do not undergo any phase transitions up to their melting point or decomposition temperature (Figure S2 in the Supplementary Materials).

Table 1. Physical and optical properties of **2,7-diBr-BTBT** and its *S*-oxides in DCM and solid state.

Compound	DCM				Solid	
	λ_{abs} [nm]	λ_{em} [nm]	Stokes Shift [nm] (cm^{-1})	ϕ	λ_{em} [nm]	T_d [°C] [1]
2,7-diBr-BTBT	337	362	25 (2049)	<1	399	256.4
2,7-diBr-BTBTDO	356	435	79 (5101)	>99	459	296.2
2,7-diBr-BTBTTO	396	486	90 (4676)	>99	495	314.5

[1] Decomposition temperature determined by TGA corresponding to 5% weight loss at 10 °C min^{-1} under nitrogen flow.

We performed UV–Vis and photoluminescence (PL) measurements on the synthesized compounds in both solution (dichloromethane (DCM)) and the solid state. The findings, detailed in Table 1 and Figure 2, demonstrate that the oxidation of sulfur atoms within the thiophene rings has a substantial impact on their optical properties. The increase in the number of oxygen atoms in the molecule results in noticeable shifts in both the absorption (λ_{abs}) and emission maxima (λ_{em}) to the longer wavelengths. Furthermore, the Stokes shift for oxidized derivatives in solution is significantly higher, ranging from 79 to 90 nm, whereas **2,7-diBr-BTBT** exhibits a much smaller shift of about 25 nm. The quantum yield (ϕ) of **2,7-diBr-BTBT** is less than 1%, which is lower than the 3.4% quantum yield observed for **BTBT** [49]. This result is in line with the expected heavy atom effect [50,51]. In contrast, for **2,7-diBr-BTBTDO** and **2,7-diBr-BTBTTO**, the quantum yields are significantly higher, surpassing 99% for both compounds. The emission maxima of the investigated compounds in the solid state are in the range of 399–495 nm and exhibit a smaller red shift of 9 to 37 nm in comparison to their emission maxima in solution.

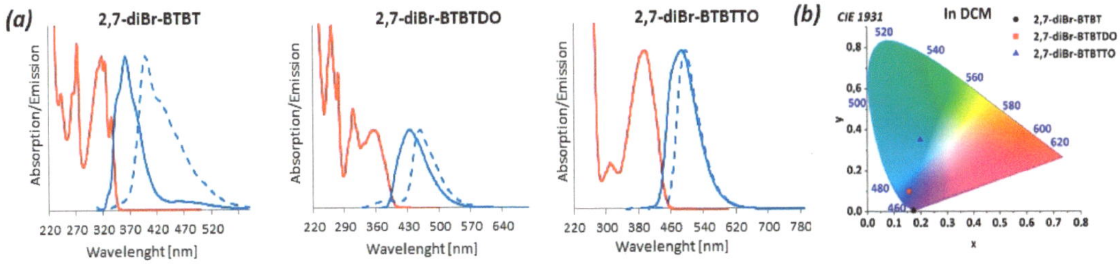

Figure 2. Optical properties of **2,7-diBr-BTBT** and its *S*-oxides; (**a**) Normalized UV–Vis spectra (red solid line—in DCM) and PL spectra (blue solid line—in DCM, blue dashed line—in solid); (**b**) Chromaticity chart.

2.4. Computational Analysis

To provide insight into the electronic changes accompanying the oxidation of sulfur atoms in **2,7-diBr-BTBT**, TDDFT calculations were performed using the PBE0/6–311+G(2d,p)//D3-M06-2X/def2-TZVP method. The results of these calculations are summarized in Table 2 and graphically presented in Figure 3. To account for solvent–solute interactions in the calculation of the electronic properties, the IEFPCM model [52–54] for DCM was applied. The calculated absorption maxima perfectly match the experimental data, with deviations ranging from 0.06 to 0.14 eV. The largest deviation was observed for **2,7-diBr-BTBTTO**, where the absorption band was red-shifted by approximately 18 nm compared to the experimental value. In all the compounds examined, the lowest energy bands were predominantly associated with the HOMO⟶LUMO transitions (>97%). During the oxidation process, the HOMO energy gradually decreased from −6.230 eV for **2,7-diBr-BTBT** to −7.048 eV for **2,7-diBr-BTBTTO**. The same trend, but to a greater extent, was observed for the LUMO energy. In this case, the energy difference between the LUMO of **2,7-diBr-BTBT** and **2,7-diBr-BTBTTO** reached 1.48 eV. As the energy of the frontier orbitals decreased, the energy band gap also reduced, ultimately reaching around 3.83 eV for **2,7-diBr-BTBTTO**. These results are in strong agreement with the findings reported for thiophene dioxides [36,55] and support our observation that the emission maxima of the investigated compounds exhibit an increasing red shift with a greater number of oxygen atoms. Hole–electron analysis [56] for individual atoms in **2,7-diBr-BTBT** revealed that during excitation to the first excited state, two sulfur atoms and two carbon atoms connecting them provide a considerable input to the hole. The sum of their contribution is almost 50%. During electron excitation, these atoms lose 0.25 electrons, which are transferred to the remaining carbon atoms. The oxidation of thiophene sulfur alters the electron-donating thienyl sulfur atom into a strong electron-accepting group, which is reflected in the electron distribution during excitation. Consequently, in **2,7-diBr-BTBTTO**, the two sulfonyl groups and their connecting carbon atoms gain 0.31 electrons, primarily from the bromine atoms. In the case of asymmetrical **2,7-diBr-BTBTDO**, electron excitation increases electron density at the sulfonyl group and the phenyl ring fused with the thiophene oxidized moiety.

Table 2. Electronic properties of **2,7-diBr-BTBT** and its *S*-oxides calculated using the TDDFT PBE0/6–311+G(2d,p)//D3-M06-2X/def2-TZVP method in DCM.

Compound	E_{HOMO} [eV]	E_{LUMO} [eV]	Energy Gap [eV]	Excited State	E_{ex} [1] [eV]	λ_{ex} [2] [nm]	Oscillator Strength	Major Transitions [%]
2,7-diBr-BTBT	−6.230	−1.734	4.496	S_1	3.762	329.6	0.765	H⟶L (97.4)
2,7-diBr-BTBTDO	−6.672	−2.447	4.225	S_1	3.422	362.3	0.605	H⟶L (97.8)
2,7-diBr-BTBTTO	−7.048	−3.218	3.830	S_1	2.990	414.6	0.560	H⟶L (98.7)

[1] E_{ex}—excitation energy; [2] λ_{ex}—excitation wavelength.

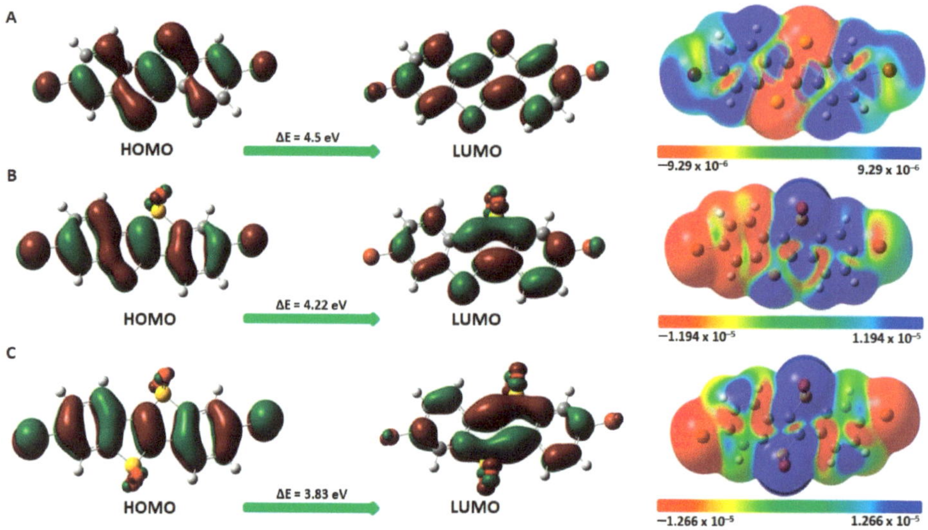

Figure 3. Graphical representation of frontier MOs (isovalue = 0.026) and changes in electron densities during excitation from the ground state to the S_1 excited state ((**A**)—**2,7-diBr-BTBT**; (**B**)—**2,7-diBr-BTBTDO**; (**C**)—**2,7-diBr-BTBTTO**).

To estimate the ambient stability of investigated compounds and their charge transport properties, the adiabatic ionization potential (IP), electron affinity (EA), and internal reorganization energy (λ) were calculated (Table 3). IP and EA are crucial for understanding the ease with which molecules can gain or lose electrons, directly influencing their electronic properties and reactivity. Furthermore, these parameters also allow us to estimate the energy barriers involved in the injection of holes and electrons into molecules, which is particularly valuable in the design and development of materials for electronic and optoelectronic applications. On the other hand, the hole (λ_h) and electron (λ_e) reorganization energies, which are associated with the process of transferring an electron or hole between two adjacent molecules, are the key parameters in the study of charge transport in organic semiconductors. Calculations at the D3-M06-2X/def2-TZVPD level revealed that the IP of the investigated compounds increased with oxidation, from 7.69 eV for unoxidized **2,7-diBr-BTBT** to 8.73 eV for **2,7-diBr-BTBTTO**. These higher IPs for the oxidized **2,7-diBr-BTBT** counterparts indicate greater oxidation stability, making these compounds less susceptible to oxidation-related degradation. The same trend was observed for the electron affinity energy, which gradually increased from 0.72 eV for **2,7-diBr-BTBT** to 2.42 eV for **2,7-diBr-BTBTTO**.

Table 3. Ionization potentials, electron affinities, and reorganization energies of **2,7-diBr-BTBT** and its *S*-oxides.

Compound	IP [eV]	EA [eV]	λ_h [eV]	λ_e [eV]	λ [eV]
2,7-diBr-BTBT	7.69	0.70	0.302	0.328	0.630
2,7-diBr-BTBTDO	8.17	1.41	0.390	0.319	0.719
2,7-diBr-BTBTTO	8.73	2.42	0.328	0.591	0.919

Thiophene sulfur oxidation also alters the internal reorganization energy (the sum of λ_h and λ_e), which increases in the order: **2,7-diBr-BTBT** < **2,7-diBr-BTBTDO** < **2,7-diBr-BTBTTO**. While the values of λ_h and λ_e are similar for **2,7-diBr-BTBT**, they differ significantly in the oxidized derivatives. For **2,7-diBr-BTBTDO**, the difference between λ_h and λ_e is about 0.07 eV, suggesting a greater propensity to conduct electrons rather than holes. The opposite situation is observed for **2,7-diBr-BTBTTO**, for which this difference is approximately −0.26 eV, indicating a preference for hole transport.

3. Materials and Methods

3.1. Synthesis of Materials

3.1.1. General Remarks

Commercial-grade reagents and solvents were used as received without further purification. The ^{1}H, ^{13}C, and 2D NMR spectra were recorded using Bruker AV Neo 400 or Bruker Avance III 500 spectrometers (Bruker, Billerica, MA, USA). Chemical shifts for ^{1}H and ^{13}C NMR spectra are reported in parts per million (ppm) relative to the residual proton signal of the deuterated solvent. High-resolution mass spectrometry (HRMS) analyses were conducted on a Waters Synapt HDMS mass spectrometer (Waters Corporation, Milford, MA, USA). The synthesis of **BTBT [46]** and **2,7-diBr-BTBT [47]** was carried out according to established literature procedures. The NMR spectra of all new compounds are included in the Supplementary Materials.

3.1.2. Synthetic Procedures

2,7-Dibromo[1]benzothieno[3,2-*b*][1]benzothiophene 5,5-dioxide (2,7-diBr-BTBTDO): A mixture of **2,7-diBr-BTBT** (100 mg, 0.25 mmol) and *m*-CPBA (248 mg, 1 mmol, 4 eq) in DCE (200 mL) was stirred at room temperature for 20 h. The reaction mixture was then washed successively with saturated aqueous solutions of Na_2SO_3 and $NaHCO_3$, followed by water. The organic layer was dried over anhydrous $MgSO_4$. Solvent evaporation, followed by the purification of the crude material by column chromatography using a dichloromethane/petroleum ether (2:1) mixture, afforded **2,7-diBr-BTBTDO** (81 mg, 76%) as a pale yellow solid. Crystals suitable for X-ray analysis were obtained by crystallization from DCE. R_f = 0.42 (DCM/petroleum ether, 2:1); ^{1}H NMR (500 MHz, $CDCl_3$) δ 8.07 (d, *J* = 1.7 Hz, 1H), 7.94–7.88 (m, 2H), 7.76 (dd, *J* = 8.1, 1.8 Hz, 1H), 7.68 (dd, *J* = 8.6, 1.7 Hz, 1H), 7.39 (d, *J* = 8.1 Hz, 1H); ^{13}C NMR (126 MHz, $CDCl_3$) δ 144.37, 144.16, 143.04, 136.82, 134.11, 130.80, 129.31, 127.09, 126.64, 125.75, 124.83, 123.51, 121.03. HRMS (APCI−) *m/z*: calcd. for $C_{14}H_6O_2S_2Br_2$ 427.8176, found 427.8181.

2,7-Dibromo[1]benzothieno[3,2-*b*][1]benzothiophene 5,5,10,10-tetraoxide (2,7-diBr-BTBTTO): To a solution of **2,7-diBr-BTBTDO** (54.8 mg, 0.128 mmol) in DCE (24 mL) stirred at 80 °C, three portions of *m*-CPBA (379 mg, 1.53 mmol, 12 eq) were added at 8 h intervals. After 32 h, the reaction mixture was cooled to room temperature. The precipitate formed during the reaction was filtered off, washed with DCM and diethyl ether, and dried, yielding **2,7-diBr-BTBTTO** (49 mg, 83%) as a yellow solid. Crystals suitable for X-ray analysis were obtained by crystallization from hot DMSO. R_f = 0.40 (DCM/petroleum ether, 2:1) ^{1}H NMR (500 MHz, CDC13) δ 7.96 (d, *J* = 1.7 Hz, 1H), 7.85 (dd, *J* = 8.1, 1.8 Hz, 1H), 7.57 (d, *J* = 8.1 Hz, 1H). Due to the very low solubility of **2,7-diBr-BTBTTO** in commonly used deuterated solvents, recording the ^{13}C NMR spectra was not feasible. HRMS (APCI−) *m/z*: calcd. for $C_{14}H_6O_4S_2Br_2$ 459.8074, found 459.8079.

3.2. X-ray Analysis

Suitable crystals of **2,7-diBr-BTBTDO**, and **2,7-diBr-BTBTTO** were selected, transferred to mineral oil, and mounted on cryo loops. The crystals were then flash-cooled directly in a stream of N_2. Diffraction intensities were recorded using a Rigaku XtaLAB Synergy-S diffractometer (Rigaku Europe SE, Neu-Isenburg, Germany) equipped with a Cu Kα radiation source (λ = 1.5418 Å) and a HyPix-6000HE hybrid photon counting detector (Rigaku Europe SE, Neu-Isenburg, Germany). The total number of runs and images was based on the strategy calculation of the CrysAlisPro program (Rigaku Oxford Diffraction, v 1.171.43.111a, 2024). The molecular models of the structures were created with the structure solution program SHELXT 2018/2 [57] using intrinsic phasing with Olex2 v.1.5 [58] as the graphical interface and refined by least squares with the 2018/3 version of SHELXL [59]. All non-hydrogen atoms were refined anisotropically. The positions of the hydrogen atoms were calculated geometrically and refined using the riding model. The structures were validated with CheckCif (http://checkcif.iucr.org, accessed on 9 July 2024) and deposited in the Cambridge Crystallographic Data Centre (CCDC) under the accession numbers 2368892 and 2368896 for **2,7-diBr-BTBTDO** and **2,7-diBr-BTBTTO**, respectively.

3.3. Optical Characterization

Electronic absorption spectra were recorded on a Shimadzu UV–VIS spectrophotometer UV 2700 (Shimadzu Scientific, Kyoto, Japan). Photoluminescence and quantum yield measurements were performed using a Horiba Scientific FluoroMax+ spectrofluorometer (Horriba Scientific, Kyoto, Japan). All measurements were conducted at room temperature in a 1 cm cuvette. Spectrophotometric grade dichloromethane (DCM) was used for measurements in solution. Photoluminescence spectra of **2,7-diBr-BTBTDO** and **2,7-diBr-BTBTTO** were recorded after excitation at their respective maximum absorption wavelengths, and a 300 nm excitation wavelength was used for **2,7-diBr-BTBT**. Photoluminescence quantum yields (ϕ) were determined using an integrating sphere.

3.4. Thermal Properties

Thermogravimetric analysis (TGA) was conducted using a TGA 2950 analyzer (TA Instruments, Eden Prairie, MN, USA). The experiments were performed under a nitrogen atmosphere in the temperature range of 50 °C to 600 °C at a heating rate of 10 °C min^{-1}. The decomposition temperature was determined at 5% weight loss. The thermal properties of the synthesized compounds were studied using differential scanning calorimetry (DSC) on a DSC 2920 (TA Instruments), heating from 0 °C to 220 °C at a rate of 5 °C min^{-1}.

3.5. Computational Details

All quantum-mechanical calculations were performed using Gaussian 09, revision D.01 [60]. For simulating optical properties, the geometries of the investigated compounds were optimized using the M06-2X [61] functional with the D3 version [62] of dispersion correction and the Karlsruhe valence triple-zeta def2-TZVP basis set [63]. Calculations were conducted in a dichloromethane (DCM) dielectric medium, applying tight convergence criteria for geometry and energy. The "ultrafine" integration grid was used for all calculations. The character of the equilibrium point on the potential energy surface was identified through frequency calculations. TDDFT calculations were performed for the first six singlet excited states using the PBE0 functional [64] in conjunction with the Polple 6–311+G(2d,p) basis set [65]. Calculations of the internal reorganization energy, adiabatic ionization potential (IP), and adiabatic electron affinity (EA) were conducted at the D3-M06-2X/def2-TZVPD level in the gas phase. Adiabatic IPs were calculated as the energy difference between the relaxed cationic state ($E^+(M^+)$) and the geometry-optimized neutral molecule ($E^0(M)$) ($IP = E^+(M^+) - E^0(M)$). EAs were determined by the energy difference between the optimized neutral molecule ($E^0(M)$) and the anionic state ($E^-(M^-)$)($EA = E^0(M) - E^-(M^-)$). The hole (λ_h) and electron (λ_h) reorganization energies were calculated using the following equations:

$$\lambda_h = \lambda_1 + \lambda_2 = [E^+(M) - E^+(M^+)] + [E^0(M^+) - E^0(M)] \quad (1)$$

$$\lambda_e = \lambda_3 + \lambda_4 = [E^0(M^-) - E^0(M)] + [E^-(M) - E^-(M^-)] \quad (2)$$

where $E^+(M)/E^-(M)$ is the energy of the cation/anion calculated at the optimized geometry of the neutral molecule, and $E^0(M^+)/E^0(M^-)$ is the energy of the neutral molecule calculated at the optimized geometry of the corresponding cation/anion. A graphical plot for the calculation of reorganization energy is presented in Figure 4.

Hole–electron analysis using Hirshfeld partitions was conducted with the Multiwfn 3.8 program [66]. Basis sets def2-TZVP [63] and def2-TZVPD [67] were downloaded from the Basis Set Exchange (BSE) [68]. The optimized geometrical parameters of the investigated compounds are provided in the Supplementary Materials.

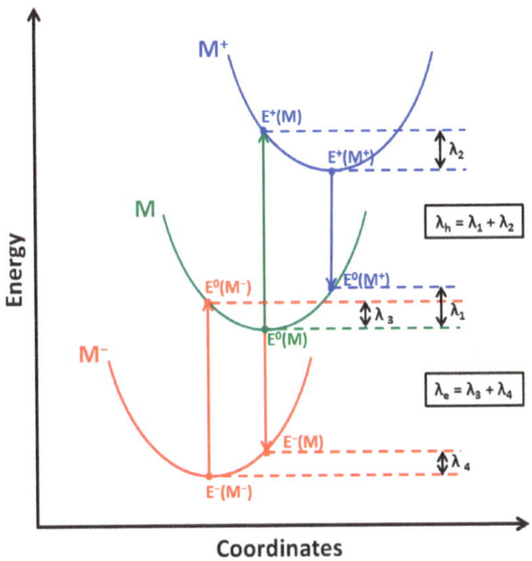

Figure 4. Diagram illustrating the calculation of the hole (λ_h) and electron (λ_e) reorganization energies.

4. Conclusions

In summary, the sulfone derivative **2,7-diBr-BTBTDO** and the disulfone derivative **2,7-diBr-BTBTTO** have been successfully synthesized via the oxidation of **2,7-diBr-BTBT** with *m*-CPBA and fully characterized using various analytical and theoretical methods. Comparative analysis of **2,7-diBr-BTBT** and its oxidized derivatives revealed the potential of sulfur oxidation as a strategy to tune the properties of **BTBT** derivatives. The oxidized compounds are characterized by red-shifted absorption and emission spectra with a Stokes shift reaching 90 nm and an impressive quantum yield above 99%. The oxidized compounds also show a decrease in both HOMO and LUMO energies, leading to a reduced energy band gap. This increased stability is further supported by their improved thermal properties, which show an increase in decomposition temperature with an increasing number of oxygen atoms. Changes in the electronic properties of the individual molecules also affect their tendency to form unique crystal patterns. Although all investigated compounds form π-stacked molecular arrangements, they differ significantly, as reflected by their different space group symmetry. All the above findings are instrumental in designing new materials based on the **BTBT** skeleton with optimized properties for organic electronic applications and fluorescence-based techniques.

Supplementary Materials: The following supporting information can be downloaded at https://www.mdpi.com/article/10.3390/molecules29153575/s1. NMR spectra of all new compounds (Section S1. NMR Spectra); crystallographic data for all investigated compounds (Section S2. X-ray analysis including Table S1. Crystal structure, data collection and refinement parameters of the **2,7-diBr-BTBT** *S*-oxides studied in this research; Table S2.1. Fractional atomic coordinates ($\times 10^4$) and equivalent isotropic displacement parameters ($Å^2 \times 10^3$) for **2,7-diBr-BTBTDO**; Table S2.2. Anisotropic displacement parameters ($Å^2 \times 10^3$) for **2,7-diBr-BTBTDO**; Table S2.3. Bond lengths for **2,7-diBr-BTBTDO** (Å); Table S2.4. Bond angles for **2,7-diBr-BTBTDO** (°); Table S2.5. Torsion angles for **2,7-diBr-BTBTDO** (°); Table S3.1. Fractional atomic coordinates ($\times 10^4$) and equivalent isotropic displacement parameters ($Å^2 \times 10^3$) for **2,7-diBr-BTBTTO**; Table S3.2. Anisotropic displacement parameters ($Å^2 \times 10^3$) for **2,7-diBr-BTBTTO**; Table S3.3. Bond lengths for **2,7-diBr-BTBTTO** (Å); Table S3.4. Bond angles for **2,7-diBr-BTBTTO** (°); Table S3.5. Torsion angles for **2,7-diBr-BTBTTO** (°); Table S4.1. List of intermolecular short contacts in the crystal structure of **2,7-diBr-BTBT**; Table S4.2. List of intermolecular short contacts in the crystal structure of **2,7-diBr-BTBTDO**; Table S4.3. List of intermolecular short contacts in the crystal structure of **2,7-diBr-BTBTTO**); thermogravimetric and differential scanning calorimetry charts (Section S3. TGA and DSC analysis including Figure S1. TGA analysis for **2,7-diBr-BTBT** *S*-oxides; Figure S2. DSC analysis for **2,7-diBr-BTBT**

S-oxides); geometrical parameters of **2,7-diBr-BTBT**, **2,7-diBr-BTBTDO**, and **2,7-diBr-BTBTTO** optimized at the TDDFT PBE0/6–311+G(2d,p)//D3-M06-2X/def2-TZVP level in DCM (IEFPCM solvation model); atomic coordinates of **2,7-diBr-BTBT** and its S-oxides (charged and neutral) optimized at the D3-M06-2X/def2-level (Section S4. Theorethical calculations).

Author Contributions: Conceptualization, R.Ż.; methodology, R.Ż.; investigation, A.R., M.M., J.K., R.D. and R.Ż.; writing—original draft preparation, A.R., R.D. and R.Ż.; writing—review and editing, R.Ż. and A.R. All authors have read and agreed to the published version of the manuscript.

Funding: This research received no external funding.

Institutional Review Board Statement: Not applicable.

Informed Consent Statement: Not applicable.

Data Availability Statement: The data presented in this study are available in Supplementary Materials.

Acknowledgments: Upgrade of the Avance III 500 NMR spectrometer used to obtain results included in this publication was supported by the funds from the EU Regional Operational Program of the Lodz Region. The Rigaku XtaLAB Synergy-S X-Ray diffractometer system used to obtain results included in this publication was supported by funds from the EU Regional Operational Program of the Lodz Region, RPLD.01.01.00-10-0008/18.

Conflicts of Interest: The authors declare no conflicts of interest.

References

1. Katz, H.E.; Bao, Z.; Gilat, S.L. Synthetic Chemistry for Ultrapure, Processable, and High-Mobility Organic Transistor Semiconductors. *Acc. Chem. Res.* **2001**, *34*, 359–369. [CrossRef] [PubMed]
2. Jiang, W.; Li, Y.; Wang, Z. Heteroarenes as high performance organic semiconductors. *Chem. Soc. Rev.* **2013**, *42*, 6113–6127. [CrossRef] [PubMed]
3. Yu, H.; Li, W.; Tian, H.; Wang, H.; Yan, D.; Zhang, J.; Geng, Y.; Wang, F. Benzothienobenzothiophene-Based Conjugated Oligomers as Semiconductors for Stable Organic Thin-Film Transistors. *ACS Appl. Mater. Interfaces* **2014**, *6*, 5255–5262. [CrossRef] [PubMed]
4. Niebel, C.; Kim, Y.; Ruzié, C.; Karpinska, J.; Chattopadhyay, B.; Schweicher, G.; Richard, A.; Lemaur, V.; Olivier, Y.; Cornil, J.; et al. Thienoacene dimers based on the thieno[3,2-*b*]thiophene moiety: Synthesis, characterization and electronic properties. *J. Mater. Chem. C* **2015**, *3*, 674–685. [CrossRef]
5. Mori, T.; Nishimura, T.; Yamamoto, T.; Doi, I.; Miyazaki, E.; Osaka, I.; Takimiya, K. Consecutive Thiophene-Annulation Approach to π-Extended Thienoacene-Based Organic Semiconductors with [1]Benzothieno[3,2-*b*][1]benzothiophene (BTBT) Substructure. *J. Am. Chem. Soc.* **2013**, *135*, 13900–13913. [CrossRef] [PubMed]
6. Izawa, T.; Mori, H.; Shinmura, Y.; Iwatani, M.; Miyazaki, E.; Takimiya, K.; Hung, H.-W.; Yahiro, M.; Adachi, C. Molecular Modification of 2,7-Diphenyl[1]benzothieno[3,2-*b*]benzothiophene (DPh-BTBT) with Diarylamino Substituents: From Crystalline Order to Amorphous State in Evaporated Thin Films. *Chem. Lett.* **2009**, *38*, 420–421. [CrossRef]
7. Keum, C.-M.; Liu, S.; Al-Shadeedi, A.; Kaphle, V.; Callens, M.K.; Han, L.; Neyts, K.; Zhao, H.; Gather, M.C.; Bunge, S.D.; et al. Tuning charge carrier transport and optical birefringence in liquid-crystalline thin films: A new design space for organic light-emitting diodes. *Sci. Rep.* **2018**, *8*, 699. [CrossRef] [PubMed]
8. Kikuchi, T.; Sasabe, H.; Watanabe, Y.; Kamata, T.; Katagiri, H.; Kido, J. A Novel Series of Thermally and Electrically Stable Hole-transporters End-capped by [1]Benzothieno[3,2-*b*][1]benzothiophenes for Organic Light-emitting Devices. *Chem. Lett.* **2019**, *48*, 219–222. [CrossRef]
9. Saito, D.; Sasabe, H.; Kikuchi, T.; Ito, T.; Tsuneyama, H.; Kido, J. Improved operational lifetime of deep-red phosphorescent organic light-emitting diodes using a benzothienobenzothiophene (BTBT)-based p-type host material. *J. Mater. Chem. C* **2021**, *9*, 1215–1220. [CrossRef]
10. Zhang, D.; Zhao, C.; Zheng, X.; Wu, L.; Xu, J.; Zhou, L.; Wong, P.K.J.; Zhang, W.; He, Y. A study on the luminescence properties of high-performance benzothieno[3,2-*b*][1]benzothiophene based organic semiconductors. *Dyes Pigm.* **2023**, *216*, 111359. [CrossRef]
11. Takimiya, K.; Ebata, H.; Sakamoto, K.; Izawa, T.; Otsubo, T.; Kunugi, Y. 2,7-Diphenyl[1]benzothieno[3,2-*b*]benzothiophene, A New Organic Semiconductor for Air-Stable Organic Field-Effect Transistors with Mobilities up to 2.0 $cm^2\ V^{-1}\ s^{-1}$. *J. Am. Chem. Soc.* **2006**, *128*, 12604–12605. [CrossRef] [PubMed]
12. Amin, A.Y.; Khassanov, A.; Reuter, K.; Meyer-Friedrichsen, T.; Halik, M. Low-Voltage Organic Field Effect Transistors with a 2-Tridecyl[1]benzothieno[3,2-*b*][1]benzothiophene Semiconductor Layer. *J. Am. Chem. Soc.* **2012**, *134*, 16548–16550. [CrossRef] [PubMed]
13. Reddy, M.R.; Kim, H.; Kim, C.; Seo, S. 2-Thiopene[1]benzothieno[3,2-*b*]benzothiophene derivatives as solution-processable organic semiconductors for organic thin-film transistors. *Synth. Met.* **2018**, *235*, 153–159. [CrossRef]
14. Iino, H.; Usui, T.; Hanna, J.-i. Liquid crystals for organic thin-film transistors. *Nat. Commun.* **2015**, *6*, 6828. [CrossRef] [PubMed]

15. Kim, H.; Rajeshkumar Reddy, M.; Kwon, G.; Choi, D.; Kim, C.; Seo, S. Synthesis and characterization of 2,7-diethynyl-benzo[b]benzo[4,5]thieno[2,3-d]thiophene derivative as organic semiconductors for organic thin-film transistors. *Synth. Met.* **2016**, *220*, 599–605. [CrossRef]
16. Yao, C.; Chen, X.; He, Y.; Guo, Y.; Murtaza, I.; Meng, H. Design and characterization of methoxy modified organic semiconductors based on phenyl[1]benzothieno[3,2-*b*][1]benzothiophene. *RSC Adv.* **2017**, *7*, 5514–5518. [CrossRef]
17. Yamamoto, T.; Takimiya, K. Facile synthesis of highly pi-extended heteroarenes, dinaphtho[2,3-b:2′,3′-f]chalcogenopheno[3,2-*b*]chalcogenophenes, and their application to field-effect transistors. *J. Am. Chem. Soc.* **2007**, *129*, 2224–2225. [CrossRef] [PubMed]
18. Kano, M.; Minari, T.; Tsukagoshi, K. All-Solution-Processed Selective Assembly of Flexible Organic Field-Effect Transistor Arrays. *Appl. Phys. Express* **2010**, *3*, 051601. [CrossRef]
19. Uemura, T.; Hirose, Y.; Uno, M.; Takimiya, K.; Takeya, J. Very High Mobility in Solution-Processed Organic Thin-Film Transistors of Highly Ordered [1]Benzothieno[3,2-*b*]benzothiophene Derivatives. *Appl. Phys. Express* **2009**, *2*, 111501. [CrossRef]
20. Ebata, H.; Izawa, T.; Miyazaki, E.; Takimiya, K.; Ikeda, M.; Kuwabara, H.; Yui, T. Highly Soluble [1]Benzothieno[3,2-*b*]benzothiophene (BTBT) Derivatives for High-Performance, Solution-Processed Organic Field-Effect Transistors. *J. Am. Chem. Soc.* **2007**, *129*, 15732–15733. [CrossRef] [PubMed]
21. Haase, K.; Teixeira da Rocha, C.; Hauenstein, C.; Zheng, Y.; Hambsch, M.; Mannsfeld, S.C.B. High-Mobility, Solution-Processed Organic Field-Effect Transistors from C8-BTBT:Polystyrene Blends. *Adv. Electron. Mater.* **2018**, *4*, 1800076. [CrossRef]
22. Liu, C.; Minari, T.; Lu, X.; Kumatani, A.; Takimiya, K.; Tsukagoshi, K. Solution-Processable Organic Single Crystals with Bandlike Transport in Field-Effect Transistors. *Adv. Mater.* **2011**, *23*, 523–526. [CrossRef] [PubMed]
23. Izawa, T.; Miyazaki, E.; Takimiya, K. Molecular Ordering of High-Performance Soluble Molecular Semiconductors and Re-evaluation of Their Field-Effect Transistor Characteristics. *Adv. Mater.* **2008**, *20*, 3388–3392. [CrossRef]
24. Qi, M.; Zhang, D.; Zhu, Y.; Zhao, C.; Li, A.; Huang, F.; He, Y.; Meng, H. Anthracene-[1]benzothieno[3,2-*b*][1]benzothiophene (BTBT) dyad and triads as p-type semiconductors for organic field-effect transistors and phototransistors. *J. Mater. Chem. C* **2024**, *12*, 6578–6587. [CrossRef]
25. Ryu, S.; Yun, C.; Ryu, S.; Ahn, J.; Kim, C.; Seo, S. Characterization of [1]Benzothieno[3,2-*b*]benzothiophene (BTBT) Derivatives with End-Capping Groups as Solution-Processable Organic Semiconductors for Organic Field-Effect Transistors. *Coatings* **2023**, *13*, 181. [CrossRef]
26. Tisovský, P.; Gáplovský, A.; Gmucová, K.; Novota, M.; Pavúk, M.; Weis, M. Synthesis and characterization of new [1]benzothieno[3,2-*b*]benzothiophene derivatives with alkyl-thiophene core for application in organic field-effect transistors. *Org. Electron.* **2019**, *68*, 121–128. [CrossRef]
27. Tayu, M.; Rahmanudin, A.; Perry, G.J.P.; Khan, R.U.; Tate, D.J.; Marcial-Hernandez, R.; Shen, Y.; Dierking, I.; Janpatompong, Y.; Aphichatpanichakul, S.; et al. Modular synthesis of unsymmetrical [1]benzothieno[3,2-*b*][1]benzothiophene molecular semiconductors for organic transistors. *Chem. Sci.* **2022**, *13*, 421–429. [CrossRef] [PubMed]
28. Zhang, F.; Qu, G.; Mohammadi, E.; Mei, J.; Diao, Y. Solution-Processed Nanoporous Organic Semiconductor Thin Films: Toward Health and Environmental Monitoring of Volatile Markers. *Adv. Funct. Mater.* **2017**, *27*, 1701117. [CrossRef]
29. Trul, A.A.; Sizov, A.S.; Chekusova, V.P.; Borshchev, O.V.; Agina, E.V.; Shcherbina, M.A.; Bakirov, A.V.; Chvalun, S.N.; Ponomarenko, S.A. Organosilicon dimer of BTBT as a perspective semiconductor material for toxic gas detection with monolayer organic field-effect transistors. *J. Mater. Chem. C* **2018**, *6*, 9649–9659. [CrossRef]
30. Pan, J.; Wu, Y.; Zhang, X.; Chen, J.; Wang, J.; Cheng, S.; Wu, X.; Zhang, X.; Jie, J. Anisotropic charge trapping in phototransistors unlocks ultrasensitive polarimetry for bionic navigation. *Nat. Commun.* **2022**, *13*, 6629. [CrossRef] [PubMed]
31. Turetta, N.; Stoeckel, M.-A.; Furlan de Oliveira, R.; Devaux, F.; Greco, A.; Cendra, C.; Gullace, S.; Gicevičius, M.; Chattopadhyay, B.; Liu, J.; et al. High-Performance Humidity Sensing in π-Conjugated Molecular Assemblies through the Engineering of Electron/Proton Transport and Device Interfaces. *J. Am. Chem. Soc.* **2022**, *144*, 2546–2555. [CrossRef] [PubMed]
32. Dong, Y.; Sun, Y.; Liu, J.; Shi, X.; Li, H.; Zhang, J.; Li, C.; Yi, Y.; Mo, S.; Fan, L.; et al. Thermally Stable Organic Field-Effect Transistors Based on Asymmetric BTBT Derivatives for High Performance Solar-Blind Photodetectors. *Adv. Sci.* **2022**, *9*, 2106085. [CrossRef] [PubMed]
33. Takimiya, K.; Bulgarevich, K.; Kawabata, K. Crystal-Structure Control of Molecular Semiconductors by Methylthiolation: Toward Ultrahigh Mobility. *Acc. Chem. Res.* **2024**, *57*, 884–894. [CrossRef] [PubMed]
34. Antolini, L.; Tedesco, E.; Barbarella, G.; Favaretto, L.; Sotgiu, G.; Zambianchi, M.; Casarini, D.; Gigli, G.; Cingolani, R. Molecular Packing and Photoluminescence Efficiency in Odd-Membered Oligothiophene S,S-Dioxides. *J. Am. Chem. Soc.* **2000**, *122*, 9006–9013. [CrossRef]
35. Barbarella, G.; Pudova, O.; Arbizzani, C.; Mastragostino, M.; Bongini, A. Oligothiophene-S,S-dioxides: a New Class of Thiophene-based Materials. *J. Org. Chem.* **1998**, *63*, 1742–1745. [CrossRef]
36. Li, P.; Cui, Y.; Song, C.; Zhang, H. Effects of Sulfur Oxidation on the Electronic and Charge Transport Properties of Fused Oligothiophene Derivatives. *J. Phys. Chem. C* **2016**, *120*, 14484–14494. [CrossRef]
37. Barbarella, G.; Favaretto, L.; Sotgiu, G.; Zambianchi, M.; Fattori, V.; Cocchi, M.; Cacialli, F.; Gigli, G.; Cingolani, R. Modified Oligothiophenes with High Photo- and Electroluminescence Efficiencies. *Adv. Mater.* **1999**, *11*, 1375–1379. [CrossRef]
38. Camaioni, N.; Ridolfi, G.; Fattori, V.; Favaretto, L.; Barbarella, G. Oligothiophene-S,S-dioxides as a class of electron-acceptor materials for organic photovoltaics. *Appl. Phys. Lett.* **2004**, *84*, 1901–1903. [CrossRef]

39. Gigli, G.; Inganäs, O.; Anni, M.; De Vittorio, M.; Cingolani, R.; Barbarella, G.; Favaretto, L. Multicolor oligothiophene-based light-emitting diodes. *Appl. Phys. Lett.* **2001**, *78*, 1493–1495. [CrossRef]
40. Varathan, E.; Subramanian, V. The role of sulfur oxidation in controlling the electronic properties of sulfur-containing host molecules for phosphorescent organic light-emitting diodes. *Phys. Chem. Chem. Phys.* **2017**, *19*, 12002–12012. [CrossRef] [PubMed]
41. Miguel, L.S.; Matzger, A.J. Regiochemical Effects of Sulfur Oxidation on the Electronic and Solid-State Properties of Planarized Oligothiophenes Containing Thieno[3,2-*b*]thiophene Units. *J. Org. Chem.* **2008**, *73*, 7882–7888. [CrossRef] [PubMed]
42. Suzuki, Y.; Okamoto, T.; Wakamiya, A.; Yamaguchi, S. Electronic Modulation of Fused Oligothiophenes by Chemical Oxidation. *Org. Lett.* **2008**, *10*, 3393–3396. [CrossRef]
43. Zherdeva, S.Y.; Zheltov, A.Y.; Kozik, T.A.; Stepanov, B.I. Investigation of the Products from the Reduction of 2,2′-Stilbenedisulfonyl Chloride by Hydriodic Acid. *J. Org. Chem. USSR (Eng. Transl.)* **1980**, *16*, 379–383.
44. Udre, V.E.; Lukevits, E.Y.; Kemme, A.A.; Bleidelis, Y.Y. New Reaction of Benzothieno[3,2-*b*]benzothiophene Disulfone. *Chem. Heterocycl. Compd.* **1980**, *16*, 234–237. [CrossRef]
45. Taki, M.; Kajiwara, K.; Yamaguchi, E.; Sato, Y.; Yamaguchi, S. Fused Thiophene-S,S-dioxide-Based Super-Photostable Fluorescent Marker for Lipid Droplets. *ACS Mater. Lett.* **2021**, *3*, 42–49. [CrossRef]
46. Saito, M.; Osaka, I.; Miyazaki, E.; Takimiya, K.; Kuwabara, H.; Ikeda, M. One-step synthesis of [1]benzothieno[3,2-*b*][1]benzothiophene from o-chlorobenzaldehyde. *Tetrahedron Lett.* **2011**, *52*, 285–288. [CrossRef]
47. Vyas, V.S.; Gutzler, R.; Nuss, J.; Kern, K.; Lotsch, B.V. Optical gap in herringbone and π-stacked crystals of [1]benzothieno[3,2-*b*]benzothiophene and its brominated derivative. *CrystEngComm* **2014**, *16*, 7389–7392. [CrossRef]
48. Tang, M.L.; Bao, Z. Halogenated Materials as Organic Semiconductors. *Chem. Mater.* **2011**, *23*, 446–455. [CrossRef]
49. Ikeda, T.; Tahara, K.; Ishimatsu, R.; Ono, T.; Cui, L.; Maeda, M.; Ozawa, Y.; Abe, M. Lewis-Pairing-Induced Electrochemiluminescence Enhancement from Electron Donor-Acceptor Diads Decorated with Tris(pentafluorophenyl)borane as an Electrochemical Protector. *Angew. Chem. Int. Ed.* **2023**, *62*, e202301109. [CrossRef] [PubMed]
50. Foley, S.; Berberan-Santos, M.N.; Fedorov, A.; Bensasson, R.V.; Leach, S.; Gigante, B. Effect of halogenated compounds on the photophysics of C70 and a monoadduct of C70: Some implications on optical limiting behaviour. *Chem. Phys.* **2001**, *263*, 437–447. [CrossRef]
51. Zakavi, S.; Naderloo, M.; Heydari-turkmani, A.; Alghooneh, L.; Eskandari, M. Effects of β-bromine substitution and core protonation on photosensitizing properties of porphyrins: Long wavelength photosensitizers. *J. Catal.* **2019**, *380*, 236–246. [CrossRef]
52. Tomasi, J.; Mennucci, B.; Cancès, E. The IEF version of the PCM solvation method: An overview of a new method addressed to study molecular solutes at the QM ab initio level. *J. Mol. Struct. THEOCHEM* **1999**, *464*, 211–226. [CrossRef]
53. Mennucci, B.; Cancès, E.; Tomasi, J. Evaluation of Solvent Effects in Isotropic and Anisotropic Dielectrics and in Ionic Solutions with a Unified Integral Equation Method: Theoretical Bases, Computational Implementation, and Numerical Applications. *J. Phys. Chem. B* **1997**, *101*, 10506–10517. [CrossRef]
54. Cancès, E.; Mennucci, B.; Tomasi, J. A new integral equation formalism for the polarizable continuum model: Theoretical background and applications to isotropic and anisotropic dielectrics. *J. Chem. Phys.* **1997**, *107*, 3032–3041. [CrossRef]
55. Santato, C.; Favaretto, L.; Melucci, M.; Zanelli, A.; Gazzano, M.; Monari, M.; Isik, D.; Banville, D.; Bertolazzi, S.; Loranger, S.; et al. Influence of the oxidation level on the electronic, morphological and charge transport properties of novel dithienothiophene S-oxide and S,S-dioxide inner core oligomers. *J. Mater. Chem.* **2010**, *20*, 669–676. [CrossRef]
56. Liu, Z.; Lu, T.; Chen, Q. An sp-hybridized all-carboatomic ring, cyclo[18]carbon: Electronic structure, electronic spectrum, and optical nonlinearity. *Carbon* **2020**, *165*, 461–467. [CrossRef]
57. Sheldrick, G. SHELXT—Integrated space-group and crystal-structure determination. *Acta Crystallogr. Sect. A* **2015**, *71*, 3–8. [CrossRef]
58. Bourhis, L.J.; Dolomanov, O.V.; Gildea, R.J.; Howard, J.A.K.; Puschmann, H. The anatomy of a comprehensive constrained, restrained refinement program for the modern computing environment—Olex2 dissected. *Acta Crystallogr. Sect. A* **2015**, *71*, 59–75. [CrossRef]
59. Sheldrick, G. Crystal structure refinement with SHELXL. *Acta Crystallogr. Sect. C* **2015**, *71*, 3–8. [CrossRef]
60. Frisch, M.J.; Trucks, G.W.; Schlegel, H.B.; Scuseria, G.E.; Robb, M.A.; Cheeseman, J.R.; Scalmani, G.; Barone, V.; Mennucci, B.; Petersson, G.A.; et al. *Gaussian 09, Revision D.01*; Gaussian, Inc.: Wallingford, CT, USA, 2009.
61. Zhao, Y.; Truhlar, D.G. The M06 suite of density functionals for main group thermochemistry, thermochemical kinetics, noncovalent interactions, excited states, and transition elements: Two new functionals and systematic testing of four M06-class functionals and 12 other functionals. *Theor. Chem. Acc.* **2008**, *120*, 215–241. [CrossRef]
62. Grimme, S.; Antony, J.; Ehrlich, S.; Krieg, H. A consistent and accurate ab initio parametrization of density functional dispersion correction (DFT-D) for the 94 elements H-Pu. *J. Chem. Phys.* **2010**, *132*, 154104. [CrossRef] [PubMed]
63. Weigend, F.; Ahlrichs, R. Balanced basis sets of split valence, triple zeta valence and quadruple zeta valence quality for H to Rn: Design and assessment of accuracy. *Phys. Chem. Chem. Phys.* **2005**, *7*, 3297–3305. [CrossRef] [PubMed]
64. Adamo, C.; Barone, V. Toward reliable density functional methods without adjustable parameters: The PBE0 model. *J. Chem. Phys.* **1999**, *110*, 6158–6170. [CrossRef]
65. Krishnan, R.; Binkley, J.S.; Seeger, R.; Pople, J.A. Self-consistent molecular orbital methods. XX. A basis set for correlated wave functions. *J. Chem. Phys.* **1980**, *72*, 650–654. [CrossRef]

66. Lu, T.; Chen, F. Multiwfn: A multifunctional wavefunction analyzer. *J. Comput. Chem.* **2012**, *33*, 580–592. [CrossRef] [PubMed]
67. Rappoport, D.; Furche, F. Property-optimized Gaussian basis sets for molecular response calculations. *J. Chem. Phys.* **2010**, *133*, 134105. [CrossRef] [PubMed]
68. Pritchard, B.P.; Altarawy, D.; Didier, B.; Gibson, T.D.; Windus, T.L. New Basis Set Exchange: An Open, Up-to-Date Resource for the Molecular Sciences Community. *J. Chem. Inf. Model.* **2019**, *59*, 4814–4820. [CrossRef] [PubMed]

molecules

MDPI

Article

Iron-Catalyzed Sulfonylmethylation of Imidazo[1,2-α]pyridines with *N,N*-Dimethylacetamide and Sodium Sulfinates

Shengnan Sun, Hexia Ye, Haibo Liu *, Junchen Li * and Xiaojing Bi *

State Key Laboratory of NBC Protection for Civilian, Beijing 102205, China; shengnansunmail@yeah.net (S.S.); yehexia6688@yeah.net (H.Y.)

* Correspondence: hbbnu@126.com (H.L.); lijch07@163.com (J.L.); xiaojingbimail@yeah.net (X.B.)

Abstract: Functionalized imidazo[1,2-α]pyridines are important scaffolds in pharmaceuticals. Herein, we present an efficient 3-sulfonylmethylation protocol for imidazo[1,2-α]pyridines by sodium sulfinates in DMA and H_2O (2:1) via an $FeCl_3$-catalyzed three-component coupling reaction. Various sulfonylmethyl imidazo[1,2-α]pyridines were thus afforded in high yields with excellent functional group tolerance. A plausible oxidation-addition mechanism was proposed.

Keywords: sulfonylmethylation; imidazo[1,2-α]pyridine; DMA; $FeCl_3$-catalyzed

Citation: Sun, S.; Ye, H.; Liu, H.; Li, J.; Bi, X. Iron-Catalyzed Sulfonylmethylation of Imidazo[1,2-α]pyridines with *N,N*-Dimethylacetamide and Sodium Sulfinates. *Molecules* **2024**, *29*, 3196. https://doi.org/10.3390/molecules29133196

Academic Editor: Evgueni Kirillov

Received: 15 May 2024
Revised: 17 June 2024
Accepted: 18 June 2024
Published: 5 July 2024

1. Introduction

As a kind of privileged scaffold, imidazo[1,2-α]pyridines are often used as the pharmacodynamic backbones of versatile drugs, such as anti-ulcer [1], anti-inflammatory [2], anti-viral [3,4], anti-bacterial [5], anti-cancer [6], anti-HIV [7] and anti-tuberculosis pharmaceutical candidates [8]. Owing to the electron-rich nature of imidazo[1,2-α]pyridine, the functionalization of imidazo[1,2-α]pyridine, especially C3-positioned C-C and C-X (O/N/S) bond construction, has attracted widespread attention [9–26]. Nevertheless, imidazo[1,2-α]pyridines containing sulfur functionality, such as sulfonylmethylated imidazo[1,2-α]pyridines, have been rarely studied before, whilst the sulfones group exists widely in drug molecules. The first typical example reported by Song employed the iron-involved tosylmethylation of imidazo[1,2-α]pyridines with *p*-toluenesulfonyl methyl isocyanide in H_2O and PEG_{400} [27]. Then, Rode described a selectfluor-mediated approach regarding the methylene-tethered arylsulfonation and benzotriazolation of imidazopyridines using DMSO as the carbon insertion synthon [28]. A very recent study reported by Tang demonstrated another efficient protocol for the direct C-3 sulfonylmethylation of imidazo[1,2-α]pyridines with glyoxylic acid and sodium sulfinates in water [29]. Some studies that have used DMA as a carbon source have been reported for the coupling of C-C bonds [30–32]. However, a new methodology enabling the C-H sulfonylmethylation of imidazo[1,2-α]pyridines is still highly desirable and valuable.

In our previous work, sodium sulfinates were found to be an effective and universal sulfur source in eco-friendly S-S and S-C bond construction [33]. In our following explorations, the unexpected sulfonylmethylation of imidazo[1,2-α]pyridines was observed when researchers tried to synthesize sulfones in DMA. Herein, we wish to present a new method for the iron-catalyzed sulfonylmethylation of imidazo[1,2-α]pyridines with DMA and sodium sulfinates to synthesize 3-sulfonylmethyl imidazo[1,2-α]pyridines (Figure 1).

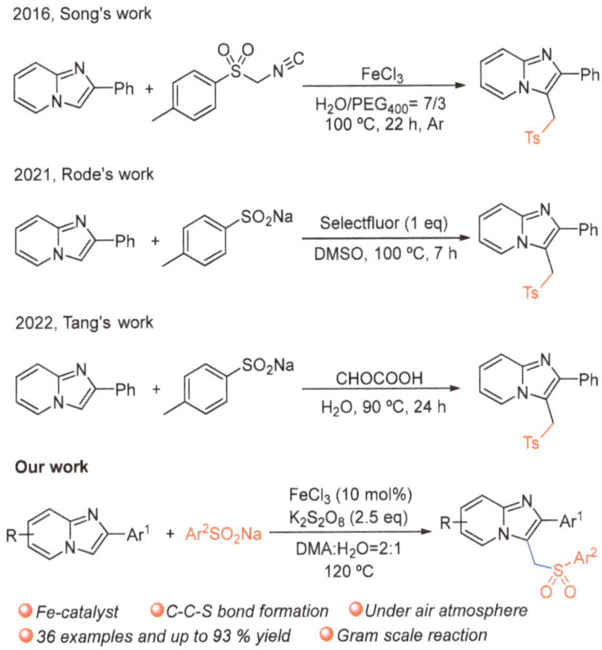

Figure 1. Comparison of recently reported sulfonylmethylation of imidazo[1,2-α]pyridines, refs. [27–29].

2. Results and Discussion

The optimization of reaction conditions was carried out by using 2-phenylimidazo[1,2-α]pyridine (**1a**) as the model substrate, sodium *p*-toluenesulfinate (**2a**) as the sulfonylation reagent and DMA as the carbon synthon. During the early stages of this work, we mainly focused on different metal catalysts using $K_2S_2O_8$ as an oxidant. The results showed that both $K_2S_2O_8$ and the catalyst were essential for the transformation (Table 1, entries 1–3). Among the metal catalysts screened, including $FeCl_3$, $FeCl_2$, Fe_3O_4, CuI, $Cu(acac)_2$, $V(acac)_2$ and $AgNO_3$, the best conversion with a 73% yield was obtained by $FeCl_3$ in the DMA and H_2O (2:1) hybrid solvent (Table 1, entries 4–10). As for the $FeCl_3$ catalyst, 10 mol% loading was effective enough to give the highest productivity of 73% product. Reducing to 5 mol% $FeCl_3$ gave comparable yields (Table 1, entries 6, 11–13). No products were obtained when the $K_2S_2O_8$ oxidant was changed to H_2O_2, TBHP, O_2 or I_2O_5 (Table 1, entries 14–17). The amount of $K_2S_2O_8$ added was also shown to have some effects on the yields, and 2.5 eq. was optimal (Table 1, entries 6, 18–20).

Further screening demonstrated that H_2O is vitally important, probably due to the fact that the low solubility of sodium sulfinate in DMA would lead to insufficient sulfur–reactant participation in a reaction without water (Table 1, entries 21–25). The best ratio of DMA and H_2O is 2:1. To our delight, when increasing the loading of **2a**, up to 90% yield was achieved (Table 1, entries 26–28). Furthermore, the reaction gave better results at higher temperatures such as 120 °C (Table 1, entries 29–32). Screening of other typical carbon source reagents, including DMF, DMSO and TMEDA, showed that DMA was the best (Table 1, entries 33–35).

Table 1. Optimization of reaction conditions [a].

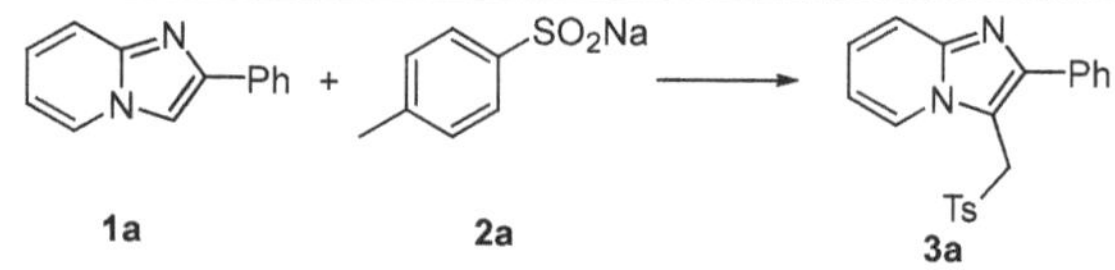

Entry	Catalyst/mol%	Oxidant/Equiv	Solvent	Yield/% [b]
1	-	-	DMA:H_2O/2:1	0
2	-	$K_2S_2O_8$/2.5	DMA:H_2O/2:1	5
3	$FeCl_3$/20	-	DMA:H_2O/2:1	0
4	CuI/20	$K_2S_2O_8$/2.5	DMA:H_2O/2:1	18
5	Cu(acac)$_2$/20	$K_2S_2O_8$/2.5	DMA:H_2O/2:1	18
6	$FeCl_3$/20	$K_2S_2O_8$/2.5	DMA:H_2O/2:1	73
7	$FeCl_2$/20	$K_2S_2O_8$/2.5	DMA:H_2O/2:1	67
8	Fe_3O_4/20	$K_2S_2O_8$/2.5	DMA:H_2O/2:1	50
9	V(acac)$_2$/20	$K_2S_2O_8$/2.5	DMA:H_2O/2:1	63
10	$AgNO_3$/20	$K_2S_2O_8$/2.5	DMA:H_2O/2:1	6
11	$FeCl_3$/5	$K_2S_2O_8$/2.5	DMA:H_2O/2:1	41
12	$FeCl_3$/10	$K_2S_2O_8$/2.5	DMA:H_2O/2:1	73
13	$FeCl_3$/30	$K_2S_2O_8$/2.5	DMA:H_2O/2:1	70
14	$FeCl_3$/10	H_2O_2/2.5	DMA:H_2O/2:1	0
15	$FeCl_3$/10	TBHP/2.5	DMA:H_2O/2:1	0
16	$FeCl_3$/10	O_2/air	DMA:H_2O/2:1	0
17	$FeCl_3$/10	I_2O_5/2.5	DMA:H_2O/2:1	0
18	$FeCl_3$/10	$K_2S_2O_8$/1.0	DMA:H_2O/2:1	57
19	$FeCl_3$/10	$K_2S_2O_8$/1.5	DMA:H_2O/2:1	64
20	$FeCl_3$/10	$K_2S_2O_8$/3.5	DMA:H_2O/2:1	73
21	$FeCl_3$/10	$K_2S_2O_8$/2.5	DMA/10 eq	0
22	$FeCl_3$/10	$K_2S_2O_8$/2.5	DMA/20 eq	0
23	$FeCl_3$/10	$K_2S_2O_8$/2.5	DMA/50 eq	43
24	$FeCl_3$/10	$K_2S_2O_8$/2.5	DMA:H_2O/1:1	43
25	$FeCl_3$/10	$K_2S_2O_8$/2.5	DMA:H_2O/5:1	70
26 [c]	$FeCl_3$/10	$K_2S_2O_8$/2.5	DMA:H_2O/2:1	80
27 [d]	$FeCl_3$/10	$K_2S_2O_8$/2.5	DMA:H_2O/2:1	90
28 [e]	$FeCl_3$/10	$K_2S_2O_8$/2.5	DMA:H_2O/2:1	90
29 [d,f]	$FeCl_3$/10	$K_2S_2O_8$/2.5	DMA:H_2O/2:1	60
30 [d,g]	$FeCl_3$/10	$K_2S_2O_8$/2.5	DMA:H_2O/2:1	70
31 [d,h]	$FeCl_3$/10	$K_2S_2O_8$/2.5	DMA:H_2O/2:1	93
32 [d,i]	$FeCl_3$/10	$K_2S_2O_8$/2.5	DMA:H_2O/2:1	93
33 [d,h]	$FeCl_3$/10	$K_2S_2O_8$/2.5	DMF:H_2O/2:1	33
34 [d,h]	$FeCl_3$/10	$K_2S_2O_8$/2.5	DMSO:H_2O/2:1	0
35 [d,h]	$FeCl_3$/10	$K_2S_2O_8$/2.5	TMEDA:H_2O/2:1	0

[a] Reaction conditions: **1a** (0.2 mmol), **2a** (0.24 mmol), catalyst, oxidant, solvent (2 mL), 110 °C, 4 h; [b] LC-MS; [c] **2a** (1.5 eq); [d] **2a** (2.0 eq); [e] **2a** (2.5 eq); [f] 90 °C; [g] 100 °C; [h] 120 °C; [i] 130 °C.

With the optimum condition in hand (Table 1, entry 31), we then evaluated the scope of structurally different sodium sulfinates with 2-phenylimidazo[1,2-α]pyridin (Table 2). As shown in Table 2, sodium sulfinates bearing electron-donating (4-Me, 3-Me, 4-iPr, 4-OMe) or electron-withdrawing (4-CF_3, 4-F, 4-Cl, 2-Cl, 4-Br) groups at the ortho, meta, or para positions were all compatible, affording good to excellent yields (Table 2, **3a–3j**). Moreover, sodium aromatic heterocyclic sulfinate also tolerated well (Table 2, **3k–3l**). To our surprise, this protocol was also applicable to sodium alkyl sulfinate, giving the target compounds in good yields (Table 2, **3m–3n**). Regrettably, no product was generated when sodium triflate and sodium difluoromethanesulfinate were used (Table 2, **3o–3p**).

Table 2. Substrate scope of sodium sulfinates [a,b].

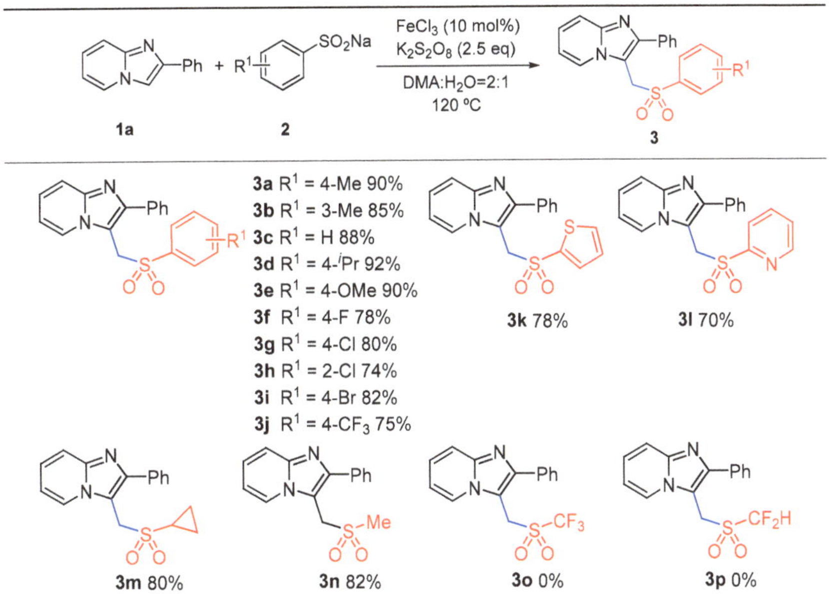

[a] Reaction conditions: **1a** (0.6 mmol), **2a** (1.2 mmol), $FeCl_3$ (10 mol%), $K_2S_2O_8$ (2.5 eq), DMA:H_2O = 2:1, 120 °C, 4 h; [b] isolated yields.

We next turned our attention to imidazo[1,2-α]pyridines substrates bearing different substituents (Table 3). To our delight, the reaction of sulfonylmethylation could be smoothly accomplished with 2-(4-bromophenyl)imidazo[1,2-α]pyridine under the standard conditions, offering compound **3qa–3qf** in good yields. This protocol can be effectively applied to 2-phenylimidazo-[1,2-α]pyridine with electron-donating groups (2-Me, 3-Me, 4-OCH_3) and electron-withdrawing groups (4-NO_2, 4-F, 4-Cl) (Table 3, **3ra–3rf**). Imidazo[1,2-α]pyridine substituted in the 6, 7, and 8 positions was also well tolerated (Table 3, **3sa–3se**). It is worth mentioning that heterocyclic substrates were amenable to this protocol, affording the desired product with good yields (Table 3, **3ta–3ud**). However, some substrates are not suitable for this protocol (Table 3, **3v–3y**).

In order to investigate the mechanism of the reaction, control experiments were explored, and the results are illustrated in Figure 2. Nearly no product was detected when 3 eq. radical scavenger (2,2,6,6-tetramethylpiperidin-1-oxyl (TEMPO) or 2,2-diphenyl-1-picrylhydrazyl (DPPH)) was added to the reaction system (Figure 2A), suggesting that a radical pathway might be involved in this protocol. Isotopic labeling experiments clearly showed that DMF provided the carbon attached to the imidazo heterocyclic backbone (Figure 2B).

According to our control experiments and the previous literature [11], a plausible mechanism of the sulfonylmethylation of imidazo[1,2-α]pyridines with DMA and sodium sulfinates is illustrated (Figure 3). DMA was firstly oxidized to iminium **B** by $FeCl_3/K_2S_2O_8$, which then reacted with electron-rich imidazopyridine to form intermediate **C** by C-C coupling. The sulfonylation of **C** by sodium *p*-toluenesulfinate generated 2-phenyl-3-(tosylmethyl)imidazo[1,2-α]pyridine (**3a**).

Table 3. Substrate scope of imidazo[1,2-α]pyridines [a,b].

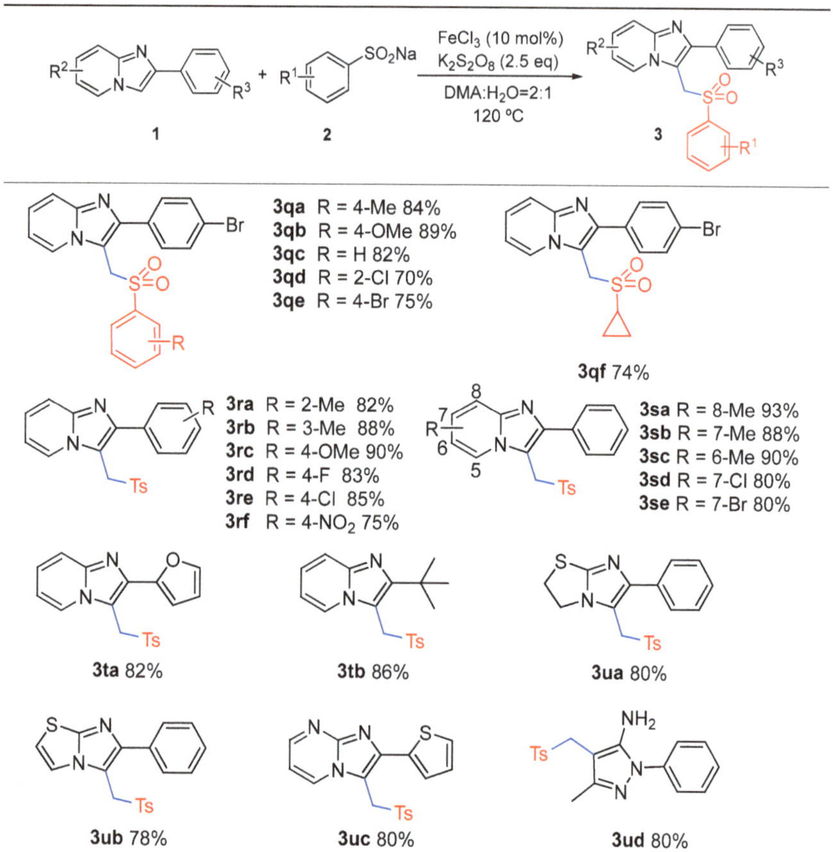

[a] Reaction conditions: **1a** (0.6 mmol), **2a** (1.2 mmol), $FeCl_3$ (10 mol%), $K_2S_2O_8$ (2.5 eq), DMA:H_2O = 2:1, 120 °C, 8 h; [b] isolated yields.

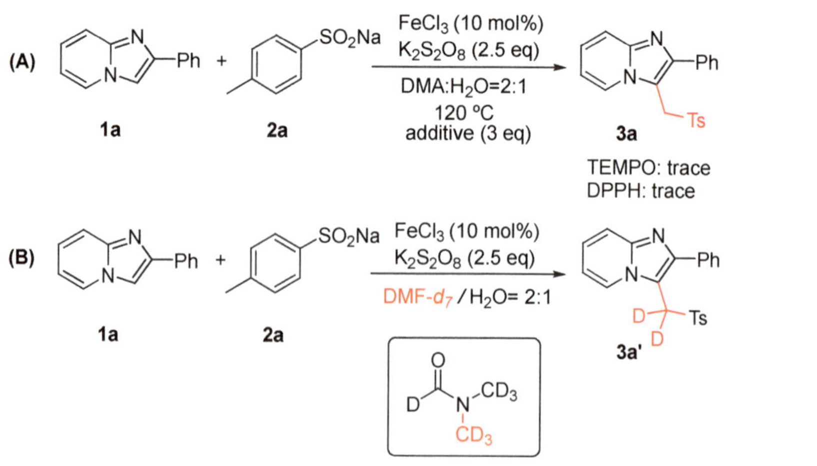

Figure 2. Control experiments. (**A**) Free radical trapping experiment; (**B**) Isotope labeling experiment.

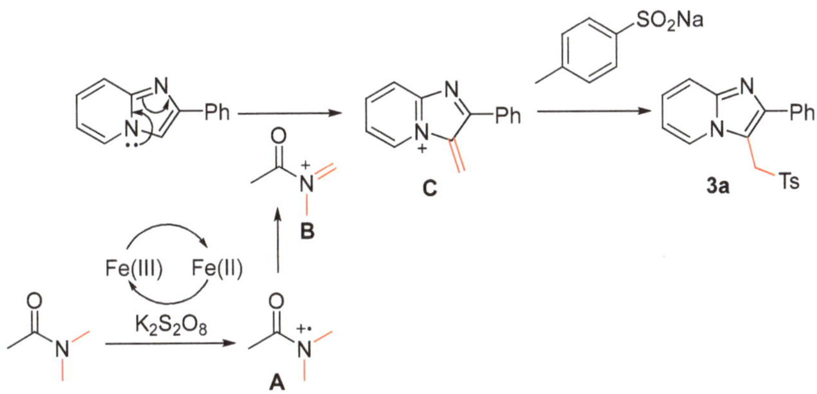

Figure 3. Proposed mechanism.

3. Experimental Section

General Information. All reactions were performed in a 10 mL tube with magnetic stirring. The imidazo[1,2-α] pyridines and sodium sulfinates are commercially available. Unless otherwise stated, all commercially available reagents were used without further purification. High-resolution mass spectra (ESI) were obtained with the Waters Xevo G2-XS QTOF. ^{1}H NMR, ^{13}C NMR and ^{19}F NMR spectra were recorded at ambient temperature on Bruker 300 M instruments. All spectra were referenced to $CDCl_3$ (^{1}H NMR δ 7.26 ppm and ^{13}C NMR δ 77.00 ppm). Data were reported as follows: chemical shift, multiplicity (s = singlet, d = doublet, t = triplet, q = quartet, dd = doublet of doublets, td = triplet of doublets, qd = quartet of doublets, m = multiplet), coupling constants (Hz) and integration.

Experimental Procedure. In a typical experiment, a 15 mL sealed tube was charged with imidazo[1,2-α]pyridine, (0.6 mmol), substituted sodium sulfinate (1.2 mmol), DMA (4 mL), H_2O (2 mL), $K_2S_2O_8$ (2.5 eq) and $FeCl_3$ (10 mol%). The mixture was allowed to stir at 120 °C and was monitored by TLC until the reaction was complete. Saturated aqueous NaCl solution was added to the reaction mixture and the aqueous phase was further extracted with ethyl acetate (3 × 10 mL). The combined organic layers were dried over anhydrous Na_2SO_4 and concentrated under a vacuum to give the crude product. The residue was purified by column chromatography on silica gel using ethyl acetate/*n*-hexane (1:1) as the eluent to provide the desired product.

2-Phenyl-3-(tosylmethyl)imidazo[1,2-α] pyridine (**3a**) [29]. White solid, 195 mg, 90% yield, ethyl acetate/*n*-hexane (1:1). ^{1}H NMR (300 MHz, $CDCl_3$) δ 8.45 (d, *J* = 6.9 Hz, 1H), 7.67 (d, *J* = 9.1 Hz, 1H), 7.38 (d, *J* = 8.3 Hz, 2H), 7.35–7.27 (m, 6H), 7.09 (d, *J* = 8.0 Hz, 2H), 6.94 (s, 1H), 4.88 (s, 2H), 2.36 (s, 3H). ^{13}C{1H} NMR (75 MHz, $CDCl_3$) δ 147.42, 146.00, 145.33, 134.22, 133.03, 129.84, 128.34, 128.23, 128.16, 128.14, 126.03, 124.98, 117.50, 112.92, 108.29, 52.69, 21.67.

2-Phenyl-3-((*m*-tolylsulfonyl)methyl)imidazo[1,2-α]pyridine (**3b**) [29]. White solid, 185 mg, 85% yield, ethyl acetate/*n*-hexane (1:1). ^{1}H NMR (300 MHz, $CDCl_3$) δ 8.45 (d, *J* = 7.0 Hz, 1H), 7.71 (d, *J* = 9.0 Hz, 1H), 7.31 (td, *J* = 7.4, 6.8, 3.9 Hz, 9H), 7.23 (d, *J* = 7.6 Hz, 1H), 6.96 (td, *J* = 6.9, 1.2 Hz, 1H), 4.88 (s, 2H), 2.26 (s, 3H). ^{13}C{1H} NMR (75 MHz, $CDCl_3$) δ 147.60, 146.02, 139.79, 137.35, 135.10, 132.97, 129.14, 128.53, 128.50, 128.34, 128.28, 126.14, 125.42, 124.99, 117.58, 113.03, 108.20, 52.76, 21.30.

2-Phenyl-3-((phenylsulfonyl)methyl)imidazo[1,2-α]pyridine (**3c**) [29]. White solid, 184 mg, 88% yield. ethyl acetate/*n*-hexane (1:1). ^{1}H NMR (300 MHz, $CDCl_3$) δ 8.47 (d, *J* = 7.0 Hz, 1H), 7.71 (d, *J* = 9.0 Hz, 1H), 7.59–7.52 (m, 3H), 7.39–7.29 (m, 8H), 6.96 (td, *J* = 6.9, 1.2 Hz, 1H), 4.91 (s, 2H). ^{13}C{1H} NMR (75 MHz, $CDCl_3$) δ 147.67, 146.04, 137.51, 134.26, 132.99, 129.28, 128.55, 128.28, 128.25, 128.19, 126.08, 124.96, 117.57, 112.97, 108.03, 52.74.

3-(((4-Isopropylphenyl)sulfonyl)methyl)-2-phenylimidazo [1,2-α]pyridine (**3d**) [29]. White solid, 215 mg, 92% yield, ethyl acetate/*n*-hexane (1:1), ^{1}H NMR (300 MHz, $CDCl_3$) δ

8.44 (d, J = 7.0 Hz, 1H), 7.68 (d, J = 9.1 Hz, 1H), 7.44 (d, J = 8.4 Hz, 2H), 7.35–7.26 (m, 6H), 7.18 (d, J = 8.3 Hz, 2H), 6.97–6.89 (m, 1H), 4.87 (s, 2H), 2.97–2.83 (p, J = 6.9 Hz, 1H), 1.22 (d, J = 6.9 Hz, 6H). ^{13}C{1H} NMR (75 MHz, $CDCl_3$) δ 155.95, 147.53, 145.99, 134.74, 133.05, 128.49, 128.39, 128.23, 128.20, 127.42, 126.01, 125.01, 117.50, 112.89, 108.31, 52.80, 34.26, 23.61.

3-(((4-Methoxyphenyl)sulfonyl)methyl)-2-phenylimidazo [1,2-α]pyridine (**3e**) [29]. White solid, 205 mg, 90% yield. ethyl acetate/n-hexane (1:1). ^{1}H NMR (300 MHz, $CDCl_3$) δ 8.36 (d, J = 7.0 Hz, 1H), 7.59 (d, J = 9.1 Hz, 1H), 7.34–7.28 (m, 2H), 7.26–7.18 (m, 6H), 6.85 (td, J = 6.8, 1.2 Hz, 1H), 6.69–6.58 (m, 2H), 4.79 (s, 2H), 3.72 (s, 3H). ^{13}C{1H} NMR (75 MHz, $CDCl_3$) δ 164.10, 147.20, 145.90, 132.99, 130.30, 128.37, 128.15, 128.10, 125.99, 124.92, 117.47, 114.34, 112.90, 108.47, 55.63, 52.68.

3-(((4-Fluorophenyl)sulfonyl)methyl)-2-phenylimidazo[1,2-α]pyridine (**3f**) [29]. White solid, 171 mg, 78% yield, ethyl acetate/n-hexane (1:1). ^{1}H NMR (300 MHz, $CDCl_3$) δ 8.36 (d, J = 6.9 Hz, 1H), 7.64 (d, J = 9.0 Hz, 1H), 7.29–7.23 (m, 5H), 7.22–7.10 (m, 5H), 6.92 (td, J = 6.9, 1.2 Hz, 1H), 4.87 (s, 2H). ^{13}C{1H} NMR (75 MHz, $CDCl_3$) δ 167.79, 164.38, 147.38, 146.08, 132.95, 131.07, 130.93, 128.52, 128.26, 128.06, 126.11, 124.84, 117.60, 116.57, 116.27, 112.99, 107.91, 52.35. ^{19}F NMR (282 MHz, $CDCl_3$) δ-102.63.

3-(((4-Chlorophenyl)sulfonyl)methyl)-2-phenylimidazo[1,2-α]pyridine (**3g**) [29]. White solid, 183 mg, 80% yield, ethyl acetate/n-hexane (1:1). ^{1}H NMR (300 MHz, $CDCl_3$) δ 8.43 (d, J = 6.8 Hz, 1H), 7.77–7.67 (m, 1H), 7.40–7.28 (m, 6H), 7.25 (s, 2H), 7.21–7.14 (m, 2H), 6.97 (d, J = 6.8 Hz, 1H), 4.93 (s, 2H). ^{13}C{1H} NMR (75 MHz, $CDCl_3$) δ 147.06, 145.93, 141.27, 135.26, 132.50, 129.60, 129.48, 128.66, 128.51, 128.14, 126.61, 124.95, 117.60, 113.40, 107.98, 52.27.

3-(((2-Chlorophenyl)sulfonyl)methyl)-2-phenylimidazo[1,2-α]pyridine (**3h**) [29]. White solid, 170 mg, 74% yield. ethyl acetate/n-hexane (1:1). ^{1}H NMR (300 MHz, $CDCl_3$) δ 8.48 (d, J = 7.0 Hz, 1H), 7.73 (dd, J = 7.9, 1.7 Hz, 1H), 7.66 (d, J = 9.1 Hz, 1H), 7.44 (d, J = 3.8 Hz, 3H), 7.35–7.27 (m, 5H), 7.24 (dd, J = 7.9, 1.2 Hz, 1H), 6.95 (td, J = 6.9, 1.2 Hz, 1H), 5.14 (s, 2H). ^{13}C{1H} NMR (75 MHz, $CDCl_3$) δ 148.07, 146.22, 135.41, 135.27, 133.08, 132.91, 132.04, 132.01, 128.67, 128.52, 128.41, 127.37, 126.24, 125.10, 117.59, 113.07, 107.48, 50.37.

3-(((4-Bromophenyl)sulfonyl)methyl)-2-phenylimidazo[1,2-α]pyridine (**3i**) [29]. White solid, 210 mg, 82% yield, ethyl acetate/n-hexane (1:1). ^{1}H NMR (300 MHz, $CDCl_3$) δ 8.39 (d, J = 6.9 Hz, 1H), 7.65 (d, J = 9.1 Hz, 1H), 7.31 (ddd, J = 7.8, 4.0, 2.3 Hz, 6H), 7.26–7.15 (m, 4H), 6.93 (td, J = 6.8, 1.2 Hz, 1H). ^{13}C{1H} NMR (75 MHz, $CDCl_3$) δ 147.37, 146.11, 135.75, 132.77, 132.36, 129.91, 129.56, 128.57, 128.32, 128.06, 126.26, 124.84, 117.63, 113.13, 107.82, 52.17.

2-Phenyl-3-(((4-(trifluoromethyl)phenyl)sulfonyl)methyl) imidazo[1,2-α]pyridine (**3j**). White solid, 188 mg, 75% yield. ethyl acetate/n-hexane (1:1), mp. 120–121 °C. ^{1}H NMR (300 MHz, $CDCl_3$) δ 8.37 (d, J = 6.9 Hz, 1H), 7.66–7.60 (m, 1H), 7.40 (d, J = 1.0 Hz, 4H), 7.33–7.26 (m, 1H), 7.21–7.12 (m, 5H), 6.92 (d, J = 6.1 Hz, 1H), 4.93 (s, 2H). ^{13}C{1H} NMR (75 MHz, $CDCl_3$) δ 147.45, 146.20, 140.44, 132.74, 128.76, 128.68, 128.32, 127.92, 126.30, 126.11, 126.06, 124.82, 121.18, 117.70, 113.15, 107.42, 51.93. HRMS (ESI) m/z: $[M + H]^+$ Calcd for $C_{21}H_{16}F_3N_2O_2S^+$ 417.0880, Found 417.0881.

2-Phenyl-3-((thiophen-2-ylsulfonyl)methyl)imidazo [1,2-α] pyridine (**3k**) [29]. White solid 166 mg, 78% yield, ethyl acetate/n-hexane (1:1). ^{1}H NMR (300 MHz, $CDCl_3$) δ 8.41 (d, J = 6.9 Hz, 1H), 7.69 (d, J = 9.1 Hz, 1H), 7.57 (dd, J = 5.0, 1.3 Hz, 1H), 7.45–7.37 (m, 2H), 7.37–7.27 (m, 5H), 6.98–6.90 (m, 2H), 4.99 (s, 2H). ^{13}C{1H} NMR (75 MHz, $CDCl_3$) δ 147.91, 146.12, 138.00, 135.41, 135.07, 132.98, 128.65, 128.41, 128.35, 128.21, 126.24, 124.95, 117.63, 113.12, 108.11, 54.07.

2-Phenyl-3-((pyridin-2-ylsulfonyl)methyl)imidazo[1,2-α] pyridine (**3l**) [29]. White solid, 146 mg, 70% yield, ethyl acetate/n-hexane (1:1), mp. 108–109 °C. ^{1}H NMR (300 MHz, $CDCl_3$) δ 8.60–8.54 (m, 1H), 8.45 (dt, J = 4.6, 1.4 Hz, 1H), 7.90–7.78 (m, 3H), 7.67–7.60 (m, 2H), 7.48–7.35 (m, 5H), 7.05 (t, J = 6.9 Hz, 1H), 5.23 (s, 2H). ^{13}C{1H} NMR (75 MHz, $CDCl_3$) δ 156.10, 150.38, 147.67, 146.20, 137.96, 133.05, 128.72, 128.59, 128.37, 127.81, 126.25, 125.00, 122.90, 117.57, 113.07, 107.15, 48.81.

3-((Cyclopropylsulfonyl)methyl)-2-phenylimidazo[1,2-α] pyridine (**3m**) [29]. White solid, 149 mg, 80% yield, ethyl acetate/n-hexane (1:1). ^{1}H NMR (300 MHz, $CDCl_3$) δ 8.40 (d, J = 7.0 Hz, 1H), 7.79 (dd, J = 8.2, 1.4 Hz, 2H), 7.71 (d, J = 9.1 Hz, 1H), 7.52–7.38 (m, 3H),

7.36–7.28 (m, 1H), 6.93 (dd, *J* = 6.9, 1.1 Hz, 1H), 4.86 (s, 2H), 2.12 (ddd, *J* = 12.8, 8.0, 4.8 Hz, 1H), 1.00 (dd, *J* = 4.7, 2.1 Hz, 2H), 0.68 (dd, *J* = 7.9, 2.2 Hz, 2H). $^{13}C\{^1H\}$ NMR (75 MHz, $CDCl_3$) δ 147.08, 145.99, 133.51, 129.00, 128.64, 128.53, 126.21, 124.91, 117.52, 113.08, 108.12, 49.91, 28.80, 4.98.

3-((Methylsulfonyl)methyl)-2-phenylimidazo[1,2-α]pyridine (**3n**) [29]. White solid, 141 mg, 82% yield. ethyl acetate/*n*-hexane (1:1). ^1H NMR (300 MHz, $CDCl_3$) δ 8.39 (d, *J* = 6.9 Hz, 1H), 7.75 (d, *J* = 6.8 Hz, 2H), 7.70 (d, *J* = 9.1 Hz, 1H), 7.54–7.47 (m, 2H), 7.47–7.41 (m, 1H), 7.36–7.30 (m, 1H), 6.94 (td, *J* = 6.9, 1.2 Hz, 1H), 4.83 (s, 2H), 2.63 (s, 3H). $^{13}C\{^1H\}$ NMR (75 MHz, $CDCl_3$) δ 147.44, 146.46, 133.54, 129.29, 128.87, 128.53, 126.30, 124.94, 117.82, 113.19, 108.02, 50.95, 39.61.

2-(4-Bromophenyl)-3-(tosylmethyl)imidazo[1,2-α]pyridine (**3qa**) [29]. White solid, 221 mg, 84% yield, ethyl acetate/*n*-hexane (1:1). ^1H NMR (300 MHz, $CDCl_3$) δ 8.42 (d, *J* = 7.0 Hz, 1H), 7.68 (d, *J* = 9.1 Hz, 1H), 7.45–7.38 (m, 4H), 7.37–7.31 (m, 1H), 7.24 (d, *J* = 8.5 Hz, 2H), 7.12 (d, *J* = 7.7 Hz, 2H), 6.95 (td, *J* = 6.9, 1.2 Hz, 1H), 4.82 (s, 2H), 2.38 (s, 3H). $^{13}C\{^1H\}$ NMR (75 MHz, $CDCl_3$) δ 146.09, 146.06, 145.65, 134.35, 132.05, 131.55, 129.95, 129.78, 128.25, 126.39, 124.93, 122.53, 117.58, 113.20, 108.48, 52.64, 21.76.

2-(4-Bromophenyl)-3-(((4-methoxyphenyl)sulfonyl)methyl) imidazo[1,2-α]pyridine (**3qb**). White solid, 243 mg, 89% yield, ethyl acetate/*n*-hexane (1:1), mp. 230–231 °C. ^1H NMR (300 MHz, $CDCl_3$) δ 8.39 (d, *J* = 7.0 Hz, 1H), 7.67–7.61 (m, 1H), 7.39 (dd, *J* = 8.7, 3.7 Hz, 4H), 7.34–7.28 (m, 1H), 7.26–7.17 (m, 2H), 6.92 (td, *J* = 6.8, 1.2 Hz, 1H), 6.73 (d, *J* = 8.9 Hz, 2H), 4.82 (s, 2H), 3.82 (s, 3H). $^{13}C\{^1H\}$ NMR (75 MHz, $CDCl_3$) δ 164.33, 146.09, 146.03, 132.19, 131.58, 130.43, 129.76, 128.38, 126.31, 124.93, 122.45, 117.61, 114.44, 113.16, 108.69, 55.82, 52.65. HRMS (ESI) m/z: $[M + H]^+$ Calcd for $C_{21}H_{18}BrN_2O_3S^+$ 457.0216, Found 457.0217.

2-(4-Bromophenyl)-3-((phenylsulfonyl)methyl) imidazo[1,2-α]pyridine (**3qc**). White solid, 209 mg, 82% yield, ethyl acetate/*n*-hexane (1:1), mp. 213–214 °C. ^1H NMR (300 MHz, $CDCl_3$) δ 8.36 (d, *J* = 7.0 Hz, 1H), 7.63 (d, *J* = 9.0 Hz, 1H), 7.52 (td, *J* = 7.0, 1.4 Hz, 3H), 7.34 (qd, *J* = 7.5, 6.7, 2.0 Hz, 4H), 7.23–7.12 (m, 3H), 6.89 (d, *J* = 6.9 Hz, 1H), 4.77 (s, 2H). $^{13}C\{^1H\}$ NMR (75 MHz, $CDCl_3$) δ 146.50, 146.15, 137.58, 134.47, 132.03, 131.76, 129.85, 129.45, 128.28, 126.42, 124.93, 122.70, 117.64, 113.24, 108.17, 52.76. HRMS (ESI) m/z: $[M + H]^+$ Calcd for $C_{20}H_{16}BrN_2O_2S^+$ 427.0111, Found 427.0111.

2-(4-Bromophenyl)-3-(((2-chlorophenyl)sulfonyl)methyl) imidazo[1,2-α]pyridine (**3qd**). White solid, 193 mg, 70% yield, ethyl acetate/*n*-hexane (1:1), mp. 227–228 °C. ^1H NMR (300 MHz, $CDCl_3$) δ 8.49 (d, *J* = 6.9 Hz, 1H), 7.80 (dd, *J* = 8.1, 1.6 Hz, 1H), 7.72 (d, *J* = 9.0 Hz, 1H), 7.52–7.40 (m, 5H), 7.40–7.31 (m, 3H), 7.01 (td, *J* = 6.9, 1.2 Hz, 1H), 5.11 (s, 2H). $^{13}C\{^1H\}$ NMR (75 MHz, $CDCl_3$) δ 146.81, 146.27, 135.59, 135.46, 133.08, 132.15, 132.08, 131.86, 130.11, 127.54, 126.62, 125.10, 122.90, 117.64, 113.38, 107.61, 50.35. HRMS (ESI) m/z: $[M–H]^-$ Calcd for $C_{20}H_{13}BrClN_2O_2S^-$ 458.9569, Found 458.9570.

2-(4-Bromophenyl)-3-(((4-bromophenyl)sulfonyl)methyl) imidazo[1,2-α]pyridine (**3qe**). White solid, 228 mg, 75% yield, ethyl acetate/*n*-hexane (1:1), mp. 244–245 °C. ^1H NMR (300 MHz, $CDCl_3$) δ 8.39 (d, *J* = 7.0 Hz, 1H), 7.70–7.66 (m, 1H), 7.50–7.44 (m, 2H), 7.44–7.38 (m, 2H), 7.38–7.32 (m, 1H), 7.30–7.26 (m, 2H), 7.20 (d, *J* = 8.5 Hz, 2H), 6.98 (td, *J* = 6.9, 1.2 Hz, 1H), 4.88 (s, 2H). $^{13}C\{^1H\}$ NMR (75 MHz, $CDCl_3$) δ 146.18, 135.92, 132.53, 131.81, 130.24, 129.69, 129.65, 126.63, 124.84, 122.81, 117.74, 113.44, 107.98, 52.27. HRMS (ESI) m/z: $[M + H]^+$ Calcd for $C_{20}H_{15}Br_2N_2O_2S^+$ 506.9195, Found 506.9196.

2-(4-Bromophenyl)-3-((cyclopropylsulfonyl)methyl) imidazo[1,2-α]pyridine (**3qf**). White solid, 173 mg, 74% yield, ethyl acetate/*n*-hexane (1:1), mp. 204–205 °C. ^1H NMR (300 MHz, $CDCl_3$) δ 8.37 (d, *J* = 7.0 Hz, 1H), 7.74–7.64 (m, 3H), 7.64–7.57 (m, 2H), 7.35–7.28 (m, 1H), 6.93 (td, *J* = 6.9, 1.2 Hz, 1H), 4.80 (s, 2H), 2.24–2.17 (m, 1H), 1.08 (dt, *J* = 6.6, 3.3 Hz, 2H), 0.81 (tt, *J* = 8.0, 3.6 Hz, 2H). $^{13}C\{^1H\}$ NMR (75 MHz, $CDCl_3$) δ 146.19, 146.09, 132.56, 132.11, 130.06, 126.32, 124.87, 122.94, 117.58, 113.17, 108.01, 50.00, 29.11, 5.12. HRMS (ESI) m/z: $[M + H]^+$ Calcd for $C_{17}H_{16}BrN_2O_2S^+$ 391.0110, Found 391.0111.

2-(*o*-Tolyl)-3-(tosylmethyl)imidazo[1,2-α]pyridine (**3ra**) [29]. White solid, 185 mg, 82%, ethyl acetate/*n*-hexane (1:1), mp.133–134 °C. ^1H NMR (300 MHz, $CDCl_3$) δ 8.44 (d, *J* = 6.9 Hz, 1H), 7.56 (d, *J* = 9.1 Hz, 1H), 7.22 (dd, *J* = 9.0, 7.1 Hz, 3H), 7.12 (td, *J* = 7.5, 1.4 Hz,

1H), 7.02 (d, J = 7.6 Hz, 3H), 6.94 (t, J = 7.3 Hz, 1H), 6.86 (td, J = 6.9, 1.2 Hz, 1H), 6.62 (dd, J = 7.5, 1.3 Hz, 1H), 4.64 (s, 2H), 2.31 (s, 3H), 1.91 (s, 3H). ^{13}C{1H} NMR (75 MHz, $CDCl_3$) δ 147.69, 145.66, 145.06, 137.38, 134.49, 131.82, 130.30, 129.87, 128.43, 128.05, 125.82, 125.29, 125.15, 117.48, 112.86, 109.31, 52.29, 21.67, 20.00.

2-(*m*-Tolyl)-3-(tosylmethyl)imidazo[1,2-α]pyridine (**3rb**) [29]. White solid, 199 mg, 88% yield, ethyl acetate/*n*-hexane (1:1). ^{1}H NMR (300 MHz, $CDCl_3$) δ 8.38 (d, J = 6.9 Hz, 1H), 7.59 (dd, J = 9.1, 1.2 Hz, 1H), 7.30 (d, J = 8.3 Hz, 2H), 7.25–7.20 (m, 1H), 7.09 (d, J = 1.3 Hz, 1H), 7.02 (d, J = 9.1 Hz, 5H), 6.84 (td, J = 6.9, 1.2 Hz, 1H), 4.79 (s, 2H), 2.27 (s, 3H), 2.23 (s, 3H). ^{13}C{1H} NMR (75 MHz, $CDCl_3$) δ 147.50, 145.92, 145.26, 138.09, 134.39, 132.84, 129.84, 128.95, 128.26, 128.21, 126.06, 125.22, 125.04, 117.46, 112.93, 108.27, 52.81, 21.70, 21.47.

2-(4-Methoxyphenyl)-3-(tosylmethyl)imidazo[1,2-α]pyridine (**3rc**) [29]. White solid, 212 mg, 90% yield, ethyl acetate/*n*-hexane (1:1). ^{1}H NMR (300 MHz, $CDCl_3$) δ 8.31 (d, J = 6.9 Hz, 1H), 7.56 (d, J = 9.0 Hz, 1H), 7.34 (d, J = 8.3 Hz, 2H), 7.24–7.18 (m, 3H), 7.03 (d, J = 7.8 Hz, 2H), 6.81 (td, J = 6.9, 1.2 Hz, 1H), 6.76–6.68 (m, 2H), 4.75 (s, 2H), 3.74 (s, 3H), 2.28 (s, 3H). ^{13}C{1H} NMR (75 MHz, $CDCl_3$) δ 159.64, 147.23, 145.81, 145.27, 134.46, 129.80, 129.44, 128.16, 125.88, 125.46, 124.80, 117.24, 113.75, 112.75, 107.63, 55.31, 52.86, 21.63.

2-(4-Fluorophenyl)-3-(tosylmethyl)imidazo[1,2-α]pyridine (**3rd**) [29]. White solid, 189 mg, 83% yield, ethyl acetate/*n*-hexane (1:1). ^{1}H NMR (300 MHz, $CDCl_3$) δ 8.32 (d, J = 7.0 Hz, 1H), 7.57 (d, J = 9.1 Hz, 1H), 7.33 (d, J = 8.3 Hz, 2H), 7.29–7.20 (m, 3H), 7.04 (d, J = 8.0 Hz, 2H), 6.94–6.80 (m, 3H), 4.74 (s, 2H), 2.33–2.26 (m, 3H). ^{13}C{1H} NMR (75 MHz, $CDCl_3$) δ 164.54, 161.25, 146.39, 145.94, 145.59, 134.49, 130.18, 130.07, 129.99, 129.21, 129.16, 128.28, 126.40, 125.04, 117.47, 115.59, 115.31, 113.18, 108.29, 52.74, 21.75.

2-(4-Chlorophenyl)-3-(tosylmethyl)imidazo[1,2-α]pyridine (**3re**) [29]. White solid, 202 mg, 85% yield. ethyl acetate/*n*-hexane (1:1). ^{1}H NMR (300 MHz, $CDCl_3$) δ 8.33 (d, J = 7.0 Hz, 1H), 7.58 (d, J = 9.1 Hz, 1H), 7.33 (d, J = 8.3 Hz, 2H), 7.26–7.16 (m, 5H), 7.04 (d, J = 8.0 Hz, 2H), 6.86 (t, J = 7.3 Hz, 1H), 4.75 (s, 2H), 2.30 (s, 3H). ^{13}C{1H} NMR (75 MHz, $CDCl_3$) δ 210.22, 146.05, 146.00, 145.62, 134.36, 134.30, 131.55, 129.94, 129.50, 128.59, 128.24, 126.39, 124.93, 117.53, 113.19, 108.46, 52.64, 21.72.

2-(4-Nitrophenyl)-3-(tosylmethyl)imidazo[1,2-α]pyridine (**3rf**). White solid, 183 mg, 75% yield, ethyl acetate/*n*-hexane (1:1), mp.133–134 °C. ^{1}H NMR (300 MHz, $CDCl_3$) δ 8.44 (d, J = 7.0 Hz, 1H), 8.23–8.11 (m, 2H), 7.74 (d, J = 9.1 Hz, 1H), 7.69–7.63 (m, 2H), 7.48 (d, J = 8.3 Hz, 2H), 7.43–7.37 (m, 1H), 7.16 (d, J = 8.0 Hz, 2H), 7.01 (td, J = 6.9, 1.1 Hz, 1H), 4.85 (s, 2H), 2.38 (s, 3H). ^{13}C{1H} NMR (75 MHz, $CDCl_3$) δ 147.47, 146.35, 145.95, 144.83, 139.70, 134.42, 130.14, 128.98, 128.33, 126.97, 125.02, 123.68, 117.89, 113.69, 109.65, 52.69, 21.74. HRMS (ESI) m/z: $[M + H]^+$ Calcd for $C_{21}H_{18}N_3O_4S^+$ 408.1013, Found 408.1013.

8-Methyl-2-phenyl-3-(tosylmethyl)imidazo[1,2-α]pyridine (**3sa**) [29]. White solid, 201 mg, 93% yield, ethyl acetate/*n*-hexane (1:1). ^{1}H NMR (300 MHz, $CDCl_3$) δ 8.21 (d, J = 6.8 Hz, 1H), 7.30 (d, J = 8.3 Hz, 2H), 7.20 (dt, J = 9.6, 2.8 Hz, 5H), 7.00 (d, J = 7.9 Hz, 3H), 6.74 (t, J = 6.9 Hz, 1H), 4.75 (s, 2H), 2.56 (s, 3H), 2.26 (s, 3H). ^{13}C{1H} NMR (75 MHz, $CDCl_3$) δ 146.96, 146.32, 145.22, 134.42, 133.25, 129.83, 128.44, 128.31, 128.19, 127.99, 127.44, 124.82, 122.74, 112.91, 108.61, 52.85, 21.68, 17.18.

7-Methyl-2-phenyl-3-(tosylmethyl)imidazo[1,2-α]pyridine (**3sb**) [29]. White solid, 199 mg, 88% yield, ethyl acetate/*n*-hexane (1:1). ^{1}H NMR (300 MHz, $CDCl_3$) δ 8.23 (d, J = 7.1 Hz, 1H), 7.34 (s, 1H), 7.27 (d, J = 8.3 Hz, 2H), 7.20–7.15 (m, 5H), 6.98 (d, J = 8.0 Hz, 2H), 6.67 (dd, J = 7.1, 1.6 Hz, 1H), 4.75 (s, 2H), 2.35 (d, J = 1.0 Hz, 3H), 2.26 (s, 3H). ^{13}C{1H} NMR (75 MHz, $CDCl_3$) δ 147.02, 146.35, 145.25, 137.28, 134.16, 133.04, 129.79, 128.28, 128.15, 128.02, 124.20, 115.84, 115.56, 107.65, 52.69, 21.66, 21.44.

6-Methyl-2-phenyl-3-(tosylmethyl)imidazo[1,2-α]pyridine (**3sc**) [29]. White solid, 203 mg, 90% yield. ethyl acetate/*n*-hexane (1:1). ^{1}H NMR (300 MHz, $CDCl_3$) δ 8.12 (s, 1H), 7.57 (d, J = 9.1 Hz, 1H), 7.38 (d, J = 8.3 Hz, 2H), 7.34–7.25 (m, 5H), 7.15 (dd, J = 9.2, 1.4 Hz, 1H), 7.07 (d, J = 8.1 Hz, 2H), 4.86 (s, 2H), 2.36 (d, J = 6.8 Hz, 6H). ^{13}C{1H} NMR (75 MHz, $CDCl_3$) δ 147.19, 145.32, 145.04, 134.35, 133.19, 129.85, 129.15, 128.33, 128.27, 128.20, 128.04, 122.69, 122.52, 116.86, 108.00, 52.81, 21.70, 18.55.

7-Chloro-2-phenyl-3-(tosylmethyl)imidazo[1,2-α]pyridine (**3sd**) [29]. White solid, 190 mg, 80% yield, ethyl acetate/*n*-hexane (1:1). ^{1}H NMR (300 MHz, $CDCl_3$) δ 8.22 (d, *J* = 7.3 Hz, 1H), 7.74 (d, *J* = 1.8 Hz, 1H), 7.26 (d, *J* = 8.3 Hz, 2H), 7.18 (s, 5H), 6.98 (d, *J* = 8.0 Hz, 2H), 6.89 (dd, *J* = 7.3, 2.0 Hz, 1H), 4.74 (s, 2H), 2.25 (s, 3H). ^{13}C{1H} NMR (75 MHz, $CDCl_3$) δ 147.76, 145.92, 145.44, 133.98, 132.33, 129.85, 128.37, 128.13, 128.03, 125.44, 120.06, 119.50, 116.76, 108.83, 52.35, 21.65.

7-Bromo-2-phenyl-3-(tosylmethyl)imidazo[1,2-α]pyridine (**3se**). White solid, 211 mg, 80% yield, ethyl acetate/*n*-hexane (1:1). mp. 231–132 °C. ^{1}H NMR (300 MHz, $CDCl_3$) δ 8.29 (d, *J* = 7.3 Hz, 1H), 7.56 (d, *J* = 1.6 Hz, 1H), 7.29–7.25 (m, 2H), 7.19 (s, 5H), 6.99 (d, *J* = 8.1 Hz, 2H), 6.79 (dd, *J* = 7.4, 2.1 Hz, 1H), 4.75 (s, 2H), 2.26 (s, 3H). ^{13}C{1H} NMR (75 MHz, $CDCl_3$) δ 147.98, 145.65, 145.45, 134.01, 132.63, 132.41, 129.86, 128.37, 128.12, 128.05, 125.50, 116.17, 114.46, 108.75, 52.39, 21.64. HRMS (ESI) m/z: [M + H]+ Calcd for $C_{21}H_{18}BrN_2O_2S^+$ 441.0267, Found 441.0267.

2-(Furan-2-yl)-3-(tosylmethyl)imidazo[1,2-α]pyridine (**3ta**) [29]. White solid, 173 mg, 82% yield, ethyl acetate/*n*-hexane (1:1). ^{1}H NMR (300 MHz, $CDCl_3$) δ 8.38 (d, *J* = 7.0 Hz, 1H), 7.60 (d, *J* = 9.2 Hz, 1H), 7.37 (d, *J* = 7.9 Hz, 2H), 7.29 (t, *J* = 8.2 Hz, 1H), 7.12 (s, 1H), 7.05 (d, *J* = 8.0 Hz, 2H), 6.92 (d, *J* = 7.0 Hz, 1H), 6.67 (d, *J* = 3.0 Hz, 1H), 6.35–6.24 (m, 1H), 5.07 (s, 2H), 2.26 (s, 3H). ^{13}C{1H} NMR (75 MHz, $CDCl_3$) δ 148.92, 146.24, 145.13, 142.10, 137.51, 133.82, 129.22, 128.31, 126.45, 124.67, 117.09, 112.98, 111.08, 108.44, 107.95, 52.55, 21.50.

2-(*t*-Butyl)-3-(tosylmethyl)imidazo[1,2-α]pyridine (**3tb**).[12] White solid, 176 mg, 86% yield, ethyl acetate/*n*-hexane (1:1). ^{1}H NMR (300 MHz, $CDCl_3$) δ 8.21 (d, *J* = 6.9 Hz, 1H), 7.60 (d, *J* = 8.2 Hz, 3H), 7.28 (d, *J* = 8.0 Hz, 2H), 7.20 (t, *J* = 8.1 Hz, 1H), 6.76 (t, *J* = 6.9 Hz, 1H), 4.89 (s, 2H), 2.41 (s, 3H), 1.30 (d, *J* = 2.1 Hz, 9H). ^{13}C{1H} NMR (75 MHz, $CDCl_3$) δ 155.80, 145.49, 144.90, 135.55, 130.04, 128.52, 125.27, 124.20, 116.92, 112.12, 106.12, 53.60, 34.18, 30.84, 21.68.

6-Phenyl-5-tosyl-2,3-dihydroimidazo[2,1-b]thiazole (**3ua**) [29]. White solid, 171 mg, 80% yield, ethyl acetate/*n*-hexane (1:1). ^{1}H NMR (300 MHz, $CDCl_3$) δ 7.40 (d, *J* = 7.8 Hz, 1H), 7.06 (d, *J* = 8.1 Hz, 1H), 4.45 (s, 1H), 4.29 (t, *J* = 7.4 Hz, 1H), 3.78 (t, *J* = 7.3 Hz, 1H). ^{13}C{1H} NMR (75 MHz, $CDCl_3$) δ 151.08, 147.55, 145.20, 133.87, 133.20, 129.67, 128.06, 128.03, 127.10, 126.68, 113.96, 53.18, 46.11, 34.77, 21.55.

6-Phenyl-5-tosylimidazo[2,1-b]thiazole (**3ub**) [29]. White solid, 165 mg, 78% yield, ethyl acetate/*n*-hexane (1:1). ^{1}H NMR (300 MHz, $CDCl_3$) δ 7.67 (d, *J* = 4.5 Hz, 1H), 7.43 (d, *J* = 8.0 Hz, 2H), 7.27–7.15 (m, 5H), 7.12 (d, *J* = 7.9 Hz, 2H), 6.86 (d, *J* = 4.4 Hz, 1H), 4.69 (s, 2H), 2.34 (s, 3H). ^{13}C{1H} NMR (75 MHz, $CDCl_3$) δ 150.77, 148.42, 145.37, 134.00, 133.01, 129.83, 128.27, 128.16, 127.78, 127.49, 119.07, 112.72, 109.77, 77.58, 77.16, 76.74, 53.45, 21.62.

2-(Thiophen-2-yl)-3-(tosylmethyl)imidazo[1,2-α] pyrimidine (**3uc**) [29]. White solid, 177 mg, 80% yield, ethyl acetate/*n*-hexane (1:1). ^{1}H NMR (300 MHz, $CDCl_3$) δ 8.72 (d, *J* = 6.7 Hz, 1H), 8.60 (s, 1H), 7.48 (d, *J* = 7.9 Hz, 2H), 7.29 (d, *J* = 5.2 Hz, 1H), 7.15 (d, *J* = 7.9 Hz, 2H), 7.07 (d, *J* = 3.2 Hz, 1H), 6.94 (s, 2H), 4.91 (s, 2H), 2.33 (s, 3H). ^{13}C{1H} NMR (75 MHz, $CDCl_3$) δ 151.32, 148.80, 145.86, 142.92, 135.04, 134.03, 132.76, 130.04, 128.37, 127.58, 127.18, 126.24, 109.20, 106.21, 52.67, 21.71.

3-Methyl-1-phenyl-4-(tosylmethyl)-1H-pyrazol-5-amine (**3ud**) [34]. Yellow solid, 167 mg, yield 82%, ethyl acetate/*n*-hexane (1:1). ^{1}H NMR (300 MHz, $CDCl_3$) δ 7.65 (d, *J* = 8.3 Hz, 2H), 7.50–7.42 (m, 4H), 7.37–7.28 (m, 3H), 4.40 (br s, 2H), 4.12 (s, 2H), 2.43 (s, 3H), 1.62 (s, 3H). ^{13}C{1H} NMR (75 MHz, $CDCl_3$) δ 148.75, 145.56, 144.99, 138.05, 134.66, 129.78, 129.58, 128.54, 127.61, 124.01, 90.17, 53.08, 21.65, 11.24.

4. Conclusions

In conclusion, we have developed a new sulfonylmethylation approach for imidazo[1,2-α]pyridines with DMA and sodium sulfinates affording diverse 3-(sulfonylmethyl)imidazo[1,2-α]pyridines in good yields with impressive simplicity. Free radical trapping and isotope labeling experiments showed that an oxidation-addition pathway might be included, and DMA provided carbon in this reaction.

Supplementary Materials: The following supporting information can be downloaded at: https://www.mdpi.com/article/10.3390/molecules29133196/s1, References [29,34] are cited in the supplementary materials.

Author Contributions: Conceptualization, X.B.; Methodology, S.S. and J.L.; Investigation, S.S. and H.Y.; Data curation, H.L. All authors have read and agreed to the published version of the manuscript.

Funding: This research received no external funding.

Institutional Review Board Statement: This study not involving humans or animals.

Informed Consent Statement: Not applicable.

Data Availability Statement: The original contributions presented in the study are included in the article/Supplementary Materials, further inquiries can be directed to the corresponding author/s.

Conflicts of Interest: There are no conflicts of interest to declare.

References

1. Gladysz, R.; Adriaenssens, Y.; De Winter, H.; Joossens, J.; Lambeir, A.-M.; Augustyns, K.; Van der Veken, P. Discovery and SAR of novel and selective inhibitors of urokinase plasminogen activator (uPA) with an imidazo[1,2-α]pyridine scaffold. *J. Med. Chem.* **2015**, *58*, 9238–9257. [CrossRef] [PubMed]
2. Kaminski, J.J.; Doweyko, A.M. Antiulcer agents. 6. Analysis of the in vitro biochemical and in vivo gastric antisecretory activity of substituted imidazo [1,2-α] pyridines and related analogues using comparative molecular field analysis and hypothetical active site lattice methodologies. *J. Med. Chem.* **1997**, *40*, 427–436. [PubMed]
3. Chen, X.; Xu, W.; Wang, K.; Mo, M.; Zhang, W.; Du, L.; Yuan, X.; Xu, Y.; Wang, Y.; Shen, J. Discovery of a novel series of imidazo[1,2-α]pyrimidine derivatives as potent and orally bioavailable lipoprotein-associated phospholipase A2 inhibitors. *J. Med. Chem.* **2015**, *58*, 8529–8541. [CrossRef] [PubMed]
4. Hamdouchi, C.; Blas, J.D.; Prado, M.D.; Gruber, J.; Heinz, B.A.; Vance, L. 2-Amino-3-substituted-6-[(E)-1-phenyl-2-(N-methylcarbamoyl) vinyl] imidazo [1,2-α] pyridines as a novel class of inhibitors of human rhinovirus: Stereospecific synthesis and antiviral activity. *J. Med. Chem.* **1999**, *42*, 50–59. [CrossRef] [PubMed]
5. Ramachandran, S.; Panda, M.; Mukherjee, K.; Choudhury, N.R.; Tantry, S.J.; Kedari, C.K.; Ramachandran, V.; Sharma, S.; Ramya, V.K.; Guptha, S.; et al. Synthesis and structure activity relationship of imidazo [1,2-α] pyridine-8-carboxamides as a novel antimycobacterial lead series. *Bioorg. Med. Chem. Lett.* **2013**, *23*, 4996–5001. [CrossRef] [PubMed]
6. Baviskar, A.T.; Madaan, C.; Preet, R.; Mohapatra, P.; Jain, V.; Agarwal, A.; Guchhait, S.K.; Kundu, C.N.; Banerjee, U.C.; Bharatam, P.V. N-fused imidazoles as novel anticancer agents that inhibit catalytic activity of topoisomerase IIα and induce apoptosis in G1/S phase. *J. Med. Chem.* **2011**, *54*, 5013–5030. [CrossRef] [PubMed]
7. Bode, M.L.; Gravestock, D.; Moleele, S.S.; van der Westhuyzen, C.W.; Pelly, S.C.; Steenkamp, P.A.; Hoppe, H.C.; Khan, T.; Nkabinde, L.A. Imidazo [1,2-α] pyridin-3-amines as potential HIV-1 non-nucleoside reverse transcriptase inhibitors. *Bioorg. Med. Chem.* **2011**, *19*, 4227–4237. [CrossRef] [PubMed]
8. Moraski, G.C.; Markley, L.D.; Cramer, J.; Hipskind, P.A.; Boshoff, H.; Bailey, M.A.; Alling, T.; Ollinger, J.; Parish, T.; Miller, M.J. Advancement of imidazo[1,2-α]pyridines with improved pharmacokinetics and nM activity vs. mycobacterium tuberculosis. *ACS Med. Chem. Lett.* **2013**, *4*, 675–679. [CrossRef] [PubMed]
9. Koubachi, J.; El Kazzouli, S.; Bousmina, M.; Guillaumet, G. Functionalization of imidazo [1,2-α] pyridines by means of metal-catalyzed cross-coupling reactions. *Eur. J. Org. Chem.* **2014**, *2014*, 5119–5123. [CrossRef]
10. Cao, H.; Zhan, H.; Lin, Y.; Lin, X.; Du, Z.; Jiang, H. Direct arylation of imidazo [1,2-α] pyridine at C-3 with aryl iodides, bro-mides, and triflates via copper (I)-catalyzed C–H bond functionali-zation. *Org. Lett.* **2012**, *14*, 1688–1692. [CrossRef]
11. Fu, H.Y.; Chen, L.; Doucet, H. Phosphine-free palladium-catalyzed direct arylation of imidazo [1,2-α] pyridines with aryl bromides at low catalyst loading. *J. Org. Chem.* **2012**, *77*, 4473–4478. [CrossRef] [PubMed]
12. Choy, P.Y.; Luk, K.C.; Wu, Y.; So, C.M.; Wang, L.-L.; Kwong, F.Y. Regioselective direct C-3 arylation of imidazo[1,2-α]pyridines with aryl tosylates and mesylates promoted by palladium-phosphine complexes. *J. Org. Chem.* **2015**, *80*, 1457–1463. [CrossRef] [PubMed]
13. Firmansyah, D.; Deperasińska, I.; Vakuliuk, O.; Banasiewicz, M.; Tasior, M.; Makarewicz, A.; Cyrański, M.K.; Kozankiewicz, B.; Gryko, D.T. Double head-to-tail direct arylation as a viable strategy towards the synthesis of the aza-analog of dihydrocyclopenta [hi] aceanthrylene-an intriguing antiaromatic heterocycle. *Chem. Commun.* **2016**, *52*, 1262–1265. [CrossRef] [PubMed]
14. Ghosh, M.; Naskar, A.; Mitra, S.; Hajra, A. Palladium-catalyzed α-selective alkenylation of imidazo[1,2-α]pyridines through aerobic cross-dehydrogenative coupling reaction. *Eur. J. Org.Chem.* **2015**, *2015*, 715–718. [CrossRef]
15. Zhan, H.; Zhao, L.; Li, N.; Chen, L.; Liu, J.; Liao, J.; Cao, H. Ruthenium-catalyzed direct C-3 oxidative olefination of imidazo [1,2-α] pyridines. *RSC Adv.* **2014**, *4*, 32013–32016. [CrossRef]
16. Cao, H.; Lei, S.; Liao, J.; Huang, J.; Qiu, H.; Chen, Q.; Qiu, S.; Chen, Y. Palladium (II)-catalyzed intermolecular oxidative C-3 alkenylations of imidazo [1,2-α] pyridines by substrate-contolled regioselective C-H functionalization. *RSC Adv.* **2014**, *4*, 50137–50140. [CrossRef]

17. Monir, K.; Bagdi, A.K.; Ghosh, M.; Hajra, A. Regioselective oxidative trifluoromethylation of imidazoheterocycles via C(sp^2)-H bond functionalization. *J. Org. Chem.* **2015**, *80*, 1332–1337. [CrossRef] [PubMed]
18. Wu, A.; Zhang, H.-R.; Jin, R.-X.; Lan, Q.; Wang, X.-S. Nickel-catalyzed C-H trifluoromethylation of electron-rich heteroarenes. *Adv. Synth. Catal.* **2016**, *358*, 3528–3533. [CrossRef]
19. Cao, H.; Lei, S.; Li, N.; Chen, L.; Liu, J.; Cai, H.; Qiu, S.; Tan, J. Cu-catalyzed selective C3-formylation of imidazo [1,2-α] pyridine C-H bonds with DMSO using molecular oxygen. *Chem. Commun.* **2015**, *51*, 1823–1825. [CrossRef]
20. Mondal, S.; Samanta, S.; Jana, S.; Hajra, A. (Diacetoxy) iodobenzene-mediated oxidative C-H amination of imidazopyridines at ambient temperature. *J. Org. Chem.* **2017**, *82*, 4504–4510. [CrossRef]
21. Lu, S.; Tian, L.-L.; Cui, T.-W.; Zhu, Y.-S.; Zhu, X.; Hao, X.-Q.; Song, M.-P. Copper-mediated C-H amination of imidazopyridines with N-fluorobenzenesulfonimide. *J. Org. Chem.* **2018**, *83*, 13991–14000. [CrossRef] [PubMed]
22. Neto, J.S.S.; Balaguez, R.A.; Franco, M.S.; Machado, V.C.S.; Saba, S.; Rafique, J.; Galetto, F.Z.; Braga, A.L. Trihaloisocyanuric acids in ethanol: An eco-friendly system for the regioselective halogenation of imidazo-heteroarenes. *Green Chem.* **2020**, *22*, 3410–3415. [CrossRef]
23. Semwal, R.; Ravi, C.; Kumar, R.; Meena, R.; Adimurthy, S. Sodium salts (NaI/NaBr/NaCl) for the halogenation of imidazo-fused heterocycles. *J. Org. Chem.* **2019**, *84*, 792–805. [CrossRef] [PubMed]
24. Guo, T.; Wei, X.-N.; Zhang, M.; Liu, Y.; Zhu, L.-M.; Zhao, Y.-H. Catalyst and additive-free oxidative dual C-H sulfenylation of imidazoheterocycles with elemental sulfur using DMSO as a solvent and an oxidant. *Chem. Commun.* **2020**, *56*, 5751–5754. [CrossRef] [PubMed]
25. Zhu, W.; Ding, Y.; Bian, Z.; Xie, P.; Xu, B.; Tang, Q.; Wu, W.; Zhou, A. One-pot three-component synthesis of alkylthio-/arylthio-substituted imidazo[1,2-α]pyridine derivatives via C(sp^2)-H functionalization. *Adv. Synth. Catal.* **2017**, *359*, 2215–2221. [CrossRef]
26. Hiebel, M.-A.; Berteina-Raboin, S. Iodine-catalyzed regioselective sulfenylation of imidazoheterocycles in PEG 400. *Green Chem.* **2015**, *17*, 937–944. [CrossRef]
27. Lu, S.; Zhu, X.-J.; Li, K.; Guo, Y.-J.; Wang, M.-D.; Zhao, X.-M.; Hao, X.-Q.; Song, M.-P. Reactivity of *p*-toluenesulfonylmethyl isocyanide: Iron-involved C-H tosylmethylation of imidazopyridines in nontoxic media. *J. Org. Chem.* **2016**, *81*, 8370–8377. [CrossRef]
28. Kalari, S.; Shinde, A.U.; Rode, H.B. Methylene-tethered arylsulfonation and benzotriazolation of aryl/heteroaryl C-H bonds with DMSO as a one-carbon surrogate. *J. Org. Chem.* **2021**, *86*, 17684–17695. [CrossRef] [PubMed]
29. Chen, J.-W.; Wen, K.-M.; Wu, Y.-R.; Shi, J.; Yao, X.-G.; Tang, X.-D. Transition metal catalyst-free C-3 sulfonylmethylation of imidazo[1,2-α]pyridines with glyoxylic acid and sodium sulfinates in water. *J. Org. Chem.* **2022**, *87*, 3780–3787. [CrossRef]
30. Lou, S.-J.; Xu, D.-Q.; Shen, D.-F.; Wang, Y.-F.; Liu, Y.-K.; Xu, Z.-Y. Highly efficient vinylaromatics generation via Iron-catalyzed sp^3 C-H bond functionalization CDC reaction: A novel approach to preparing substituted benzo[α]phenazines. *Chem. Commun.* **2012**, *48*, 11993–11995. [CrossRef]
31. Modi, A.; Ali, W.; Patela, B.K. *N,N*-dimethylacetamide (DMA) as a methylene synthon for regioselective linkage of imidazo[1,2-α]pyridine. *Adv. Synth. Catal.* **2016**, *358*, 2100–2107. [CrossRef]
32. Kaswan, P.; Nandwana, N.K.; DeBoef, B.; Kumar, A. Vanadyl acetylacetonate catalyzed methylenation of imidazo[1,2-α]pyridines by using dimethylacetamide as a methylene source: Direct access to bis(imidazo[1,2-α]pyridin-3-yl)methanes. *Adv. Synth. Catal.* **2016**, *358*, 2108–2115. [CrossRef]
33. Sun, S.-N.; Li, J.-C.; Gou, Y.-B.; Gao, Z.-H.; Bi, X.-J. Controllable synthesis of disulfides and thiosulfonates from sodium sulfinates mediated by hydroiodic acid using ethanol and H_2O as solvents. *Org. Biomol. Chem.* **2022**, *20*, 8885–8892. [CrossRef] [PubMed]
34. Chen, J.-W.; Tian, J.-H.; Wen, K.-M.; Gao, Q.-W.; Shi, J.; Yao, X.-G.; Wu, T.; Tang, X.-D. A new method for C (sp^2)-H sulfonylmethylation with glyoxylic acid and sodium sulfinates. *Org. Biomol. Chem.* **2022**, *20*, 1652–1655. [CrossRef]

molecules

MDPI

Article

The Influence of the Comonomer Ratio and Reaction Temperature on the Mechanical, Thermal, and Morphological Properties of Lignin Oil–Sulfur Composites

Katelyn A. Tisdale, Nawoda L. Kapuge Dona and Rhett C. Smith *

Department of Chemistry and Center for Optical Materials Science and Engineering Technology, Clemson University, Clemson, SC 29634, USA

* Correspondence: rhett@clemson.edu

Abstract: Although lignin is a plentiful biomass resource, it continually exists as an underutilized component of biomass material. Elemental sulfur is another abundant yet underutilized commodity produced as a by-product resulting from the refining of fossil fuels. The current study presents a strategy for preparing five durable composites via a simple one-pot synthesis involving the reaction of lignin oil and elemental sulfur. These lignin oil–sulfur composites $LOS_x@T$ (where x = wt. % sulfur, ranging from 80 to 90, and T represents the reaction temperature in °C) were prepared via the reaction of elemental sulfur and lignin oil (LO) with elemental sulfur. The resulting composites could be remelted and reshaped several times without the loss of mechanical strength. Mechanical, thermal, and morphological studies showed that $LOS_x@T$ possesses properties competitive with some mechanical properties of commercial building materials, exhibiting favorable compressive strengths (22.1–35.9 MPa) and flexural strengths (5.7–6.5 MPa) exceeding the values required for many construction applications of ordinary Portland cement (OPC) and brick formulations. While varying the amount of organic material did not result in a notable difference in mechanical strength, increasing the reaction temperature from 230 to 300 °C resulted in a significant increase in compressive strength. The results reported herein reveal potential applications of both lignin and waste sulfur during the ongoing effort toward developing recyclable and sustainable building materials.

Keywords: sustainable composites; sulfur; lignin; lignin oil; compressive strength

check for updates

Citation: Tisdale, K.A.; Kapuge Dona, N.L.; Smith, R.C. The Influence of the Comonomer Ratio and Reaction Temperature on the Mechanical, Thermal, and Morphological Properties of Lignin Oil–Sulfur Composites. *Molecules* **2024**, *29*, 4209. https://doi.org/10.3390/molecules29174209

Academic Editor: Ming Wang

Received: 3 July 2024
Revised: 30 August 2024
Accepted: 31 August 2024
Published: 5 September 2024

1. Introduction

Lignin holds about 30% of the Earth's carbon [1–4] and is a widely available potential feedstock for the chemical industry. Biomass processing produces over 5 billion mt of waste lignin annually, and paper production contributes a supplementary 600,000 tons/year of lignin waste [5]. Despite the high abundance of lignin, its complex structure (Figure 1a) [6–10] introduces many challenges when developing methods for its valorization [11–16]. Lignin is composed of mixtures of coniferyl alcohol, 4-*p*-coumaryl alcohol, and sinapyl alcohol units linked together, forming the subunits *p*-hydroxyphenyl, guaiacyl, and syringyl, correspondingly, in lignin (Figure 1b) [17–28]. The magnitude of the contribution of each component varies with both the plant species and the part of the plant from which the lignin is derived [1,29,30].

Both "upstream" and "downstream" processes contribute to efficient lignin valorization. Upstream processes involve separating and isolating lignin, lignin bioengineering, and catalytic conversion, whereas downstream methods involve the depolymerization and upgrading of lignin [31–33]. Depolymerization yields bio-oils or lignin oils comprising mixtures of small molecular fragments of lignin and is among the most promising routes to lignin employment. Green methods of lignin depolymerization are therefore being hotly pursued. A few "green" solvents employed in extraction processes involve supercritical fluids, liquified gases, and bio-based solvents. Supercritical fluid extraction (SFE) requires an extraction solvent that can be removed easily, a shorter extraction time, and enhanced selectivity and efficiency [34]. Bio-based solvents are derivative of a broad range of biomass sources, including the extraction of vegetable oils, the fermentation of carbohydrates, and the steam distillation of wood [35–37]. Liquified gases are another green solvent. They require low temperature for easy evaporation, allowing for room-temperature liquid gas extractions to be performed at room temperature, consuming minimal energy and extracts containing an insignificant amount residual solvents [38,39]. A recent report [40] discussing lignin's mild thermolytic solvolysis of yielding a solubilized form of lignin (termed lignin oil) garnered particular interest for the present study due to the process being performed at temperatures ranging from 100 to 350 °C by means of various alcohols as solvents [41]. By heating in an ethanol solvent at a low temperature of 100 °C, this procedure accessibly accomplished the conversion of lignin (64 wt. %) to the desired product, lignin oil (LO). Ethanol is a preferred green solvent and is advantageous because of its low cost and availability [42–45]. Lignin oil produced by this reported procedure was used in the studies described herein.

Another, existing abundant, and underutilized by-product is elemental sulfur. Over 80 Mt/y of elemental sulfur is produced from fossil fuel refining [46–62]. Multiple efforts have been previously reported for preparing composites from lignin or lignin derivatives and elemental sulfur [63–67].

Early examples of high-sulfur-content materials (HSMs) were made using the inverse vulcanization [57,68–82] of olefins [64,83–92]. The preparation of lignin–sulfur composites thus required the modification of lignin with olefin-bearing substituents before the reaction with sulfur could take place [63,65]. The previously reported LS_x (where x = wt. % sulfur, varying between 80 and 99), for example, was prepared using allyl-derivatized lignin (10 wt. %) [63], while ELS_x@T (where x = wt. % sulfur, varying between 80 and 90, and T represents reaction temperature in °C) was prepared by reactions of oleic-esterified lignin with sulfur [65]. Two reaction temperatures were used for the preparation of ELS_x@T due to the fact that the reaction at 180 °C strictly produces S–C bonds at olefinic sites, while reacting at a higher temperature of 230 °C can produce S–C bonds at both olefinic sites and aryl sites [65]. Although LS_x and ELS_x@T proved to be durable lignin–sulfur composites, the multistep synthesis required the use of solvents and separation methods that detracted from the desired greenness, affordability, and process atom economy related to these materials.

The preparation of CLS_{90} (90 wt. % S, 10 wt. % chlorolignin, a paper/pulping waste product) [93] via the reaction of sulfur (90 wt. %) with chlorolignin (10 wt. %) through radical-induced aryl halide–sulfur polymerization (RASP) [94] presented a more proficient one-pot synthesis to produce a durable composite [95]. However, this process yields 0.5 equivalents of toxic S_2Cl_2 for each C–S bond formed, making it less environmentally friendly and less atom-economical compared to composites prepared via inverse vulcanization. A wide range of composites incorporating lignin have also been reported that do not require sulfur [96–105].

The search for a more sustainable S–C bond forming reaction to allow the direct utilization of lignin without the need for olefination has simplified the synthetic process toward lignin–sulfur composites and has improved the atom economy of the processes used for their preparation. Recently reported C–S bond-forming mechanisms during the reaction of sulfur with anisole derivatives (Scheme 1) such as *O,O′*-dimethylbisphenol A [106], guaiacol [64], and others [107] suggest that the direct reaction of lignin oil mixtures with sulfur was possible, as was validated in a recent proof-of-principle study detailing the crosslinking that takes place between sulfur and small lignin derivatives present in lignin oil [108].

Herein, five composites were prepared by heating lignin oil (LO) with elemental sulfur for 2 h. The composites, including LOS_x@T (where x = wt. % sulfur, varying from 80 to 90, and T represents the reaction temperature in °C), were of interest to exploit the features of previously reported ELS_{90}@180 and ELS_{90}@230 due to S–C bond-forming at both olefinic and aryl sites after an increase in temperature. Previously studied model compounds with functional groups found within thermal degradation products or lignin subunits were reacted with elemental sulfur, implementing matching conditions [107] as reported to enhance the understanding of S–C bond-forming reactions within the lignin–sulfur composites prepared herein (Scheme 1a). The thermal, morphological, and mechanical properties were analyzed using powder X-ray diffraction (XRD), scanning electron microscopy accompanied by elemental mapping via energy-dispersive X-ray analysis (SEM-EDX), thermogravimetric analysis (TGA), flexural strength analysis, mechanical test stand analysis, and differential scanning calorimetry (DSC).

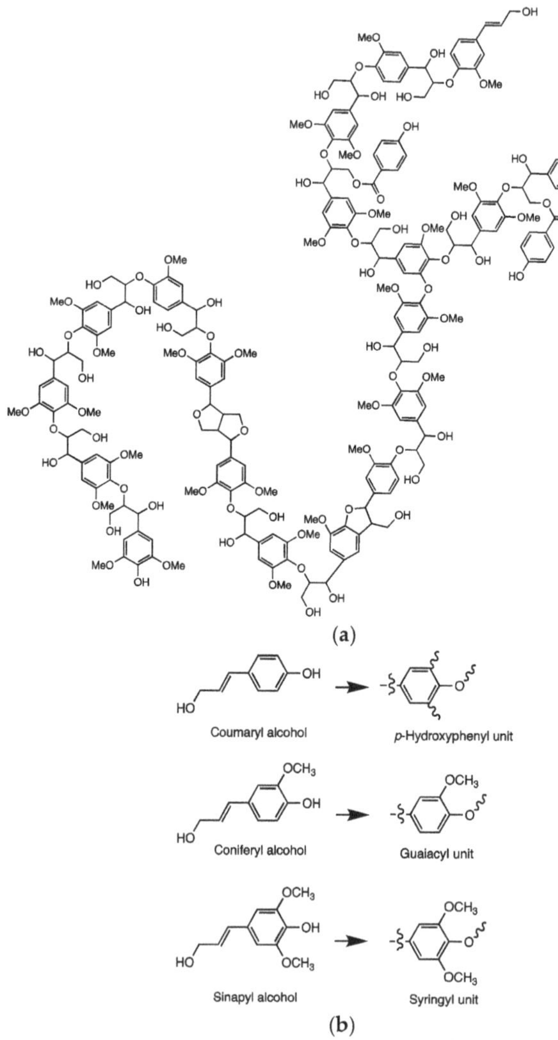

Figure 1. Structure displaying a lignin polymer (**a**) and lignin biosynthetic precursors (**b**).

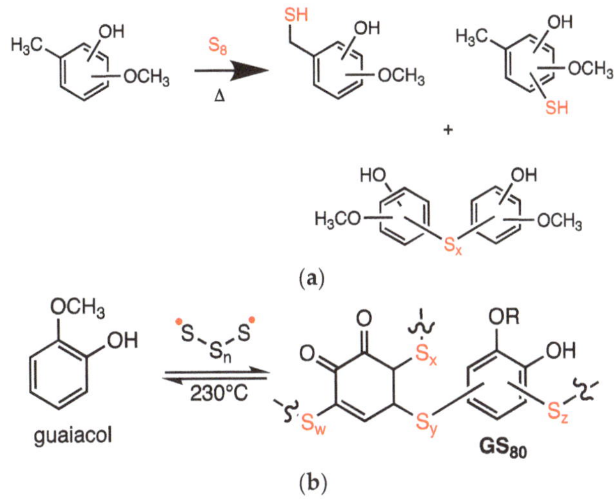

Scheme 1. (**a**) S–C bond-forming reactions for lignin-derived small molecules bearing some combination of $-CH_3$, $-OH$, and $-OCH_3$ substituents [107] and (**b**) the major S–C bond forming reaction of composite GS_{80} from sulfur and guaiacol. Polymeric sulfur is represented by $-S_x-$ with subscript letters w, x, y, and z signifying some amount of sulfur in the polymer chain.

2. Results and Discussion

2.1. Synthesis and Chemical Characterization of Composites

The lignin oil utilized herein was prepared and characterized as previously reported [108].

Composites LOS_x@T (where x = wt. % sulfur, varying from 80 to 90, and T represents the reaction temperature in °C) were prepared by heating lignin oil (10, 15, or 20 wt. %) and sulfur (90, 85, or 80 wt. %) to the requisite temperature in a sealed vessel for 2 h. The composites LOS_{80}@230 and LOS_{85}@230, and LOS_{90}@230 [108] were prepared in thick-walled glass pressure vessels slowly heated to 230 °C and then held at 230 °C for 2 h. Composites LOS_{90}@250 and LOS_{90}@300 were prepared in a stainless steel autoclave reactor The composites were mechanically stirred and slowly heated to either 250 °C or 300 °C, respectively, and held at that temperature for 2 h. After cooling to room temperature, all the materials solidified into a dark brown, remeltable solid. After melting, the composite material was transferred into molds for shaping and cooled to room temperature (Figures 2 and 3).

The sulfur present in high sulfur-content materials (HSMs) exists as both $-S_x-$ chains that are linked covalently to the organic comonomers and entrapped sulfur species such as *cyclo*-S_8, known as "dark sulfur", which is not linked covalently to the organic comonomers [109–111]. Dark sulfur's relative quantity within an HSM can influence both its thermal and its mechanical properties. In this study, dark sulfur was extracted from each HSM using ethyl acetate. The wt. % of each sample that was soluble (dark sulfur) and the insoluble fraction (organic crosslinked network) are provided in Table 1. The majority of sulfur in the reported lignin–sulfur composites was stabilized as covalently attached crosslinking catenates, with all of the composites showing a similar contribution of dark sulfur over a narrow range of 22–31 wt. %. Elemental analysis confirmed that very little organic material was extractable, with soluble portions consisting of 94–99% sulfur. It is important to note that the extraction experiment was only conducted to assess the relative contributions of the dark sulfur and of the crosslinked network to the overall structure. All subsequent thermal, morphological, and mechanical testing was conducted on the complete composites, comprising the network with the entrapped dark sulfur.

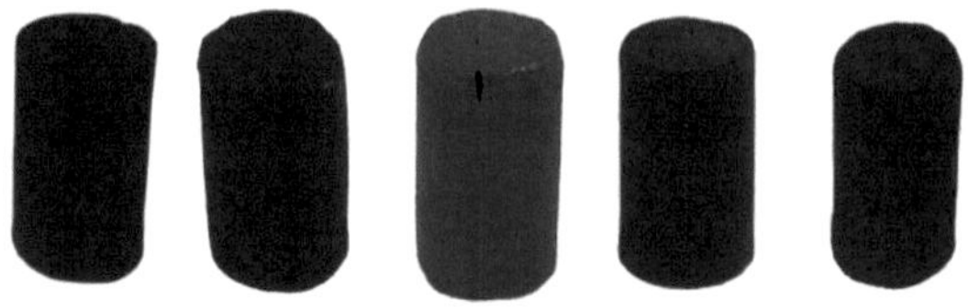

Figure 2. Representative photos of compressional cylinders of (from left to right) LOS_{80}@230, LOS_{85}@230, LOS_{90}@230, LOS_{90}@250, and LOS_{90}@300.

Figure 3. Representative photos of rectangular prisms for flexural strength tests of LOS_{90}@230 (**left**) and LOS_{85}@230 (**right**).

Table 1. Soluble (dark sulfur) and insoluble (crosslinked material) wt. % fractions of lignin–sulfur composites.

Materials	Dark Sulfur [a] (wt. %)	Crosslinked Material [a] (wt. %)
LOS_{80}@230	22	82
LOS_{85}@230	31	77
LOS_{90}@230	21	82
LOS_{90}@250	24	78
LOS_{90}@300	22	81

[a] Sum of soluble and insoluble fractions is greater than 100% due to rounding and errors associated with the measurements.

Powder X-ray diffraction (PXRD) (Figures 4–6 and S1–S5, Supplementary Materials) was used for the initial qualitative assessment of crystallinity and species contributing to the crystalline domains of LOS_x@T composites. As the amount of organic material increased, the PXRD data reveal the anticipated corresponding reduction in crystallinity: LOS_{90}@230 is a crystalline polymer, LOS_{85}@230 is a partially amorphous polymer, and LOS_{80}@230 is primarily amorphous. This trend reflects the diminished capacity for the crystalline packing of poly/oligosulfur crosslinking chains as they become progressively shorter in response to the availability of more organic crosslinkable sites. The sharp peaks are the crystalline sulfur, while broad baselines between 20 and 30 2θ, most noticeable for the amorphous LOS_{80}@230 (Figure 6), are characteristic of the polymer. LOS_{90}@230 is almost entirely crystalline sulfur within the polymer composite, as seen by the very tall, sharp peaks (Figure 4). While PXRD provides qualitative confirmation of a decreasing percentage of crystallinity with increasing organic comonomer, it only provides a semiquantitative estimate of the percentage of crystallinity. Due to the nature of the composite having numerous diffractions of crystalline sulfur, it is difficult to identify lignin's contribution within the pattern. Therefore, differential scanning calorimetry (DSC) data were used to calculate a more accurate percentage crystallinity for the composites (Table 2). We find that the percentage crystallinity values correlate quantitatively well with the PXRD patterns, which show increasing intensity of broad features about 20–30 2θ.

FT-IR spectra revealed further evidence of the formation of C–S bonds between the polymeric sulfur and lignin oil components (Figure 7). At 660 cm^{-1}, a visible peak indicates a C–S stretch [112,113]. The IR spectra (Figures S6 and S7, Supplementary Materials) also revealed that LOS_x@T composites preserved peaks characteristic of LO such as O–H stretches (3200–3500 cm^{-1}), C–H stretches (2930 cm^{-1}), and C–O stretches

(1030–1200 cm^{-1}) [114,115]. Other important peaks to note are those characteristic of lignin bands such as the C=O aromatic skeletal vibration observed from 1600 to 1620 cm^{-1} and the unconjugated carbonyl groups observed from 1700 to 1720 cm^{-1}. As previously reported, the C–S stretch for LOS_{90}@230 is predictably broadened because it has the lowest organic content and reaction temperature of the composites and therefore the longest average polysulfur chains between organic sites, leading to the greatest polydispersity of the possible bond stretch energies. The displacement of the bands observed can also be attributed to the reaction temperatures of the different composites. For example, slightly sharper peaks were observed in LOS_{80}@230, LOS_{85}@230, and LOS_{90}@230 from 600 to 1200 cm^{-1} (Figure S8, Supplementary Materials) when compared to LOS_{90}@250 and LOS_{90}@300, which were reacted at higher temperatures.

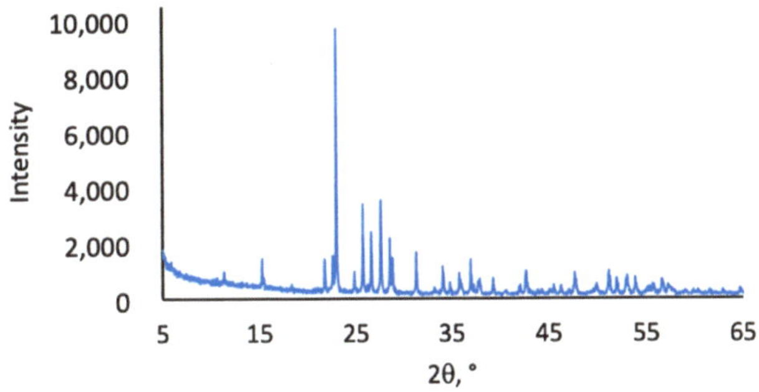

Figure 4. Powder XRD data for LOS_{90}@230.

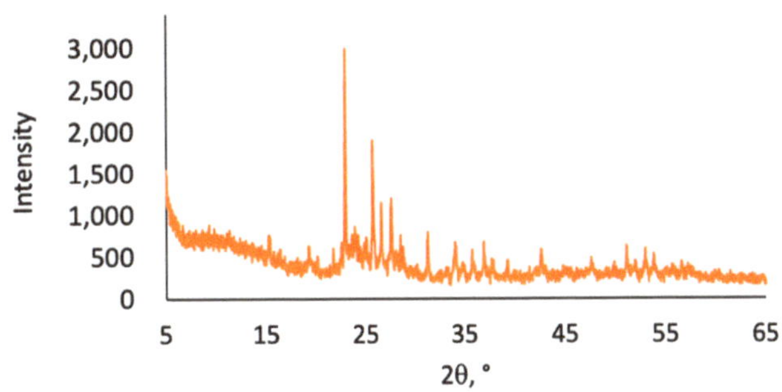

Figure 5. Powder XRD data for LOS_{85}@230.

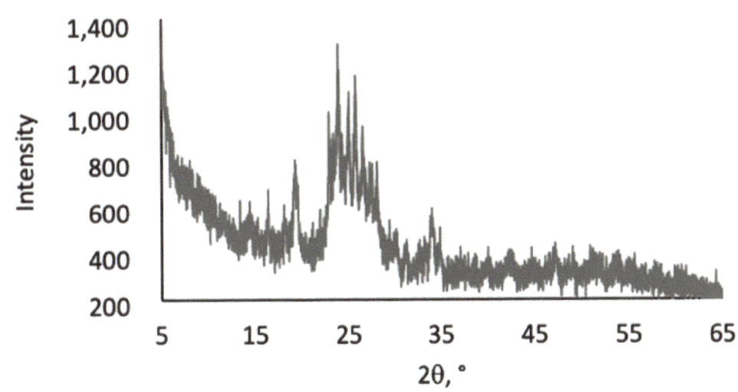

Figure 6. Powder XRD trace for LOS_{80}@230.

Table 2. Thermal and morphological properties of lignin oil–sulfur composites, elemental sulfur, and other composites.

Materials	$T_d^{[a]}$/°C	$T_m^{[b]}$/°C	$T_{g,DSC}^{[c]}$/°C	ΔH_m J/g	ΔH_{cc} J/g	Percentage Crystallinity [d]
LOS_{90}@230 [e]	229	114	−36	26	ND [j]	57
LOS_{85}@230 [e]	230	106	−35	35	26	19
LOS_{80}@230 [e]	230	105	−35	32	32	1
LOS90@250 [e]	231	117	−36	30	16	30
LOS90@300 [e]	230	119	ND [j]	ND [j]	ND [j]	ND [j]
LS_{80} [f]	236	118	−34	30	25	8
LS_{85} [f]	235	118	−35	34	13	40
LS_{90} [f]	237	117	ND [j]	34	ND [j]	67
LS_{95} [f]	235	107	ND [j]	37	ND [j]	63
LS_{99} [f]	233	113	ND [j]	47	ND [j]	91
GS_{80} [g]	264	ND [j]	−30	ND [j]	ND [j]	ND [j]
ELS_{90}@180 [h]	230	117	−37	ND [j]	ND [j]	ND [j]
ELS_{90}@230 [h]	234	117	−37	ND [j]	ND [j]	ND [j]
ELS_{80}@180 [h]	231	117	−37	ND [j]	ND [j]	ND [j]
S_8 [i]	228/229	119	ND [j]	45	ND [j]	100

[a] The temperature at which a mass loss of 5% is observed. [b] The peak temperature at the peak maximum of the endothermic melting. [c] Glass transition temperature. [d] Percentage crystallinity reduction for each composite sample calculated with respect to sulfur (normalized to 100%). [e] Lignin oil and elemental sulfur (where x = wt. % sulfur, varying from 80 and 90, and T represents the reaction temperature in °C). [f] LS_x composites comprising 10 wt. % allyl-derivatized lignin and elemental sulfur, where x = wt. % sulfur in monomer feed, ranging from 80 to 99. [g] Composites consisting of guaiacol (20 wt. %) and elemental sulfur (80 wt. %). [h] ELS_x@T composites comprising esterified lignin and elemental sulfur (where x = wt. % sulfur in the reaction mixture, varying from 80 to 90, and T represents the reaction temperature in °C). [i] Elemental sulfur. [j] Not determined.

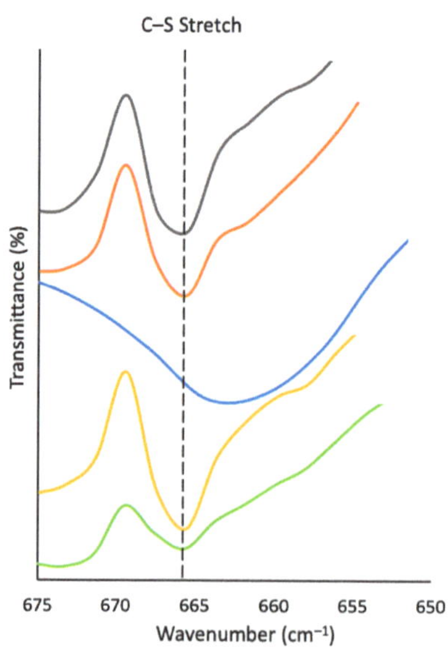

Figure 7. Portion of the FT-IR spectrum revealing the S–C stretch for LOS_{80}@230 (grey trace), LOS_{85}@230 (orange trace), LOS_{90}@230 (blue trace), LOS_{90}@250 (yellow trace), and LOS_{90}@300 (green trace). The entire set of IR spectra are presented in the Supplementary Materials file.

Scanning electron microscopy (SEM, Figures S8–S12) imaging accompanied by element mapping by energy-dispersive X-ray analysis (EDX) was utilized to evaluate the dispersal of elements in the $LOS_x@T$ composites (Figure 8). SEM/EDX data revealed a homogeneous distribution of sulfur, carbon, and oxygen.

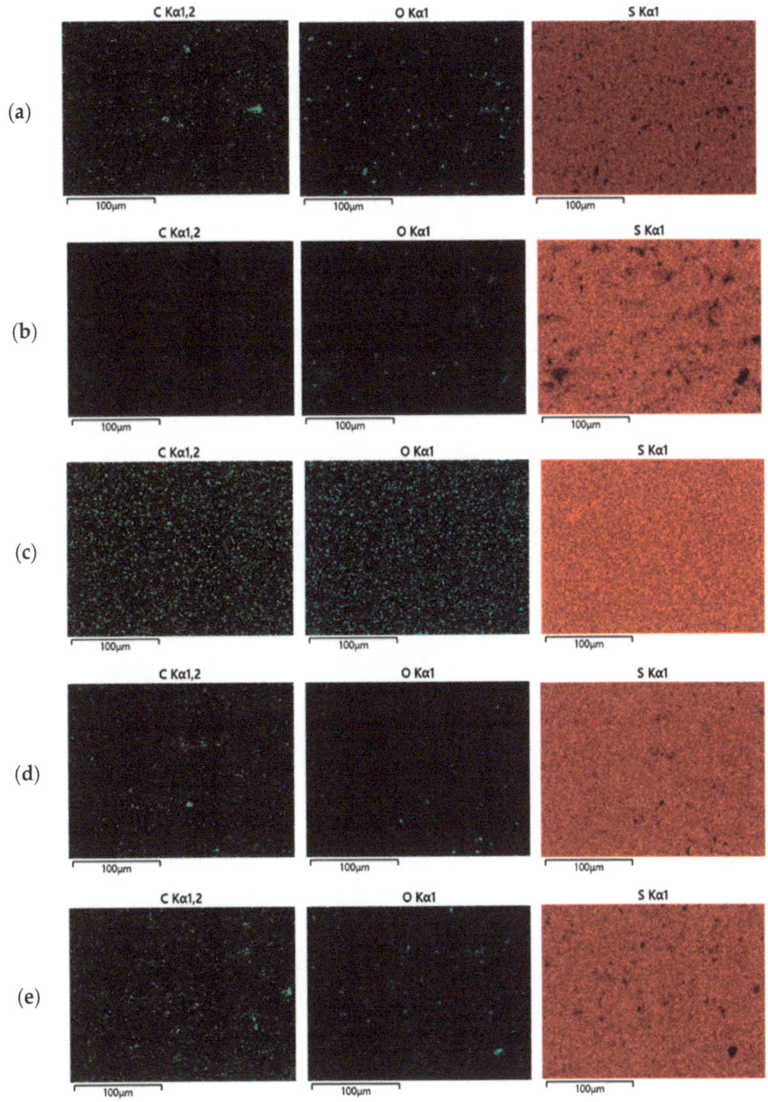

Figure 8. Elemental mapping of the elements carbon (green, left), oxygen (blue, middle), and sulfur (red, right) in (**a**) $LOS_{80}@230$, (**b**) $LOS_{85}@230$, (**c**) $LOS_{90}@230$, (**d**) $LOS_{90}@250$, and (**e**) $LOS_{90}@300$ by energy-dispersive X-ray analysis (EDX).

2.2. Thermal and Morphological Properties of $LOS_x@T$

The thermal stability of the reported $LOS_x@T$ composites was assessed using thermogravimetric analysis (Figure 9). The TGA data for lignin oil (LO) prior to its reaction with elemental sulfur are also provided in Figure 9, and the origin of the decomposition events in LO have been previously delineated [116]. The TGA trace for each of the composites showed a single decomposition temperature (T_d) ranging from 229 to 231 °C. These temperatures are characteristic of such HSMs due to the fact that elemental sulfur has a T_d of

229 °C, which is attributable to the sublimation of sulfur in the LOS_x@T composites. The smaller decomposition event at >300 °C in LOS_{90}@230 and LOS_{90}@250 is likely attributable to similar mechanisms leading to the corresponding decomposition events observed in lignin oil. The T_d values for LOS_x@T and other previously reported lignin-containing HSMs are summarized in Table 2.

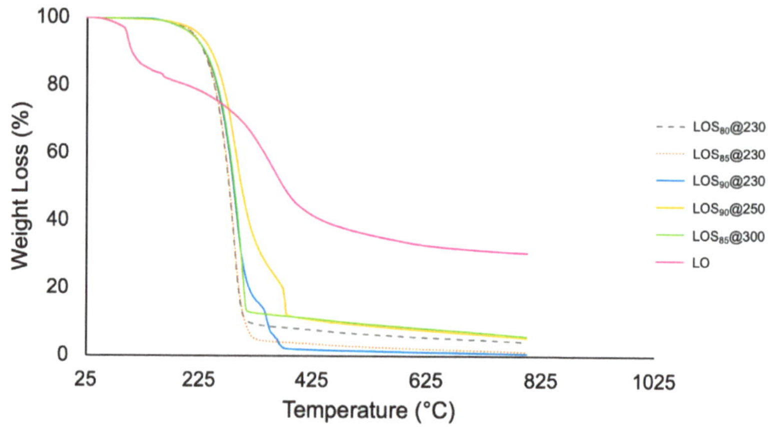

Figure 9. Thermogravimetric analysis (TGA) traces for LOS_{80}@230 (grey dashed line), LOS_{85}@230 (orange dotted line), LOS_{90}@230(blue solid line), LOS_{90}@250 (yellow solid line), LOS_{90}@300 (green solid line), and lignin oil (LO) (pink solid line).

The analysis of LOS_x@T composites by differential scanning calorimetry (DSC) (Figures S13–S27, Supplementary Materials) showed thermal transitions that were concomitant with the presence of polymeric sulfur and *cyclo*-S_8. LOS_{90}@230, LOS_{90}@250, and LOS_{90}@300 exhibited the melting peaks (114 °C, 117 °C, and 119 °C) expected for *cyclo*-S_8. However, LOS_{85}@230 and LOS_{80}@230 exhibited lower melting peaks at 106 °C and 105 °C, indicating an α-to-β transition of sulfur. A glass transition temperature (T_g) was also witnessed (Figure 10) at −35 °C or −36 °C for all composites other than LOS_{90}@300, diagnostic for polymeric sulfur [117–119]. Similar to the previously studied $LS_{90–99}$ composites, LOS_{90}@300 did not exhibit a glass transition temperature. This is likely due to the higher temperature allowing for more crosslinking, leading to shorter sulfur chains.

Melting and cold crystallization enthalpies determined from DSC data were employed for the calculation of the percentage crystallinity of the composites, further indicating the presence of *cyclo*-S_8, using Equation (1).

$$\Delta\chi_c = 1 - \left\{ \frac{\Delta H_m - \Delta H_{cc}}{\Delta H_{m(S)} - \Delta H_{cc(S)}} \right\} \times 100\% \quad (1)$$

where ΔX_c signifies the change in the percentage crystallinity with respect to sulfur, ΔH_m signifies the melting enthalpy of the composite, ΔH_{cc} represents the cold crystallization enthalpy of the composite, $\Delta H_{m(S)}$ is sulfur's melting enthalpy, and $\Delta H_{cc(S)}$ is sulfur's cold crystallization enthalpy.

The percentage crystallinity of LOS_{90}@230 is 57% relative to the crystallinity of pure elemental sulfur. Lower percentage crystallinity results in the reduced brittleness of HSMs. Compared to LS_x composites (Table 1), LOS_{90}@230 possesses a lower percentage crystallinity than any of the composites made from allylated lignin, for example, $LS_{90–99}$ (percentage crystallinity of 67–91%). However, LOS_{90}@230 possesses a higher percentage crystallinity than that of composites LS_{80} (8%) and LS_{85} (40%), reflecting the trend more qualitatively revealed by PXRD data (vide supra). These trends are summarized in Figure 11. LOS_{85}@230 also has a lower percentage crystallinity (19%) than that of

LOS_{90}@230 (57%) and $LS_{85–99}$ (40–91%). However, LOS_{85}@230 possesses a higher percentage crystallinity than that of composite LS_{80} (8%). The reaction temperature does not seem to have a significant effect on the percentage crystallinity of lignin–sulfur materials.

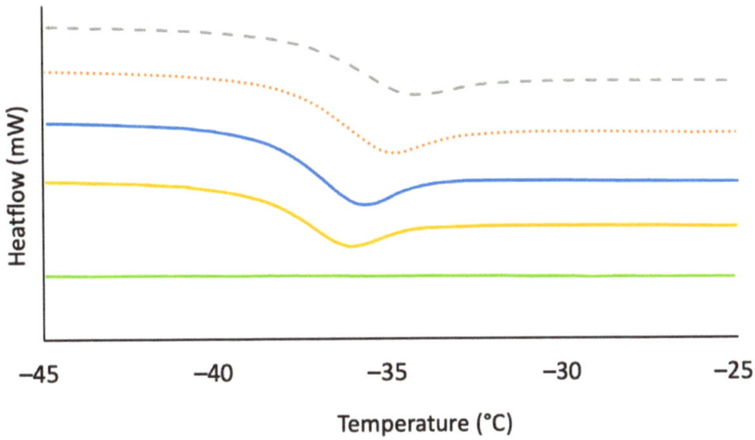

Figure 10. Inset of differential scanning calorimetry (DSC) thermograms emphasizing the glass transition temperature (T_g) of LOS_{90}@230 (blue solid line), LOS_{85}@230 (orange dotted line), LOS_{80}@230 (grey dashed line), LOS_{90}@250 (yellow solid line), and LOS_{90}@300 (green solid line). Full DSC thermographs from −60 to 140 C° are provided in the Supplementary Information file.

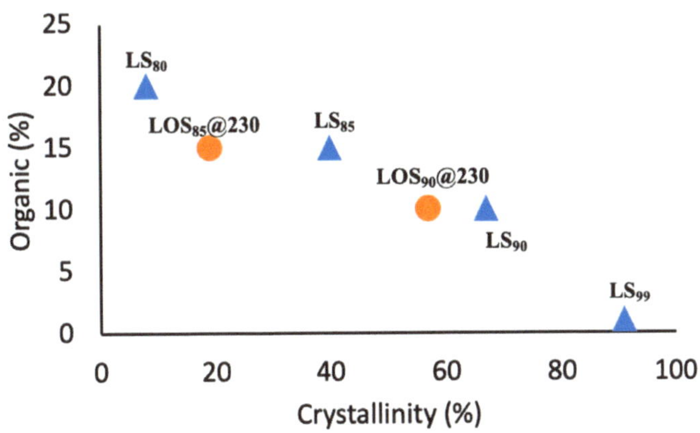

Figure 11. A graphical representation of the general trend of the effect increasing the percentage of organic material has on the percentage crystallinity of LS_x (blue triangles) and LOS_x@T (orange circles).

2.3. *Mechanical Properties and Environmental Impact Considerations*

The LOS_x@T composites studied herein must meet the specified mechanical strength requirements to be considered as potential replacements for less sustainable materials. For example, OPC requires a compressive strength greater than 17 MPa and a flexural strength greater than 3 MPa. The compressive strength of the LOS_x@T composites reported exceeded that required of OPC with strengths ranging from 22.1 to 35.9 MPa. Other lignin-containing HSMs have been reported and compared to the LOS_x@T composites. The composites ELS_x@T (where x = wt. % sulfur in the reaction mixture and T represents the reaction temperature in °C), for example, were prepared by reacting sulfur with oleic-esterified lignin. ELS_{80}@180 has a notably lower compressive strength (10.9 MPa) than LOS_x@T composites.

ELS_{80}@180 comprises twice the amount of organic component as LOS_{90}, LOS_{90}@250, and LOS_{90}@300. ELS_{80}@180 was also reacted at a much lower temperature in comparison to LOS_x@T composites, whose reaction temperatures ranged from 230 to 300 °C. A larger amount of organic material and lower reaction temperature possibly contributes to a lower sulfur rank, resulting in a lower compressive strength. The ELS_{90}@180 and ELS_{90}@230 composites exhibited similar compressive strengths as the LOS_{90}@230, LOS_{85}@230, LOS_{80}@230, and LOS_{90}@250 composites. The LOS_{90}@300 exhibited a much higher compressive strength (35.9 MPa) than all other composites reported in Table 3 and compared graphically in Figure 12 (stress–strain plots are presented in Figures S28–S32, Supplementary Materials). This can possibly be attributed to the higher reaction temperature of 300 °C, which can lead to a higher sulfur rank and a higher compressive strength [65]. While there was not a clear trend displaying how an increase in organic material affects the compressive strength, an increase in compressive strength is observed as the reaction temperature is increased (Figure 13). The preparation of the oleic-esterified lignin (the precursor for ELS_x@T) is a much more time- and energy-intensive process compared to the process for preparing the lignin oil required for preparing LOS_x@T composites. Therefore, preparing lignin-oil-containing HSMs is a more realistic concept for acquiring comparable compressive strength features.

Table 3. Mechanical properties of the lignin–sulfur composites LOS_x@T with other previously reported lignin-containing HSMs composites for comparison.

Materials	Compressive Strength (MPa)	Flexural Strength/Modulus (MPa)	Compressive Strength (% of OPC)
LOS_{90}@230 [a]	22.1 ± 2.5	5.7/186	130
LOS_{85}@230 [a]	26.0 ± 0.3	6.5/236	153
LOS_{80}@230 [a]	22.6 ± 1.7	ND [k]	133
LOS90@250 [a]	22.1 ± 1.3	ND [k]	130
LOS90@300 [a]	35.9 ± 1.5	ND [k]	211
LS_{80} [b]	ND [k]	2.1/87	ND [k]
LS_{85} [b]	ND [k]	1.5/76	ND
LS_{90} [b]	ND [k]	1.7/57	ND
LS_{95} [b]	ND [k]	ND [k]	ND
LS_{99} [b]	ND [k]	ND [k]	ND
GS_{80} [c]	ND [k]	ND [k]	ND
ELS_{90}@180 [d]	20.1 ± 2.3	3.3/ND [k]	118
ELS_{90}@230 [d]	26.8 ± 0.5	3.9/ND [k]	158
ELS_{80}@180 [d]	10.9 ± 0.85	2.7/ND [k]	64
CLS_{80} [e]	ND [k]	3.6/ND [k]	ND [k]
$mAPS_{95}$ [f]	17.0	5.6/ND [k]	100
OPC [g]	17.0	3.7/580	100
S-DCPD (1:1) [h]	ND	6.0/3700	ND
S-DCPD–linseed oil (2:1:1) [i]	ND	4.7/1250	ND
S-DCPD–limonene (2:1:1) [j]	ND	1.9/1750	ND

[a] LOS_x@T composites made from lignin oil and elemental sulfur (where x = wt. % sulfur, varying from 80 to 90, and T represents the reaction temperature in °C). [b] LS_x composites comprising 10 wt. % allyl-derivatized lignin and elemental sulfur, where x = wt. % sulfur in monomer feed, varying from 80 to 99. [c] Composites comprising guaiacol (20 wt. %) and elemental sulfur (80 wt. %). [d] ELS_x@T composites comprising esterified lignin and elemental sulfur (where x = wt. % sulfur in the reaction mixture, varying from 80 to 90 and T represents the reaction temperature in °C). [e] Composites comprising chlorolignin (20 wt. %) and elemental sulfur (80 wt. %). [f] Composites comprising allyl lignin (2 wt. %), allyl cellulose (3 wt. %) and elemental sulfur (95 wt. %). [g] Ordinary Portland cement. [h] Composites comprising elemental sulfur (50 wt. %) and dicyclopentadiene (DCPD) (50 wt. %). [i] Composites comprising elemental sulfur (50 wt. %), DCPD (25 wt. %), and linseed oil (25 wt. %). [j] Composites comprising elemental sulfur (50 wt. %), DCPD (25 wt. %), and limonene (25 wt. %). [k] Not determined.

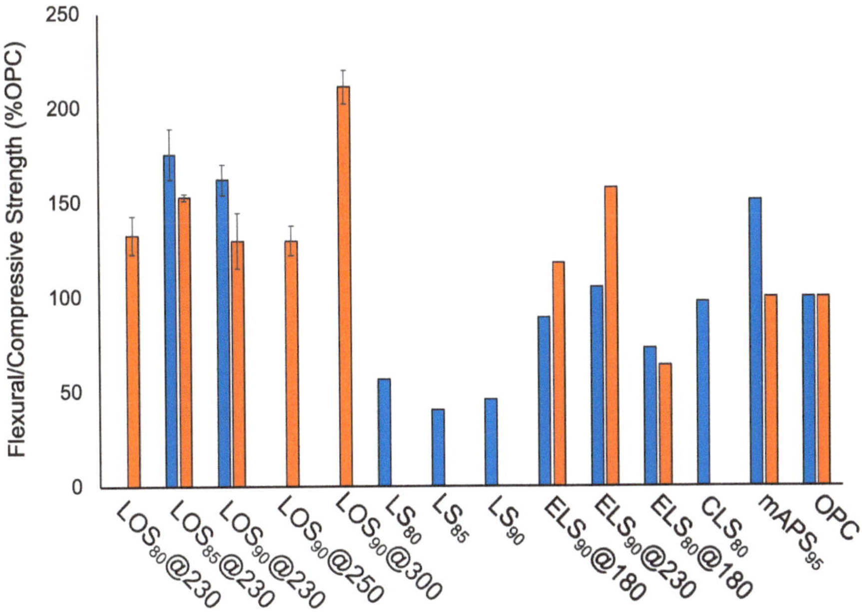

Figure 12. Graphical representation comparing the flexural strength (blue bars) and compressive strength (orange bars) of LOS_x@T composites with lignin–sulfur composites (LS_x), lignin/cellulose-containing sulfur composites (ELS_x@T, CLS_{80}, $mAPS_{95}$), and ordinary Portland cement (OPC). Bars not displayed signify that the attendant value was not previously reported for the listed material or could not be obtained for the current research.

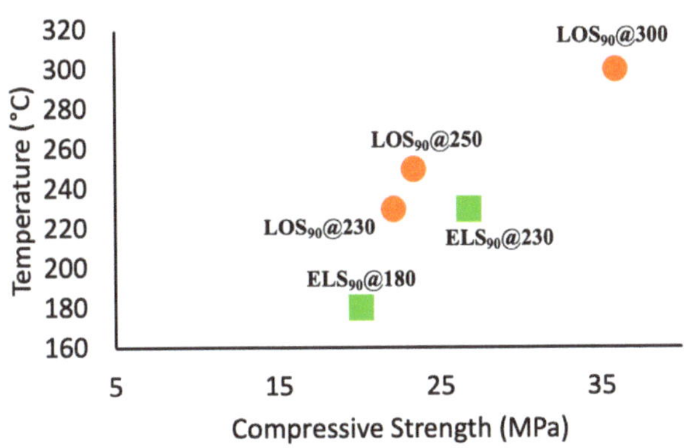

Figure 13. A graphical representation of the general trend of the effect increasing reaction temperature has on the compressive strength of ELS_x@T (green squares) and LOS_x@T (orange circles).

The flexural strengths/moduli of LOS_{90}@230 and LOS_{85}@230 composites were tested and assessed to reveal high flexural strengths (5.7 MPa and 6.5 MPa) and moduli (186 and 236 MPa) at room temperature (Table 3, stress–strain curves are provided in Figures S33 and S34, Supplementary Materials). These flexural strengths and moduli are much higher than the flexural strengths of previously reported LS_x (where x = wt. % sulfur, prepared by the inverse vulcanization of allylated lignin) composites with flexural strengths ranging from 1.5 to 2.1 MPa and moduli ranging from 57 to 87 MPa. LOS_{90}@230 and LOS_{85}@230 also exhibited higher flexural strengths than ELS_x@T composites (2.7–3.3 MPa), as well as

CLS_{80} (3.6 MPa), a chlorolignin–sulfur composite comprising 80 wt. % sulfur. However, LOS_{90}@230 exhibited a similar flexural strength to that of $mAPS_{95}$ (5.6 MPa), a composite prepared by the inverse vulcanization of allyl lignin (2 wt. %), allyl cellulose (3 wt. %), and 95 wt. % sulfur. The flexural strengths/moduli were previously reported for additional HSM polymers containing organic crosslinkers. For example, dicyclopentadiene (DCPD), an industrial, inexpensive by-product, forms S-DCPD (50 wt. % dicyclopentadiene and 50 wt. % sulfur) after reaction with elemental sulfur. LOS_{85}@230 (6.5 MPa) and LOS_{90}@230 (5.7 MPa) compare moderately to S-DCPD (6.0 MPa). However, S-DCPD experiences a decrease in flexural strength when organic crosslinkers are incorporated. Linseed oil reacts with DCPD and sulfur to yield S-DCPD–linseed (50 wt. % sulfur, 25 wt. % dicyclopentadiene, and 25 wt. % linseed oil) and exhibits a faintly lower flexural strength (4.7 MPa). However, when a cheap biomass material known as limonene reacts with DCPD and sulfur to yield S-DCPD–limonene (25 wt. % dicyclopentadiene, 25 wt. % limonene, and 50 wt. % sulfur), a considerably lower flexural strength is revealed (1.9 MPa). Conversely, the flexural moduli of S-DCPD–linseed (1250 MPa), S-DCPD (3700 MPa), and S-DCPD–limonene (1750 MPa) [120] considerably surpass those of LOS_{85}@230 (236 MPa) and LOS_{90}@230 (186 MPa). Another important observation to note is that the flexural strengths of both the LOS_{90}@230 and LOS_{85}@230 composites exceed OPC's flexural strength (3.7 MPa), therefore giving competition for commercial building materials. The mechanical strength characteristics of the LOS_x@T and previously reported lignin-containing HSMs are summarized in Table 3 and compared graphically in Figure 12.

3. Materials and Methods

3.1. Materials

Sulfur powder (99.5%) was bought from Alfa Aesar (Haverhill, MA, USA). The chemicals employed did not undergo any further purification. The lignin oil used herein was produced via a reported method implementing the thermal solvolysis of kraft lignin (Sigma Aldrich, St. Louis, MO, USA) in the solvent ethanol following the reported procedure [40]. A detailed description of the preparation and characterization of lignin oil was previously reported for the synthesis and analysis of LOS_{90}@230 [108]. The same lignin oil was used to prepare the remaining LOS_x@T composites reported herein.

3.2. General Considerations and Instrumentation

Fourier transform infrared spectra were acquired using an IR instrument (Shimadzu IRAffinity-1S, Shimadzu Corporation, Columbia, MD, USA) equipped with an ATR attachment. Scans were collected over the range of 400–4000 cm^{-1} at ambient temperature with a resolution of 8 cm^{-1}.

SEM was attained using a Schottky Field Emission Scanning Electron Microscope SU5000 (Hitachi High-Tech, Tokyo, Japan) operating in variable pressure mode while using an accelerating voltage of 15 kV.

Compressional strength measurements were commenced with cylinders by means of a Mark-10 ES30 (Mark-10 Corporation, Copiague, NY, USA) Manual Test Stand equipped with a Mark 10 M3-200 Force Gauge (USA). The composites studied herein were melted and formed in molds (Smooth-On Oomoo® 25 tin-cure, Oomoo Corp., Richmond, BC, Canada) and then allowed to cool to room temperature. The cylinders stood at room temperature for 4 d before compressive strength was tested. Tests for each composite were performed in triplicates, and the strength reported is the average of the three runs. The long-term stability of the reported materials is unknown.

Flexural strength analyses were performed using a Mettler Toledo DMA 1 STARe System (Mettler Toledo, Columbus, OH, USA) in single-cantilever mode. The samples were formed in silicon resin molds (Smooth-On Oomoo® 25 tin-cure). The sample dimensions were 1.5 × 10.7 × 5.0 mm. The temperature was 25 °C with a clamping force of 1 cN m. Samples were assessed in triplicates, and the observed flexural strengths/moduli are an average of the three trials.

TGA data were recorded (Mettler Toledo TGA 2 STARe System, TA Instruments, New Castle, DE, USA) across the temperature range of 20–800 °C at a heating rate of 10 $°C·min^{-1}$ under a flow of N_2 (100 $mL·min^{-1}$).

DSC data were obtained by means of a Mettler Toledo DSC 3 STARe System (Mettler Toledo, Columbus, OH, USA) across a temperature range of −60 °C to 140 °C while implementing a heating rate of 10 $°C·min^{-1}$ under a flow of N_2 (200 $mL·min^{-1}$). DSC measurements were conducted over three cooling and heating cycles. Data from the third heating cycle are reported herein, with the first cycle removing solvent impurities. During the third cycle, the glass-transition temperature was observed. DSC data were utilized to calculate the percentage crystallinity using Equation (1).

UV-vis data were collected on Agilent Technologies Cary 60 UV-vis (Agilent Technologies, Inc., Santa Clara, CA, USA) using Simple Reads software (Cary WinUV Scan Application Version 5.1.0.1016) over the range of 400–600 nm. Dark sulfur was quantified by using the extinction coefficient of sulfur at 275 nm along with the absorbance of ethyl acetate soluble fraction of each composite.

3.3. Synthesis of LOS_x@T Composites

LOS_{80}@230 and LOS_{85}@230 were prepared by adding elemental sulfur (8.005 g, 8.500 g) and lignin oil (2.001 g, 1.503 g) to a heavy-walled pressure flask sealed by a Viton O-ring and a PTFE stopper along with a PTFE stir bar and then placed in an oil bath at 180 °C under continuous magnetic stirring. Next, the temperature was increased to 230 °C. The mixture was heated under rapid continuous stirring for a duration of 2 h. During the reaction time, the reaction mixture appeared homogeneous and turned dark brown in color. Stirring was stopped, and the reaction was removed from the heat. The material was cooled to room temperature, forming a solid dark brown substance. Forming the compressive strength cylinders (Figure 2) and flexural strength rectangular prisms (Figure 3) required remelting the product at 180 °C and then pouring the product into molds, followed by cooling to room temperature. This method resulted in a 99.7% yield for LOS_{80}@230 and a 99.8% yield for LOS_{85}@230. For LOS_{80}@230 elemental analysis, the theoretical values were C 12.86, H 1.22, and S 80; the actual values found were C 10.94, H 0.46, and S 84.85. For LOS_{85}@230 elemental analysis, the theoretical were C 9.64, H 0.91, and S 85; the actual values found were C 3.76, H 0.05, and S 94.49. LOS_{90}@230 was prepared as previously reported [108].

LOS90@250 and LOS90@300 were prepared by adding elemental sulfur (13.500 g, 13.505 g) and lignin oil (1.501 g, 1.500 g) to a stainless steel autoclave reactor and heated to 250 °C under rapid mechanical stirring. After 2 h, the heat and stirring was stopped, and the autoclave vessel cooled to room temperature. After cooling, the homogeneous, dark brown solid was removed from the reaction vessel. Forming the compressive strength cylinders (Figure 2) required remelting the product at 180 °C and then pouring the product into molds, followed by cooling to room temperature. For LOS_{90}@250 elemental analysis, the theoretical values were C 6.43, H 0.61, and S 90; the actual values found were C 7.33, H 0.14, and S 91.24. For LOS_{90}@300 elemental analysis, the theoretical values were C 6.43, H 0.61, and S 90; the actual values found were C 5.43, H 0.05, and S 93.06.

3.4. General Method for Dark Sulfur Quantification

The method used herein to determine the quantity of dark sulfur is based on a method previously reported by Hasell's group [109,110]. This method uses absorbance to quantify dark sulfur in HSMs. A small fraction of the composite (6–7 mg) was weighed to +/−0.0001 g precision using a microbalance and placed in a 250 mL volumetric flask with ethyl acetate and stirred for 30 min. This duration of agitation was chosen due to the solubility of elemental sulfur. Ethyl acetate has a lower solubility for sulfur and not oligomeric fractions; therefore, it will extracts dark sulfur within the composite [110]. The solution was then measured using a Carry-UV. A scan was taken over the range of 400–600 nm, and the absorbance at 275 nm was used in the calibration curve that was used to determine the concentration.

3.5. *General Method for Ethyl Acetate Extractions for Dark Sulfur Quantification*

A small amount of each composite (~50 mg) was placed in a vial along with 10 mL of ethyl acetate. This was then stirred for a duration of 1 h at room temperature. After stirring was finished, the vial was left standing still to allow the material to settle, and the ethyl acetate was then pipetted into a different vial. The ethyl acetate was then allowed to evaporate, with the soluble portion remaining in the vial.

4. Conclusions

Herein, we have demonstrated that lignin oil containing a mixture of oligomers can readily react with sulfur diradicals at varying amounts (80–90 wt. % S) and relatively low temperatures (230–300 °C). The resulting material formed durable, remeltable composites exhibiting compressive strengths (22.1–35.9 MPa) and flexural strengths (5.7–6.5 MPa) exceeding that of OPC (17 MPa). Increasing the amount of organic material did not result in a drastic change in properties; however, it was shown that increasing the reaction temperature improved the mechanical strength of the composites and decreased the crystallinity, resulting in a less brittle substance. These results show the great potential lignin oil has in the efforts to valorize the abundant yet underutilized lignin biomass.

Supplementary Materials: The following supporting information can be downloaded at https://www.mdpi.com/article/10.3390/molecules29174209/s1. Figure S1. Powder XRD trace of LOS_{80}@230 (top) compared to that of alpha sulfur (bottom); Figure S2. Powder XRD trace of LOS_{85}@230 (top) compared to that of alpha sulfur (bottom); Figure S3. Powder XRD trace of LOS_{90}@230 (top) compared to that of alpha sulfur (bottom); Figure S4. Powder XRD trace of LOS_{90}@250 (top) compared to that of alpha sulfur (bottom); Figure S5. Powder XRD trace of LOS_{90}@300 (top) compared to that of alpha sulfur (bottom); Figure S6. Portion of the FT-IR spectrum of LOS_{80}@230 (grey trace), LOS_{85}@230 (orange trace), LOS_{90}@230 (blue trace), LOS_{90}@250 (yellow trace), and LOS_{90}@300 (green trace); Figure S7. Portion of the FT-IR spectrum of LOS_{80}@230 (grey trace), LOS_{85}@230 (orange trace), LOS_{90}@230 (blue trace), LOS_{90}@250 (yellow trace), and LOS_{90}@300 (green trace); Figure S8. Scanning electron microscopy (SEM) image of LOS_{80}@230; Figure S9. Scanning electron microscopy (SEM) image of LOS85@230; Figure S10. Scanning electron microscopy (SEM) image of LOS90@230; Figure S11. Scanning electron microscopy (SEM) image of LOS90@250; Figure S12. Scanning electron microscopy (SEM) image of LOS90@300; Figure S13. Differential scanning calorimetry (DSC) traces (endothermic down) of the first heating (blue line) and first cooling (orange line) cycle for LOS80@230; Figure S14. Differential scanning calorimetry (DSC) traces (endothermic down) of the second heating (blue line) and second cooling (orange line) cycle for LOS80@230; Figure S15. Differential Scanning calorimetry (DSC) traces (endothermic down) of the third heating (blue line) and third cooling (orange line) cycle for LOS80@230; Figure S16. Differential scanning calorimetry (DSC) traces (endothermic down) of the first heating (blue line) and first cooling (orange line) cycle for LOS85@230; Figure S17. Differential scanning calorimetry (DSC) traces (endothermic down) of the second heating (blue line) and second cooling (orange line) cycle for LOS85@230; Figure S18. Differential scanning calorimetry (DSC) traces (endothermic down) of the third heating (blue line) and third cooling (orange line) cycle for LOS85@230; Figure S19. Differential scanning calorimetry (DSC) traces (endothermic down) of the first heating (blue line) and first cooling (orange line) cycle for LOS90@230; Figure S20. Differential scanning calorimetry (DSC) traces (endothermic down) of the second heating (blue line) and second cooling (orange line) cycle for LOS90@230; Figure S21. Differential scanning calorimetry (DSC) traces (endothermic down) of the third heating (blue line) and third cooling (orange line) cycle for LOS90@230; Figure S22. Differential scanning calorimetry (DSC) traces (endothermic down) of the first heating (blue line) and first cooling (orange line) cycle for LOS90@250; Figure S23. Differential scanning calorimetry (DSC) traces (endothermic down) of the second heating (blue line) and second cooling (orange line) cycle for LOS90@250; Figure S24. Differential scanning calorimetry (DSC) traces (endothermic down) of the third heating (blue line) and third cooling (orange line) cycle for LOS90@250; Figure S25. Differential scanning calorimetry (DSC) traces (endothermic down) of the first heating (blue line) and first cooling (orange line) cycle for LOS90@300; Figure S26. Differential scanning calorimetry (DSC) traces (endothermic down) of the second heating (blue line) and second cooling (orange line) cycle for LOS_{90}@300; Figure S27. Differential scanning calorimetry (DSC) traces (endothermic down) of the third heating (blue line) and third cooling (orange line) cycle for LOS_{90}@300; Figure S28. Representative stress–strain plot for the compressive-strength measurements of LOS_{80}@230; Figure S29. Representative stress–strain plot for the compressive-strength measurements of LOS_{85}@230; Figure S30. Representative stress–strain plot for the compressive-strength measurements of LOS_{90}@230; Figure S31. Representative stress–strain plot for the compressive-strength measurements of LOS_{90}@250; Figure S32. Representative stress–strain plot for the compressive-strength measurements of LOS_{90}@300; Figure S33. Stress–strain curves of LOS_{85}@230 determined during flexural strength testing. The orange line represents the propagations of the linear region of the stress–strain curve; Figure S34. P Stress–strain curves of LOS_{90}@230 determined during flexural strength testing. The orange line represents the propagations of the linear region of the stress–strain curve.

Author Contributions: K.A.T. and N.L.K.D.: data curation, formal analysis, investigation, validation. K.A.T. and R.C.S.: writing—original draft. R.C.S.: conceptualization, funding acquisition, methodology, resources, supervision. N.L.K.D.: writing—review and editing. All authors have read and agreed to the published version of the manuscript.

Funding: This research is funded by the National Science Foundation grant number CHE-2203669 awarded to RCS.

Institutional Review Board Statement: Not applicable.

Informed Consent Statement: Not applicable.

Data Availability Statement: Data are contained within the article and Supplementary Materials.

Conflicts of Interest: The authors declare no conflicts of interest.

References

1. Wang, C.; Kelley, S.S.; Venditti, R.A. Lignin-Based Thermoplastic Materials. *ChemSusChem* **2016**, *9*, 770–783. [CrossRef]
2. Boerjan, W.; Ralph, J.; Baucher, M. Lignin Biosynthesis. *Annu. Rev. Plant Biol.* **2003**, *54*, 519–546. [CrossRef] [PubMed]
3. Argyropoulos, D.S.; Jurasek, L.; Krištofová, L.; Xia, Z.; Sun, Y.; Paluš, E. Abundance and Reactivity of Dibenzodioxocins in Softwood Lignin. *J. Agric. Food Chem.* **2002**, *50*, 658–666. [CrossRef] [PubMed]
4. Xie, S.; Ragauskas, A.J.; Yuan, J.S. Lignin Conversion: Opportunities and Challenges for the Integrated Biorefinery. *Ind. Biotechnol.* **2016**, *12*, 161–167. [CrossRef]
5. Zoia, L.; Salanti, A.; Frigerio, P.; Orlandi, M. Exploring allylation and Claisen rearrangement as a novel chemical modification of lignin. *BioResources* **2014**, *9*, 6540–6561. [CrossRef]
6. Ralph, J.; Lapierre, C.; Boerjan, W. Lignin structure and its engineering. *Curr. Opin. Biotechnol.* **2019**, *56*, 240–249. [CrossRef]
7. Katahira, R.; Elder, T.J.; Beckham, G.T. A Brief Introduction to Lignin Structure. In *Lignin Valorization: Emerging Approaches*; Beckham, G.T., Ed.; The Royal Society of Chemistry: Piccadilly, London, 2018; pp. 1–20.
8. Hatakeyama, H.; Hatakeyama, T. Lignin Structure, Properties, and Applications. In *Biopolymers: Lignin, Proteins, Bioactive Nanocomposites*; Abe, A., Dusek, K., Kobayashi, S., Eds.; Springer: Berlin/Heidelberg, Germany, 2010; pp. 1–63.
9. Collins, M.N.; Nechifor, M.; Tanasă, F.; Zănoagă, M.; McLoughlin, A.; Stróżyk, M.A.; Culebras, M.; Teacă, C.-A. Valorization of lignin in polymer and composite systems for advanced engineering applications—A review. *Int. J. Biol. Macromol.* **2019**, *131*, 828–849. [CrossRef]
10. Petridis, L.; Schulz, R.; Smith, J.C. Simulation Analysis of the Temperature Dependence of Lignin Structure and Dynamics. *J. Am. Chem. Soc.* **2011**, *133*, 20277–20287. [CrossRef]
11. Ragauskas, A.J.; Beckham, G.T.; Biddy, M.J.; Chandra, R.; Chen, F.; Davis, M.F.; Davison, B.H.; Dixon, R.A.; Gilna, P.; Keller, M.; et al. Lignin Valorization: Improving Lignin Processing in the Biorefinery. *Science* **2014**, *344*, 1246843. [CrossRef]
12. Zakzeski, J.; Bruijnincx, P.C.A.; Jongerius, A.L.; Weckhuysen, B.M. The Catalytic Valorization of Lignin for the Production of Renewable Chemicals. *Chem. Rev.* **2010**, *110*, 3552–3599. [CrossRef]
13. Sethupathy, S.; Murillo Morales, G.; Gao, L.; Wang, H.; Yang, B.; Jiang, J.; Sun, J.; Zhu, D. Lignin valorization: Status, challenges and opportunities. *Bioresour. Technol.* **2022**, *347*, 126696. [CrossRef] [PubMed]
14. Sun, R.-C. Lignin Source and Structural Characterization. *ChemSusChem* **2020**, *13*, 4385–4393. [CrossRef] [PubMed]
15. Reshmy, R.; Athiyaman Balakumaran, P.; Divakar, K.; Philip, E.; Madhavan, A.; Pugazhendhi, A.; Sirohi, R.; Binod, P.; Kumar Awasthi, M.; Sindhu, R. Microbial valorization of lignin: Prospects and challenges. *Bioresour. Technol.* **2022**, *344*, 126240. [CrossRef] [PubMed]
16. Singhvi, M.; Kim, B.S. Lignin valorization using biological approach. *Biotechnol. Appl. Biochem.* **2021**, *68*, 459–468. [CrossRef]

17. Crestini, C.; Lange, H.; Sette, M.; Argyropoulos, D.S. On the structure of softwood kraft lignin. *Green Chem.* **2017**, *19*, 4104–4121. [CrossRef]
18. Suota, M.J.; da Silva, T.A.; Zawadzki, S.F.; Sassaki, G.L.; Hansel, F.A.; Paleologou, M.; Ramos, L.P. Chemical and structural characterization of hardwood and softwood LignoForce™ lignins. *Ind. Crops Prod.* **2021**, *173*, 114138. [CrossRef]
19. Figueiredo, P.; Lintinen, K.; Hirvonen, J.T.; Kostiainen, M.A.; Santos, H.A. Properties and chemical modifications of lignin: Towards lignin-based nanomaterials for biomedical applications. *Prog. Mater. Sci.* **2018**, *93*, 233–269. [CrossRef]
20. Rohella, R.S.; Sahoo, N.; Chakravortty, V. Lignin macromolecule. *Resonance* **1997**, *2*, 60–66. [CrossRef]
21. Vanholme, R.; Demedts, B.; Morreel, K.; Ralph, J.; Boerjan, W. Lignin Biosynthesis and Structure. *Plant Physiol.* **2010**, *153*, 895–905. [CrossRef]
22. Adler, E. Structural elements of lignin. *Ind. Eng. Chem.* **1957**, *49*, 1377–1383. [CrossRef]
23. Xia, M.; Valverde-Barrantes, O.J.; Suseela, V.; Blackwood, C.B.; Tharayil, N. Characterizing natural variability of lignin abundance and composition in fine roots across temperate trees: A comparison of analytical methods. *New Phytol.* **2022**, *236*, 2358–2373. [CrossRef] [PubMed]
24. Ponnusamy, V.K.; Nguyen, D.D.; Dharmaraja, J.; Shobana, S.; Banu, J.R.; Saratale, R.G.; Chang, S.W.; Kumar, G. A review on lignin structure, pretreatments, fermentation reactions and biorefinery potential. *Bioresour. Technol.* **2019**, *271*, 462–472. [CrossRef]
25. Sette, M.; Wechselberger, R.; Crestini, C. Elucidation of Lignin Structure by Quantitative 2D NMR. *Chem.—A Eur. J.* **2011**, *17*, 9529–9535. [CrossRef]
26. Vanholme, R.; Morreel, K.; Ralph, J.; Boerjan, W. Lignin engineering. *Curr. Opin. Plant Biol.* **2008**, *11*, 278–285. [CrossRef] [PubMed]
27. Martínez, Á.T.; Rencoret, J.; Marques, G.; Gutiérrez, A.; Ibarra, D.; Jiménez-Barbero, J.; del Río, J.C. Monolignol acylation and lignin structure in some nonwoody plants: A 2D NMR study. *Phytochemistry* **2008**, *69*, 2831–2843. [CrossRef] [PubMed]
28. Li, M.; Pu, Y.; Ragauskas, A.J. Current Understanding of the Correlation of Lignin Structure with Biomass Recalcitrance. *Front. Chem.* **2016**, *4*, 45. [CrossRef]
29. Bonawitz, N.D.; Chapple, C. The Genetics of Lignin Biosynthesis: Connecting Genotype to Phenotype. *Annu. Rev. Genet.* **2010**, *44*, 337–363. [CrossRef]
30. Novaes, E.; Kirst, M.; Chiang, V.; Winter-Sederoff, H.; Sederoff, R. Lignin and Biomass: A Negative Correlation for Wood Formation and Lignin Content in Trees. *Plant Physiol.* **2010**, *154*, 555–561. [CrossRef]
31. Rinaldi, R.; Jastrzebski, R.; Clough, M.T.; Ralph, J.; Kennema, M.; Bruijnincx, P.C.; Weckhuysen, B.M. Paving the Way for Lignin Valorisation: Recent Advances in Bioengineering, Biorefining and Catalysis. *Angew. Chem. (Int. Ed.)* **2016**, *55*, 8164–8215. [CrossRef]
32. Wong, S.S.; Shu, R.; Zhang, J.; Liu, H.; Yan, N. Downstream processing of lignin derived feedstock into end products. *Chem. Soc. Rev.* **2020**, *49*, 5510–5560. [CrossRef]
33. Sun, Z.; Cheng, J.; Wang, D.; Yuan, T.-Q.; Song, G.; Barta, K. Downstream Processing Strategies for Lignin-First Biorefinery. *ChemSusChem* **2020**, *13*, 5199–5212. [CrossRef] [PubMed]
34. Yao, Y.; Farac, N.F.; Azimi, G. Supercritical Fluid Extraction of Rare Earth Elements from Nickel Metal Hydride Battery. *ACS Sustain. Chem. Eng.* **2018**, *6*, 1417–1426. [CrossRef]
35. Janicka, P.; Płotka-Wasylka, J.; Jatkowska, N.; Chabowska, A.; Fares, M.Y.; Andruch, V.; Kaykhaii, M.; Gębicki, J. Trends in the new generation of green solvents in extraction processes. *Curr. Opin. Green Sustain. Chem.* **2022**, *37*, 100670. [CrossRef]
36. Cañadas, R.; González-Miquel, M.; González, E.J.; Núñez de Prado, A.; Díaz, I.; Rodríguez, M. Sustainable Recovery of High Added-Value Vanilla Compounds from Wastewater Using Green Solvents. *ACS Sustain. Chem. Eng.* **2021**, *9*, 4850–4862. [CrossRef]
37. Cañadas, R.; González-Miquel, M.; González, E.J.; Díaz, I.; Rodríguez, M. Evaluation of bio-based solvents for phenolic acids extraction from aqueous matrices. *J. Mol. Liq.* **2021**, *338*, 116930. [CrossRef]
38. Rapinel, V.; Rombaut, N.; Rakotomanomana, N.; Vallageas, A.; Cravotto, G.; Chemat, F. An original approach for lipophilic natural products extraction: Use of liquefied n-butane as alternative solvent to n-hexane. *LWT-Food Sci. Technol.* **2017**, *85*, 524–533. [CrossRef]
39. Rapinel, V.; Breil, C.; Makerri, C.; Jacotet-Navarro, M.; Rakotomanomana, N.; Vallageas, A.; Chemat, F. Feasibility of using liquefied gas HFO-1234ze (trans-1,3,3,3-tetrafluoroprop-1-ene) as an alternative to conventional solvents for solid–liquid extraction of food ingredients and natural products. *LWT-Food Sci. Technol.* **2017**, *83*, 225–234. [CrossRef]
40. Kouris, P.D.; van Osch, D.J.G.P.; Cremers, G.J.W.; Boot, M.D.; Hensen, E.J.M. Mild thermolytic solvolysis of technical lignins in polar organic solvents to a crude lignin oil. *Sustain. Energy Fuels* **2020**, *4*, 6212–6226. [CrossRef]
41. Jessop, P.G. Searching for green solvents. *Green Chem.* **2011**, *13*, 1391–1398. [CrossRef]
42. Tekin, K.; Hao, N.; Karagoz, S.; Ragauskas, A.J. Ethanol: A Promising Green Solvent for the Deconstruction of Lignocellulose. *ChemSusChem* **2018**, *11*, 3559–3575. [CrossRef]
43. Schutyser, W.; Renders, T.; Van den Bosch, S.; Koelewijn, S.F.; Beckham, G.T.; Sels, B.F. Chemicals from lignin: An interplay of lignocellulose fractionation, depolymerisation, and upgrading. *Chem. Soc. Rev.* **2018**, *47*, 852–908. [CrossRef]
44. Mabee, W.E.; McFarlane, P.N.; Saddler, J.N. Biomass availability for lignocellulosic ethanol production. *Biomass Bioenergy* **2011**, *35*, 4519–4529. [CrossRef]
45. Capello, C.; Fischer, U.; Hungerbühler, K. What is a green solvent? A comprehensive framework for the environmental assessment of solvents. *Green Chem.* **2007**, *9*, 927–934. [CrossRef]

46. Zhang, X.; Tang, Y.; Qu, S.; Da, J.; Hao, Z. H2S-Selective Catalytic Oxidation: Catalysts and Processes. *ACS Catal.* **2015**, *5*, 1053–1067. [CrossRef]
47. Lim, J.; Pyun, J.; Char, K. Recent approaches for the direct use of elemental sulfur in the synthesis and processing of advanced materials. *Angew. Chem. Int. Ed.* **2015**, *54*, 3249–3258. [CrossRef]
48. Crockett, M.P.; Evans, A.M.; Worthington, M.J.; Albuquerque, I.S.; Slattery, A.D.; Gibson, C.T.; Campbell, J.A.; Lewis, D.A.; Bernardes, G.J.; Chalker, J.M. Sulfur-Limonene Polysulfide: A Material Synthesized Entirely from Industrial By-Products and Its Use in Removing Toxic Metals from Water and Soil. *Angew. Chem. Int. Ed.* **2016**, *55*, 1714–1718. [CrossRef] [PubMed]
49. Worthington, M.J.H.; Kucera, R.L.; Chalker, J.M. Green chemistry and polymers made from sulfur. *Green Chem.* **2017**, *19*, 2748–2761. [CrossRef]
50. Yan, P.; Zhao, W.; Tonkin, S.J.; Chalker, J.M.; Schiller, T.L.; Hasell, T. Stretchable and Durable Inverse Vulcanized Polymers with Chemical and Thermal Recycling. *Chem. Mater.* **2022**, *34*, 1167–1178. [CrossRef]
51. Davis, A.E.; Sayer, K.B.; Jenkins, C.L. A comparison of adhesive polysulfides initiated by garlic essential oil and elemental sulfur to create recyclable adhesives. *Polym. Chem.* **2022**, *13*, 4634–4640. [CrossRef]
52. Eder, M.L.; Call, C.B.; Jenkins, C.L. Utilizing Reclaimed Petroleum Waste to Synthesize Water-Soluble Polysulfides for Selective Heavy Metal Binding and Detection. *ACS Appl. Polym. Mater.* **2022**, *4*, 1110–1116. [CrossRef]
53. Herrera, C.; Ysinga, K.J.; Jenkins, C.L. Polysulfides Synthesized from Renewable Garlic Components and Repurposed Sulfur Form Environmentally Friendly Adhesives. *ACS Appl. Mater. Interfaces* **2019**, *11*, 35312–35318. [CrossRef]
54. Orme, K.; Fistrovich, A.H.; Jenkins, C.L. Tailoring Polysulfide Properties through Variations of Inverse Vulcanization. *Macromolecules* **2020**, *53*, 9353–9361. [CrossRef]
55. Westerman, C.R.; Jenkins, C.L. Dynamic Sulfur Bonds Initiate Polymerization of Vinyl and Allyl Ethers at Mild Temperatures. *Macromolecules* **2018**, *51*, 7233–7238. [CrossRef]
56. Grimm, A.P.; Scheiger, J.M.; Roesky, P.W.; Théato, P. Inverse vulcanization of trimethoxyvinylsilane particles. *Polym. Chem.* **2022**, *13*, 5852–5860. [CrossRef]
57. Gomez, I.; Leonet, O.; Blazquez, J.A.; Mecerreyes, D. Inverse Vulcanization of Sulfur using Natural Dienes as Sustainable Materials for Lithium-Sulfur Batteries. *ChemSusChem* **2016**, *9*, 3419–3425. [CrossRef]
58. Wagenfeld, J.-G.; Al-Ali, K.; Almheiri, S.; Slavens, A.F.; Calvet, N. Sustainable applications utilizing sulfur, a by-product from oil and gas industry: A state-of-the-art review. *Waste Manag.* **2019**, *95*, 78–89. [CrossRef]
59. Wujcik, K.H.; Wang, D.R.; Raghunathan, A.; Drake, M.; Pascal, T.A.; Prendergast, D.; Balsara, N.P. Lithium Polysulfide Radical Anions in Ether-Based Solvents. *J. Phys. Chem. C* **2016**, *120*, 18403–18410. [CrossRef]
60. Diez, S.; Hoefling, A.; Theato, P.; Pauer, W. Mechanical and Electrical Properties of Sulfur-Containing Polymeric Materials Prepared via Inverse Vulcanization. *Polymers* **2017**, *9*, 59. [CrossRef]
61. Hoefling, A.; Lee, Y.J.; Theato, P. Sulfur-Based Polymer Composites from Vegetable Oils and Elemental Sulfur: A Sustainable Active Material for Li–S Batteries. *Macromol. Chem. Phys.* **2017**, *218*, 1600303. [CrossRef]
62. Westerman, C.R.; Walker, P.M.; Jenkins, C.L.; Westerman, C.R.; Walker, P.M. Synthesis of Terpolymers at Mild Temperatures Using Dynamic Sulfur Bonds in Poly(S-Divinylbenzene). *J. Vis. Exp.* **2019**, *147*, e59620. [CrossRef]
63. Karunarathna, M.S.; Lauer, M.K.; Thiounn, T.; Smith, R.C.; Tennyson, A.G. Valorization of waste to yield recyclable composites of elemental sulfur and lignin. *J. Mater. Chem. A* **2019**, *7*, 15683–15690. [CrossRef]
64. Karunarathna, M.S.; Lauer, M.K.; Smith, R.C. Facile route to an organosulfur composite from biomass-derived guaiacol and waste sulfur. *J. Mater. Chem. A* **2020**, *8*, 20318–20322. [CrossRef]
65. Karunarathna, M.S.; Maladeniya, C.P.; Lauer, M.K.; Tennyson, A.G.; Smith, R.C. Durable composites by vulcanization of oleyl-esterified lignin. *RSC Adv.* **2023**, *13*, 3234–3240. [CrossRef] [PubMed]
66. Mohamed, A.-M.O.; El Gamal, M. Sulfur based hazardous waste solidification. *Environ. Geol.* **2007**, *53*, 159–175. [CrossRef]
67. Rauchfuss, T. Under sulfur's spell. *Nat. Chem.* **2011**, *3*, 648. [CrossRef]
68. Chung, W.J.; Griebel, J.J.; Kim, E.T.; Yoon, H.; Simmonds, A.G.; Ji, H.J.; Dirlam, P.T.; Glass, R.S.; Wie, J.J.; Nguyen, N.A.; et al. The use of elemental sulfur as an alternative feedstock for polymeric materials. *Nat. Chem.* **2013**, *5*, 518–524. [CrossRef]
69. Griebel, J.J.; Nguyen, N.A.; Namnabat, S.; Anderson, L.E.; Glass, R.S.; Norwood, R.A.; Mackay, M.E.; Char, K.; Pyun, J. Dynamic Covalent Polymers via Inverse Vulcanization of Elemental Sulfur for Healable Infrared Optical Materials. *ACS Macro Lett.* **2015**, *4*, 862–866. [CrossRef]
70. Zhang, Y.; Konopka, K.M.; Glass, R.S.; Char, K.; Pyun, J. Chalcogenide hybrid inorganic/organic polymers (CHIPs) via inverse vulcanization and dynamic covalent polymerizations. *Polym. Chem.* **2017**, *8*, 5167–5173. [CrossRef]
71. Tonkin, S.J.; Gibson, C.T.; Campbell, J.A.; Lewis, D.A.; Karton, A.; Hasell, T.; Chalker, J.M. Chemically induced repair, adhesion, and recycling of polymers made by inverse vulcanization. *Chem. Sci.* **2020**, *11*, 5537–5546. [CrossRef]
72. Smith, J.A.; Wu, X.; Berry, N.G.; Hasell, T. High sulfur content polymers: The effect of crosslinker structure on inverse vulcanization. *J. Polym. Sci. Part A Polym. Chem.* **2018**, *56*, 1777–1781. [CrossRef]
73. Yan, P.; Zhao, W.; Zhang, B.; Jiang, L.; Petcher, S.; Smith, J.A.; Parker, D.J.; Cooper, A.I.; Lei, J.; Hasell, T. Inverse Vulcanized Polymers with Shape Memory, Enhanced Mechanical Properties, and Vitrimer Behavior. *Angew. Chem. Int. Ed.* **2020**, *59*, 13371–13378. [CrossRef]
74. Parker, D.J.; Chong, S.T.; Hasell, T. Sustainable inverse-vulcanised sulfur polymers. *RSC Adv.* **2018**, *8*, 27892–27899. [CrossRef] [PubMed]

75. Yan, P.; Wang, H.; Dodd, L.J.; Hasell, T. Processable crosslinked terpolymers made from elemental sulfur with wide range of thermal and mechanical properties. *Commun. Mater.* **2023**, *4*, 89. [CrossRef]
76. Zhang, Y.; Pavlopoulos, N.G.; Kleine, T.S.; Karayilan, M.; Glass, R.S.; Char, K.; Pyun, J. Nucleophilic Activation of Elemental Sulfur for Inverse Vulcanization and Dynamic Covalent Polymerizations. *J. Polym. Sci. Part A Polym. Chem.* **2019**, *57*, 7–12. [CrossRef]
77. Wu, X.; Smith, J.A.; Petcher, S.; Zhang, B.; Parker, D.J.; Griffin, J.M.; Hasell, T. Catalytic inverse vulcanization. *Nat. Commun.* **2019**, *10*, 647. [CrossRef] [PubMed]
78. Lundquist, N.A.; Tikoalu, A.D.; Worthington, M.J.H.; Shapter, R.; Tonkin, S.J.; Stojcevski, F.; Mann, M.; Gibson, C.T.; Gascooke, J.R.; Karton, A.; et al. Reactive Compression Molding Post-Inverse Vulcanization: A Method to Assemble, Recycle, and Repurpose Sulfur Polymers and Composites. *Chem.—A Eur. J.* **2020**, *26*, 10035–10044. [CrossRef] [PubMed]
79. Zhang, B.; Gao, H.; Yan, P.; Petcher, S.; Hasell, T. Inverse vulcanization below the melting point of sulfur. *Mater. Chem. Front.* **2020**, *4*, 669–675. [CrossRef]
80. Yan, P.; Zhao, W.; McBride, F.; Cai, D.; Dale, J.; Hanna, V.; Hasell, T. Mechanochemical synthesis of inverse vulcanized polymers. *Nat. Commun.* **2022**, *13*, 4824. [CrossRef]
81. Jia, J.; Liu, J.; Wang, Z.-Q.; Liu, T.; Yan, P.; Gong, X.-Q.; Zhao, C.; Chen, L.; Miao, C.; Zhao, W.; et al. Photoinduced inverse vulcanization. *Nat. Chem.* **2022**, *14*, 1249–1257. [CrossRef]
82. Alex, A.; Singha, N.K.; Choudhury, S. Exploring inverse vulcanization in lithium–sulfur batteries. *Curr. Opin. Electrochem.* **2023**, *39*, 101271. [CrossRef]
83. Karunarathna, M.S.; Lauer, M.K.; Tennyson, A.G.; Smith, R.C. Copolymerization of an aryl halide and elemental sulfur as a route to high sulfur content materials. *Polym. Chem.* **2020**, *11*, 1621–1628. [CrossRef]
84. Lopez, C.V.; Maladeniya, C.P.; Smith, R.C. Lithium-Sulfur Batteries: Advances and Trends. *Electrochem* **2020**, *1*, 226–259. [CrossRef]
85. Lauer, M.K.; Tennyson, A.G.; Smith, R.C. Green Synthesis of Thermoplastic Composites from a Terpenoid-Cellulose Ester. *ACS Appl. Polym. Mater.* **2020**, *2*, 3761–3765. [CrossRef]
86. Lauer, M.K.; Karunarathna, M.S.; Tennyson, A.G.; Smith, R.C. Recyclable, sustainable, and stronger than portland cement: A composite from unseparated biomass and fossil fuel waste. *Mater. Adv.* **2020**, *1*, 590–594. [CrossRef]
87. Lauer, M.K.; Karunarathna, M.S.; Tennyson, A.G.; Smith, R.C. Robust, remeltable and remarkably simple to prepare biomass–sulfur composites. *Mater. Adv.* **2020**, *1*, 2271–2278. [CrossRef]
88. Maladeniya, C.P.; Karunarathna, M.S.; Lauer, M.K.; Lopez, C.V.; Thiounn, T.; Smith, R.C. A role for terpenoid cyclization in the atom economical polymerization of terpenoids with sulfur to yield durable composites. *Mater. Adv.* **2020**, *1*, 1665–1674. [CrossRef]
89. Smith, A.D.; Tennyson, A.G.; Smith, R.C. Sulfur-Containing Polymers Prepared from Fatty Acid-Derived Monomers: Application of Atom-Economical Thiol-ene/Thiol-yne Click Reactions and Inverse Vulcanization Strategies. *Sustain. Chem.* **2020**, *1*, 209–237. [CrossRef]
90. Lauer, M.K.; Tennyson, A.G.; Smith, R.C. Inverse vulcanization of octenyl succinate-modified corn starch as a route to biopolymer–sulfur composites. *Mater. Adv.* **2021**, *2*, 2391–2397. [CrossRef]
91. Lopez, C.V.; Smith, A.D.; Smith, R.C. High strength composites from low-value animal coproducts and industrial waste sulfur. *RSC Adv.* **2022**, *12*, 1535–1542. [CrossRef]
92. Thiounn, T.; Karunarathna, M.S.; Slann, L.M.; Lauer, M.K.; Smith, R.C. Sequential crosslinking for mechanical property development in high sulfur content composites. *J. Polym. Sci.* **2020**, *58*, 2943–2950. [CrossRef]
93. Lindholm-Lehto, P.C.; Knuutinen, J.S.; Ahkola, H.S.J.; Herve, S.H. Refractory organic pollutants and toxicity in pulp and paper mill wastewaters. *Environ. Sci. Pollut. Res.* **2015**, *22*, 6473–6499. [CrossRef] [PubMed]
94. Lee, J.M.; Noh, G.Y.; Kim, B.G.; Yoo, Y.; Choi, W.J.; Kim, D.-G.; Yoon, H.G.; Kim, Y.S. Synthesis of Poly(phenylene polysulfide) Networks from Elemental Sulfur and p-Diiodobenzene for Stretchable, Healable, and Reprocessable Infrared Optical Applications. *ACS Macro Lett.* **2019**, *8*, 912–916. [CrossRef] [PubMed]
95. Karunarathna, M.S.; Tennyson, A.G.; Smith, R.C. Facile new approach to high sulfur-content materials and preparation of sulfur-lignin copolymers. *J. Mater. Chem. A* **2020**, *8*, 548–553. [CrossRef]
96. Melro, E.; Duarte, H.; Eivazi, A.; Costa, C.; Faleiro, M.L.; da Costa, A.M.R.; Antunes, F.E.; Valente, A.J.M.; Romano, A.; Norgren, M.; et al. Poly(butylene succinate)-Based Composites with Technical and Extracted Lignins from Wood Residues. *ACS Appl. Polym. Mater.* **2024**, *6*, 1169–1181. [CrossRef]
97. Li, X.; Li, J.; Shen, X.; Cao, M.; Wang, Y.; Zhang, W.; Xu, Y.; Ling, Z.; Chen, S.; Xu, F. Transparent Cellulose/Lignin Composite Films with Adjustable Haze and UV-Blocking Performance for Light Management. *ACS Sust. Chem. Eng.* **2024**, *12*, 5427–5435. [CrossRef]
98. Liu, S.; Li, C.; Sun, Y.; Qiu, X.; Li, X.; Sun, C.; Liu, Y.; Yu, Q.; Yu, B.; Cai, M.; et al. BTA-P4444-Lig-Functionalized MXene to Prepare Anticorrosion and Wear-Resistant Integrated Waterborne Epoxy Composite Coating. *ACS Sust. Chem. Eng.* **2024**, *12*, 8247–8260. [CrossRef]
99. Singh, K.; Yuvrajsinh, G.M.; Mehra, S.; Mishra, A.; Kumar, A. Bioionic Liquid-Assisted Transparent Sodium Alginate-κ-Carrageenan-Lignin Composite Films with Excellent Ultraviolet Shielding, Antioxidant, and Antibacterial Properties. *ACS Sustain. Chem. Eng.* **2024**, *12*, 4314–4327. [CrossRef]
100. Wang, L.; Wang, Y.; Zhang, J.; Wang, F.; Liu, Z.; Jiang, J. Investigation on self-healing polyurethane coating doped with lignin composites for protecting cementitious materials. *Constr. Build. Mater.* **2024**, *411*, 134368. [CrossRef]

101. Fazeli, M.; Mukherjee, S.; Baniasadi, H.; Abidnejad, R.; Mujtaba, M.; Lipponen, J.; Seppala, J.; Rojas, O.J. Lignin beyond the status quo: Recent and emerging composite applications. *Green Chem.* **2024**, *26*, 593–630. [CrossRef]
102. Yang, W.; Wang, D.; Feng, S.; He, S.; Xiao, H.; Dai, H.; Han, J. Heteroatom-doped hierarchically porous thick bulk carbon derived from a Pleurotus eryngii/lignin composite: A free-standing and high mass loading electrode for high-energy-density storage. *Green Chem.* **2024**, *26*, 4633–4644. [CrossRef]
103. Chen, Y.; Ma, Q.; Wang, K.; Cho, M.; Jiang, L. Thermoplastic starch nanocomposites derived directly from cornmeal. *J. Compos. Mater.* **2024**, *58*, 1077–1088. [CrossRef]
104. Campos, G.N.; da Rocha, E.B.D.; Furtado, C.R.G.; de Figueiredo, M.A.G.; de Sousa, A.M.F. Using carboxyl groups to improve the compatibility of XNBR/lignin composites. *Polym. Compos.* **2024**, *45*, 4124–4137. [CrossRef]
105. Montazeri, M.; Norouzbeigi, R. Investigation of synergistic effects incorporating esterified lignin and guar gum composite aerogel for sustained oil spill cleanup. *Sci. Rep.* **2024**, *14*, 13892. [CrossRef]
106. Thiounn, T.; Karunarathna, M.S.; Lauer, M.K.; Tennyson, A.G.; Smith, R.C. Detoxification of bisphenol A via sulfur-mediated carbon–carbon σ-bond scission. *RSC Sustain.* **2023**, *1*, 535–542. [CrossRef]
107. Kapuge Dona, N.L.; Maladeniya, C.P.; Smith, R.C. Reactivity of Biomass-Derived Olefins with Elemental Sulfur: Mechanistic Insight. *Eur. J. Org. Chem.* **2024**, *27*, e202301269. [CrossRef]
108. Tisdale, K.A.; Dona, N.L.K.; Maladeniya, C.P.; Smith, R.C. Green and Atom Economical Route to High Compressive Strength Lignin Oil-Sulfur Composites. *J. Polym. Environ.* **2024**, 1–13. [CrossRef]
109. Dale, J.J.; Petcher, S.; Hasell, T. Dark Sulfur: Quantifying Unpolymerized Sulfur in Inverse Vulcanized Polymers. *ACS Appl. Polym. Mater.* **2022**, *4*, 3169–3173. [CrossRef]
110. Dale, J.J.; Stanley, J.; Dop, R.A.; Chronowska-Bojczuk, G.; Fielding, A.J.; Neill, D.R.; Hasell, T. Exploring Inverse Vulcanisation Mechanisms from the Perspective of Dark Sulfur. *Eur. Polym. J.* **2023**, *195*, 112198. [CrossRef]
111. Shankarayya Wadi, V.K.; Jena, K.K.; Khawaja, S.Z.; Yannakopoulou, K.; Fardis, M.; Mitrikas, G.; Karagianni, M.; Papavassiliou, G.; Alhassan, S.M. NMR and EPR Structural Analysis and Stability Study of Inverse Vulcanized Sulfur Copolymers. *ACS Omega* **2018**, *3*, 3330–3339. [CrossRef]
112. Rao, C.N.R.; Venkataraghavan, R.; Kasturi, T.R. Contribution to the Infrared Spectra of Organosulphur Compounds. *Can. J. Chem.* **1964**, *42*, 36–42. [CrossRef]
113. Bastian, E.J., Jr.; Martin, R.B. Disulfide vibrational spectra in the sulfur-sulfur and carbon-sulfur stretching region. *J. Phys. Chem.* **1973**, *77*, 1129–1133. [CrossRef]
114. Stark, N.M.; Yelle, D.J.; Agarwal, U.P. Techniques for characterizing lignin. *Lignin Polym. Compos.* **2016**, *2016*, 49–66.
115. Collier, W.E.; Schultz, T.P.; Kalasinsky, V.F. Infrared Study of Lignin: Reexamination of Aryl-Alkyl Ether C—O Stretching Peak Assignments. *Holzforschung* **1992**, *46*, 523–528. [CrossRef]
116. Brebu, M.; Vasile, C. Thermal degradation of lignin—A review. *Cellul. Chem. Technol.* **2010**, *44*, 353–363.
117. Griebel, J.J.; Glass, R.S.; Char, K.; Pyun, J. Polymerizations with elemental sulfur: A novel route to high sulfur content polymers for sustainability, energy and defense. *Prog. Polym. Sci.* **2016**, *58*, 90–125. [CrossRef]
118. Tobolsky, A.V.; MacKnight, W.; Beevers, R.B.; Gupta, V.D. The glass transition temperature of polymeric sulphur. *Polymer* **1963**, *4*, 423–427. [CrossRef]
119. Bandzierz, K.; Reuvekamp, L.; Dryzek, J.; Dierkes, W.; Blume, A.; Bielinski, D. Influence of Network Structure on Glass Transition Temperature of Elastomers. *Materials* **2016**, *9*, 607. [CrossRef]
120. Smith, J.A.; Green, S.J.; Petcher, S.; Parker, D.J.; Zhang, B.; Worthington, M.J.H.; Wu, X.; Kelly, C.A.; Baker, T.; Gibson, C.T.; et al. Crosslinker Copolymerization for Property Control in Inverse Vulcanization. *Chem.—A Eur. J.* **2019**, *25*, 10433–10440. [CrossRef]

molecules

Article

New Sulfenate Sources for Double Pallado-Catalyzed Cross-Coupling Reaction: Application in Symmetrical Biarylsulfoxide Synthesis, and Evidence of TADF Properties

Valentin Magné [1], Iulia Cretoiu [1], Sonia Mallet-Ladeira [2], Eddy Maerten [1] and David Madec [1,*]

[1] Laboratoire Hétérochimie Fondamentale et Appliquée (UMR 5069), Université de Toulouse, CNRS, 118 Route de Narbonne, CEDEX 09, 31062 Toulouse, France; valentin.magne@univ-tlse3.fr (V.M.); iulia.cretoiu@univ-tlse3.fr (I.C.); eddy.maerten@univ-tlse3.fr (E.M.)

[2] Institut de Chimie de Toulouse (UAR 2599), 118 Route de Narbonne, CEDEX 09, 31062 Toulouse, France; sonia.ladeira@univ-tlse3.fr

* Correspondence: david.madec@univ-tlse3.fr

Abstract: Tetrahydro-4*H*-thiopyran-4-one 1-oxide **1** and sulfinyl-di-*tert*-butylpropionate **2** were reported as sources of bis-sulfenate anion and applied in a double pallado-catalyzed cross-coupling reaction for the synthesis of symmetrical biarylsulfoxides, tolerating a large array of electronic properties and bulkiness. The photophysical properties of a biarylsulfoxide have been explored, demonstrating an unreported TADF phenomenon on sulfoxide-containing scaffolds.

Keywords: sulfoxides; sulfenate; palladium; TADF

check for updates

Citation: Magné, V.; Cretoiu, I.; Mallet-Ladeira, S.; Maerten, E.; Madec, D. New Sulfenate Sources for Double Pallado-Catalyzed Cross-Coupling Reaction: Application in Symmetrical Biarylsulfoxide Synthesis, and Evidence of TADF Properties. *Molecules* **2024**, *29*, 4809. https://doi.org/10.3390/molecules29204809

Academic Editor: Ming Wang

Received: 11 September 2024
Revised: 4 October 2024
Accepted: 9 October 2024
Published: 11 October 2024

1. Introduction

Sulfoxides are ubiquitous scaffolds in organic chemistry. The simplest of the sulfoxides, dimethyl sulfoxide (DMSO) is a solvent of choice for synthetic chemists and biochemists because of its physicochemical properties and biocompatibility [1–4]. Natural sulfoxides or precursors of sulfoxides are involved in various biological processes [5,6] such as alliin bio-transformed into allicin, an antifungal agent, Sparsomycin with antibiotic activities or Oxisuran with immunosuppressive properties. Aromatic sulfoxides are often used in synthetic therapeutic chemistry, with, for example, Sulindac with anti-inflammatory activity, Sulmazole in the treatment of cardiovascular anomalies or Esomeprazole involved in proton pump inhibition (Figure 1). Aromatic sulfoxides are also important in coordination chemistry for the stabilization of organometallic complexes and their applications in catalysis [7–10], for the stabilization of low-valent species of main group elements [11–15], and in materials science [16,17]. Aromatic sulfoxides are mainly prepared by oxidation of the corresponding thioethers, or by nucleophilic substitution reaction of sulfinate amides or esters [18]. Each of these two methods suffers from limitations such as the formation of over-oxidation products, or the incompatibility of nucleophilic reagents with different functional groups.

Recently, a new synthetic strategy involving a pallado-catalyzed cross coupling with sulfenate anions was developed [19–23]. This approach allows perfect control of the oxidation state of the sulfoxide synthesized, as well as good compatibility with a wide range of functional groups (Scheme 1).

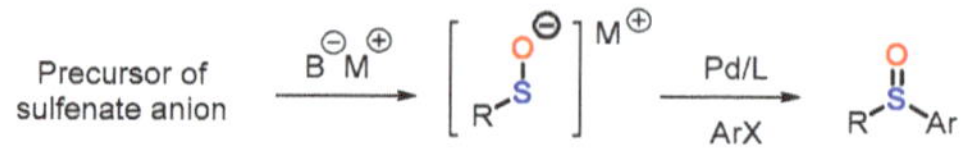

Scheme 1. Palladium-catalyzed arylation of sulfenate anions.

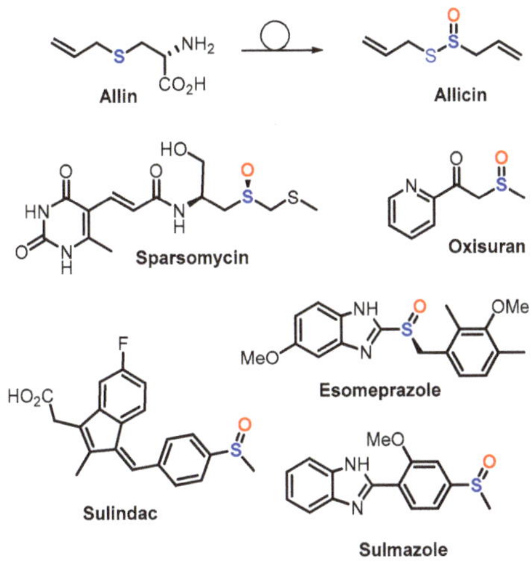

Figure 1. Examples of bioactive sulfoxides or precursors of sulfoxides.

This pallado-catalyzed coupling reaction allows the synthesis of various unsymmetrical aromatic sulfoxides from different sulfenate anion precursors (Figure 2) [24–34], and enantioselective versions were also developed [25,34–36].

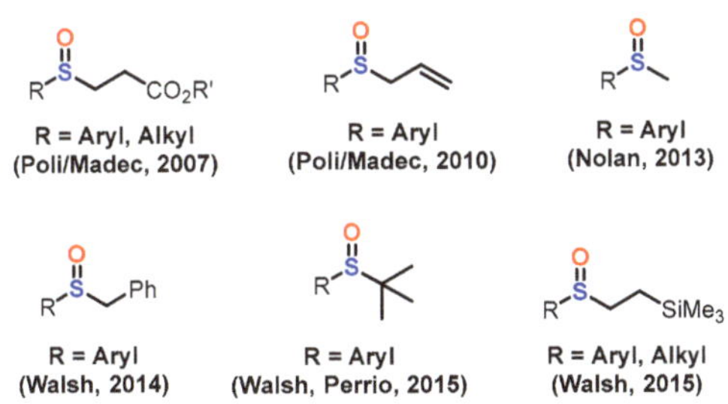

Figure 2. Sources of sulfenate anions in palladium-catalyzed arylation [19,24–34].

In contrast, the use of sulfenate ions for the synthesis of symmetrical sulfoxides by palladium-catalyzed cross-coupling reaction remains sporadic [28,29,32]. Interested in the preparation of various symmetrical sulfoxides, in particular with highly hindered groups, we decided to explore alternative sources of bis-sulfenate synthon [37].

2. Results and Discussion

Thanks to our team experience with the β-H elimination systems for generation of sulfenate anions, we have rapidly identified two candidates as new sources of bis-sulfenate synthon: the commercially available tetrahydro-4*H*-thiopyran-4-one 1-oxide **1** as a cyclic precursor and sulfinyl-di-*tert*-butylpropionate **2** as a linear precursor (Figure 3).

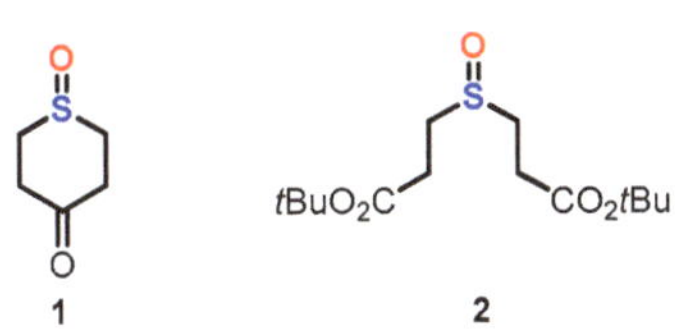

Figure 3. New sources of bis-sulfenate synthons.

Tetrahydro-4*H*-thiopyran-4-one 1-oxide **1** and sulfinyl-di-*tert*-butylpropionate **2** appear as white crystalline solids that are stable for months when kept in dry conditions (Figure 4). Chemical **2** was synthesized by esterification reaction of inexpensive thiodipropionic acid with *tert*-butyl alcohol in acidic conditions, then oxidation. This two-step procedure is efficient and can be performed on a multigram scale (14.5 g). However, when exposed to moisture, tetrahydro-4*H*-thiopyran-4-one 1-oxide **1** degrades into the corresponding gem diol in the solid state and in solution. Importantly, the latter is unreactive in the catalytic system described herein.

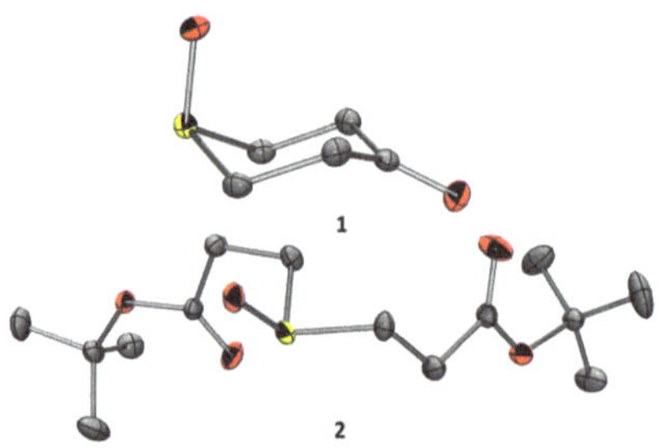

Figure 4. ORTEP view of X-ray structure of **1** and **2** at 50% probability; hydrogen atoms and solvent molecule have been omitted for clarity.

We began our catalytic studies with **1** using 4-iodotoluene **3a** as the model substrate (Table 1). Pleasingly, upon using classic sulfenate cross-coupling conditions, bis(dibenzylidenea cetone)palladium(0) as the palladium source, Xantphos as ligand and cesium carbonate as the base in toluene at 110 °C, we obtained the desired product **4a** with 48% yield (entry 1). Noteworthy, 10% of scrambling product (4-tolylsulfinyl-benzene) was obtained as well. This side product, arising from the aryl exchange on Xantphos palladium oxidative addition complexes, is well known and typically observed in sulfenate cross coupling [24]. Surprisingly, $Pd_2(dba)_3$, often used as a palladium(0) source in sulfenate cross-couplings, led to no conversion (entry 2). Different ligands such as SPhos, XPhos or DPEPhos (entries 3–5) or palladium catalysts such as CX21 [Allylchloro[1,3-bis(2,6-diisopropylphenyl)-imidazol-2-ylidene] palladium(II)] (entry 6) were thus investigated, not providing any product and evidencing that Xantphos is the ligand of choice for this cross-coupling reaction. Upon lowering the catalytic charge to 2 mol%, the reaction became sluggish (entries 7–8), convincing us to further optimize the reaction conditions with 5 mol%. The optimization of the base then proved critical in this reaction (entries 9–12). Indeed, when DBU was used, a drastic improvement of the yield to 61% was obtained along with only a very small amount of scrambling product (entry 11). Remarkably, lowering the temperature to 30 °C did not affect the reaction yield, providing the product with 62% yield (entry 12). Noteworthy, to the best of our knowledge, those reaction conditions are the smoothest and most selective scrambling-wise reported to date.

Nonetheless, despite allowing the cross-coupling in smooth conditions along with small amounts of scrambling product, the isolated yield remains quite low using **1** as a bis-sulfenate source, presumably because of its degradation in basic media happening too quickly for the catalytic system. We then naturally turned to the use of sulfinyl-di-*tert*-butylpropionate **2** as a more robust reactant under the reaction conditions.

We first used the reaction conditions previously developed for **1** using 4-iodotoluene **3a** [$Pd(dba)_2$/Xantphos, DBU as base at 30 °C], which provided the desired biarylsulfoxide **4a** with 71% yield. However, the reaction needed 72 h for completion at this temperature. We then carried out the reaction at 80 °C and obtained the product **4a** with 83% yield in 24 h. Encouraged by these results, we decided to continue the optimization of the reaction conditions on the more challenging sterically hindered 1-iodonaphatalene **3t** (Table 2). Indeed, the desired product was not observed within the previously developed conditions, even upon increasing the reaction temperature to 80 °C (entry 1). We thus reasoned that the sulfinyl-di-*tert*-butylpropionate **2** was being degraded too quickly in the presence of DBU.

Gratifyingly, changing the base to Cs_2CO_3 furnished the desired 1,1′-sulfinyldinaphthalene **4t** in 80% of isolated yield (entry 2). Finally, using the Buchwald precatalyst XantPhosPdG3, which is easy to handle and known to furnish highly active catalytic Pd(0) species under basic conditions, provided a major improvement of the reaction both in catalytic charge (2 mol%) and isolated yield (95%) (entry 3).

Table 1. Optimization of the reaction conditions [a].

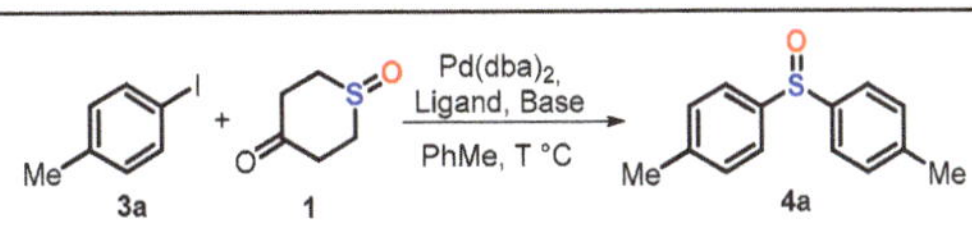

Entry	Ligand	Base	Yield (%) [b]
1	Xantphos (12%)	Cs_2CO_3	48 (10) [c]
2	Xantphos (12%)	Cs_2CO_3	n.r. [d]
3	SPhos (10%)	Cs_2CO_3	n.r. [c]
4	XPhos (10%)	Cs_2CO_3	n.r. [c]
5	DPEPhos (5%)	Cs_2CO_3	n.r. [c]
6	CX21 (10%) [e]	Cs_2CO_3	n.r.
7	Xantphos (5%)	Cs_2CO_3	33 (7)
8	Xantphos (2%) [f]	Cs_2CO_3	16 (3)
9	Xantphos (5%)	K_3PO_4	12 (4)
10	Xantphos (5%)	DIPEA	n.r.
11	Xantphos (5%)	DBU	61 (2)
12	Xantphos (5%)	DBU	62 (1) [g]

[a] **3a** (0.3 mmol, 3.0 equiv.), **1** (0.1 mmol, 1.0 equiv.), $Pd(dba)_2$ (5 mol%), base (0.4 mmol, 4 equiv.), PhMe (1 mL), 6 h, 110 °C. [b] isolated yield, yield of scrambling product within brackets, n.r. stands for no reaction. [c] 10 mol% of $Pd(dba)_2$ used. [d] $Pd_2(dba)_3$ (5 mol%) used. [e] CX21 stands for Allylchloro[1,3-bis(2,6-diisopropylphenyl)-imidazol-2-ylidene]palladium(II). [f] $Pd(dba)_2$ (2 mol%). [g] Temperature = 30 °C.

Table 2. Optimization of the reaction conditions [a].

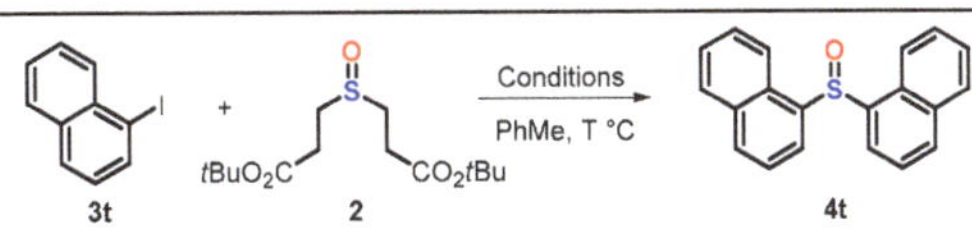

Entry	Conditions	Yield (%) [b]
1	$Pd(dba)_2$ (5%), Xantphos (5%), DBU, 80 °C	n.r.
2	$Pd(dba)_2$ (5%), Xantphos (5%), Cs_2CO_3, 80 °C	80 (6)
3	XantphosPdG3 (2%), Cs_2CO_3, 80 °C	95 (3)

[a] **3t** (0.3 mmol, 3.0 equiv.), **2** (0.1 mmol, 1.0 equiv.), base (0.4 mmol, 4 equiv.), PhMe (1 mL), 18 h. [b] isolated yield, yield of scrambling product within brackets, n.r. stands for no reaction.

After optimizing the reaction conditions with both reactants **1** and **2**, we turned our attention towards exploring their reaction scope and limitations. A large span of electronic effects was thus explored (Scheme 2). Electron-rich substrates are well tolerated using both catalytic systems, with method B using the reactant **2** generally providing the biarylsulfoxides **4** with higher yields. Noteworthy, obtaining **4h** in 70% yield in one step is a striking example of the advantage of the present methodology, the synthesis of this compound by an oxidation approach being really complicated. The reaction method A using **1** showed a drastic loss of activity upon investigating electron-poor iodoarenes with no conversion observed for groups more electron-withdrawing than halogens. Method B, though, allowed the cross-coupling reaction furnishing the highly electron-poor products such as pentafluorosulfanyl **4o** or nitro **4p** derivatives in 89% and 82% yield, respectively. Finally, we turned our attention to the study of steric effects. Although method A proved completely inefficient in the case of ortho-substituted substrates, method B showed robustness providing the desired products in good to excellent yields. Moreover, the cross-coupling could be adapted to accommodate aryl bromides. For example, the sterically demanding 9-bromoanthracene could be engaged in the reaction, increasing both the temperature to 110 °C and the catalyst loading to 5 mol%, furnishing the corresponding biarylsulfoxide **4u** in 49% yield.

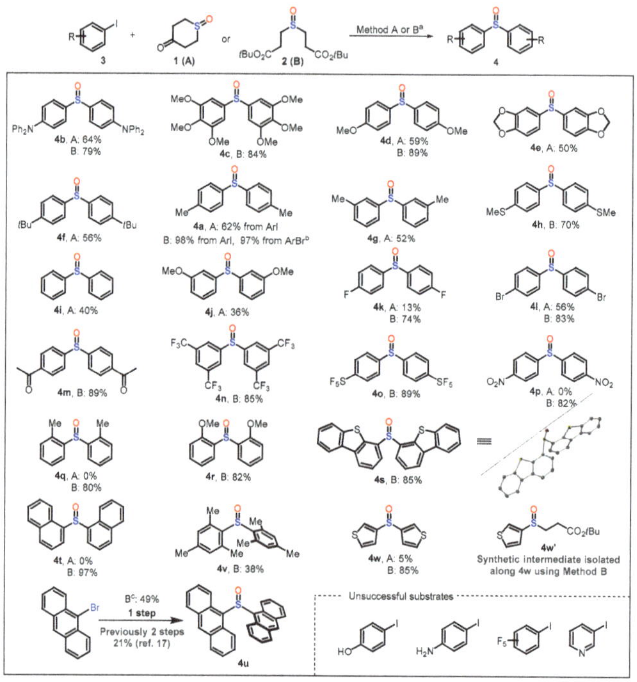

Scheme 2. Reaction scope of the pallado-catalyzed cross coupling of sulfenates with aryl iodides. [a] Reaction method A: iodoarene (0.9 mmol, 3.0 equiv.), 1 (0.3 mmol, 1.0 equiv.), $Pd(dba)_2$ (5 mol%), Xantphos (5 mol%), DBU (1.2 mmol, 4 equiv.), PhMe (3 mL), 30 °C, 6 h. Reaction method B: iodoarene (0.9 mmol, 3.0 equiv.), 2 (0.3 mmol, 1.0 equiv.), XantphosPdG3 (2 mol%), Cs_2CO_3 (1.2 mmol, 4 equiv.), PhMe (3 mL), 80 °C, 18 h. [b] THF was used instead of PhMe. [c] Reaction was carried out at 110 °C using 5 mol% of XantphosPdG3.

Noteworthy, the only known synthesis of this compound required two steps and furnished the product with 21% overall yield [38]. In addition, this product **4u** is particularly interesting for the study of SO extrusion under irradiation [38,39]. Finally, 3-iodothiophene was engaged in the reaction conditions, once again highlighting the superiority of method B over A, providing the desired sulfoxide at 85% and 5% yield, respectively. Interestingly, the compound **4w**, synthesized in method B, was obtained as a 90/10 mixture with **4w′**, thereby confirming the anticipated stepwise cross-coupling mechanism (Scheme 3).

The results presented herein clearly show the superiority of method B using the reactant **2** over method A. The difference in reactivity presumably arises on one hand from the degradation rate of the bis-sulfenate in their respective reaction conditions. On the other hand, the temperature may have an impact on the cross-coupling kinetics itself. The best conditions thus lay in the best balance between the generation rate of the sulfenate and its consumption in the catalytic cross-coupling reaction.

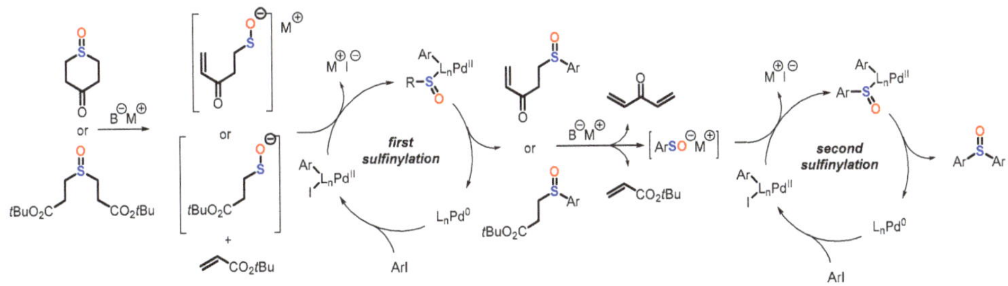

Scheme 3. Proposed reaction mechanism for the double pallado-catalyzed cross-coupling of aryl iodides with the sulfenate anion.

Intrigued by the strong fluorescence observed upon handling **4b**, we decided to study its photophysical studies (Figure 5). Noteworthy, its parent sulfone **5** and some closely related derivatives have received much attention in the last decade because of their thermally activated delayed fluorescence (TADF), which makes them promising in a number of applications such as OLEDs [40–42]. In a toluene solution, **4b** presents a maximum of absorption at 333 nm and an emission band at 391 nm, while **5**'s maximum of absorption is at 352 nm and emits at 401 nm (Figure 5). Additionally, for both compounds, a strong solvatochromic fluorescence is observed, ranging from 372 nm to 461 nm in cyclohexane and methanol respectively for **4b**, and 383 nm to 462 nm for **5** (see Supplementary Materials). The wider range of **4b**'s solvatochromic emission (89 nm) compared to **5** (79 nm) probably arises from the higher polarizability of the sulfoxide moiety. Finally, we turned our attention towards the lifetime of the excited states (Table 3). We first focused on reproducing the studies on **5** and obtained comparable results with the literature, showing first a fast fluorescence decay component with a lifetime on 2.43 ns followed by a second, slower decay with a 111 μs lifetime, characteristic of the TADF phenomenon. Pleasingly, in a similar fashion, **4b** shown two fluorescence decay components with 0.93 ns and 94 μs of lifetime, thus showing that **4b** possess comparable TADF properties with its parent sulfone **5**. Finally, the luminescence quantum yields were measured, **5** displaying a high 69.1% whereas **4b** furnished a lower yield of 25.7%.

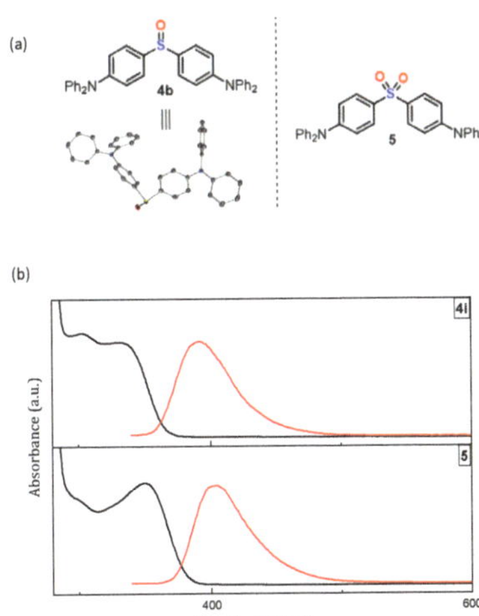

Figure 5. (**a**) Molecular structures of **4b** and **5** and ORTEP view of X-ray structure of **4b** at 50% probability. (**b**) Normalized emission and absorption spectra of **4b** and **5** in toluene at 298 K. black line: absorbance, red line: pl intensity.

Table 3. Photophysical properties for compounds **4b** and **5** in toluene at 298 K.

Compounds	λ_{max} (nm)	Φ (%)	τ (ns/μs)
4b	391	25.7	0.93/94
5	401	69.1	2.43/111

3. Materials and Methods

3.1. Reagents and Solvents

Unless otherwise noted, reagents were purchased from commercial suppliers and used directly without further purification. Unless indicated, technical grade solvents were purchased from commercial suppliers and used without further purification. Toluene was dried by passing over two columns of activated alumina, kept over activated 4A molecular sieves, and degassed by thorough argon sparging. All water was deionized before use. Unless stated, all reactions were carried out in conventional glassware. 'Room temperature' varied between 18 °C and 25 °C.

3.2. Analysis and Characterization

Analytical Thin Layer Chromatography (TLC) was performed on Merck aluminum-backed silica gel 60 F254 plates (Darmstadt, Germany). Developed TLC plates were visualized by ultraviolet (UV) irradiation (254 nm) or by staining with a solution of potassium permanganate. Column chromatography was carried using Merk silica gel 60 Å, 220–440 mesh. Fourier Transform Infrared Spectrometry (FTIR) was carried out using a Thermo Nicolet 6700 (Waltham, MA, USA) using an Attenuated Total Reflection (ATR) attachment and peaks are reported in terms of frequency of absorption (cm^{-1}). High Resolution Mass Spectrometry (HRMS) data were acquired using a GCT Premier CAB109 TOF mass spectrometer (Milford, MA, USA) equipped with DCI-CH_4 ionization. HRMS data were quoted to four decimal places (0.1 mDa). All NMR spectra were recorded on either a Bruker AV 300 or Bruker AV 500 (Billerica, MA, USA) and were internally referenced to residual solvent signals (residual $CHCl_3$ was referenced at δ 7.26 and $CDCl_3$ at δ 77.16 for ^1H and ^{13}C NMR, respectively, while residual $CHDCl_2$ is referenced at δ 5.32 and CD_2Cl_2 at δ 53.84 for ^1H and ^{13}C NMR, respectively). All NMR chemical shifts (δ) were reported in parts per million (ppm) and coupling constants (*J*) are given in hertz (Hz). The ^1H NMR spectra are reported as follows: δ (multiplicity, coupling constant *J*, number of protons), "app" stands for apparent.

The solution samples (0.1 mM) for the luminescence studies were prepared in an argon-filled glove-box using dry and degassed toluene. UV-VIS absorption spectra of the compounds in toluene were measured on an Agilent Cary 60 UV-Vis spectrophotometer in the range of 280–600 nm. The photoluminescence spectra were recorded on an Agilent Cary Eclipse spectrofluorometer (Santa Clara, CA, USA). The absolute PL quantum yield (PLQY) was obtained by using Quantaurus-QY Plus (HAMAMATSU, Shizuoka, Japan). The phosphorescence decay characteristics of the solution samples was recorded using a Horiba Fluorolog 3-2iHR 320 spectrofluorometer (Kyoto, Japan), equipped with a UV xenon flash lamp (FL-1035, 0.05–25 Hz flash rate, 3 μs pulse at FWHM). Excitation wavelengths were selected thanks to the double-monochromator of the spectrophotometer. The fast decay components were recorded in TCSPC mode thanks to a Horiba NanoLED module (Kyoto, Japan), piloting a pulsed diode NanoLED (371 nm). TADF measurements were recorded and analyzed using FluorEssence V3.8 from Horiba. Lifetime measurements were recorded with Data Station 2.7.0.4 and analyzed with DAS-6 6.8.0.10, from Horiba.

Single-crystal X-ray data were collected at low temperature (193(2)K) on a Bruker APEX II Quazar diffractometer (Billerica, MA, USA) equipped with a 30 W air-cooled microfocus source (**1**) or on a Bruker D8 VENTURE diffractometer equipped with a PHOTON III detector (**2**, **4b** and **4s**), using MoKα radiation (λ = 0.71037 Å). The structures were solved by intrinsic phasing method [43] and refined by full-matrix least- squares method on F2 [44]. All non-H atoms were refined with anisotropic displacement parameters and all the hydrogen atoms were refined isotropically at calculated positions using a riding model.

3.3. Synthesis of the Sulfenate Sources

Tetrahydro-4*H*-thiopyran-4-one 1-oxide 1:

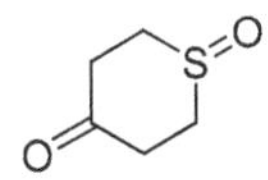

Prepared following a reported procedure [45]. Oxone® (15.35 g, 50 mmol, 1 equiv.) was added to a stirred suspension of wet alumina (50 g, prepared from 50 g of neutral alumina mixed with 10 mL of water) in DCM (500 mL). Tetrahydro-4*H*-thiopyran-4-one (5.8 g, 50 mmol, 1 equiv.) was added in one portion, and the reaction mixture was stirred for 24 h at room temperature. The reaction mixture was filtered over a sintered funnel, rinsing with DCM (2×100 mL), before evaporating the solvent under reduced pressure. The crude mixture was then purified over silica gel chromatography (gradient from 30 to 100% EtOAc:Pentane) yielding the desired crystalline product (2.00 g, 15.2 mmol, 30%).

^{1}H NMR ($CDCl_3$, 300 MHz) δ_H 3.47–3.23 (m, 4H), 2.98–2.78 (m, 2H), 2.61–2.48 (m, 2H). $^{13}C\{^{1}H\}$ NMR ($CDCl_3$, 75 MHz) δ_C 204.9, 47.5, 32.4.

Data are in agreement with the literature [46].

Di(*tert*-butyl)-3,3′-thiodipropionate:

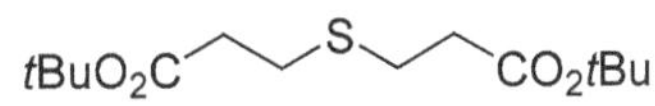

Prepared following a reported procedure [47]. Conc. H_2SO_4 (11 mL, 200 mmol, 2 equiv.) was added over a vigorously stirred suspension of anhydrous $MgSO_4$ (96.2 g, 800 mmol, 8 eq.) in 500 mL of dichloromethane in a 1 L flask. After 15 min of stirring, thiodipropionic acid (17.8 g, 100 mmol, 1 equiv.) and *tert*-butanol (94 mL, 1.0 mol, 10 equiv.) were added before quickly sealing the flask with a new rubber septum and the reaction mixture was stirred for 65 h at room temperature. A volume of 300 mL of saturated aq. $NaHCO_3$ was carefully added in portions before transferring the reaction mixture in a separatory funnel. The organic phase was separated and treated once with brine, then dried over Na_2SO_4. The volatiles were then removed in a rotavapor to obtain an oily residue and purified over silica gel column chromatography (gradient from 0 to 3% EtOAc:Pentane), yielding the desired product as a colorless oil (17.2 g, 59 mmol, 59% yield)

^{1}H NMR ($CDCl_3$, 300 MHz) δ_H 2.75 (t, J = 7.5 Hz, 4H), 2.51 (t, J = 7.4 Hz, 4H), 1.45 (s, 18H). $^{13}C\{^{1}H\}$ NMR (75 MHz, $CDCl_3$) δ_c 171.3, 81.0, 36.2, 28.2, 27.3. HRMS (DCI-CH_4) Calc'd for $C_{14}H_{26}O_4S$ $[M]^+$ 290.1552, found 290.1561. FTIR (neat) ν_{max}/cm^{-1} 2978, 1724, 1366, 1245, 1139, 1046, 844, 755.

Di(*tert*-butyl)-3,3′-thiodipropionate S-oxide 2:

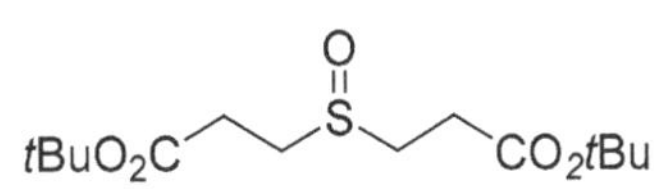

Prepared following a reported procedure [45]. Di(*tert*-butyl)-3,3′-thiodipropionate (17.2 g, 59 mmol, 1 equiv.) was added to a vigorously stirred suspension of wet alumina (60 g, prepared from 50 g of neutral alumina mixed with 10 mL of water) and Oxone® (18.1 g, 59 mmol, 1 equiv.) in 300 mL of CH_2Cl_2. The mixture was left stirring for 18 h and filtrated over a sintered funnel. The solids were washed with EtOAc and the filtrate was evaporated under reduced pressure. The crude mixture was then purified over silica gel chromatography (gradient from 0 to 60% EtOAc;Pentane) yielding a pure colorless product, crystallizing upon standing (14.6 g, 47.6 mmol, 81%).

^{1}H NMR ($CDCl_3$, 300 MHz) δ_H 2.97 (ddd, *J* = 12.9, 8.2, 7.4 Hz, 2H), 2.83 (app dt, *J* = 12.9, 6.6 Hz, 2H), 2.73–2.66 (m, 4H), 1.41 (s, 18H). $^{13}C\{^{1}H\}$ NMR (75 MHz, $CDCl_3$) δ_C 170.4, 81.7, 47.4, 28.2, 28.1. HRMS (DCI-CH_4) Calc'd for $C_{14}H_{26}O_5S$ $[M]^+$ 306.1501, found 315.1509. FTIR (neat) ν_{max}/cm^{-1} 2980, 2932, 1702, 1364, 1226, 1154, 1032, 950, 844, 754.

3.4. General Procedures for the Pallado-Catalyzed Cross Coupling Reactions

Method A:

Dry and degassed toluene (3 mL) was added to a Schlenk tube containing $Pd(dba)_2$ (8.6 mg, 0.015 mmol, 5 mol%) and Xantphos (8.7 mg, 0.015 mmol, 5 mol%) under inert atmosphere. The solution was stirred for 2 min, then degassed DBU (180 µL, 1.2 mmol, 4 equiv.) was added, inducing a color change of the solution from dark red to clear orange-yellow. A second Schlenk tube containing tetrahydro-4*H*-thiopyran-4-one 1-oxide (39.6 mg, 0.3 mmol, 1 equiv.) and solid aryl iodide (0.9 mmol, 3 equiv.) was evacuated and back-filled with argon three times. Liquid aryl iodides (0.9 mmol, 3 equiv.) were added subsequently to evacuation and back-filling. The palladium solution was then cannulated in the second Schlenk tube, and the resulting reaction mixture was stirred at 30 °C for 6 h. Then, the solution was evaporated to dryness and purified over silica gel column chromatography.

Method B:

Sulfinyl-di-*tert*-butylpropionate (91.9 mg, 0.3 mmol, 1 equiv.), Xantphos-Phos-Pd-G3 (5.7 mg, 0.006 mmol, 2 mol%), aryl iodide (0.9 mmol, 3 equiv.) and Cs_2CO_3 (391 mg, 1.2 mmol, 4 equiv.) were weighted in a Schlenk tube. The Schlenk tube was capped with a rubber septum before being evacuated and backfilled with argon three times. A volume of 3 mL of dry and degassed toluene was added and the reaction mixture was heated at 80 °C for 16 h. A volume of 20 mL of distilled water and 20 mL of dichloromethane were added, and the organic phase separated; the aqueous phase was subsequently extracted with dichloromethane twice. The combined organic phases were dried over Na_2SO_4 before removing the volatiles using a rotavapor. The crude product was then purified over silica gel column chromatography.

3.5. Biarylsulfoxide Characterization

4,4′-sulfinylbis(N,N-diphenylaniline) 4b:

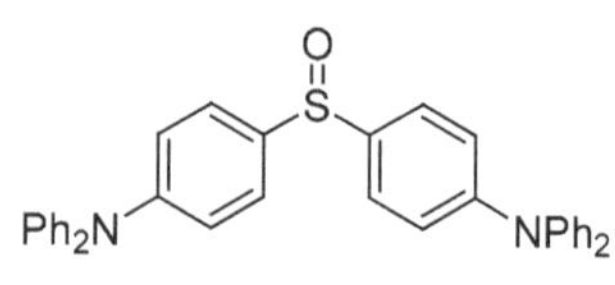

Synthesized according to the general procedure using 4-iodo-*N,N*-diphenylaniline (334 mg, 0.9 mmol). Purification by flash column chromatography (gradient from 0 to 25% EtOAc:Pentane) afforded the title compound as a white foam (Method A: 102.8 mg, 0.19 mmol, 64%: Method B: 127.3 mg, 0.24 mmol, 79%).

^{1}H NMR ($CDCl_3$, 300 MHz) δ_H 7.44 (d, *J* = 8.8 Hz, 4H), 7.29 (dd, *J* = 8.5, 6.8 Hz, 8H), 7.15–7.08 (m, 12H), 7.06 (d, *J* = 8.9 Hz, 4H). $^{13}C\{^{1}H\}$ NMR ($CDCl_3$, 75 MHz) δ_C 150.6, 146.9, 136.8, 129.7, 126.7, 125.7, 124.4, 121.7. HRMS (DCI-CH_4) Calc'd for $C_{36}H_{29}N_2OS$ $[M + H]^+$ 537.2001, found 537.2009. FTIR (neat) ν_{max}/cm^{-1} 3057, 2923, 1580, 1486, 1315, 1272, 1090, 1045, 754, 696.

5,5′-sulfinylbis(1,2,3-trimethoxybenzene) 4c:

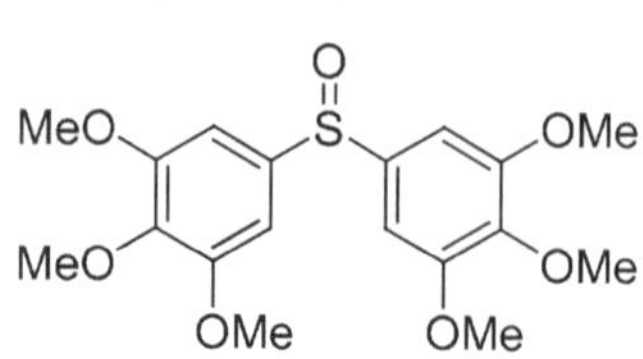

Synthesized according to the general procedure using 5-iodo-1,2,3-trimethoxybenzene (265 mg, 0.9 mmol). Purification by flash column chromatography (gradient from 0 to 80% EtOAc:Pentane) afforded the title compound as a pale yellow amorphous solid (Method B: 96.2 mg, 0.252 mmol, 84%).

^{1}H NMR ($CDCl_3$, 300 MHz) δ_H 6.87 (s, 4H), 3.88 (s, 12H), 3.86 (s, 6H). ^{13}C{^{1}H} NMR ($CDCl_3$, 75 MHz) δ_C 154.1, 140.4, 140.2, 101.9, 61.1, 56.6. HRMS (DCI-CH_4) Calc'd for $C_{18}H_{23}O_7S$ [M + H]$^+$ 383.1165 found 383.1161. FTIR (neat) ν_{max}/cm^{-1} 3084, 3053, 2939, 1590, 1496, 1462, 1409, 1307, 1235, 1127, 1102, 1059, 1004.

4,4′-sulfinylbis(methoxybenzene) 4d:

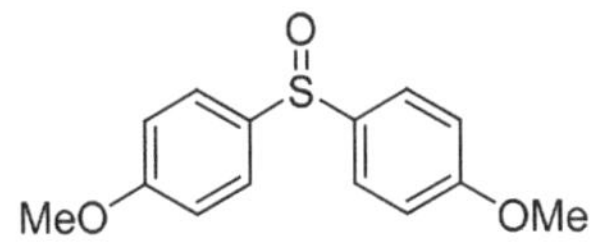

Synthesized according to the general procedure using 4-iodo-anisole (211 mg, 0.9 mmol). Purification by flash column chromatography (gradient from 0 to 25% EtOAc:Pentane) afforded the title compound as a beige amorphous solid (Method A: 46.3 mg, 0.18 mmol, 59% Method B: 69.8 mg, 0.27 mmol, 89%).

^{1}H NMR ($CDCl_3$, 300 MHz) δ_H 7.52 (d, J = 8.8 Hz, 4H), 6.94 (d, J = 8.8 Hz, 4H), 3.79 (s, 6H). ^{13}C{^{1}H} NMR ($CDCl_3$, 75 MHz) δ_C 161.9, 137.1, 126.9, 114.8, 55.6.

Data are in agreement with the literature [48].

5,5′-sulfinylbis(benzo[d][1,3]dioxole) 4e:

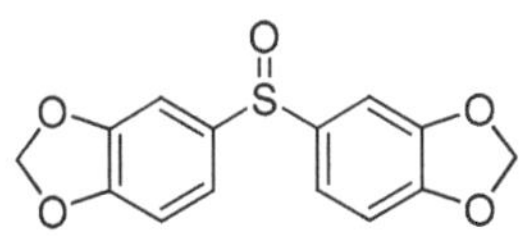

Synthesized according to the general procedure using 5-iodo-1,3-benzodioxole (114 µL, 0.9 mmol). Purification by flash column chromatography (gradient from 0 to 30% EtOAc:Pen tane) afforded the title compound as a beige amorphous solid (Method A: 43.9 mg, 0.15 mmol, 50%).

^{1}H NMR ($CDCl_3$, 300 MHz) δ_H 7.16 (dd, J = 8.0, 1.7 Hz, 2H), 7.00 (d, J = 1.7 Hz, 2H), 6.85 (d, J = 8.0 Hz, 2H), 5.99 (app q, J = 1.4 Hz, 4H). ^{13}C{^{1}H} NMR ($CDCl_3$, 75 MHz) δ_C 150.4, 148.8, 139.1, 120.0, 108.7, 104.9, 102.0.

Data are in agreement with the literature [48].

4,4′-sulfinylbis(tert-butylbenzene) 4f:

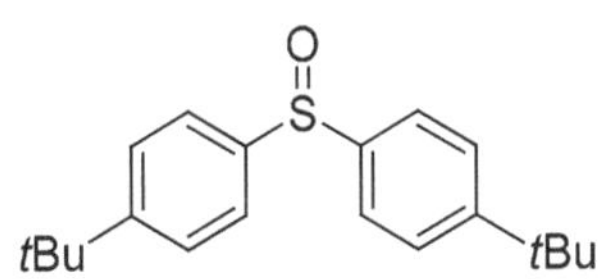

Synthesized according to the general procedure using 4-tert-butyliodobenzene (160 µL, 0.9 mmol). Purification by flash column chromatography (gradient from 0 to 15% EtOAc:Pen tane) afforded the title compound as a beige amorphous solid (Method A: 52.5 mg, 0.17 mmol, 56%).

^{1}H NMR ($CDCl_3$, 300 MHz) δ_H 7.57 (d, J = 8.6 Hz, 4H), 7.47 (d, J = 8.6 Hz, 4H), 1.30 (s, 18H). ^{13}C{^{1}H} NMR ($CDCl_3$, 75 MHz) δ_C 154.6, 142.5, 126.4, 125.0, 35.1, 31.3. HRMS (DCI-CH_4) Calc'd for $C_{20}H_{27}OS$ [M + H]$^+$ 315.1783, found 315.1776. FTIR (neat) ν_{max}/cm^{-1} 3058, 3023, 2961, 2929, 2868, 1650, 2592, 1397, 1267, 1083, 1042, 1009, 840, 829, 589, 564.

4,4′-sulfinylbis(methylbenzene) 4a:

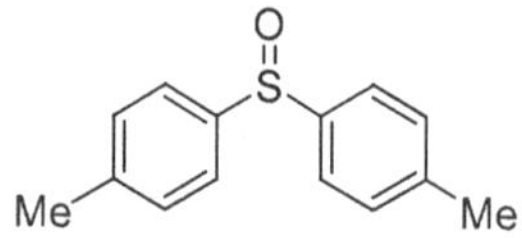

Synthesized according to the general procedure using 4-iodotoluene (196 mg, 0.9 mmol). Purification by flash column chromatography (gradient from 0 to 20% EtOAc:Pentane) afforded the title compound as a beige amorphous solid (Method A: 42.4 mg, 0.18 mmol, 55%; Method B: 67.9 mg, 0.295 mmol, 98%).

Synthesized according to the general procedure using 4-bromotoluene (154 mg, 0.9 mmol), using THF in place of PhMe (Method B: 67.0 mg, 0.29 mmol, 97%).

^{1}H NMR ($CDCl_3$, 300 MHz) δ_H 7.51 (d, J = 8.2 Hz, 4H), 7.25 (d, J = 8.1 Hz, 4H), 2.36 (s, 6H). $^{13}C\{^1H\}$ NMR ($CDCl_3$, 75 MHz) δ_C 142.8, 141.5, 130.1, 125.0, 21.5.

Data are in agreement with the literature [24].

3,3′-sulfinylbis(methylbenzene) 4g:

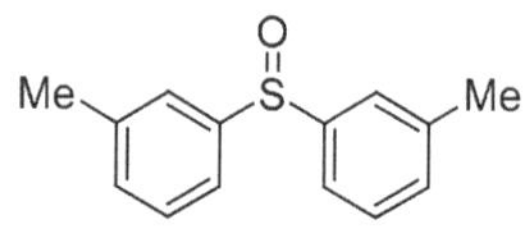

Synthesized according to the general procedure using 3-iodotoluene (115 µL, 0.9 mmol). Purification by flash column chromatography (gradient from 0 to 20% EtOAc:Pentane) afforded the title compound as a beige amorphous solid (Method A: 36.2 mg, 0.16 mmol, 52%).

^{1}H NMR ($CDCl_3$, 300 MHz) δ_H 7.47 (s, 2H), 7.41 (d, J = 7.7 Hz, 2H), 7.32 (t, J = 7.6 Hz, 2H), 7.22 (d, J = 7.4 Hz, 2H), 2.37 (s, 6H). $^{13}C\{^1H\}$ NMR ($CDCl_3$, 75 MHz) δ_C 145.5, 139.6, 131.9, 129.2, 125.1, 122.1, 21.5.

Data are in agreement with the literature [49].

(sulfinylbis(4,1-phenylene))bis(methylsulfane) 4h:

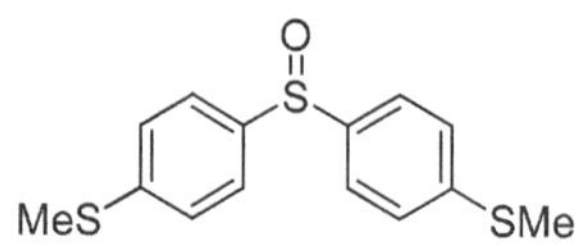

Synthesized according to the general procedure using (4-iodophenyl)(methyl)sulfane (225 mg, 0.9 mmol). Purification by flash column chromatography (gradient from 0 to 30% EtOAc:Pentane) afforded the title compound as beige amorphous solid (Method B: 62.2 mg, 0.21 mmol, 70%).

^{1}H NMR ($CDCl_3$, 300 MHz) δ_H 7.51 (d, J = 8.6 Hz, 4H), 7.27 (d, J = 8.6 Hz, 4H), 2.47 (s, 6H). $^{13}C\{^1H\}$ NMR ($CDCl_3$, 75 MHz) δ_C 143.4, 141.7, 126.3, 125.4, 15.2. HRMS (DCI-CH_4) Calc'd for $C_{14}H_{15}OS$ $[M+H]^+$ 295.0285, found 295.0279. FTIR (neat) $\nu max/cm^{-1}$ 3051, 2985, 2920, 1575, 1476, 1435, 1391, 1095, 1071, 1046, 811, 744, 554.

Sulfinyldibenzene 4i:

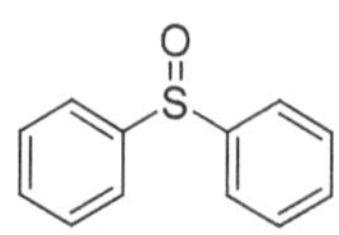

Synthesized according to the general procedure using iodobenzene (100 µL, 0.9 mmol). Purification by flash column chromatography (gradient from 0 to 20% EtOAc:Pentane) afforded the title compound as a beige amorphous solid (Method A: 27.4 mg, 0.12 mmol, 40%).

^{1}H NMR ($CDCl_3$, 300 MHz) δ_H 7.71–7.59 (m, 4H), 7.54–7.34 (m, 6H). ^{13}C{^{1}H} NMR ($CDCl_3$, 75 MHz) δ_C 145.7, 131.2, 129.5, 124.9.

Data are in agreement with the literature [48].

3,3′-sulfinylbis(methoxybenzene) 4j:

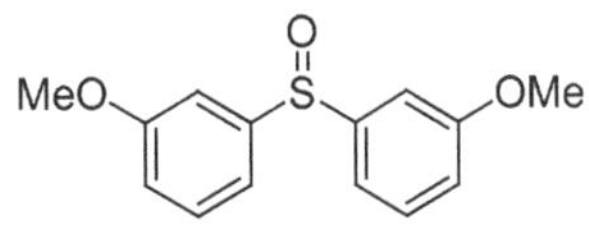

Synthesized according to the general procedure using 3-iodoanisole (107 μL, 0.9 mmol). Purification by flash column chromatography (gradient from 0 to 30% EtOAc:Pentane) afforded the title compound as a beige amorphous solid (Method A: 28.2 mg, 0.11 mmol, 36%).

^{1}H NMR ($CDCl_3$, 300 MHz) δ_H 7.34 (t, J = 7.9 Hz, 2H), 7.23 (dd, J = 2.4, 1.6 Hz, 2H), 7.22–7.12 (m, 2H), 3.81 (s, 6H). ^{13}C{^{1}H} NMR ($CDCl_3$, 75 MHz) δ_C 160.5, 147.0, 130.4, 117.5, 117.1, 109.2, 55.7. HRMS (DCI-CH_4) Calc'd for $C_{14}H_{15}O_3S$ $[M+H]^+$ 263.0742, found 263.0733. FTIR (neat) νmax/cm^{-1} 3062, 2957, 2924, 2854, 1725, 1593, 1578, 1477, 1247, 1037, 781, 689.

4,4′-sulfinylbis(fluorobenzene) 4k:

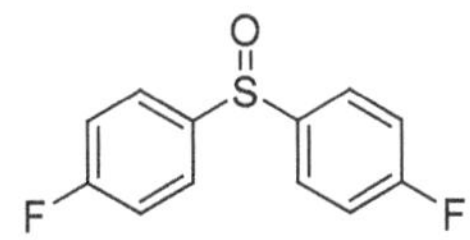

Synthesized according to the general procedure using 4-fluoroiodobenzene (104 μL, 0.9 mmol). Purification by flash column chromatography (gradient from 0 to 20% EtOAc:Pen tane) afforded the title compound as a beige amorphous solid (Method A: 12.9 mg, 0.054 mmol, 13%, Method B: 53.2 mg, 0.22 mmol, 74%).

^{1}H NMR ($CDCl_3$, 300 MHz) δ_H 7.61 (dd, J = 8.9, 5.1 Hz, 4H), 7.15 (t, J = 8.6 Hz, 4H). ^{13}C{^{1}H} NMR ($CDCl_3$, 75 MHz) δ_C δ 164.5 (d, $J_{C\text{-}F}$ = 252.2 Hz), 141.1 (d, $J_{C\text{-}F}$ = 2.3 Hz), 127.2 (d, $J_{C\text{-}F}$ = 9.0 Hz), 116.9 (d, $J_{C\text{-}F}$ = 22.6 Hz).

Data are in agreement with the literature [48].

4,4′-sulfinylbis(bromobenzene) 4l:

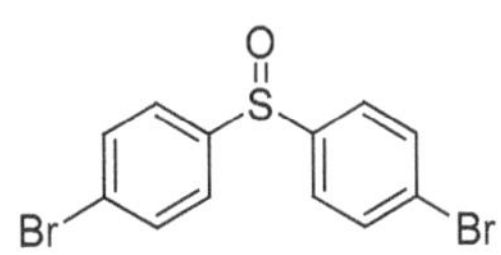

Synthesized according to the general procedure using 1-bromo-4-iodobenzene (255 mg, 0.9 mmol). Purification by flash column chromatography (gradient from 0 to 20% EtOAc:Pen tane) afforded the title compound as a beige amorphous solid (Method A: 60.4 mg, 0.17 mmol, 56%, Method B: 89.8 mg, 0.25 mmol, 83%).

^{1}H NMR ($CDCl_3$, 300 MHz) δ_H 7.66–7.55 (m, 4H), 7.55–7.44 (m, 4H). ^{13}C{^{1}H} NMR ($CDCl_3$, 75 MHz) δ_C 144.5, 132.9, 126.3, 126.1.

Data are in agreement with the literature [50].

1,1′-(sulfinylbis(4,1-phenylene))bis(ethan-1-one) 4m:

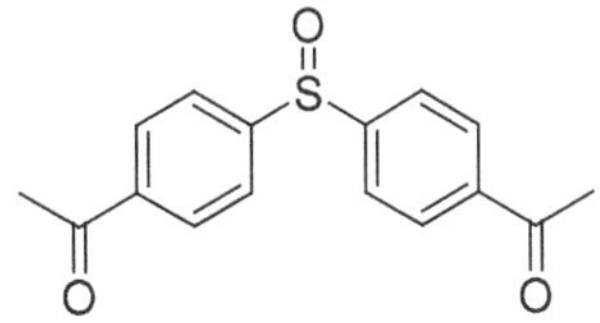

Synthesized according to the general procedure using 1-(4-iodophenyl)ethan-1-one (221.4 mg, 0.9 mmol). Purification by flash column chromatography (gradient from 0 to 70% EtOAc:Pentane) afforded the title compound as a pale yellow amorphous solid (Method B: 76.8 mg, 0.27 mmol, 89%).

^{1}H NMR ($CDCl_3$, 300 MHz) δ_H 8.04 (d, J = 8.6 Hz, 4H), 7.77 (d, J = 8.6 Hz, 4H), 2.60 (s, 6H). ^{13}C{^{1}H} NMR ($CDCl_3$, 75 MHz) δ_C 197.0, 150.1, 139.4, 129.5, 124.8, 26.9.

Data are in agreement with the literature [51].

5,5′-sulfinylbis(1,3-bis(trifluoromethyl)benzene) 4n:

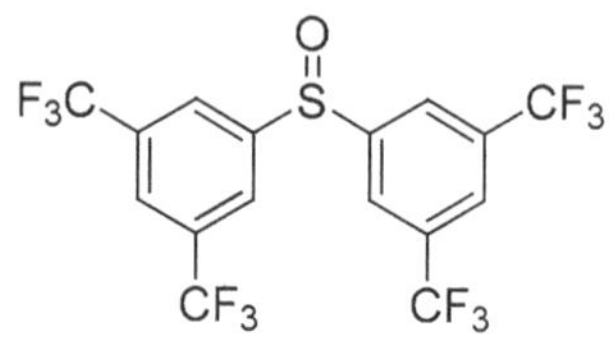

Synthesized according to the general procedure using 1-iodo-3,5-bis(trifluoromethyl)benzene (159 μL, 0.9 mmol). Purification by flash column chromatography (gradient from 0 to 10% EtOAc:Pentane) afforded the title compound as a pale yellow amorphous solid (Method B: 120.9 mg, 0.26 mmol, 85%).

^{1}H NMR ($CDCl_3$, 600 MHz) δ_H 8.16 (s, 4H), 8.02 (s, 2H). ^{13}C{^{1}H} NMR ($CDCl_3$, 151 MHz) δ_C 147.8, 133.8 (q, J = 34.6 Hz), 126.0 (p, J = 3.6 Hz), 124.7 (d, J = 3.7 Hz), 122.5 (q, J = 273.7 Hz).

Data are in agreement with the literature [52].

(sulfinylbis(4,1-phenylene))bis(pentafluoro-λ^6-sulfane) 4o:

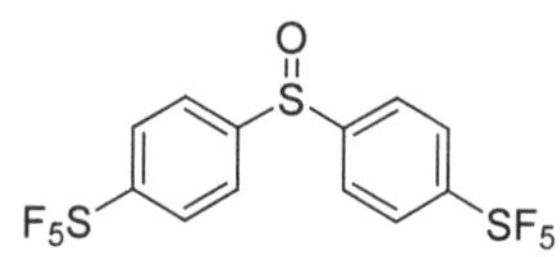

Synthesized according to the general procedure using pentafluoro(4-iodophenyl)-λ^6-sulfane (297 mg, 0.9 mmol). Purification by flash column chromatography (gradient from 0 to 20% EtOAc:Pentane) afforded the title compound as a pale yellow amorphous solid (Method B: 121.1 mg, 0.27 mmol, 89%).

^{1}H NMR ($CDCl_3$, 300 MHz) δ_H 7.90 (d, J = 8.9 Hz, 4H), 7.79 (d, J = 9.1 Hz, 4H). ^{13}C{^{1}H} NMR ($CDCl_3$, 151 MHz) δ_C 156.1 (p, J = 18.4 Hz), 148.7, 127.6 (p, J = 4.8 Hz), 125.0. ^{19}F{^{1}H} NMR ($CDCl_3$, 282 MHz) δ_F −62.97, -102.95—103.26 (m). HRMS (DCI-CH_4) Calc'd for $C_{12}H_8OF_{10}S_3$ $[M]^+$ 453.9578, found 453.9588. FTIR (neat) νmax/cm^{-1} 3103, 3068, 2928, 1605, 1480, 1397, 1103, 1057, 840, 724, 666, 601.

4,4′-sulfinylbis(nitrobenzene) 4p:

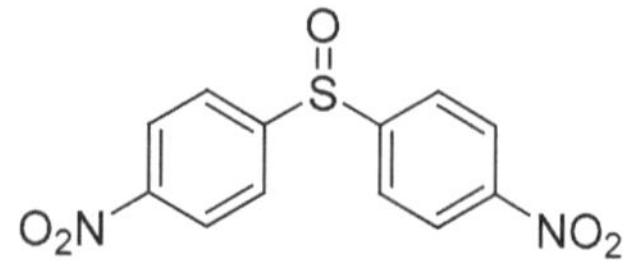

Synthesized according to the general procedure using 1-iodo-4-nitrobenzene (224.1 mg, 0.9 mmol). Purification by flash column chromatography (gradient from 0 to 10% EtOAc:Pen tane) afforded the title compound as a pale yellow amorphous solid (Method B: 71.7 mg, 0.25 mmol, 82%).

^{1}H NMR ($CDCl_3$, 300 MHz) δ_H 8.36 (d, J = 8.8 Hz, 4H), 7.90 (d, J = 8.8 Hz, 4H). ^{13}C{^{1}H} NMR ($CDCl_3$, 75 MHz) δ_C 151.6, 149.9, 125.5, 125.1.

Data are in agreement with the literature [50].

2,2′-sulfinylbis(methylbenzene) 4q:

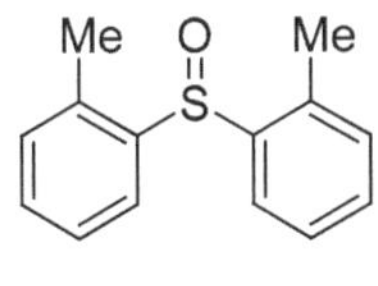

Synthesized according to the general procedure using 2-iodotoluene (114 μL, 0.9 mmol). Purification by flash column chromatography (gradient from 0 to 20% EtOAc:Pentane) afforded the title compound as a beige amorphous solid (Method B: 55.6 mg, 0.24 mmol, 80%).

^{1}H NMR ($CDCl_3$, 300 MHz) δ_H 7.75–7.64 (m, 2H), 7.42–7.30 (m, 4H), 7.24–7.15 (m, 2H), 2.42 (s, 6H). ^{13}C{^{1}H} NMR ($CDCl_3$, 75 MHz) δ_C 141.9, 136.7, 131.2, 131.0, 127.2, 126.1, 18.7.

Data are in agreement with the literature [49].

2,2′-sulfinylbis(methoxybenzene) 4r:

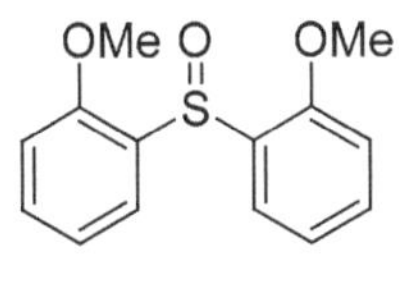

Synthesized according to the general procedure using 1-iodo-2-methoxybenzene (117 μL, 0.9 mmol). Purification by flash column chromatography (gradient from 0 to 60% EtOAc:Pentane) afforded the title compound as a beige amorphous solid (Method B: 64.9 mg, 0.25 mmol, 82%).

^{1}H NMR ($CDCl_3$, 300 MHz) δ_H 7.65 (dd, J = 7.7, 1.7 Hz, 2H), 7.41 (ddd, J = 8.3, 7.4, 1.7 Hz, 2H), 7.07 (app. td, J = 7.5, 1.0 Hz, 2H), 6.89 (dd, J = 8.3, 1.0 Hz, 2H), 3.80 (s, 6H). ^{13}C{^{1}H} NMR ($CDCl_3$, 75 MHz) δ_C 156.8, 132.5, 132.4, 127.0, 121.4, 111.3, 56.0.

Data are in agreement with the literature [49].

4,4′-sulfinyldidibenzo[*b,d*]thiophene 4s:

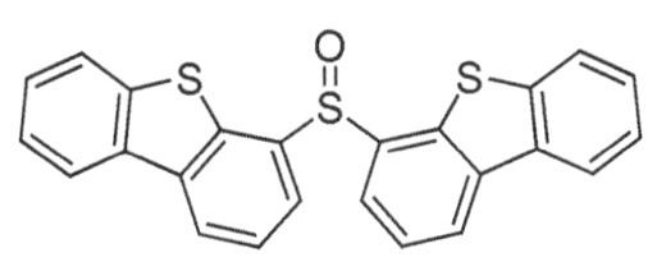

Synthesized according to the general procedure using 4-iododibenzo[*b,d*]thiophene (465 mg, 1.5 mmol, 3 equiv. Purification by flash column chromatography (gradient from 0 to 20% EtOAc:Pentane) afforded the title compound as a beige amorphous solid (Method B: 181.8 mg, 0.438 mmol, 85%).

^{1}H NMR ($CDCl_3$, 300 MHz) δ_H 8.22 (dd, J = 7.9, 1.1 Hz, 2H), 8.18 (dd, J = 7.6, 1.1 Hz, 2H), 8.14–8.10 (m, 2H), 7.87–7.83 (m, 2H), 7.61 (app t, J = 7.7 Hz, 2H), 7.52–7.41 (m, 4H). ^{13}C{^{1}H} NMR ($CDCl_3$, 75 MHz) δ_C 140.1, 137.9, 136.4, 136.2, 134.1, 127.7, 125.1, 124.9, 124.9, 124.6, 122.9, 121.9. HRMS (DCI-CH_4) Calc'd for $C_{24}H_{15}OS_3$ [M+H]$^+$ 415.0285 found 415.0268. FTIR (neat) νmax/cm^{-1} 3053, 2922, 1434, 1398, 1389, 1056, 1028, 749, 703.

1,1′-sulfinyldinaphthalene 4t:

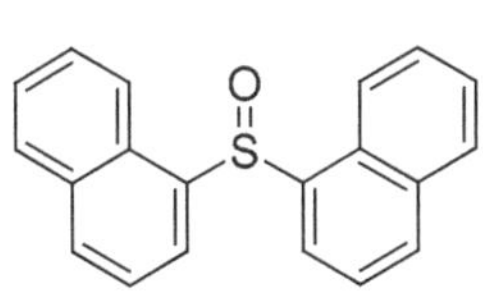

Synthesized according to the general procedure using 1-iodonaphthalene (131 μL, 0.9 mmol). Purification by flash column chromatography (gradient from 0 to 25% EtOAc:Pentane) afforded the title compound as a pale yellow amorphous solid (Method B: 87.7 mg, 0.29 mmol, 97%).

^{1}H NMR (CDCl$_3$, 300 MHz) δ_H 8.42–8.30 (m, 2H), 8.06 (dd, J = 7.3, 1.2 Hz, 2H), 7.97 (app dt, J = 8.2, 1.1 Hz, 2H), 7.96–7.85 (m, 2H), 7.61–7.51 (m, 6H). ^{13}C{^{1}H} NMR (CDCl$_3$, 75 MHz) δ_C 139.9, 133.8, 132.1, 130.1, 129.1, 127.8, 126.9, 125.8, 125.8, 122.7.

Data are in agreement with the literature [53].

9,9′-sulfinyldianthracene 4u:

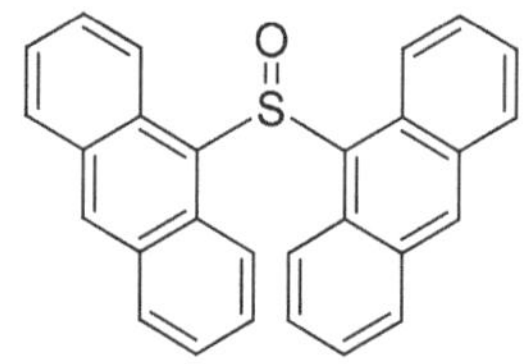

Synthesized according to the general procedure using 9-bromoanthracene (386 mg, 1.5 mmol, 3 equiv.), and carrying the reaction at 110 °C. Purification by flash column chromatography (gradient from 0 to 20% EtOAc:Pentane) afforded the title compound as a bright yellow solid (Method B: 98 mg, 0.24 mmol, 49%).

^{1}H NMR (CD$_2$Cl$_2$, 300 MHz) δ_H 9.32–9.13 (m, 4H), 8.51 (s, 2H), 8.05–7.90 (m, 4H), 7.52–7.32 (m, 8H). ^{13}C{^{1}H} NMR (CD$_2$Cl$_2$, 75 MHz) δ_C 134.5, 132.8, 131.6, 131.0, 129.9, 128.0, 125.8, 123.5.

Data are in agreement with the literature [38].

2,2′-sulfinylbis(1,3,5-trimethylbenzene) 4v:

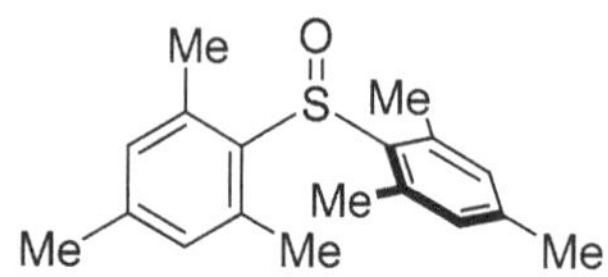

Synthesized according to the general procedure using 2-iodo-1,3,5-trimethylbenzene (221.5 mg, 0.9 mmol). Purification by flash column chromatography (gradient from 0 to 20% EtOAc:Pentane) afforded the title compound as a beige amorphous solid (Method B: 32.8 mg, 0.12 mmol, 38%).

^{1}H NMR (CDCl$_3$, 300 MHz) δ_H 6.86–6.77 (m, 4H), 2.42 (s, 12H), 2.27 (s, 6H). ^{13}C{^{1}H} NMR (CDCl$_3$, 75 MHz) δ_C 140.6, 138.6, 136.6, 131.3, 21.0, 19.6.

Data are in agreement with the literature [54].

3,3′-sulfinyldithiophene 4w:

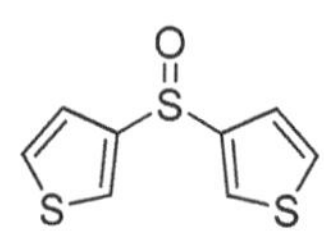

Synthesized according to the general procedure using 3-iodothiophene (91 µL, 0.9 mmol). Purification by flash column chromatography (gradient from 0 to 30% EtOAc:Pentane) afforded the title compound as a yellow oily residue (Method A: 3.4 mg, 0.02 mmol, 5%; Method B: 54.8 mg consisting of a 90/10 mixture of the titled compound and tert-butyl 3-(thiophen-3-ylsulfinyl)propanoate, 0.26 mmol, 85%).

^{1}H NMR (CDCl$_3$, 300 MHz) δ_H 7.80 (dd, J = 3.0, 1.3 Hz, 2H), 7.43 (dd, J = 5.1, 3.0 Hz, 2H), 7.11 (dd, J = 5.2, 1.3 Hz, 2H). ^{13}C{^{1}H} NMR (CDCl$_3$, 75 MHz) δ_C 144.3, 128.5, 127.0, 124.2. HRMS (DCI-CH$_4$) Calc'd for $C_8H_7OS_3$ [M+H]$^+$ 214.9659 found 215.9660. FTIR (neat) vmax/cm^{-1} 3090, 3066, 1398, 1204, 1096, 1072, 1040, 791, 633.

tert-butyl 3-(thiophen-3-ylsulfinyl)propanoate 4w′:

The titled compound was isolated as a co-eluting byproduct upon isolation of 3,3′-sulfinyldithiophene using method B (10%).

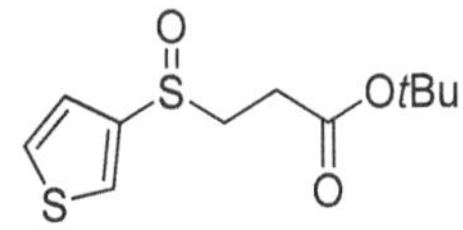

^{1}H NMR ($CDCl_3$, 300 MHz) δ_H 7.76 (dd, J = 3.0, 1.3 Hz, 1H), 7.51 (dd, J = 5.1, 3.0 Hz, 1H), 7.23 (dd, J = 5.1, 1.3 Hz, 1H), 3.20 (ddd, J = 13.4, 8.4, 6.9 Hz, 1H), 3.06 (ddd, J = 13.4, 8.2, 6.0 Hz, 1H), 2.75 (ddd, J = 17.1, 8.2, 6.8 Hz, 1H), 2.50 (ddd, J = 17.2, 8.4, 6.0 Hz, 1H), 1.43 (s, 9H).

3.6. Photophysical Properties of 4,4′-Sulfinylbis(N,N-Diphenylaniline) 4b and 4,4′-Sulfonylbis(N,N-Diphenylaniline) 5

Solutions of various concentrations of the desired compound were prepared in dry and degassed PhMe in an argon filled glove-box. The solution was transferred in a 3 mL quartz cuvette before sealing it tightly with a PTFE cap. The cuvette was removed from the glove-box and rapidly studied.

3.7. X-ray Data

CCDC-2364220 (**1**), CCDC-2364221 (**2**), CCDC-2364222 (**4b**) and CCDC-2364223 (**4s**) contain the supplementary crystallographic data for this paper (Sections S1–S3). These data can be obtained free of charge from The Cambridge Crystallographic Data Centre via www.ccdc.cam.ac.uk/structures (accessed on 10 October 2024).

4. Conclusions

In summary, we designed two new reactants, smoothly furnishing the bis-sulfenate anion synthon in basic conditions. These reactants were then applied in the development of a pallado-catalyzed methodology for the synthesis of symmetrical biarylsulfoxides tolerating a large array of electronic properties and bulkiness. Finally, the photophysical properties of the biarylsulfoxide **4b** have been explored, demonstrating an unreported TADF phenomenon on sulfoxide-containing scaffolds.

Supplementary Materials: The following supporting information can be downloaded at: https://www.mdpi.com/article/10.3390/molecules29204809/s1, S1: Photophysical studies data. S2: NMR Spectra. S3: X-ray diffraction studies.

Author Contributions: D.M. and V.M. designed the experiments. V.M. and I.C. conducted the chemical reactions described in the manuscript. S.M.-L. carried the X-ray diffraction analysis. D.M., V.M. and E.M. wrote the manuscript and all authors contributed to the reading and revising of the manuscript. All authors have read and agreed to the published version of the manuscript.

Funding: This work was supported by the Centre National de la Recherche Scientifique (CNRS) and the Universite' de Toulouse (UPS). We thank the ANR funding agency for financial support of the CROSS project (ANR-21-CE09-002).

Institutional Review Board Statement: Not applicable.

Informed Consent Statement: Not applicable.

Data Availability Statement: Data are available on request from the authors.

Conflicts of Interest: The authors declare no conflicts of interest.

References

1. David, N.A. The Pharmacology of Dimethyl Sulfoxide. *Annu. Rev. Pharmacol.* **1972**, *12*, 353–374. [CrossRef] [PubMed]
2. Rubin, L.F. Toxicity of dimethyl sulfoxide, alone and in combination. *Ann. N. Y. Acad. Sci.* **1975**, *243*, 98–103. [CrossRef] [PubMed]
3. Brayton, C.F. Dimethyl sulfoxide (DMSO), a review. *Cornell Vet.* **1986**, *76*, 61–90. [PubMed]
4. Aronson, J.K. (Ed.) *Meyler's Side Effects of Drugs*, 16th ed.; Elsevier: Amsterdam, The Netherlands, 2016; pp. 992–993.
5. Feng, M.; Tang, B.; Liang, S.H.; Jiang, X. Sulfur Containing Scaffolds in Drugs: Synthesis and Application in Medicinal Chemistry. *Curr. Top. Med. Chem.* **2016**, *16*, 1200–1216. [CrossRef]

6. Bentley, R. Role of sulfur chirality in the chemical processes of biology. *Chem. Soc. Rev.* **2005**, *34*, 609–624. [CrossRef]
7. Calligaris, M. Structure and bonding in metal sulfoxide complexes. *Coord. Chem. Rev.* **1996**, *153*, 83–154. [CrossRef]
8. Calligaris, M. Structure and bonding in metal sulfoxide complexes: An update. *Coord. Chem. Rev.* **2004**, *248*, 351–375. [CrossRef]
9. Mellah, M.; Voituriez, A.; Schulz, E. Chiral Sulfur Ligands for Asymmetric Catalysis. *Chem. Rev.* **2007**, *107*, 5133–5209. [CrossRef]
10. Sipos, G.; Drinkel, E.E.; Dorta, R. The emergence of sulfoxides as efficient ligands in transition metal catalysis. *Chem. Soc. Rev.* **2015**, *44*, 3834–3860. [CrossRef]
11. Lentz, N.; Mallet-Ladeira, S.; Baceiredo, A.; Kato, T.; Madec, D. Germylene–sulfoxide as a potential hemilabile ligand: Application in coordination chemistry. *Dalton Trans.* **2018**, *47*, 15751–15756. [CrossRef]
12. Deak, N.; du Boullay, O.T.; Moraru, I.-T.; Mallet-Ladeira, S.; Madec, D.; Nemes, G. A non-symmetric sulfur-basedO,C,O-chelating pincer ligand leading to chiral germylene and stannylene. *Dalton Trans.* **2019**, *48*, 2399–2406. [CrossRef] [PubMed]
13. Deak, N.; du Boullay, O.T.; Mallet-Ladeira, S.; Moraru, I.; Madec, D.; Nemes, G. Synthesis and Characterization of a Novel Bis-Sulfoxide and Its Evaluation as a Ligand in p-Block Chemistry. *Eur. J. Inorg. Chem.* **2020**, *2020*, 3729–3737. [CrossRef]
14. Lentz, N.; Cuevas-Chavez, C.; Mallet-Ladeira, S.; Sotiropoulos, J.-M.; Baceiredo, A.; Kato, T.; Madec, D. Germylene-β-sulfoxide Hemilabile Ligand in Coordination Chemistry. *Inorg. Chem.* **2020**, *60*, 423–430. [CrossRef]
15. Authesserre, U.; Hameury, S.; Dajnak, A.; Saffon-Merceron, N.; Baceiredo, A.; Madec, D.; Maerten, E. Complexes of Dichlorogermylene with Phosphine/Sulfoxide-Supported Carbone as Ligand. *Molecules* **2021**, *26*, 2005. [CrossRef] [PubMed]
16. Watanabe, S.; Takayama, T.; Nishio, H.; Matsushima, K.; Tanaka, Y.; Saito, S.; Sun, Y.; Oyaizu, K. Synthesis of colorless and high-refractive-index sulfoxide-containing polymers by the oxidation of poly(phenylene sulfide) derivatives. *Polym. Chem.* **2022**, *13*, 1705–1711. [CrossRef]
17. Li, M.; Berritt, S.; Wang, C.; Yang, X.; Liu, Y.; Sha, S.-C.; Wang, B.; Wang, R.; Gao, X.; Li, Z.; et al. Sulfenate anions as organocatalysts for benzylic chloromethyl coupling polymerization via C=C bond formation. *Nat. Commun.* **2018**, *9*, 1754. [CrossRef]
18. Kaiser, D.; Klose, I.; Oost, R.; Neuhaus, J.; Maulide, N. Bond-Forming and -Breaking Reactions at Sulfur(IV): Sulfoxides, Sulfonium Salts, Sulfur Ylides, and Sulfinate Salts. *Chem. Rev.* **2019**, *119*, 8701–8780. [CrossRef]
19. Maitro, G.; Prestat, G.; Madec, D.; Poli, G. An escapade in the world of sulfenate anions: Generation, reactivity and applications in domino processes. *Tetrahedron Asymmetry* **2010**, *21*, 1075–1084. [CrossRef]
20. Yang, L.; Wang, B.; Yin, X.; Zeng, Q. Advances of Sulfenate Anions in Catalytic Asymmetric Synthesis of Sulfoxides. *Chem. Rec.* **2021**, *22*, e202100242. [CrossRef]
21. Riddell, A.B.; Smith, M.R.A.; Schwan, A.L. The generation and reactions of sulfenate anions. An update. *J. Sulfur Chem.* **2022**, *43*, 540–592. [CrossRef]
22. Yin, X.; Zhang, Q.; Zeng, Q. Advance in the Synthesis of Sulfoxides and Sulfinamides from β-Sulfinyl Esters. *Organics* **2023**, *4*, 173–185. [CrossRef]
23. Saito, F. Recent Developments on the Synthesis of Sulfoxides via Sulfenate Anions. *Synthesis* **2023**, *56*, 220–228. [CrossRef]
24. Maitro, G.; Vogel, S.; Prestat, G.; Madec, D.; Poli, G. Aryl Sulfoxides via Palladium-Catalyzed Arylation of Sulfenate Anions. *Org. Lett.* **2006**, *8*, 5951–5954. [CrossRef]
25. Maitro, G.; Vogel, S.; Sadaoui, M.; Prestat, G.; Madec, D.; Poli, G. Enantioselective Synthesis of Aryl Sulfoxides via Palladium-Catalyzed Arylation of Sulfenate Anions. *Org. Lett.* **2007**, *9*, 5493–5496. [CrossRef]
26. Bernoud, E.; Le Duc, G.; Bantreil, X.; Prestat, G.; Madec, D.; Poli, G. Aryl Sulfoxides from Allyl Sulfoxides via [2,3]-Sigmatropic Rearrangement and Domino Pd-Catalyzed Generation/Arylation of Sulfenate Anions. *Org. Lett.* **2009**, *12*, 320–323. [CrossRef]
27. Izquierdo, F.; Chartoire, A.; Nolan, S.P. Direct S-Arylation of Unactivated Arylsulfoxides Using [Pd(IPr*)(cin)Cl]. *ACS Catal.* **2013**, *3*, 2190–2193. [CrossRef]
28. Gelat, F.; Lohier, J.; Gaumont, A.; Perrio, S. tert-Butyl Sulfoxides: Key Precursors for Palladium-Catalyzed Arylation of Sulfenate Salts. *Adv. Synth. Catal.* **2015**, *357*, 2011–2016. [CrossRef]
29. Zhang, M.; Jia, T.; Sagamanova, I.K.; Pericás, M.A.; Walsh, P.J. tert-Butyl Phenyl Sulfoxide: A Traceless Sulfenate Anion Precatalyst. *Org. Lett.* **2015**, *17*, 1164–1167. [CrossRef] [PubMed]
30. Jia, T.; Zhang, M.; Jiang, H.; Wang, C.Y.; Walsh, P.J. Palladium-Catalyzed Arylation of Alkyl Sulfenate Anions. *J. Am. Chem. Soc.* **2015**, *137*, 13887–13893. [CrossRef]
31. Jiang, H.; Jia, T.; Zhang, M.; Walsh, P.J. Palladium-Catalyzed Arylation of Aryl Sulfenate Anions with Aryl Bromides under Mild Conditions: Synthesis of Diaryl Sulfoxides. *Org. Lett.* **2016**, *18*, 972–975. [CrossRef]
32. Jia, T.; Bellomo, A.; Montel, S.; Zhang, M.; EL Baina, K.; Zheng, B.; Walsh, P.J. Diaryl Sulfoxides from Aryl Benzyl Sulfoxides: A Single Palladium-Catalyzed Triple Relay Process. *Angew. Chem. Int. Ed.* **2013**, *53*, 260–264. [CrossRef] [PubMed]
33. Jia, T.; Zhang, M.; Sagamanova, I.K.; Wang, C.Y.; Walsh, P.J. Palladium Catalyzed Diaryl Sulfoxide Generation from Aryl Benzyl Sulfoxides and Aryl Chlorides. *Org. Lett.* **2015**, *17*, 1168–1171. [CrossRef]
34. Jia, T.; Zhang, M.; McCollom, S.P.; Bellomo, A.; Montel, S.; Mao, J.; Dreher, S.D.; Welch, C.J.; Regalado, E.L.; Williamson, R.T.; et al. Palladium-Catalyzed Enantioselective Arylation of Aryl Sulfenate Anions: A Combined Experimental and Computational Study. *J. Am. Chem. Soc.* **2017**, *139*, 8337–8345. [CrossRef]
35. Wu, C.; Berritt, S.; Liang, X.; Walsh, P.J. Palladium-Catalyzed Enantioselective Alkenylation of Sulfenate Anions. *Org. Lett.* **2019**, *21*, 960–964. [CrossRef] [PubMed]
36. Wang, L.; Chen, M.; Zhang, P.; Li, W.; Zhang, J. Palladium/PC-Phos-Catalyzed Enantioselective Arylation of General Sulfenate Anions: Scope and Synthetic Applications. *J. Am. Chem. Soc.* **2018**, *140*, 3467–3473. [CrossRef]

37. Zhang, J.; Suzuki, K.; Ohmori, K. Total Syntheses of Sparsomycin and Sparoxomycins A1 and A2 via Sulfenate-Anion-Mediated Iterative C–S Bond Formation. *Org. Lett.* **2023**, *25*, 9036–9040. [CrossRef]
38. Christensen, P.R.; Patrick, B.O.; Caron, É.; Wolf, M.O. Oxidation-State-Dependent Photochemistry of Sulfur-Bridged Anthracenes. *Angew. Chem. Int. Ed.* **2013**, *52*, 12946–12950. [CrossRef] [PubMed]
39. Magné, V.; Lenk, R.; Mallet-Ladeira, S.; Maerten, E.; Madec, D. Pentafluorophenyl Copper–Biarylsulfoxide Complexes: Synthesis and Photoreactivity. *Molecules* **2024**, *29*, 3332. [CrossRef] [PubMed]
40. Zhang, Q.; Li, J.; Shizu, K.; Huang, S.; Hirata, S.; Miyazaki, H.; Adachi, C. Design of Efficient Thermally Activated Delayed Fluorescence Materials for Pure Blue Organic Light Emitting Diodes. *J. Am. Chem. Soc.* **2012**, *134*, 14706–14709. [CrossRef]
41. Wong, M.Y.; La-Placa, M.-G.; Pertegas, A.; Bolink, H.J.; Zysman-Colman, E. Deep-blue thermally activated delayed fluorescence (TADF) emitters for light-emitting electrochemical cells (LEECs). *J. Mater. Chem. C* **2017**, *5*, 1699–1705. [CrossRef]
42. Hepguler, A.; Ulukan, P.; Catak, S. The photophysical properties of sulfone-based TADF emitters in relation to their structural properties. *Phys. Chem. Chem. Phys.* **2023**, *25*, 31457–31470. [CrossRef] [PubMed]
43. Sheldrick, G.M. SHELXT—Integrated space-group and crystal-structure determination. *Acta Crystallogr. Sect. A Found. Adv.* **2015**, *71*, 3–8. [CrossRef] [PubMed]
44. Sheldrick, G.M. Crystal structure refinement with SHELXL. *Acta Crystallogr. Sect. C Struct. Chem.* **2015**, *71*, 3–8. [CrossRef] [PubMed]
45. Greenhalgh, R.P. Selective Oxidation of Phenyl Sulphides to Sulphoxides or Sulphones Using Oxone® and Wet Alumina. *Synlett* **1992**, *1992*, 235–236. [CrossRef]
46. Johnson, J.A.; Zhang, X.; Reeson, T.C.; Chen, Y.-S.; Zhang, J. Facile Control of the Charge Density and Photocatalytic Activity of an Anionic Indium Porphyrin Framework via in Situ Metalation. *J. Am. Chem. Soc.* **2014**, *136*, 15881–15884. [CrossRef]
47. Wright, S.W.; Hageman, D.L.; Wright, A.S.; McClure, L.D. Convenient preparations of t-butyl esters and ethers from t-butanol. *Tetrahedron Lett.* **1997**, *38*, 7345–7348. [CrossRef]
48. Li, W.-X.; Yang, B.-W.; Ying, X.; Zhang, Z.-W.; Chu, X.-Q.; Zhou, X.; Ma, M.; Shen, Z.-L. Nickel-Catalyzed Direct Cross-Coupling of Diaryl Sulfoxide with Aryl Bromide. *J. Org. Chem.* **2022**, *87*, 11899–11908. [CrossRef]
49. Zhu, Y.; Li, Y.; Zhang, B.; Zhang, F.; Yang, Y.; Wang, X. Palladium-Catalyzed Enantioselective C−H Olefination of Diaryl Sulfoxides through Parallel Kinetic Resolution and Desymmetrization. *Angew. Chem. Int. Ed.* **2018**, *57*, 5129–5133. [CrossRef]
50. Chun, J.-H.; Morse, C.L.; Chin, F.T.; Pike, V.W. No-carrier-added [18F]fluoroarenes from the radiofluorination of diaryl sulfoxides. *Chem. Commun.* **2013**, *49*, 2151–2153. [CrossRef]
51. Jeon, H.B.; Kim, K.T.; Kim, S.H. Selective oxidation of sulfides to sulfoxides with cyanuric chloride and urea–hydrogen peroxide adduct. *Tetrahedron Lett.* **2014**, *55*, 3905–3908. [CrossRef]
52. Huang, M.; Wu, Z.; Krebs, J.; Friedrich, A.; Luo, X.; Westcott, S.A.; Radius, U.; Marder, T.B. Ni-Catalyzed Borylation of Aryl Sulfoxides. *Chem.—A Eur. J.* **2021**, *27*, 8149–8158. [CrossRef] [PubMed]
53. Hampel, T.; Ruppenthal, S.; Sälinger, D.; Brückner, R. Desymmetrization of Prochiral Diaryl Sulfoxides by an Asymmetric Sulfoxide–Magnesium Exchange. *Chem.—A Eur. J.* **2012**, *18*, 3136–3140. [CrossRef] [PubMed]
54. Bandgar, B.P.; Makone, S.S. Lithium/Sodium Perchlorate Catalyzed Synthesis of Symmetrical Diaryl Sulfoxides. *Synth. Commun.* **2004**, *34*, 743–750. [CrossRef]

molecules

MDPI

Article

Oxysulfonylation of Alkynes with Sodium Sulfinates to Access β-Keto Sulfones Catalyzed by $BF_3 \cdot OEt_2$

Shi-Wei Yu, Zu-Jia Chen, Huan-Qing Li, Wen-Xi Li, Yun Li, Zong Li and Zhao-Yang Wang *

School of Chemistry, South China Normal University, Guangzhou Key Laboratory of Analytical Chemistry for Biomedicine, GDMPA Key Laboratory for Process Control and Quality Evaluation of Chiral Pharmaceuticals, Key Laboratory of Theoretical Chemistry of Environment, Ministry of Education, Guangzhou 510006, China; 2021022671@m.scnu.edu.cn (S.-W.Y.); 2022022607@m.scnu.edu.cn (Z.-J.C.); 2022022611@m.scnu.edu.cn (H.-Q.L.); 2023022646@m.scnu.edu.cn (W.-X.L.); m15192479187@163.com (Y.L.); lzscnu@outlook.com (Z.L.)

* Correspondence: wangzy@scnu.edu.cn; Tel.: +86-020-3931-0258; Fax: +86-020-3931-0187

Abstract: An efficient and operationally simple method for the synthesis of β-keto sulfones through the $BF_3 \cdot OEt_2$-promoted reaction of alkynes and sodium sulfinates is developed. With its facile and selective access to the targets, it features good functional group compatibility, mild conditions, easily available starting materials, and good yields. Notably, the reaction does not require metal catalysts or chemical reagents with pungent odors.

Keywords: sodium sulfinates; alkynes; sulfonylation reaction; $BF_3 \cdot OEt_2$; β-Keto sulfones

Citation: Yu, S.-W.; Chen, Z.-J.; Li, H.-Q.; Li, W.-X.; Li, Y.; Li, Z.; Wang, Z.-Y. Oxysulfonylation of Alkynes with Sodium Sulfinates to Access β-Keto Sulfones Catalyzed by $BF_3 \cdot OEt_2$. *Molecules* **2024**, *29*, 3559. https://doi.org/10.3390/molecules29153559

Academic Editor: Georg Manolikakes

Received: 11 July 2024
Revised: 26 July 2024
Accepted: 26 July 2024
Published: 28 July 2024

1. Introduction

Sulfone compounds are of considerable importance in synthetic and medicinal chemistry [1,2]. For example, the introduction of sulfonyl groups into medicines can substantially influence their polarity, acidity, aqueous solubility, and other properties [3,4], so the efficient synthesis of sulfone compounds has attracted extensive attention recently [5,6]. Among them, β-keto sulfone as a kind of unique sulfone compound has been extensively used in biomedicine, such as for anti-schistosomal, anti-analgesic, and antibacterial effects (Figure 1) [7–9]. At the same time, it can also be used as a synthon and is widely used in synthetic chemistry [10–12].

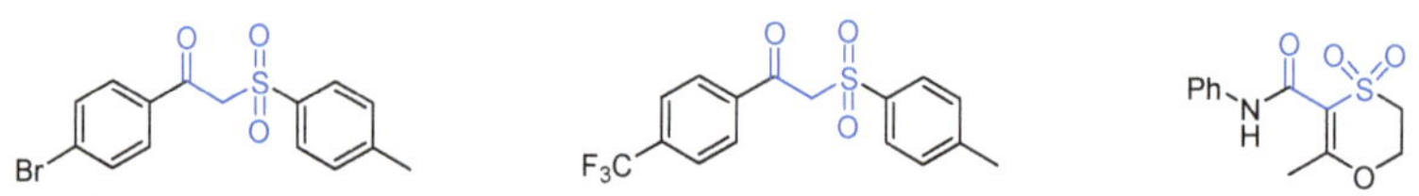

Figure 1. Some bioactive β-keto sulfone derivatives.

Traditionally, β-keto sulfones can be synthesized through the nucleophilic alkylation of sodium sulfinates with acyl halides [13]. In recent years, with the rise of green synthetic chemistry and the development of various sulfur reagents [14–17], the coupling reaction of unsaturated bonds (e.g., alkynes) and sulfur reagents has gradually become the main way to synthesize β-keto sulfones [18–20]. Among them, being bench-stable, commercially or readily available, and easy to handle [21,22], sodium sulfinates/sulfinic acid is widely used in the synthesis of β-keto sulfones [23–26].

For instance, Lei's group [27] reported the oxysulfonylation of terminal alkynes with sulfinic acids catalyzed by pyridine (Scheme 1a). Afterward, Kumar's group [28] also reported a similar reaction (Scheme 1b). Recently, Sun's group [29] reported the oxysulfonylation of arylpropiolic acids and sodium sulfnates to generate β-keto sulfones using only hexafluoroisopropanol (HFIP) as a solvent and oxygen as a green oxidant (Scheme 1c).

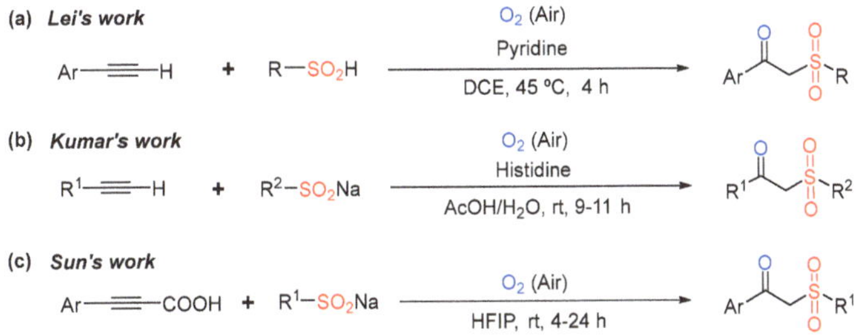

Scheme 1. Examples of access to β-keto sulfones starting from sodium sulfinates or sulfinic acids [27–29].

On the other hand, $BF_3{\cdot}OEt_2$ exhibits strong Lewis acidity [30,31], and its excellent catalytic activity is widely used in various organic synthesis reactions [32,33]. In our previous work [34], we reported a $BF_3{\cdot}OEt_2$-mediated difunctionalization reaction of sodium sulfinates and alkynes to obtain β-sulfinyl alkenylsulfone (Scheme 2b). In fact, during the condition optimization for the above-mentioned work, we also unexpectedly found that if the amount of $BF_3{\cdot}OEt_2$ was below 1.0 equiv., the main product changed from β-sulfinyl alkenylsulfones to β-keto sulfones.

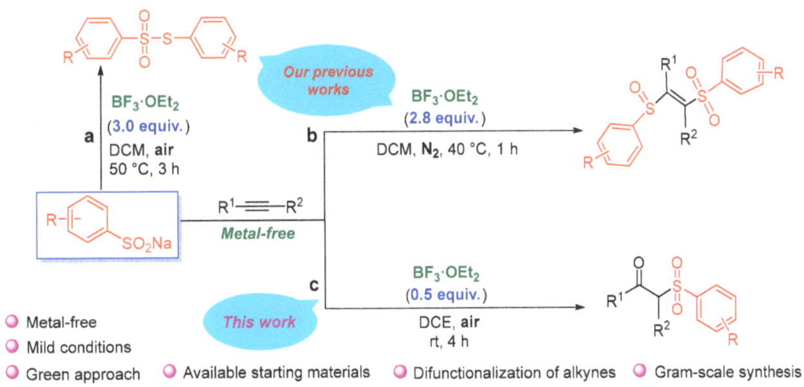

Scheme 2. Serial synthesis reactions of sodium sulfinates catalyzed by $BF_3{\cdot}OEt_2$ [34,35].

In view of this, based on our previous research on C-S bond construction [21,36], especially the sodium sulfinate reaction with $BF_3{\cdot}OEt_2$ as a catalyst [34,35], herein, we hope to report a new reaction under air atmosphere to efficiently obtain β-keto sulfones via the radical pathway (Scheme 2c). This reaction is designed to synthesize β-keto sulfones without any metal catalysts and does not require the use of chemical reagents with irritating odors such as pyridine and acetic acid. Therefore, this method is more gentle and greener and its successful development is also conducive to expanding the application of $BF_3{\cdot}OEt_2$ as a catalyst. In particular, the regulation and control through this organosulfur reagent and $BF_3{\cdot}OEt_2$ catalyst under different reaction conditions can synthesize different products, such as thiosulfonates, β-sulfinyl alkenylsulfones, and β-keto sulfones. Thus, this strategy is of great significance in the fields of organic sulfur chemistry, BF_3 chemistry, and alkyne conversion.

2. Results and Discussion

2.1. Optimization of Reaction Conditions

We attempted the reaction of phenylacetylene (**1a**) and sodium *p*-tolylsulfinate (**2a**) as a model in dichloromethane (DCM) to systematically investigate the influences of different factors to optimize the best conditions. The results are summarized in Table 1.

Table 1. Optimization of reaction conditions [a].

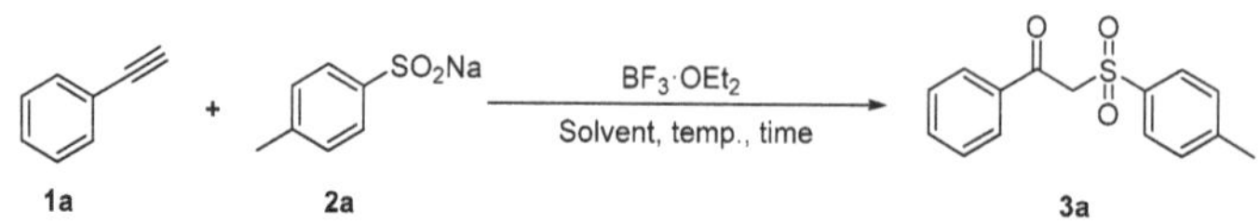

Entry	Acid (equiv.)	1a:2a	Temp. (°C)	Time (h)	Solvent	Yield (%) [b]
1	$BF_3{\cdot}OEt_2$ (0.5)	1:2.4	rt	4	DCM	38
2	$BF_3{\cdot}OEt_2$ (0.5)	1:2.4	rt	6	DCM	33
3	$BF_3{\cdot}OEt_2$ (0.5)	1:2.4	rt	2	DCM	15
4	$BF_3{\cdot}OEt_2$ (0.5)	1:2.4	40	4	DCM	32
5	$BF_3{\cdot}OEt_2$ (0.5)	1:2.4	0	4	DCM	Trace
6	$BF_3{\cdot}OEt_2$ (0.25)	1:2.4	rt	4	DCM	22
7	$BF_3{\cdot}OEt_2$ (0.75)	1:2.4	rt	4	DCM	27
8	**$BF_3{\cdot}OEt_2$ (0.5)**	**1:2.4**	**rt**	**4**	**DCE**	**56**
9	$BF_3{\cdot}OEt_2$ (0.5)	1:2.4	rt	4	EA	24
10	$BF_3{\cdot}OEt_2$ (0.5)	1:2.4	rt	4	DMSO	N.D. [c]
11	$BF_3{\cdot}OEt_2$ (0.5)	1:2.4	rt	4	CH_3OH	N.D.
12	$BF_3{\cdot}OEt_2$ (0.5)	1:2.4	rt	4	C_2H_5OH	N.D.
13	$BF_3{\cdot}OEt_2$ (0.5)	1:2.4	rt	4	THF	N.D.
14	$BF_3{\cdot}OEt_2$ (0.5)	1:2.4	rt	4	CH_3CN	18
15	$BF_3{\cdot}OEt_2$ (0.5)	1:2.4	rt	4	Toluene	N.D.
16	$BF_3{\cdot}OEt_2$ (0.5)	1:2.4	rt	4	Acetone	N.D.
17	$BF_3{\cdot}OEt_2$ (0.5)	1:2.4	rt	4	DMF	N.D.
18	$BF_3{\cdot}OEt_2$ (0.5)	1:2.4	rt	4	AcOH	N.D.
19	$BF_3{\cdot}OEt_2$ (0.5)	1:2.4	rt	4	1,4-Dioxane	N.D.
20	$BF_3{\cdot}OEt_2$ (0.5)	1:2.4	rt	4	$CHCl_3$	12
21	$BF_3{\cdot}OEt_2$ (0.5)	1:2.4	rt	4	Ethyl ether	N.D.
22	$BF_3{\cdot}OEt_2$ (0.5)	1:2.4	rt	4	Dimethoxyethane	N.D.
23	$BF_3{\cdot}OEt_2$ (0.5)	1:2.0	rt	4	DCE	43
24	$BF_3{\cdot}OEt_2$ (0.5)	1:2.8	rt	4	DCE	42

[a] Reaction conditions: Usually, **1a** (0.3 mmol), **2a** (0.72 mmol), $BF_3{\cdot}OEt_2$ (0.15 mmol), and DCE (4 mL) at room temperature (rt) for 4 h under air atmosphere. [b] Isolated yield. [c] N.D. = Not Detected.

Initially, we examined the reaction time (Entries 1–3). When the reaction time increased to 6 h, there was no improvement. When the reaction time was shortened, the yield decreased sharply.

In addition, lowering or increasing the reaction temperature did not increase the yield (Entries 4–5 vs. Entry 1).

Similarly, we also examined the amount of $BF_3{\cdot}OEt_2$ used. It was found that increasing or decreasing the amount of Lewis acid did not further increase the yield (Entries 6–7 vs. Entry 1).

At the same time, the effect of solvents on the model reaction was investigated (Entries 1, 8–22). It was found that the solvent plays a significant role in the success of this reaction. Among different solvents, e.g., ethyl acetate (EA), dimethyl sulfoxide (DMSO), tetrahydrofuran (THF), and *N*,*N*-dimethylformamide (DMF), using 1,2-dichloroethane (DCE) as the solvent was more favorable to the formation of **3a** (Entry 8 vs. Entry 1). Therefore, we decided to choose DCE as the best solvent.

Furthermore, we investigated the ratio of reactants **1a** and **2a**. When the ratio of **1a**:**2a** was 1:2.4, the best effect was achieved (Entry 8 vs. Entries 23–24).

Thus, the optimized reaction conditions were identified as using **1a** (0.3 mmol), **2a** (0.72 mmol), and $BF_3{\cdot}OEt_2$ (0.15 mmol) as the catalyst and 4.0 mL of DCE as the solvent at room temperature (rt) for 4 h.

2.2. *Scope of Substrates*

Under the optimized conditions, a range of alkynes **1** and sodium sulfinates **2** were applied in the transformation to establish the scope and generality of this reaction.

We first investigated various alkyne substrates **1** to explore the scope of this reaction (Table 2).

Table 2. Substrate scope of various alkyne compounds **1** sodium sulfinates **2** [a, b].

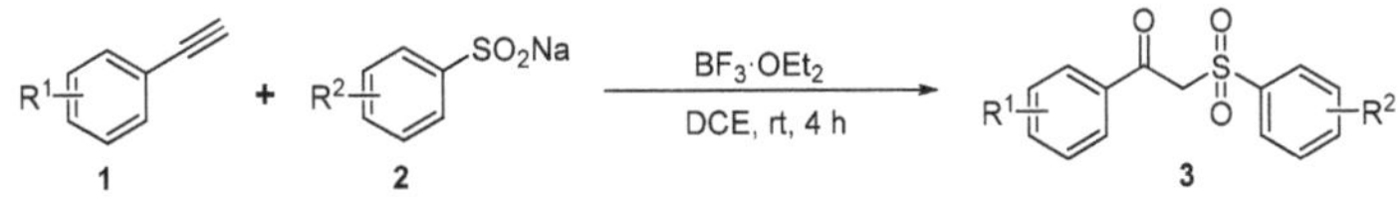

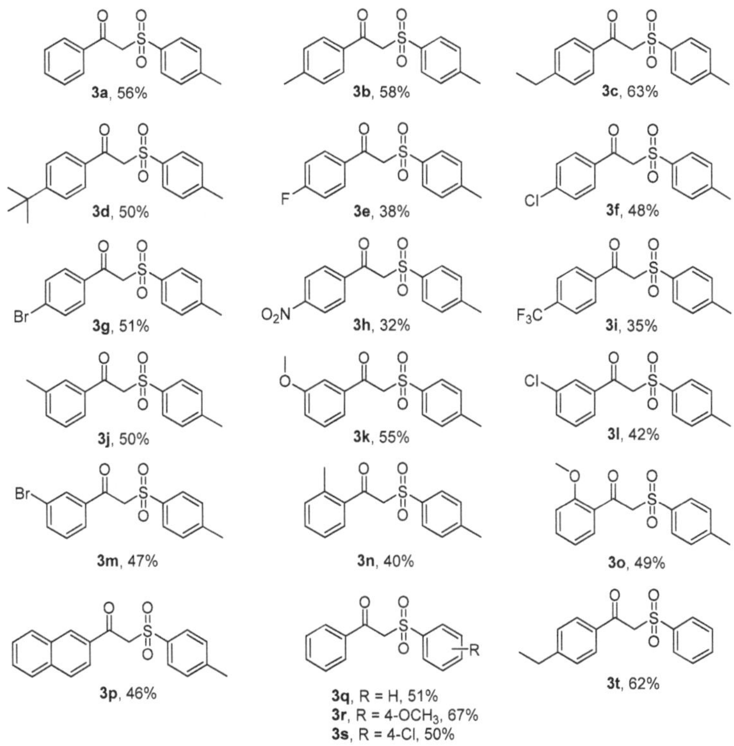

[a] Reaction conditions: Usually, **1** (0.30 mmol), **2** (0.72 mmol), $BF_3{\cdot}OEt_2$ (0.15 mmol), and DCE (4 mL) were added and stirred at room temperature (rt) for 4 h. [b] Isolated yield.

Pleasingly, many arylalkynes **1** bearing electron-donating substituents (e.g., methyl, ethyl, and *tert*-butyl) at the para-position of the aryl group were successfully employed with moderate yields (**3a-3d**, 50–63%). For other substituents, e.g., halogen, there was a slight decrease in the yield of the desired product (**3e-3g**, 38–51%). In particular, when the strong electron-withdrawing groups (e.g., $-CF_3$, $-NO_2$) were bearing in **1**, the yield decreased further (**3h-3i**, 32–35%).

In the next step, the effect of different positions of the arylalkyne ring on the yield was investigated. Due to the steric effect, for the same substituent group, when it was a *meta*-substituted group in the arylalkyne substrate **1**, the yield was lower than that of the *para*-substituted group, such as **3b** (58%) vs. **3j** (50%), **3f** (48%) vs. **3l** (42%), and **3g** (51%) vs. **3m** (47%).

As anticipated, when the arylalkyne substrate **1** was bearing *ortho*-substituted, the yield was further reduced, such as **3b** (58%) vs. **3j** (50%) vs. **3n** (40%), and **3k** (55%) vs. **3o** (49%). Furthermore, polycyclic substituted alkynes **1** also can be transformed into the corresponding products (e.g., **3p**, 46%).

Importantly, the scope of sodium arylsulfinate substrates **2** was also examined. The results are also summarized in Table 2. Obviously, the substituted **2** was able to react with **1a** to produce the corresponding products with moderate yields (**3q-3t**, 51–67%). Among them, the substrate **2** containing electron-donating groups performed better than those with electron-withdrawing groups, such as **3r** (67%) vs. **3a** (56%) vs. **3q** (51%) vs. **3s** (50%), and **3c** (63%) vs. **3t** (62%).

2.3. Structural Characterization Analysis

From the ^{1}H NMR spectra of the target compounds, it can be seen that the ^{1}H NMR data of the twenty compounds **3a-3t** are consistent with the simulated data of the hydrogen atom in the target products. For example, the ^{1}H NMR spectra of the synthesized compounds **3a-3t** show that there was a single peak with the integration of two units near 4.72 ppm, which was a characteristic peak caused by methylene hydrogen in the structure of the target products.

At the same time, in the ^{13}C NMR spectra, the single peak of the chemical shift value near 64.0 ppm was also a characteristic peak of methylene carbon. In addition, regardless of whether the product contains one fluorine atom or trifluoromethyl group, the corresponding chemical shift can be found in ^{19}F NMR, as anticipated.

In a word, these test results indicate that the characterization results of NMR are consistent with expectations. The tested data of the synthesized compounds **3a-3t**, including the data of the melting point (m.p.), are also consistent with data in the references reported before [17,19,23,25].

Although NMR tests have proven that compounds **3a-3t** have the expected structure, in order to further determine the structure of the product, we also conducted single-crystal cultivation of compound **3a**. Its crystal resolution data are shown in Table 3, successfully affirming that **3a** does have the expected structure [37].

Table 3. X-ray crystal data and structure refinement for compound **3a**.

Compound	3a
Empirical formula	$C_{15}H_{14}O_3S$
Formula weight	274.32
Temperature (K)	297
Wavelength (Å)	0.71073
Crystal system	monoclinic
Space group	$P2_1/n$
Unit cell dimensions (Å, °)	$a = 7.7540(5)$, $b = 11.5019(8)$, $c = 15.3137(10)$
	$\alpha = 90$, $\beta = 98.746(7)$, $\gamma = 90$
Volume (Å^3)	1349.88(16)
Z	4
Density (calculated) (g/cm^3)	1.350
Absorption coefficient (mm^{-1})	0.240
F(000)	576.0
Theta range for data collection	2.691 to 29.072
Index ranges	$-10 \leq h \leq 10$, $-15 \leq k \leq 13$, $-20 \leq l \leq 20$

Table 3. *Cont.*

Compound	3a
Reflections collected	6338
Independent reflections	3117 [R(int) = 0.0218, R(sigma) = 0.0379]
Completeness to theta = 29.072°	86.4 %
Absorption correction	Multi-Scan
Max. and min. transmission	1.000 and 0.885
Refinement method	Least Squares minimisation
Data / restraints / parameters	3117 / 0 / 173
Goodness-of-fit on F^2	1.027
Final R indices [I>2sigma(I)]	R_1 = 0.0489, wR_2 = 0.1141
R indices (all data)	R_1 = 0.0668, wR_2 = 0.1229
Largest diff. peak and hole	0.26 and −0.26 e.Å-3

Accordingly, the X-ray single-crystal diffraction results of compound **3a** in Figure 2 indicate that the molecule contains two benzene rings, which are connected to a carbonyl group and a sulfonyl group, respectively. Moreover, the carbonyl and sulfonyl groups are connected by a methylene group. Thus, the structure of compound **3a** can also be aptly confirmed through the single crystal structure.

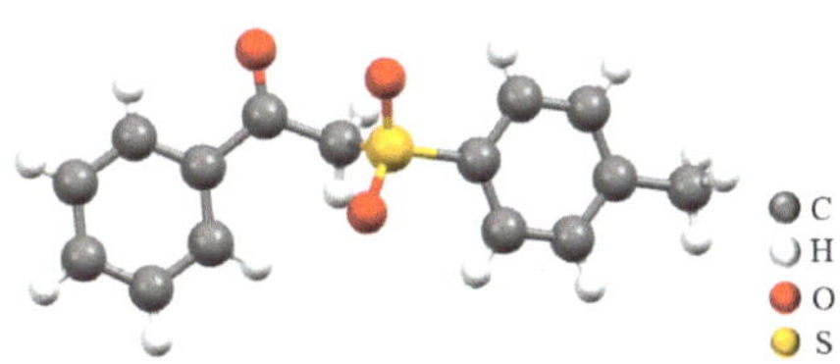

Figure 2. The crystal structure of **3a** (**CCDC**: 2367445).

On the other hand, the molecular weights of these radicals in control experiments were obtained by high-resolution mass spectrometry (HR-MS), and the error between the tested value and the calculated value was within a reasonable range.

2.4. Mechanism Investigation

In order to understand the reaction mechanism, some control experiments were carried out. The results are shown in Scheme 3.

Initially, once 2,2,6,6-tetramethyl-1-piperidinyloxy (TEMPO) as a free radical scavenger was added under the standard reaction conditions, the formation of **3a** was obviously suppressed (Scheme 3a). Importantly, the capture of the sulfonyl radical by TEMPO could be detected by HR-MS (Figure S2), indicating that a free radical pathway may be involved in this process.

Similarly, when using 1,1-diphenylethylene as a free radical scavenger, there was no normal reaction for the formation of **3a** (Scheme 3b), and the capture of the sulfonyl radical by 1,1-diphenylethylene was also confirmed by HR-MS. In Figure 3, it can be seen that the theoretical calculated value of $[M+H]^+$ of intermediates is 335.1100, and the actual test value is 335.1096 with an error value of 0.0004, which is indeed within the reasonable range. This also indicates that there is a free radical pathway.

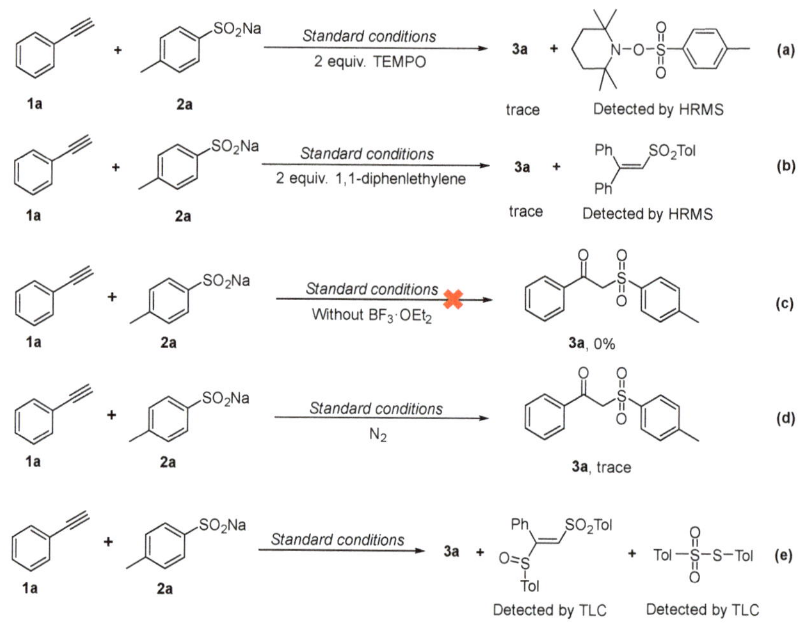

Scheme 3. Control experiments.

What is more, the control experiment also implies that Lewis acid $BF_3{\cdot}OEt_2$ could be essential for this reaction (Scheme 3c).

At the same time, in order to determine the source of the carboxyl oxygen atom in compound **3**, an experiment in an oxygen-free atmosphere was investigated. When the reaction was carried out in a N_2 atmosphere, the target product could not be obtained (Scheme 3d). This result shows that O_2 in the air may be the source of the carboxyl oxygen atom in the product [28].

Furthermore, when the reaction was carried out under standard conditions, β-sulfinyl alkenylsulfone and thiosulfonate were easily observed by TLC (Scheme 3e), which may be the byproducts of the reaction.

Thus, on the basis of the above control experiments and the previously reported studies [14,16,23,25,27–29,34,35,38,39], a possible reaction pathway is proposed as Scheme 4.

Firstly, $BF_3{\cdot}H_2O$ is produced in situ from $BF_3{\cdot}OEt_2$ in the case of a trace amount of water [34,35]. Then, $BF_3{\cdot}H_2O$ reacts with sodium sulfinate **2**, giving sulfinic acid **A** and Na[BF_3OH]. Subsequently, sulfinyl sulfone **B** is generated from sulfinic acid **A**, and it is easy to produce sulfonyl radical **I** and sulfinyl radical **II** from intermediate **B** under heating conditions [38,39].

Secondly, sulfonyl radical **I** is added to alkyne **1** to give intermediate **C** [14,23], which is then trapped by oxygen to generate intermediate **D** [25]. Finally, intermediate **D** forms intermediate **E** [16] via the hydrogen ion and radical **II**, while causing radical **II** to generate free radical **I**. Intermediate **E** is prone to tautomerism [27–29], giving product **3** (Scheme 4).

Considering the preliminary experimental results (Scheme 2a and 2b) [34,35], we speculate that the amount of $BF_3{\cdot}OEt_2$ determines the amount of sodium sulfite involved in the reaction and less $BF_3{\cdot}OEt_2$ is beneficial to the reaction of less sodium sulfite (Scheme 2c). At the same time, the reaction atmosphere [35] and the presence [34] or absence [35] of acetylene substrate determine which radicals are easier to generate and react. This is why different products can be produced in similar $BF_3{\cdot}OEt_2$ catalytic systems (Scheme 2).

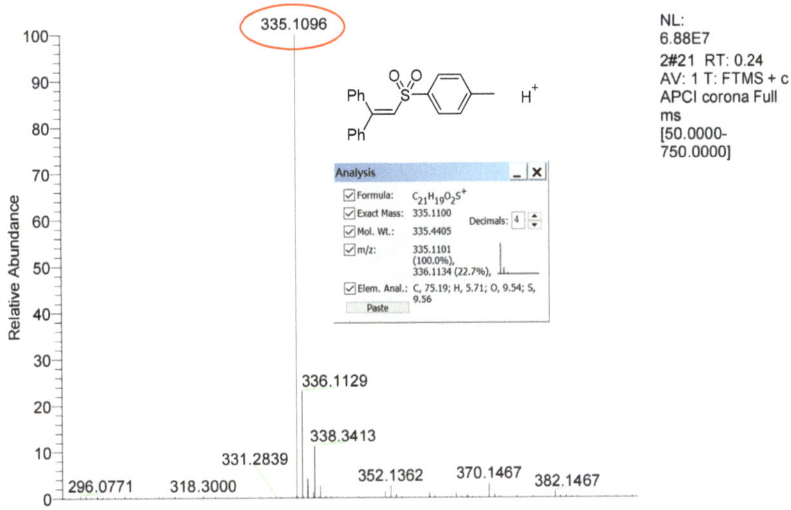

Figure 3. The HR-MS of the capture of sulfonyl radical by scavenger 1,1-diphenylethylene in the control experiment.

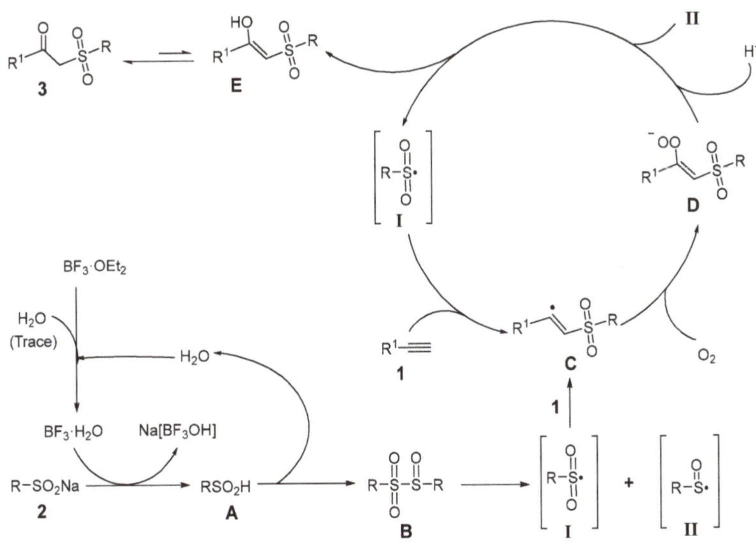

Scheme 4. Proposed mechanism.

In addition, as shown in Scheme 5, sulfinyl radicals **II** are extremely unstable in an air atmosphere and are prone to disproportionation to form sulfonyl radicals **I** and thiyl radicals **III** [40,41], which easily couple to form thiosulfonates **F** (**Minor path a** from intermediate **II**) [42,43]. Similarly, a small amount of β-sulfinyl alkenylsulfones **G** (**Minor path b** from intermediate **II**) [34] as a by-product can also be observed, which may be caused by the incomplete disproportionation of sulfinyl radicals [44].

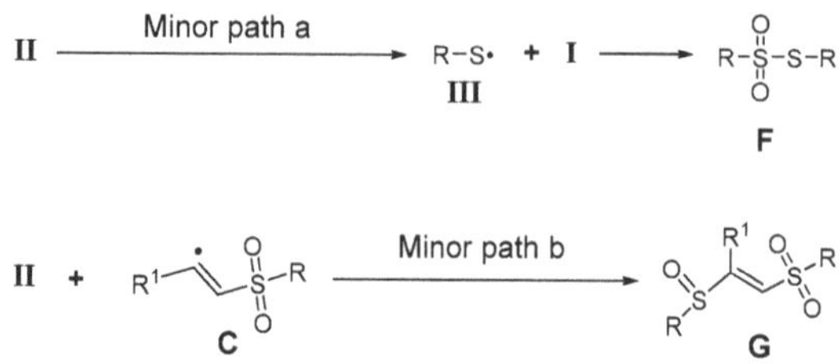

Scheme 5. Proposed mechanism for the by-products.

2.5. Gram-scale Reaction

Considering that there is a wide application of β-keto sulfones in organic synthesis and biomedicine [45], in order to demonstrate the practicability of this reaction, further gram-scale research was carried out by selecting the synthesis of the target compound **3c** as an example (Scheme 6).

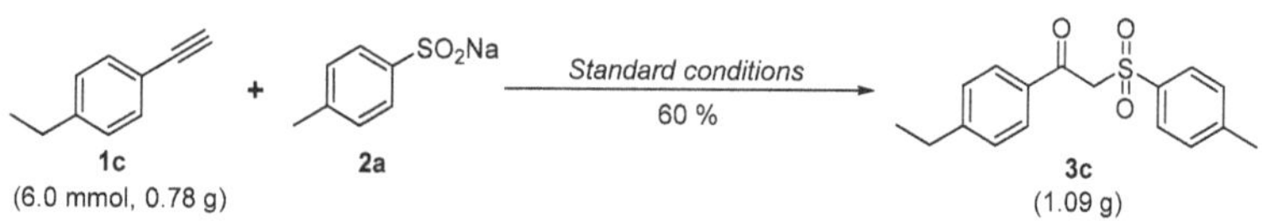

Scheme 6. Gram-scale reaction of **3c**.

Obviously, the gram-scale synthesis of **3c** is satisfactory. With the increase in the amount of reactant **1c** from 0.3 to 6.0 mmol, the yield of **3c** is only slightly decreased (60% vs. 63%), and the target compound **3c** can be smoothly synthesized in gram-scale with a good yield.

3. Materials and Methods

3.1. General Information

^{1}H and ^{13}C NMR spectra were collected on an AVANCE NEO-600 in $CDCl_3$ using tetramethylsilane (TMS) as an internal standard. Mass spectra were recorded on a Thermo Scientific ISQ gas chromatograph-mass spectrometer. High-resolution mass spectra (HR-MS) were obtained with a MAT 95XP mass spectrometer. The melting point (m.p.) was measured with a WRS-1B melting point instrument. Single-crystal X-ray analysis was performed using Agilent Gemini E. Reactions were monitored using thin-layer chromatography (TLC) and visualized with a UV light at 254 nm.

All reagents and solvents were purchased from commercial sources and used without further purification.

3.2. Experimental Procedure for Sodium Sulfinates 2

Different sodium sulfinates **2** were synthesized as Scheme 7 according to the procedure in the literature [22,26,46].

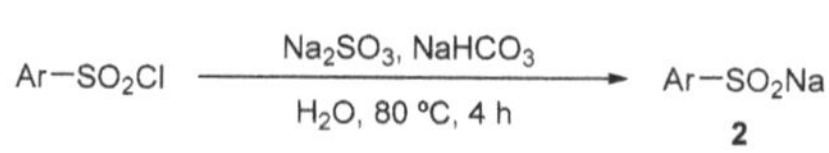

Scheme 7. Synthesis route of different sodium sulfinates **2**.

According to the literature [22,26,46], the mixture of arylsulfonyl chloride (10 mmol), sodium sulfite **2** (20 mmol), and sodium bicarbonate (20 mmol) in H_2O (15 mL) was stirred at 80 °C for 4 h. Water was removed by a rotary evaporator. Then, the remaining solid was extracted and recrystallized by ethanol to obtain the required compound **2**.

3.3. Experimental Procedure for Compounds **3a-3t**

As shown in Scheme 8, the mixture of alkyne compound **1** (0.30 mmol, 1.0 equiv.), sodium sulfinate **2** (0.72 mmol, 2.4 equiv.), and $BF_3 \cdot OEt_2$ (0.15 mmol, 0.5 equiv.) in DCE (4 mL) under an air atmosphere was stirred at room temperature for 4 h. After the completion of the reaction, ethyl acetate (EA) (15 mL) was poured into the reaction mixture. The organic layers were extracted with a saturated sodium chloride solution (3 × 15 mL). Then, the organic layer was dried over anhydrous Na_2SO_4. Finally, after the filtration and evaporation of the solvents under reduced pressure, the crude product was purified by column chromatography on silica gel to afford the desired product **3**.

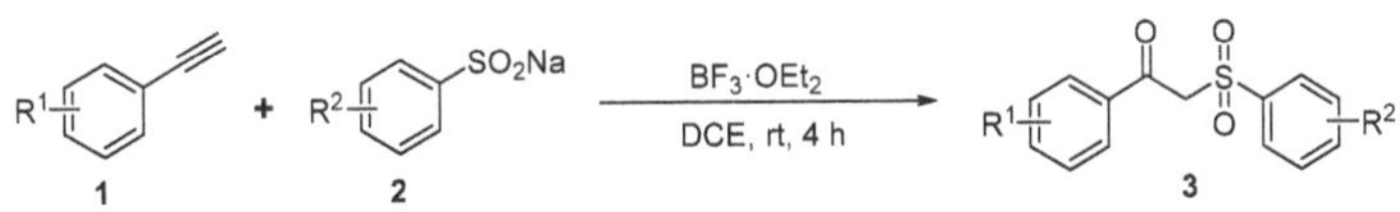

Scheme 8. Synthesis route for Compounds **3a-3t**.

3.4. Characterization Data for All Products **3a-3t**

The structures of the serial compounds **3a-3t** were systematically characterized via NMR, m.p., etc., and the corresponding data are summarized as follows:

(1) 1-Phenyl-2-tosylethan-1-one (**3a**), white solid (43.0 mg, 56%); m.p.: 106–108 °C (103–105 °C [23]); ^{1}H NMR (600 MHz, $CDCl_3$), *δ*, ppm: 2.44 (*s*, 3H, CH_3), 4.72 (*s*, 2H, CH_2), 7.33 (*d*, *J* = 8.4 Hz, 2H, ArH), 7.46–7.49 (*m*, 2H, ArH), 7.61 (*t*, *J* = 7.2 Hz, 1H, ArH), 7.76 (*d*, *J* = 8.4 Hz, 2H, ArH), 7.94 (*d*, *J* = 7.2 Hz, 2H); ^{13}C NMR (150 MHz, $CDCl_3$), *δ*, ppm: 21.9, 63.8, 128.8, 129.0, 129.5, 130.0, 134.5, 135.8, 135.9, 145.5, 188.3.

(2) 1-(*p*-Tolyl)-2-tosylethan-1-one (**3b**), white solid (50.1 mg, 58%); m.p.: 109–111 °C (119–121 °C [47]); ^{1}H NMR (600 MHz, $CDCl_3$), *δ*, ppm: 2.42 (*s*, 3H, CH_3), 2.44 (*s*, 3H, CH_3), 4.69 (*s*, 2H, CH_2), 7.27 (*d*, *J* = 8.4 Hz, 2H, ArH), 7.33 (*d*, *J* = 8.4 Hz, 2H, ArH), 7.75 (*d*, *J* = 8.4 Hz, 2H, ArH), 7.85 (*d*, *J* = 8.4 Hz, 2H, ArH); ^{13}C NMR (150 MHz, $CDCl_3$), *δ*, ppm: 21.9, 21.9, 63.7, 128.7, 129.6, 129.7, 129.9, 133.5, 135.9, 145.4, 145.7, 187.8.

(3) 1-(4-Ethylphenyl)-2-tosylethan-1-one (**3c**), white solid (57.1 mg, 63%); m.p.: 95–97 °C (84% [48]); ^{1}H NMR (600 MHz, $CDCl_3$), *δ*, ppm: 1.26 (*t*, *J* = 7.2 Hz, 3H, CH_3), 2.44 (*s*, 3H, CH_3), 2.72 (*q*, *J* = 7.2 Hz, 2H, CH_2), 4.69 (*s*, 2H, CH_2), 7.30 (*d*, *J* = 8.4 Hz, 2H, ArH), 7.33 (*d*, *J* = 8.4 Hz, 2H, ArH), 7.76 (*d*, *J* = 8.4 Hz, 2H, ArH), 7.87 (*d*, *J* = 8.4 Hz, 2H, ArH); ^{13}C NMR (150 MHz, $CDCl_3$), *δ*, ppm: 15.2, 21.8, 29.2, 63.7, 128.5, 129.7, 129.7, 129.9, 133.7, 135.9, 145.4, 151.8, 187.8.

(4) 1-(4-(*tert*-Butyl)phenyl)-2-tosylethan-1-one (**3d**), white solid (49.5 mg, 50%); m.p.: 93–94 °C (95–97 °C [49]); ^{1}H NMR (600 MHz, $CDCl_3$), *δ*, ppm: 1.34 (*s*, 9H, CH_3), 2.43 (*s*, 3H CH_3), 4.69 (*s*, 2H, CH_2), 7.32 (*d*, *J* = 8.4 Hz, 2H, ArH), 7.48 (*d*, *J* = 8.4 Hz, 2H, ArH), 7.76 (*d*, *J* = 8.4 Hz, 2H, ArH), 7.88 (*d*, *J* = 8.4 Hz, 2H, ArH); ^{13}C NMR (150 MHz, $CDCl_3$), *δ*, ppm: 21.8, 31.1, 35.4, 63.6, 126.0, 128.7, 129.5, 129.9, 133.4, 135.9, 145.4, 158.5, 187.8.

(5) 1-(4-Fluorophenyl)-2-tosylethan-1-one (**3e**), white solid (33.3 mg, 38%); m.p.: 134–135 °C (125–127 °C [47]); ^{1}H NMR (600 MHz, $CDCl_3$), *δ*, ppm: 2.45 (*s*, 3H, CH_3), 4.68 (*s*, 2H CH_2), 7.14–7.17 (*m*, 2H, ArH), 7.34 (*d*, *J* = 8.4 Hz, 2H, ArH), 7.75 (*d*, *J* = 8.4 Hz, 2H, ArH), 7.99–8.01 (*m*, 2H, ArH); ^{13}C NMR (150 MHz, $CDCl_3$), *δ*, ppm: 21.9, 63.9, 116.2 (*d*, *J* = 22.5 Hz) 128.7, 130.0, 132.3, 132.4 (*d*, *J* = 10.5 Hz), 135.7, 145.7, 166.6 (*d*, *J* = 256.5 Hz), 186.7; ^{19}F NMR (564 MHz, $CDCl_3$), δ, ppm: -102.4.

(6) 1-(4-Chlorophenyl)-2-tosylethan-1-one (**3f**), white solid (44.4 mg, 48%); m.p.: 132–134 °C (138–139 °C [46]); ^{1}H NMR (600 MHz, $CDCl_3$), *δ*, ppm: 2.45 (*s*, 3H, CH_3), 4.68 (*s*, 2H, CH_2), 7.34 (*d*, *J* = 7.8 Hz, 2H, ArH), 7.46 (*d*, *J* = 8.4 Hz, 2H, ArH), 7.74 (*d*, *J* = 7.8 Hz, 2H, ArH), 7.90 (*d*, *J* = 8.4 Hz, 2H, ArH); ^{13}C NMR (150 MHz, $CDCl_3$), *δ*, ppm: 21.9, 63.9, 128.7, 129.3, 130.0, 130.9, 134.2, 135.7, 141.2, 145.7, 187.2.

(7) 1-(4-Bromophenyl)-2-tosylethan-1-one (**3g**), white solid (53.9 mg, 51%); m.p.: 140–141 °C (143–145 °C [23]); ^{1}H NMR (600 MHz, $CDCl_3$), *δ*, ppm: 2.45 (*s*, 3H, CH_3), 4.67 (*s*, 2H, CH_2), 7.34 (*d*, *J* = 8.4 Hz, 2H, ArH), 7.63 (*d*, *J* = 9.0 Hz, 2H, ArH), 7.74 (*d*,

J = 8.4 Hz, 2H, ArH), 7.82 (*d*, *J* = 9.0 Hz, 2H, ArH); ^{13}C NMR (150 MHz, $CDCl_3$), *δ*, ppm: 21.9, 63.9, 128.7, 130.0, 130.1, 131.0, 132.4, 134.6, 135.7, 145.7, 187.4.

(8) 1-(4-Nitrophenyl)-2-tosylethan-1-one (**3h**), white solid (30.6 mg, 32%); m.p.: 140–142 °C (145–147 °C [19]); ^{1}H NMR (600 MHz, $CDCl_3$), *δ*, ppm: 2.47 (*s*, 3H, CH_3), 4.75 (*s*, 2H, CH_2), 7.37 (*d*, *J* = 8.4 Hz, 2H, ArH), 7.74 (*d*, *J* = 8.4 Hz, 2H, ArH), 8.15 (*d*, *J* = 9.0 Hz, 2H, ArH), 8.33 (*d*, *J* = 9.0 Hz, 2H, ArH); ^{13}C NMR (150 MHz, $CDCl_3$), *δ*, ppm: 21.9, 64.3, 124.1, 128.7, 130.2, 130.7, 135.5, 140.1, 146.1, 151.0, 187.1.

(9) 2-Tosyl-1-(4-trifluoromethylphenyl)ethan-1-one (**3i**), white solid (35.9 mg, 35%); m.p.: 134–136 °C (129–131 °C [23]); ^{1}H NMR (600 MHz, $CDCl_3$), *δ*, ppm: 2.45 (*s*, 3H, CH_3), 4.73 (*s*, 2H, CH_2), 7.34 (*d*, *J* = 8.4 Hz, 2H, ArH), 7.73–7.76 (*m*, 4H, ArH), 8.07 (*d*, *J* = 8.4 Hz, 2H, ArH); ^{13}C NMR (150 MHz, $CDCl_3$), *δ*, ppm: 21.9, 64.0, 123.5 (*q*, *J* = 271.5 Hz), 126.0 (*q*, *J* = 4.5 Hz128.7, 129.9, 130.1, 135.5 (*q*, *J* = 33.0 Hz), 135.6, 138.4, 145.9, 187.6; ^{19}F NMR (564 MHz, $CDCl_3$), δ, ppm: -63.3.

(10) 1-(*m*-Tolyl)-2-tosylethan-1-one (**3j**), white solid (43.2 mg, 50%); m.p.: 97–98 °C (97–99 °C [47]); ^{1}H NMR (600 MHz, $CDCl_3$), *δ*, ppm: 2.40 (*s*, 3H, CH_3), 2.44 (*s*, 3H, CH_3), 4.70 (*s*, 2H, CH_2), 7.33 (*d*, *J* = 8.4 Hz, 2H, ArH), 7.35–7.38 (*m*, 1H, ArH), 7.42 (*d*, *J* = 7.8 Hz, 1H, ArH), 7.71 (*s*, 1H, ArH), 7.74 (*d*, *J* = 7.8 Hz, 1H, ArH), 7.76 (*d*, *J* = 8.4 Hz, 2H, ArH); ^{13}C NMR (150 MHz, $CDCl_3$), *δ*, ppm: 21.4, 21.9, 63.7, 126.8, 128.8, 128.9, 129.8, 129.9, 135.3, 136.0, 138.9, 145.5, 188.4.

(11) 1-(3-Methoxyphenyl)-2-tosylethan-1-one (**3k**), colorless oil (50.2 mg, 55%) (54% [47]); ^{1}H NMR (600 MHz, $CDCl_3$), *δ*, ppm: 2.44 (*s*, 3H, CH_3), 3.84 (*s*, 3H, OCH_3), 4.70 (*s*, 2H, CH_2), 7.16 (*d*, *J* = 7.8 Hz, 1H, ArH), 7.34 (*d*, *J* = 8.4 Hz, 2H, ArH), 7.37–7.40 (*m*, 1H, ArH), 7.43 (*s*, 1H, ArH), 7.52 (*d*, *J* = 7.8 Hz, 1H, ArH), 7.77 (*d*, *J* = 8.4 Hz, 2H, ArH); ^{13}C NMR (150 MHz, $CDCl_3$), *δ*, ppm: 21.9, 55.6, 63.8, 113.1, 121.3, 122.3, 128.8, 130.0, 135.9, 137.2, 145.5, 160.1, 188.4.

(12) 1-(3-Chlorophenyl)-2-tosylethan-1-one (**3l**), white solid (38.8 mg, 42%); m.p.: 71–73 °C (71–73 °C [47]); ^{1}H NMR (600 MHz, $CDCl_3$), *δ*, ppm: 2.45 (*s*, 3H, CH_3), 4.68 (*s*, 2H, CH_2), 7.34 (*d*, *J* = 7.8 Hz, 2H, ArH), 7.42–7.46 (*m*, 1H, ArH), 7.58 (*d*, *J* = 7.2 Hz, 1H, ArH), 7.75 (*d*, *J* = 7.8 Hz, 2H, ArH), 7.84–7.86 (*m*, 2H, ArH); ^{13}C NMR (150 MHz, $CDCl_3$), *δ*, ppm: 21.9, 63.9, 127.7, 128.7, 129.3, 130.1, 130.3, 134.4, 135.4, 135.7, 137.4, 145.8, 187.2.

(13) 1-(3-Bromophenyl)-2-tosylethan-1-one (**3m**), white solid (49.6 mg, 47%); m.p.: 140–142 °C (140–141 °C [50]); ^{1}H NMR (600 MHz, $CDCl_3$), *δ*, ppm: 2.45 (*s*, 3H, CH_3), 4.68 (*s*, 2H, CH_2), 7.34 (*d*, *J* = 8.4 Hz, 2H, ArH), 7.36–7.39 (*m*, 1H, ArH), 7.72–7.75 (*m*, 3H, ArH), 7.89 (*d*, *J* = 7.8 Hz, 1H, ArH), 8.00 (*s*, 1H, ArH); ^{13}C NMR (150 MHz, $CDCl_3$), *δ*, ppm: 21.9, 63.8, 123.3, 128.2, 128.7, 130.1, 130.5, 132.2, 135.6, 137.3, 137.5, 145.8, 187.1.

(14) 1-(*o*-Tolyl)-2-tosylethan-1-one (**3n**), white solid (34.6 mg, 40%); m.p.: 99–101 °C (95–97 °C [51]); ^{1}H NMR (600 MHz, $CDCl_3$), *δ*, ppm: 2.43 (*s*, 3H, CH_3), 2.44 (*s*, 3H, CH_3), 4.69 (*s*, 2H, CH_2), 7.24–7.26 (*m*, 1H, ArH), 7.28 (*d*, *J* = 7.8 Hz, 1H, ArH), 7.33 (*d*, *J* = 8.4 Hz, 2H, ArH), 7.41–7.43 (*m*, 1H, ArH), 7.73 (*d*, *J* = 7.8 Hz, 1H, ArH), 7.75 (*d*, *J* = 8.4 Hz, 2H, ArH); ^{13}C NMR (150 MHz, $CDCl_3$), *δ*, ppm: 21.7, 21.9, 65.7, 126.0, 128.6, 130.0, 130.6, 132.4, 132.9, 135.8, 136.1, 140.2, 145.4, 190.6.

(15) 1-(2-Methoxyphenyl)-2-tosylethan-1-one (**3o**), white solid (44.7 mg, 49%); m.p.: 124–126 °C (128–130 °C [51]); ^{1}H NMR (600 MHz, $CDCl_3$), *δ*, ppm: 2.42 (*s*, 3H, CH_3), 3.88 (*s*, 3H, OCH_3), 4.92 (*s*, 2H, CH_2), 6.90 (*d*, *J* = 8.4 Hz, 1H, ArH), 6.97–7.00 (*m*, 1H, ArH), 7.29 (*d*, *J* = 8.4 Hz, 2H, ArH), 7.47–7.50 (*m*, 1H, ArH), 7.65 (*d*, 1H, *J* = 7.8 Hz, ArH), 7.74 (*d*, *J* = 8.4 Hz, 2H, ArH); ^{13}C NMR (150 MHz, $CDCl_3$), *δ*, ppm: 21.7, 55.7, 67.5, 111.7, 121.0, 126.3, 128.5, 129.5, 131.3, 135.2, 136.7, 144.8, 159.0, 189.2.

(16) 1-(Naphthalen-2-yl)-2-tosylethan-1-one (**3p**), white solid (44.7 mg, 46%); m.p.: 150–151 °C (150–152 °C [17]); ^{1}H NMR (600 MHz, $CDCl_3$), *δ*, ppm: 2.43 (*s*, 3H, CH_3), 4.86 (*s*, 2H, CH_3), 7.34 (*d*, *J* = 8.4 Hz, 2H, ArH), 7.59–7.62 (*m*, 1H, ArH), 7.65–7.68 (*m*, 1H, ArH), 7.80 (*d*, *J* = 8.4 Hz, 2H, ArH), 7.90–7.93 (*m*, 2H, ArH), 7.98–8.00 (*m*, 2H, ArH), 8.48 (*s*, 1H, ArH); ^{13}C NMR (150 MHz, $CDCl_3$), *δ*, ppm: 21.8, 64.0, 124.1, 127.3, 128.0, 128.8, 129.0, 129.5, 130.0, 130.1, 132.3, 132.4, 133.3, 135.8, 135.2, 145.6, 188.4

(17) 1-Phenyl-2-(phenylsulfonyl)ethan-1-one (**3q**), white solid (39.8 mg, 51%); m.p.: 86–88 °C (85–87 °C [23]); ^{1}H NMR (600 MHz, $CDCl_3$), δ, ppm: 4.74 (*s*, 2H, CH_2), 6.46–7.49 (*m*, 2H, ArH), 7.53–7.56 (*m*, 2H, ArH-11,13), 7.62 (*t*, *J* = 7.8 Hz, 1H, ArH), 7.66 (*t*, *J* = 7.8 Hz, 1H, ArH), 7.90 (*d*, *J* = 7.2 Hz, 2H, ArH), 7.93 (*d*, *J* = 7.2 Hz, 2H, ArH); ^{13}C NMR (150 MHz, $CDCl_3$), δ, ppm: 63.6, 128.7, 129.0, 129.3, 129.4, 134.4, 134.5, 135.8, 138.8, 188.1.

(18) 2-(4-Methoxypheny)sulfonyl)-1-phenylethan-1-one (**3r**), white solid (50.6 mg, 53%); m.p.: 100–102 °C (100–102 °C [25]); ^{1}H NMR (600 MHz, $CDCl_3$), δ, ppm: 3.86 (*s*, 3H, OCH_3), 4.71 (*s*, 2H, CH_2), 6.97 (*d*, *J* = 9.0 Hz, 2H, ArH), 7.46–7.48 (*m*, 2H, ArH), 7.61 (*t*, *J* = 7.8 Hz, 1H, ArH), 7.79 (*d*, *J* = 9.0 Hz, 2H, ArH), 7.93 (*d*, *J* = 7.2 Hz, 2H, ArH); ^{13}C NMR (150 MHz, $CDCl_3$), δ, ppm: 55.8, 63.8, 114.5, 128.9, 129.4, 130.3, 131.0, 134.4, 135.9, 164.2, 188.4.

(19) 2-(4-Chlorophenylsulfonyl)-1-phenylethan-1-one (**3s**), white solid (44.1 mg, 50%); m.p.: 131–132 °C (132–134 °C [17]); ^{1}H NMR (600 MHz, $CDCl_3$), δ, ppm: 4.74 (*s*, 2H, CH_2), 7.48–7.51 (*m*, 2H, ArH), 7.52 (*d*, *J* = 8.4 Hz, 2H, ArH), 7.64 (*t*, *J* = 7.8 Hz, 1H, ArH), 7.83 (*d*, *J* = 8.4 Hz, 2H, ArH), 7.94 (*d*, *J* = 7.8 Hz, 2H, ArH); ^{13}C NMR (150 MHz, $CDCl_3$), δ, ppm: 63.5, 129.1, 129.4, 129.7, 130.3, 134.7, 135.7, 137.2, 141.3, 188.0.

(20) 1-(4-Ethylphenyl)-2-phenylsulfonylethan-1-one (**3t**), white solid (53.6 mg, 62%); m.p.: 128–129 °C (127–129 °C [25]); ^{1}H NMR (600 MHz, $CDCl_3$), δ, ppm: 1.25 (*t*, *J* = 7.8 Hz, 3H, CH_3), 2.71 (*q*, *J* = 7.8 Hz, 2H, CH_2), 4.72 (s, 2H, CH_2), 7.29 (*d*, *J* = 8.4 Hz, ArH), 7.52–7.55 (*m*, 2H, ArH), 7.65 (*t*, *J* = 7.8 Hz, 1H, ArH), 7.86 (*d*, *J* = 8.4 Hz, 2H, ArH), 7.89 (*d*, *J* = 7.2 Hz, 2H, ArH); ^{13}C NMR (150 MHz, $CDCl_3$), δ, ppm: 15.1, 29.1, 63.5, 128.5, 128.7, 129.3, 129.7, 133.6, 134.3, 138.9, 151.8, 187.6.

The detailed ^{1}H, ^{13}C, and ^{19}F NMR spectra for all compounds **3a-3t** are provided in the Supplementary Materials.

4. Conclusions

In summary, we successfully found a $BF_3{\cdot}OEt_2$-mediated oxysulfonylation reaction of sodium sulfinates and alkynes. The reaction does not need any metal catalysts, and the simple and mild conditions make it a convenient procedure for the synthesis of β-keto sulfones. A possible radical mechanism was also proposed on the basis of control studies. Importantly, this reaction uses oxygen in the air as an oxidant and does not need to use chemical reagents with pungent odors such as pyridine and acetic acid. It meets the requirements of the development of green chemistry.

Supplementary Materials: The following supporting information can be downloaded at: https://www.mdpi.com/article/10.3390/molecules29153559/s1, and contains the ^{1}H, ^{13}C, and ^{19}F NMR spectra for all compounds **3a-3t.**

Author Contributions: Conceptualization, Z.-Y.W.; methodology, S.-W.Y. and Z.-J.C.; formal analysis, S.-W.Y. and H.-Q.L.; data curation, Y.L. and W.-X.L.; writing—original draft preparation, S.-W.Y.; writing—review and editing, S.-W.Y., Z.L. and Z.-Y.W.; project administration, Z.-Y.W.; funding acquisition, Z.-Y.W. All authors have read and agreed to the published version of the manuscript.

Funding: This research was supported by the Guangdong Basic and Applied Basic Research Foundation (no. 2021A1515012342).

Institutional Review Board Statement: Not applicable.

Informed Consent Statement: Not applicable.

Data Availability Statement: All data supporting the findings of this study are available within the paper and within its Supplementary Materials published online.

Conflicts of Interest: The authors declare no conflicts of interest.

References

1. Corpas, J.; Kim-Lee, S.-H.; Mauleón, P.; Arrayás, R.G.; Carretero, J.C. Beyond classical sulfone chemistry: Metal- and photocatalytic approaches for C-S bond functionalization of sulfones. *Chem. Soc. Rev.* **2022**, *51*, 6774–6823. [CrossRef]
2. Wu, H.; Chen, S.G.; Liu, C.N.; Zhao, Q.S.; Wang, Z.; Jin, Q.R.; Sun, S.J.; Guo, J.; He, X.W.; Walsh, P.J.; et al. Construction of C-S and C-Se bonds from unstrained ketone precursors under photoredox catalysis. *Angew. Chem. Int. Ed.* **2024**, *63*, e202314790. [CrossRef]
3. Markham, A.; Keam, S.J. Danoprevir: First global approval. *Drugs* **2018**, *78*, 1271–1276. [CrossRef] [PubMed]
4. Wu, Y.R.; Chen, J.W.; Li, L.; Wen, K.M.; Yao, X.G.; Pang, J.X.; Wu, T.; Tang, X.D. Copper-mediated decarboxylative sulfonylation of arylacetic acids with sodium sulfinates. *Org. Lett.* **2020**, *22*, 7164–7168. [CrossRef]
5. Wasilewska-Rosa, A.; Kisiel, K.; Tkaczyk, A.; Loska, R. Stereospecific synthesis of allenes from β-ketosulfones. *Adv. Synth. Catal.* **2023**, *365*, 704–708. [CrossRef]
6. Lu, L.J.; Luo, J.; Milstein, D. Manganese(I)-pincer catalyzed α-alkylation of sulfones by alcohols. *ACS Catal.* **2023**, *13*, 5949–5954. [CrossRef]
7. Inanaga, K.; Fukuyama, T.; Kubota, M.; Komatsu, Y.; Chiba, H.; Kayano, A.; Tagami, K. Novel and efficient chromium(II)-mediated desulfonylation of α-sulfonyl ketone. *Org. Lett.* **2015**, *17*, 3158–3161. [CrossRef] [PubMed]
8. Ghosh, S.; Samanta, S.; Ghosh, A.K.; Neogi, S.; Hajra, A. Advances in oxosulfonylation reaction. *Adv. Synth. Catal.* **2020**, *362*, 4552–4578. [CrossRef]
9. Reddy, R.J.; Kumari, A.H.; Kumar, J.J. Recent advances in the synthesis and applications of β-keto sulfones: New prospects for the synthesis of β-keto thiosulfones. *Org. Biomol. Chem.* **2021**, *19*, 3087–3118. [CrossRef]
10. Mulina, O.M.; Ilovaisky, A.I.; Parshin, V.D.; Terent'ev, A.O. Oxidative sulfonylation of multiple carbon-carbon bonds with sulfonyl hydrazides, sulfinic acids and their salts. *Adv. Synth. Catal.* **2020**, *362*, 4579–4654. [CrossRef]
11. Yang, W.H.; Zhou, Y.; Tong, Y.; Wang, Z.Y.; Wu, H.L.; Mei, L.; Li, Q.; Xie, L.Y.; Zhang, J.Q.; Wei, W.T. Nickel-catalyzed reaction between vinyl azides and an alkyl sulfonyl radical generated from DMSO: Rapid access to β-keto sulfones. *ACS Sustainable Chem. Eng.* **2024**, *12*, 5046–5051. [CrossRef]
12. Abdukerem, D.; Chen, H.; Mao, Z.; Xia, K.; Zhu, W.; Liu, C.; Yu, Y.; Abdukader, A. Transition metal-free C(sp^3)-H selenation of β-ketosulfones. *Org. Biomol. Chem.* **2024**, *22*, 2075–2080. [CrossRef] [PubMed]
13. Pospísil, J.; Sato, H. Practical synthesis of β-acyl and β-alkoxycarbonyl heterocyclic sulfones. *J. Org. Chem.* **2011**, *76*, 2269–2272. [CrossRef] [PubMed]
14. Ni, B.Q.; Zhang, B.B.; Han, J.E.; Peng, B.H.; Shan, Y.L.; Niu, T.F. Heterogeneous carbon nitrides photocatalysis multicomponent hydrosulfonylation of alkynes to access β-keto sulfones with the insertion of sulfur dioxide in aerobic aqueous medium. *Org. Lett.* **2020**, *22*, 670–674. [CrossRef] [PubMed]
15. Pampana, V.K.K.; Charpe, V.P.; Sagadevan, A.; Das, D.K.; Lin, C.C.; Hwu, J.R.; Hwang, K.C. Oxy-sulfonylation of terminal alkynes via C-S coupling enabled by copper photoredox catalysis. *Green Chem.* **2021**, *23*, 3569–3574. [CrossRef]
16. Song, T.; Zhang, Y.P.; Wang, C.; Li, Y.F.; Yang, Y. Photocatalytic aerobic oxysulfonylation of alkynes to access β-keto sulfones catalyzed by OVs-N-Nb_2O_5. *Chin. J. Chem.* **2022**, *40*, 2618–2624. [CrossRef]
17. Dam, B.; Sahoo, A.K.; Patel, B.K. Visible-light-mediated synthesis of β-keto sulfones using g-C_3N_4 as a recyclable photocatalyst under sustainable conditions. *Green Chem.* **2022**, *24*, 7122–7130. [CrossRef]
18. Wang, Y.J.; Zhao, Y.H.; Cai, C.Q.; Wang, L.Y.; Gong, H. Dioxygen-triggered oxosulfonylation/sulfonylation of terminal olefins toward β-keto sulfones/sulfones. *Org. Lett.* **2021**, *23*, 8296–8301. [CrossRef]
19. Chikunova, E.I.; Kukushkin, V.Y.; Dubovtsev, A.Y. Atom-economic synthesis of β-ketosulfones based on gold-catalyzed highly regioselective hydration of alkynylsulfones. *Green Chem.* **2022**, *24*, 3314–3320. [CrossRef]
20. Tyagi, A.; Taneja, N.; Khan, J.; Hazra, C.K. I_2-Catalyzed/mediated C-S and C-I bond formation: Solvent- and metal-free approach for the synthesis of β-keto sulfones and branched sulfones. *Adv. Synth. Catal.* **2023**, *365*, 1247–1254. [CrossRef]
21. Wang, B.-W.; Jiang, K.; Li, J.-X.; Luo, S.-H.; Wang, Z.-Y.; Jiang, H.-F. 1,1-Diphenylvinylsulfide as functional AIEgen derived from the aggregation-caused-quenching molecule 1,1-diphenylvinylsulfide through simple thioetherification. *Angew. Chem. Int. Ed.* **2020**, *59*, 2338–2343. [CrossRef]
22. Tian, Z.B.; Gong, Q.H.; Huang, T.Z.; Liu, L.; Chen, T.Q. Practical electro-oxidative sulfonylation of phenols with sodium arenesulfinates generating arylsulfonate esters. *J. Org. Chem.* **2021**, *86*, 15914–15926. [CrossRef] [PubMed]
23. Handa, S.; Fennewald, J.C.; Lipshutz, B.H. Aerobic oxidation in nanomicelles of aryl alkynes, in water at room temperature. *Angew. Chem. Int. Ed.* **2014**, *53*, 3432–3435. [CrossRef] [PubMed]
24. Peng, Z.; Hong, Y.Y.; Peng, S.; Xu, X.Q.; Tang, S.S.; Yang, L.H.; Xie, L.Y. Photo-sensitizer-free synthesis of β-keto sulfones via visible-light-induced oxysulfonylation of alkenes with sulfonic acids. *Org. Biomol. Chem.* **2021**, *19*, 4537–4541. [CrossRef] [PubMed]
25. Chen, X.Y.; Lu, S.X.; Zheng, Y.Y.; Wang, J.G.; Yang, L.; Sun, P. Synthesis of β-keto sulfones by oxy-sulfonylation of alkynes in HFIP. *Adv. Synth. Catal.* **2022**, *364*, 1305–1312. [CrossRef]
26. Saxena, B.; Patel, R.I.; Sharma, S.; Sharma, A. Mechanochemical-assisted decarboxylative sulfonylation of α,β-unsaturated carboxylic acids with sodium sulfinate salts. *Green Chem.* **2024**, *26*, 2721–2729. [CrossRef]
27. Lu, Q.Q.; Zhang, J.; Zhao, G.L.; Qi, Y.; Wang, H.M.; Lei, A.W. Dioxygen-triggered oxidative radical reaction: Direct aerobic difunctionalization of terminal alkynes toward β-keto sulfones. *J. Am. Chem. Soc.* **2013**, *135*, 11481–11484. [CrossRef] [PubMed]
28. Kumar, N.; Kumar, A. Amino acid-catalyzed direct synthesis of β-keto sulfones via aerobic difunctionalization of terminal alkynes in an aqueous medium. *ACS Sustain. Chem. Eng.* **2019**, *7*, 9182–9188. [CrossRef]

29. Chen, X.Y.; Chang, X.Q.; Zhang, S.C.; Lu, S.X.; Yang, L.; Sun, P. Air-triggered, catalyst-free decarboxylative oxysulfonylation of arylpropiolic acids with sodium sulfinates. *Environ. Chem. Lett.* **2022**, *20*, 2773–2779. [CrossRef]
30. Chen, W.; Ma, Y.H.; He, W.Y.; Wu, Y.X.; Huang, Y.C.; Zhang, Y.P.; Tian, H.C.; Wei, K.; Yang, X.D.; Zhang, H.B. Structure units oriented approach towards collective synthesis of sarpagine-ajmaline-koumine type alkaloids. *Nat. Commun.* **2022**, *13*, 908. [CrossRef]
31. Chen, Z.-J.; Yu, S.-W.; Zhou, Y.-J.; Li, H.-Q.; Qiu, Q.-W.; Li, M.-X.; Wang, Z.-Y. Application of $BF_3{\cdot}OEt_2$ in organic synthesis as a catalyst or synthon. *Chin. J. Org. Chem.* **2023**, *43*, 3107–3118. [CrossRef]
32. Li, A.Q.; Zhao, J.; Zhang, C.; Jiang, Q.X.; Zhu, B.F.; Cao, H. Lewis acid-promoted three-component cyclization for the construction of functionalized oxazoles. *J. Org. Chem.* **2023**, *88*, 27–38. [CrossRef] [PubMed]
33. Yu, X.L.; Hu, J.W.; Cao, J.; Xu, L.W. Intramolecular Hosomi-Sakurai reaction for the synthesis of benzoxasiloles. *J. Org. Chem.* **2024**, *89*, 9027–9030. [CrossRef] [PubMed]
34. Yu, S.-W.; Chen, Z.-J.; Chen, Z.-H.; Chen, S.-H.; Yang, K.; Xu, W.-J.; Wang, Z.-Y. Trace water in a $BF_3{\cdot}OEt_2$ system: A facile access to sulfinyl alkenylsulfones from alkynes and sodium sulfinates. *Org. Biomol. Chem.* **2023**, *21*, 7776–7781. [CrossRef] [PubMed]
35. Cao, L.; Luo, S.-H.; Jiang, K.; Hao, Z.-F.; Wang, B.-W.; Pang, C.-M.; Wang, Z.-Y. Disproportionate coupling reaction of sodium sulfinates mediated by $BF_3{\cdot}OEt_2$: An approach to symmetrical/unsymmetrical thiosulfonates. *Org. Lett.* **2018**, *20*, 4754–4758. [CrossRef] [PubMed]
36. Wu, H.-Q.; Yang, K.; Chen, X.-Y.; Arulkumar, M.; Wang, N.; Chen, S.-H.; Wang, Z.-Y. A 3,4-dihalo-2(5*H*)-furanone initiated ring-opening reaction of DABCO in the absence of a metal catalyst and additive and its application in a one-pot two-step reaction. *Green Chem.* **2019**, *21*, 3782–3788. [CrossRef]
37. CCDC 2367445 (for 3a) Contains the Supplementary Crystallographic Data for This Paper. These Data Can Be Obtained Free of Charge from the Cambridge Crystallographic Data Centre. Available online: www.ccdc.cam.ac.uk/data_request/cif (accessed on 3 July 2024).
38. Wang, Z.K.; Zhang, Z.S.; Zhao, W.J.; Sivaguru, P.; Zanoni, G.; Wang, Y.Y.; Anderson, E.A.; Bi, X.H. Synthetic exploration of sulfinyl radicals using sulfinyl sulfones. *Nat. Commun.* **2021**, *12*, 5244. [CrossRef]
39. Zhang, Z.; Song, Q.; Feng, C.; Wang, Z.; Zhao, W.; Ning, Y.; Wu, Y. The sulfinylsulfonation of alkynes for β-sulfinyl alkenylsulfone. *Chem. –Asian J.* **2022**, *17*, e202200299. [CrossRef] [PubMed]
40. Freeman, F. *vic*-Disulfoxides and *OS*-sulfenyl sulfinates. *Chem. Rev.* **1984**, *84*, 117–135. [CrossRef]
41. Nguyen, V.D.; Haug, G.C.; Greco, S.G.; Trevino, R.; Karki, G.B.; Arman, H.D.; Larionov, O.V. Decarboxylative sulfinylation enables a direct, metal-free access to sulfoxides from carboxylic acids. *Angew. Chem. Int. Ed.* **2022**, *61*, e202210525. [CrossRef]
42. Mampuys, P.; McElroy, C.R.; Clark, J.H.; Orru, R.V.A.; Maes, B.U.W. Thiosulfonates as emerging reactants: Synthesis and applications. *Adv. Synth. Catal.* **2020**, *362*, 3–64. [CrossRef]
43. Lv, Y.F.; Luo, J.Y.; Ma, Y.C.; Dong, Q.; He, L. Visible-light-promoted sulfonylation of thiols with aryldiazonium and sodium metabisulphite leading to unsymmetrical thiosulfonates. *Org. Chem. Front.* **2021**, *8*, 2461–2467. [CrossRef]
44. Zhang, Z.X.; Wang, X.R.; Sivaguru, P.; Wang, Z.K. Exploring the synthetic application of sulfinyl radicals. *Org. Chem. Front.* **2022**, *9*, 6063–6076. [CrossRef]
45. Chen, X.W.; Luo, Z.L.; Hu, Z.J.; Sun, D.H.; He, Y.Y.; Lu, J.N.; Chen, L.L.; Liu, S.Y. Discovery of potent thiazolidin-4-one sulfone derivatives for inhibition of proliferation of osteosarcoma in vitro and in vivo. *Eur. J. Med. Chem.* **2024**, *266*, 116082. [CrossRef] [PubMed]
46. Du, B.; Qian, P.; Wang, Y.; Mei, H.; Han, J.; Pan, Y. Cu-Catalyzed deoxygenative C2-sulfonylation reaction of quinoline *N*-Oxides with sodium sulfinate. *Org. Lett.* **2016**, *18*, 4144–4147. [CrossRef] [PubMed]
47. Liu, C.L.; Liang, Z.; Jialingbieke, A.; Gao, J.; Du, D. NHC-Catalyzed Synthesis of α-sulfonyl ketones via radical- mediated sulfonyl methylation of aldehydes. *Org. Lett.* **2023**, *25*, 2657–2662. [CrossRef] [PubMed]
48. Jadhao, A.R.; Gaikwad, S.S.; Patil, L.R.; Waghmode, S.B. Sonication-assisted one pot, metal-free synthesis of β-keto sulfones from styrenes, NBS and aromatic sodium sulfinate salts. *Chem. Pap.* **2021**, *75*, 4959–4968. [CrossRef]
49. Deng, S.Q.; Liang, E.; Wu, Y.R.; Tang, X.D. Efficient sulfonylation of ketones with sodium sulfinates for the synthesis of β-keto sulfones. *Tetrahedron Lett.* **2018**, *59*, 3955–3957. [CrossRef]
50. Chen, W.T.; Liu, X.Y.; Chen, E.; Chen, B.H.; Shao, J.A.; Yu, Y.P. KI-Mediated radical multi-functionalization of vinyl azides: A one-pot and efficient approach to β-keto sulfones and α-halo-β-keto sulfones. *Org. Chem. Front.* **2017**, *4*, 1162–1166. [CrossRef]
51. Yavari, I.; Shaabanzadeh, S. Electrochemical synthesis of β-ketosulfones from switchable starting materials. *Org. Lett.* **2020**, *22*, 464–467. [CrossRef]

molecules

MDPI

Article

Controllable Synthesis of Thioacetals/Thioketals and β-Sulfanyl Ketones Mediated by Methanesulfonic Anhydride and Sulfuric Acid Sulfuric Acid from Aldehyde/Acetone and Thiols

Hexia Ye [1], Xinyao Zhao [1], Yajie Fu [1,2], Haibo Liu [1], Junchen Li [1,*] and Xiaojing Bi [1,*]

1 State Key Laboratory of NBC Protection for Civilian, Beijing 102205, China; yehexia6688@yeah.net (H.Y.); xy01253@yeah.net (X.Z.); yajief2022@163.com (Y.F.); hbbnu@126.com (H.L.)

2 School of Chemistry and Environmental Engineering, Sichuan University of Science &Engineering, 180 Xueyuan Street, Huixing Lu, Zigong 643000, China

* Correspondence: lijch07@163.com (J.L.); xiaojingbimail@yeah.net (X.B.)

Abstract: A novel and controllable synthesis of thioacetals/thioketals and β-sulfanyl ketones mediated by the reaction of aldehyde/acetone with thiols has been developed. In this protocol, β-sulfanyl ketones can be generated without the prior preparation of α, β-unsaturated carbonyl compounds. A variety of thiols reacted with aldehyde/acetone and provided the corresponding thioacetals/thioketals and β-sulfanyl ketones in good to excellent yields, respectively. This protocol is operationally simple, mild, and atom-economical, providing controllable access to thioacetals/thioketals and thia-Michael addition products under mild conditions.

Keywords: thia-Michael addition; thioacetal; thioketal; methanesulfonic anhydride; H_2SO_4

Citation: Ye, H.; Zhao, X.; Fu, Y.; Liu, H.; Li, J.; Bi, X. Controllable Synthesis of Thioacetals/Thioketals and β-Sulfanyl Ketones Mediated by Methanesulfonic Anhydride and Sulfuric Acid Sulfuric Acid from Aldehyde/Acetone and Thiols. *Molecules* **2024**, *29*, 4785. https://doi.org/10.3390/molecules29204785

Academic Editor: Antonio Massa

Received: 4 September 2024
Revised: 6 October 2024
Accepted: 8 October 2024
Published: 10 October 2024

1. Introduction

Sulfides have garnered significant attention from scientists due to their diverse chemical valence states and the resulting rich chemical structure and stereo conformation. Thioether, sulfoxide, sulfone, and sulfonamide are widely utilized pharmacophores in commercially available pharmaceuticals [1]. Furthermore, owing to the high reactivity of sulfides, they are frequently employed as a key intermediate in the total synthesis of natural products [2] and various synthetic reactions [3]. Thioacetals and thioketals, among a wide variety of sulfur-containing compounds, are commonly utilized in organic synthesis as precursors for fluorination and alkylation and for olefin formation [4–6]. Furthermore, thioketals exhibit diverse biological activities and are employed as drugs with numerous pharmacological effects (Figure 1) [7–9]. The construction of C-S bonds through the Michael addition reaction between mercaptan with α, β-unsaturated carbonyl compounds holds significant value in both chemistry and biology [10].

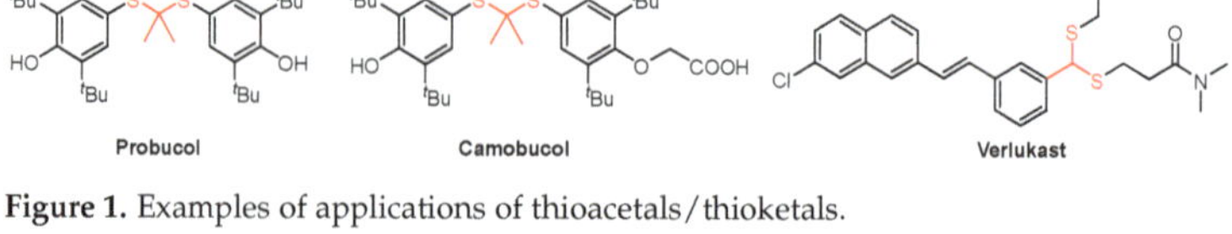

Figure 1. Examples of applications of thioacetals/thioketals.

Thioacetals and thioketals are typically synthesized through the condensation of carbonyl compounds with thiols or dithiols, employing protic acids, Lewis acids, or photocatalysis (Figure 2a) [11–18]. The addition of thiol to aryl and alkyl substituted α, β-unsaturated carbonyl compounds represents the primary method for synthesizing β-sulfanyl ketones

(Figure 2b) [19–21]. Herein, we investigated a novel and controllable synthesis of thioacetals/thioketals and β-sulfanyl ketones from acetone and thiols. With this synthetic method, β-sulfanyl ketones can be formed without preparation of α, β-unsaturated carbonyl compounds in advance (Figure 2c).

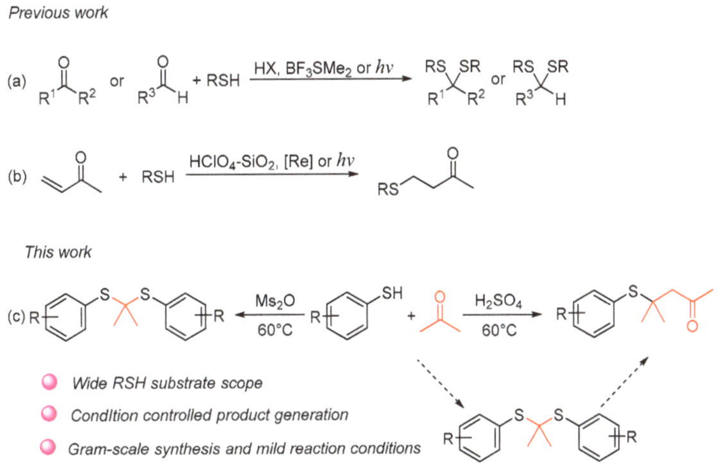

Figure 2. Overview of synthetic strategies to access thioacetals/thioketals and β-sulfanyl ketones. (**a**) The synthesis strategies of thioacetals and thioketals; (**b**) The synthesis strategies of β-sulfanyl ketones; (**c**) Our work.

2. Results and Discussion

The Michael addition reaction has been extensively studied due to its high application value. In the previous investigations, we discovered that mercaptan can be transformed into a thia-Michael addition product when mediated by p-toluene sulfonic anhydride in acetone as the solvent. Thioketals and α, β-unsaturated ketone were also monitored during the reaction (Table 1). According to preliminary results, a multitude of experiments were conducted to optimize the reaction conditions.

We initiated our investigation using 4-methylbenzenethiol (**1a**) and acetone (**2**) as the model substrates and then allowed them to react at 60 °C for 24 h (Table 1, entry 1). Initially, various additives (Ts_2O, Ms_2O, TsOH, HCl, and H_2SO_4) were added to the model reaction to identify suitable accelerators. Both Ts_2O and TsOH exhibited comparable facilitation abilities in this reaction (Table 1, entries 1 and 3). Compared to Ts_2O, the addition of TsOH makes the reaction proceed more completely (Table 1, entry 3). There are no intermediates remaining after 24 h. Notably, it was observed that the addition of Ms_2O mainly generates thioketal for an extended period, with only a limited number of products being generated over time.

Inorganic acids can also promote the reaction, and sulfuric acid displayed superior efficacy among them (Table 1, entry 5). This enhancement may be attributed to the high concentration of hydrogen ions dissociated from sulfuric acid relative to other acids used in equivalent amounts. Decreasing the amount of acetone provides better results (Table 1, entry 6), whereas reducing the amount of H_2SO_4 is not conducive to the reaction (Table 1, entries 7–8). Enhanced yields were achieved using DCE, DCM, and toluene as solvents (Table 1, entries 6, 9–10), while polar solvents, such as water and acetonitrile, hindered progress (Table 1, entries 11–14). Neither lower nor higher temperatures provided optimal outcomes for this process (Table 1, entries 15–18). According to the above experiments, we established optimal reaction conditions, which yielded product **4a** with an isolated yield of 86% (Table 1, entry 6).

Table 1. Optimization of reaction conditions [a].

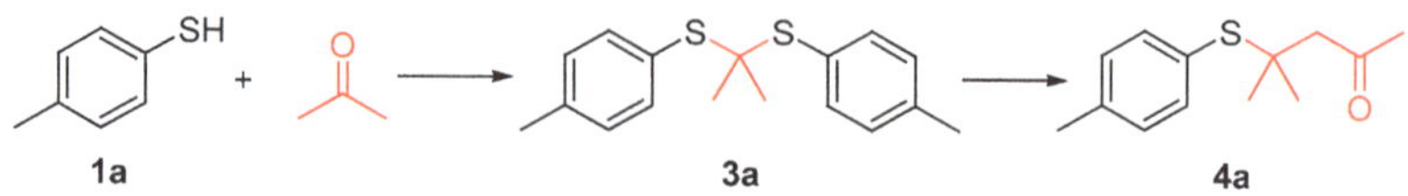

Entry	Acid (eq.)	T (°C)	Solvent	Yield of 3a (%) [b]	Yield of 4a (%) [c]
1	Ts_2O (1)	60	Acetone	7	72
2	Ms_2O (1)	60	Acetone	62	3
3	TsOH (1)	60	Acetone	-	78
4	Hydrochloric acid (1)	60	Acetone	35	4
5	H_2SO_4 (1)	60	Acetone	-	88
6	H_2SO_4 (1)	60	Acetone + DCE	-	95 (86) [d]
7	H_2SO_4 (0.1)	60	Acetone + DCE	--	21
8	H_2SO_4 (0.5)	60	Acetone + DCE	-	74
9	H_2SO_4 (1)	60	Acetone + DCM	-	91
10	H_2SO_4 (1)	60	Acetone + Toluene	-	92
11	H_2SO_4 (1)	60	Acetone + H_2O	-	1
12	H_2SO_4 (1)	60	Acetone + CH_3CN	-	26
13	H_2SO_4 (1)	60	Acetone + EtOAc	-	49
14	H_2SO_4 (1)	60	Acetone + THF	-	0
15	H_2SO_4 (1)	rt	Acetone + DCE	-	15
16	H_2SO_4 (1)	40	Acetone + DCE	-	42
17	H_2SO_4 (1)	50	Acetone + DCE	-	78
18	H_2SO_4 (1)	70	Acetone + DCE	-	92

[a] Reaction conditions. Entries 1–5: **1a** (0.2 mmol), acid (0.2 mmol), acetone (2 mL), 60 °C, 24 h; entries 6–14: **1a** (0.2 mmol), H_2SO_4, acetone:solvent = 1:5 (2 mL), 60 °C, 24 h; [b] GC-MS; [c] LC-MS; [d] isolated yield.

With the optimized reaction conditions established, we investigated the scope and generality of thiol substrates in conjunction with ketones (Table 2). Overall, both electron-donating and electron-withdrawing substituted substrates provided the desired products in good yields. A diverse array of functional groups, such as Me, *i*-Pr, OMe, F, Cl, Br, I, CF_3, NO_2, and COOMe, were compatible with this reaction and provided satisfactory product yields (Table 2, **4a–4m**). Substituted benzyl mercaptan and naphthalene-2-thiol were well-tolerated under the optimal reaction conditions (Table 2, **4n–4q**, **4t**). However, alkyl mercaptans and heteroaryl mercaptans afforded the corresponding products with lower yields (Table 2, **4r–4s**, **4u**). Motivated by the above results, we further explored the substrate scope of ketones, such as acetophenone and 2-pentanone. Unfortunately, no desired products were obtained due to the steric hindrance.

In the process of condition optimization, it was observed that when methanesulfonic anhydride was introduced into the reaction mixture, thioketal emerged as the main product, rather than β-sulfanyl ketone (Table 1, entry 2). The results prompted us to carry out an extended study of the reaction with methanesulfonic anhydride as the accelerator (Table 3). It was found that the thiophenol and benzyl mercaptan substituted with Me or OMe can obtain the target substances in medium to good yields (Table 3, **3a–3g**), while thiophenol substituted with other substituents (NO_2, Cl, Br, etc.) has a complex reaction system, and very little thioketal is generated, which is difficult to be separated and purified.

Table 2. H_2SO_4 facilitated thia-Michael addition between thiols and ketones [a,d].

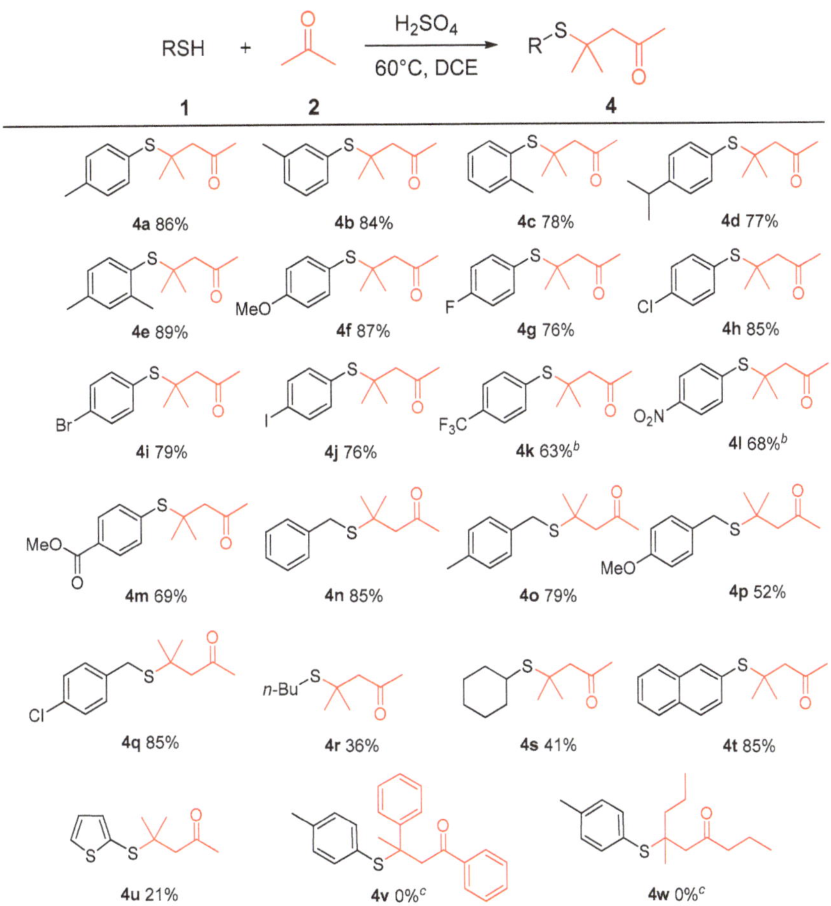

[a] Reaction conditions: **1** (0.6 mmol), H_2SO_4 (0.6 mmol), acetone:DCE = 1:5 (6 mL), 60 °C, 24 h, [b] 28 h; [c] reaction conditions: **1** (0.6 mmol), acetophenone/2-pentanone (0.9 mmol), H_2SO_4 (0.6 mmol), DCE (6 mL), 60 °C, 24 h. [d] Isolated yield.

Table 3. Ms_2O facilitated thioacetalization of acetone to thioacetals [a,b].

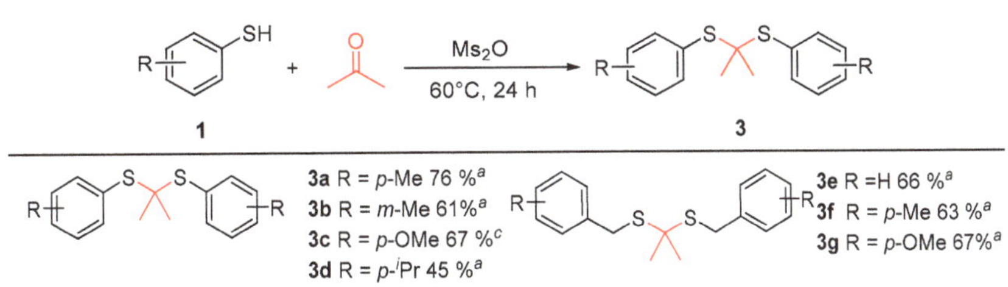

[a] Reaction conditions: **1** (0.6 mmol), Ms_2O (0.6 mmol), acetone: DCE = 1:5 (6 mL), 60 °C, 20 h; [b] isolated yield; [c] 8 h.

In addition to acetone, the applicability of aromatic aldehydes as substrates was also discussed. The results indicated that the target product could be obtained with medium to good yield from different substituted aromatic aldehydes (Table 4). Notably, even thiophenol substituted with NO_2 achieved a yield of approximately 70% for the target material (Table 4, **3′o**).

Table 4. Ms_2O facilitated thioacetalization of aldehyde to thioacetals [a,b].

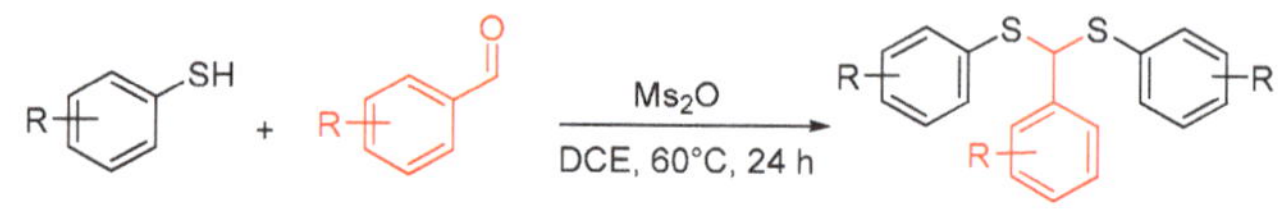

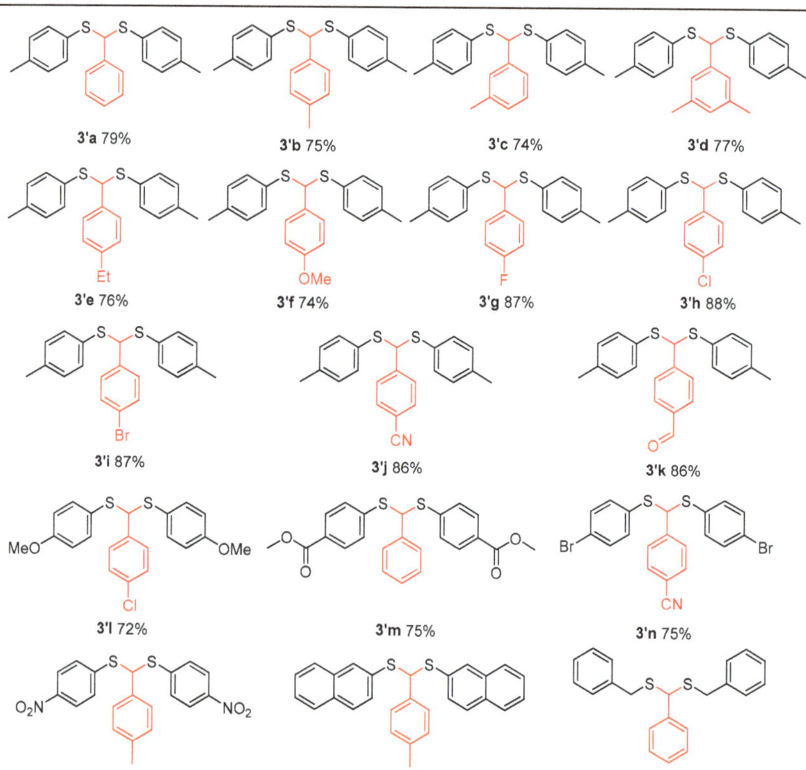

3'a 79%, **3'b** 75%, **3'c** 74%, **3'd** 77%, **3'e** 76%, **3'f** 74%, **3'g** 87%, **3'h** 88%, **3'i** 87%, **3'j** 86%, **3'k** 86%, **3'l** 72%, **3'm** 75%, **3'n** 75%, **3'o** 70%, **3'p** 71%, **3'q** 73%

[a] Reaction conditions: **1** (0.6 mmol), **2′** (0.9 mmol), Ms_2O (0.6 mmol), DCE (5 mL), 60 °C, 20 h; [b] isolated yield.

During the reaction process, α, β-unsaturated ketone and thioketal were identified alongside substrates and desired products at various stages of the reaction. In order to clarify the reaction mechanism, several control experiments were performed. In the absence of 4-methylbenzenethiol, acetone was converted into α, β-unsaturated ketone **A** under current reaction conditions (Figure 3a). When **A** was introduced into the reaction instead of acetone, a yield of 42% for product **4a** was obtained (Figure 3b). Thioketal reacted with acetone under the optimal conditions to produce **4a** in a yield of 71% (Figure 3c). Additionally, with thioketals **3a** and **A** as substrates, only 35% of **4a** was obtained (Figure 3d). The reaction process was monitored via GC-MS, and it was surprising that at the initial stage of the reaction, only thioketal was detected in the reaction mixture without a product or α, β-unsaturated ketone **A**. The presence of compound **A** became evident in later stages. Thus, we assume that the reaction (b), (c), and (d) exist at the same time, and reaction (c) is the main process (Figure 3).

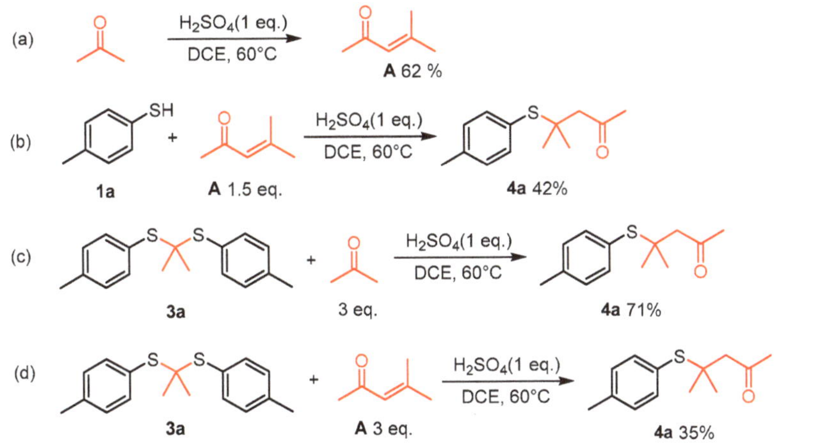

Figure 3. Control experiments. (**a**) Acetone converted into α, β-unsaturated ketone in the absence of 4-methylbenzenethiol under optimal reaction conditions; (**b**) 4-Methylbenzenethiol reacted with **A** to form **4a**; (**c**) Thioketal **3a** reacted with 3 eq acetone under optimal reaction conditions to produce **4a** in a yield of 71%; (**d**) Thioketal **3a** reacted with **A** to produce the target compound **4a** with a yield of 35%.

To demonstrate the practicality of the present method, a gram-scale experiment was performed under the standard reaction conditions, yielding the product **4a** in a 78% yield (Figure 4). The gram-scale experiment shows the promising application of this method.

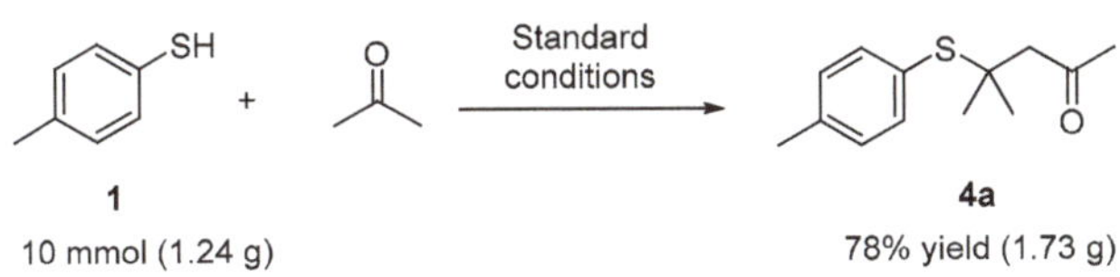

Figure 4. Gram-scale experiment.

Based on the above results and the relevant literature [22–25], a possible mechanism was proposed, as shown in Figure 5. The reaction products (**4**) are generated more rapidly via Path A. Thioketals are formed uniformly and quickly, which is reversible under acidic conditions, leading to positively charged sulfur species that subsequently react with acetone to generate the final products. In contrast, only a minimal quantity of products is produced through Path B. The α, β-unsaturated ketone **A** was detected later using GC-MS. The conversion of acetone into diketone alcohol occurs at a slow rate. Diacetone alcohol eliminates one molecule of H_2O and directly reacts with thiols to produce the final products. By comparing these two reaction mechanisms, we propose that β-sulfanyl ketones are preferentially formed through reactions with highly reactive charged sulfur species (Path A).

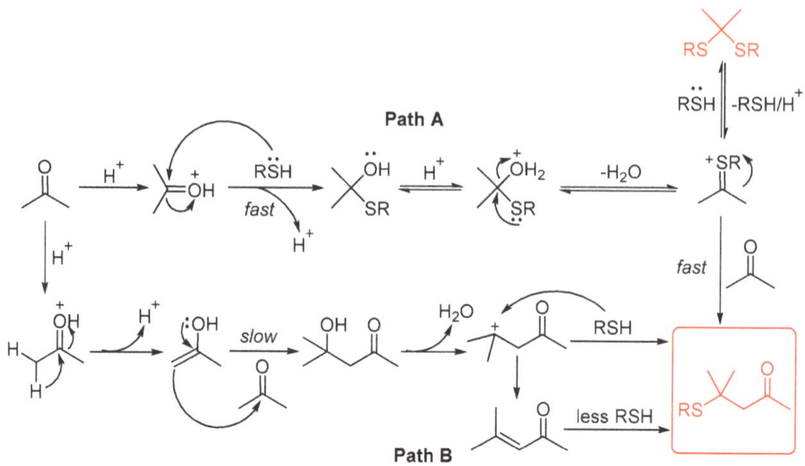

Figure 5. Plausible mechanism.

3. Materials and Methods

General Information.

All reactions were performed in a sealed tube with magnetic stirring. Unless otherwise stated, all commercially available reagents (innochem, Beijing, China) were used without further purification. Reactions were monitored using thin-layer chromatography (TLC), GC/MS, or LC/MS. NMR spectra were recorded on Bruker DRX-300 instruments (Bruker, Rheinstetten, Germany) and were calibrated using residual undeuterated solvent ($CHCl_3$ at 7.26 ppm for ^1H NMR and 77.16 ppm for ^{13}C NMR). Data were reported as follows: chemical shift, multiplicity (s = singlet, d = doublet, t = triplet, q = quartet, dd = doublet of doublets, td = triplet of doublets, qd = quartet of doublets, m = multiplet), coupling constants (Hz) and integration. High-resolution mass spectra (HRMS) were recorded on an Agilent LC/Xevo G_2-XS QTOF mass spectrometer (Agilent, Palo Alto, CA, USA) using electrospray ionization time of flight reflectron experiments.

Experimental Procedure.

A 10 mL sealed tube was charged with substituted various thiols (0.6 mmol), methanesulfonic anhydride (Ms_2O, 0.6 mmol), acetone (1 mL), and DCE (5 mL). The resulting solution was stirred at 60 °C and monitored using TLC until the reaction was complete. Saturated aqueous Na_2CO_3 solution (10 mL) and EtOAc (10 mL) was added to the mixture. The layers were separated, and the aqueous layer was washed with EtOAc (2 × 10 mL). Combined organic layers were washed with brine (10 mL), dried with anhydrous Na_2SO_4, filtered, and concentrated. The residue was purified using column chromatography on silica gel, with hexane/ethyl acetate = 20:1 as the eluent to provide the desired products **3**.

A 10 mL sealed tube was charged with thiols (0.6 mmol), methanesulfonic anhydride (Ms_2O, 0.6 mmol), **2′** (0.9 mmol), and DCE (5 mL). The resulting solution was stirred at 60 °C and monitored by TLC until the reaction was complete. Saturated aqueous Na_2CO_3 solution (10 mL) and EtOAc (10 mL) were added to the resulting mixture. The layers were separated, and the aqueous layer was washed with EtOAc (2 × 10 mL). The combined organic layers were washed with brine (10 mL), dried with anhydrous Na_2SO_4, filtered, and concentrated. The residue was purified using column chromatography on silica gel, with hexane/ethyl acetate = 50:1 as the eluent to provide the desired products **3′**.

A 10 mL sealed tube was charged with thiols (0.6 mmol), H_2SO_4 (0.6 mmol), acetone (1 mL), and DCE (5 mL). The resulting solution was stirred at 60 °C and monitored using TLC until the reaction was complete. Saturated aqueous Na_2CO_3 solution (10 mL) and EtOAc (10 mL) were added to the resulting mixture. The layers were separated, and the aqueous layer was washed with EtOAc (2 × 10 mL). The combined organic layers were washed with brine (10 mL), dried with anhydrous Na_2SO_4, filtered, and concentrated. The residue was purified using column chromatography on silica gel, with hexane/ethyl acetate = 10:1 as the eluent to provide the desired products **4**.

The gram-scale experiment procedure. A 250 mL round-bottomed flask was charged with 4-methylbenzenethiol (**1a**) (10 mmol,1.2432 g), H_2SO_4 (10 mmol), acetone (16 mL), and DCE (84 mL). The bottle was sealed with a rubber stopper, and the resulting solution was stirred at 60 °C for 24 h. Saturated aqueous Na_2CO_3 solution (100 mL) and EtOAc (100 mL) were added to the resulting mixture. The layers were separated, and the aqueous layer was washed with EtOAc (2 × 100 mL). The combined organic layers were washed with brine (100 mL), dried with anhydrous Na_2SO_4, filtered, and concentrated. The residue was purified using column chromatography on silica gel, with hexane/ethyl acetate = 10:1 as the eluent to provide the desired product **4a**. Characterization data for the products can be found in the Supplementary Materials, along with references of previous reports which provide support for the identities of the products [20,26–31].

Propane-2,2-diylbis(*p*-tolylsulfane) (**3a**) [26]. White solid, 76% yield. ^{1}H NMR (300 MHz, $CDCl_3$) δ 7.53 (d, *J* = 8.0 Hz, 4H), 7.16 (d, *J* = 7.9 Hz, 4H), 2.37 (s, 6H), 1.49 (s, 6H). ^{13}C NMR (75 MHz, $CDCl_3$) δ 139.37, 137.16, 129.48, 128.92, 59.24, 30.77, 21.43.

Propane-2,2-diylbis(*m*-tolylsulfane) (**3b**). White solid, 61% yield. ^{1}H NMR (300 MHz, Chloroform-d) δ 7.45 (d, J = 6.5 Hz, 4H), 7.20 (dd, J = 12.1, 7.7 Hz, 4H), 2.35 (s, 6H), 1.51 (s, 6H). ^{13}C NMR (75 MHz, $CDCl_3$) δ 138.35, 137.66, 134.08, 132.14, 129.98, 128.44, 59.34, 30.99, 21.42.

Propane-2,2-diylbis((4-methoxyphenyl)sulfane) (**3c**). White solid, 67% yield. ^{1}H NMR (300 MHz, $CDCl_3$) δ 7.54 (d, *J* = 8.6 Hz, 4H), 6.88 (d, *J* = 8.7 Hz, 4H), 3.82 (s, 6H), 1.45 (s, 6H). ^{13}C NMR (75 MHz, $CDCl_3$) δ = 160.62, 138.82, 123.28, 114.14, 59.33, 55.41, 30.52.

Propane-2,2-diylbis((4-isopropylphenyl)sulfane) (**3d**). White solid, 45% yield. ^{1}H NMR (300 MHz, Chloroform-*d*) δ 7.56 (d, *J* = 8.1 Hz, 4H), 7.21 (d, *J* = 8.0 Hz, 4H), 2.93 (hept, *J* = 6.8 Hz, 2H), 1.50 (s, 6H), 1.26 (d, *J* = 6.9 Hz, 12H). ^{13}C NMR (75 MHz, $CDCl_3$) δ 150.14, 137.19, 129.31, 126.83, 59.51, 34.02, 30.88, 24.01.

Propane-2,2-diylbis(benzylsulfane) (**3e**). White solid, 66% yield. ^{1}H NMR (300 MHz, $CDCl_3$) δ 7.34 (dt, *J* = 14.4, 7.5 Hz, 10H), 3.93 (s, 4H), 1.67 (s, 6H). ^{13}C NMR (75 MHz, $CDCl_3$) δ 137.87, 129.23, 128.62, 127.03, 57.30, 35.19, 30.85.

Propane-2,2-diylbis((4-methylbenzyl)sulfane) (**3f**). White solid, 63% yield. ^{1}H NMR (300 MHz, $CDCl_3$) δ 7.45 (d, *J* = 7.8 Hz, 4H), 7.33 (d, *J* = 7.8 Hz, 4H), 4.07 (s, 4H), 2.54 (s, 6H), 1.83 (s, 6H). ^{13}C NMR (75 MHz, $CDCl_3$) δ 136.62, 134.72, 129.31, 129.12, 57.18, 34.86, 30.84, 21.22.

Propane-2,2-diylbis((4-methoxybenzyl)sulfane) (**3g**). White solid, 67% yield. ^{1}H NMR (300 MHz, $CDCl_3$) δ 7.37 (d, *J* = 8.5 Hz, 4H), 6.95 (d, *J* = 8.5 Hz, 4H), 3.95 (s, 4H), 3.89 (s, 6H), 1.72 (s, 6H). ^{13}C NMR (75 MHz, $CDCl_3$) δ 158.66, 130.28, 129.71, 114.03, 57.04, 55.36, 34.52, 30.85.

(Phenylmethylene)bis(*p*-tolylsulfane) (**3′a**) [27]. White solid, 79% yield. ^{1}H NMR (300 MHz, $CDCl_3$) δ 7.30 (dd, *J* = 7.6, 1.7 Hz, 2H), 7.25–7.20 (m, 7H), 7.02 (d, *J* = 7.9 Hz, 4H), 5.29 (s, 1H), 2.27 (s, 6H). ^{13}C NMR (75 MHz, $CDCl_3$) δ 140.08, 138.13, 133.25, 130.95, 129.69, 128.48, 128.00, 61.39, 21.31.

(*p*-Tolylmethylene)bis(*p*-tolylsulfane) (**3′b**). White solid, 75% yield. ^{1}H NMR (300 MHz, $CDCl_3$) δ 7.35 (d, *J* = 8.0 Hz, 6H), 7.16 (t, *J* = 7.9 Hz, 6H), 5.43 (s, 1H), 2.42 (s, 3H), 2.40 (s, 6H). ^{13}C NMR (75 MHz, $CDCl_3$) δ 137.91, 137.69, 137.14, 133.03, 131.29, 129.64, 129.17, 127.85, 61.18, 21.25.

(*m*-Tolylmethylene)bis(*p*-tolylsulfane) (**3′c**). White solid, 74% yield. ^{1}H NMR (300 MHz, $CDCl_3$) δ 7.29 (d, *J* = 8.1 Hz, 4H), 7.22–7.15 (m, 3H), 7.09 (d, *J* = 7.9 Hz, 5H), 5.34 (s, 1H), 2.34 (s,9H). ^{13}C NMR (75 MHz, $CDCl_3$) δ 139.91, 138.13, 138.00, 133.12, 131.10, 129.63, 128.78, 128.56, 128.31, 125.00, 61.44, 21.47, 21.26.

((3, 5-Dimethylphenyl)methylene)bis(*p*-tolylsulfane) (**3′d**). White solid, 77% yield. ^{1}H NMR (300 MHz, $CDCl_3$) δ 7.29 (d, *J* = 8.1 Hz, 4H), 7.09 (d, *J* = 7.9 Hz, 4H), 7.00 (s, 2H), 6.91

(s, 1H), 5.31 (s, 1H), 2.34 (s, 6H), 2.30 (s, 6H). ^{13}C NMR (75 MHz, $CDCl_3$) δ 139.82, 137.98, 137.92, 133.05, 131.27, 129.73, 129.61, 125.64, 61.54, 21.35, 21.26.

((4-Ethylphenyl)methylene)bis(*p*-tolylsulfane) (**3′e**). White solid, 76% yield. ^{1}H NMR (300 MHz, $CDCl_3$) δ 7.31 (t, *J* = 8.5 Hz, 6H), 7.12 (dd, *J* = 16.9, 8.0 Hz, 6H), 5.38 (s, 1H), 2.66 (q, *J* = 7.6 Hz, 2H), 2.34 (s, 6H), 1.26 (t, *J* = 7.6 Hz, 3H). ^{13}C NMR (75 MHz, $CDCl_3$) δ 144.03, 137.89, 137.24, 132.99, 131.21, 129.62, 127.96, 127.84, 61.17, 28.61, 21.25, 15.52.

((4-Methoxyphenyl)methylene)bis(*p*-tolylsulfane) (**3′f**). White solid, 74% yield. ^{1}H NMR (300 MHz, $CDCl_3$) δ 7.24–7.20 (m, 6H), 7.03 (d, *J* = 7.9 Hz, 4H), 6.77 (d, *J* = 8.7 Hz, 2H), 5.29 (s, 1H), 3.77 (s, 3H), 2.29 (s, 6H). ^{13}C NMR (75 MHz, $CDCl_3$) δ 159.27, 138.02, 133.12, 132.14, 131.15, 129.68, 129.22, 113.84, 60.70, 55.40, 21.31.

((4-Fluorophenyl)methylene)bis(*p*-tolylsulfane) (**3′g**). White solid, 87% yield. ^{1}H NMR (300 MHz, $CDCl_3$) δ 7.39–7.27 (m, 6H), 7.10 (d, *J* = 8.1 Hz, 4H), 6.97 (t, *J* = 8.6 Hz, 2H), 5.38 (s, 1H), 2.35 (s, 6H). ^{13}C NMR (75 MHz, $CDCl_3$) δ 162.20 (d, *J* = 246.9 Hz), 138.27, 135.85 (d, *J* = 3.2 Hz), 133.35, 130.53, 129.72, 129.61, 115.28 (d, *J* = 21.7 Hz), 60.42, 21.25. ^{19}F NMR (282 MHz, $CDCl_3$) δ -113.83 (tt, *J* = 8.6, 5.3 Hz).

((4-Chlorophenyl)methylene)bis(*p*-tolylsulfane) (**3′h**). White solid, 88% yield. ^{1}H NMR (300 MHz, $CDCl_3$) δ 7.36–7.27 (m, 8H), 7.12 (d, *J* = 7.9 Hz, 4H), 5.37 (s, 1H), 2.37 (s, 6H). ^{13}C NMR (75 MHz, $CDCl_3$) δ 138.62, 138.32, 133.51, 133.36, 130.38, 129.74, 129.31, 128.53, 60.51, 21.25.

((4-Bromophenyl)methylene)bis(*p*-tolylsulfane) (**3′i**). White solid, 87% yield. ^{1}H NMR (300 MHz, $CDCl_3$) δ 7.41 (d, *J* = 8.5 Hz, 2H), 7.26 (m, 6H), 7.10 (d, *J* = 8.0 Hz, 4H), 5.33 (s, 1H), 2.35 (s, 6H). ^{13}C NMR (75 MHz, $CDCl_3$) δ 139.14, 138.34, 133.36, 131.49, 130.34, 129.75, 129.63, 121.72, 60.58, 21.28.

4-(Bis(*p*-tolylthio)methyl)benzonitrile (**3′j**). White solid, 86% yield. ^{1}H NMR (300 MHz, $CDCl_3$) δ 7.44 (d, *J* = 8.2 Hz, 2H), 7.31 (d, *J* = 8.2 Hz, 2H), 7.18 (d, *J* = 8.0 Hz, 4H), 7.01 (d, *J* = 8.0 Hz, 4H), 5.27 (s, 1H), 2.25 (s, 6H). ^{13}C NMR (75 MHz, $CDCl_3$) δ 145.29, 138.64, 133.54, 132.02, 129.74, 129.50, 128.56, 118.57, 111.30, 60.55, 21.15.

4-(Bis(*p*-tolylthio)methyl)benzaldehyde (**3′k**). White solid, 86% yield. ^{1}H NMR (300 MHz, $CDCl_3$) δ 9.70 (s, 1H), 7.51 (d, *J* = 8.1 Hz, 2H), 7.21 (d, *J* = 8.1 Hz, 2H), 7.01 (d, *J* = 8.0 Hz, 4H), 6.81 (d, *J* = 7.9 Hz, 4H), 5.12 (s, 1H), 2.06 (s, 6H). ^{13}C NMR (75 MHz, $CDCl_3$) δ 191.66, 146.82, 138.53, 135.66, 133.51, 129.87, 129.74, 128.54, 60.90, 21.19.

((4-Chlorophenyl)methylene)bis((4-methoxyphenyl)sulfane) (**3′l**). White solid, 72% yield. ^{1}H NMR (300 MHz, $CDCl_3$) δ 7.73–7.66 (m, 4H), 7.65–7.53 (m, 4H), 7.23–7.17 (m, 4H), 5.54 (s, 1H), 4.19 (s, 6H). ^{13}C NMR (75 MHz, $CDCl_3$) δ 160.16, 138.83, 136.15, 133.45, 129.34, 128.50, 124.33, 114.50, 62.11, 55.39.

Dimethyl 4,4′-((phenylmethylene)bis(sulfanediyl))dibenzoate (**3′m**). White solid, 75% yield. ^{1}H NMR (300 MHz, $CDCl_3$) δ 7.89 (d, *J* = 8.5 Hz, 4H), 7.46 (dd, *J* = 7.7, 1.5 Hz, 2H), 7.35 (d, *J* = 8.5 Hz, 4H), 7.29 (d, *J* = 7.4 Hz, 2H), 5.69 (s, 1H), 3.87 (s, 6H). ^{13}C NMR (75 MHz, $CDCl_3$) δ 166.49, 140.94, 138.13, 130.01, 129.97, 128.85, 128.71, 128.66, 127.93, 57.59, 52.18.

4-(Bis((4-bromophenyl)thio)methyl)benzonitrile (**3′n**). White solid, 75% yield. ^{1}H NMR (300 MHz, $CDCl_3$) δ 7.55 (d, *J* = 8.3 Hz, 2H), 7.38 (d, *J* = 8.4 Hz, 6H), 7.17 (dd, *J* = 8.8, 2.1 Hz, 4H), 5.33 (s, 1H). ^{13}C NMR (75 MHz, $CDCl_3$) δ 144.25, 134.79, 132.43, 132.30, 131.97, 128.59, 123.20, 118.39, 112.07, 60.03.

(*p*-tolylmethylene)bis((4-nitrophenyl)sulfane) (**3′o**). White solid, 70% yield. ^{1}H NMR (300 MHz, $CDCl_3$) δ 8.08 (d, *J* = 8.8 Hz, 4H), 7.42 (dd, *J* = 8.3, 6.5 Hz, 6H), 7.17 (d, *J* = 7.9 Hz, 2H), 5.80 (s, 1H), 2.34 (s, 3H). ^{13}C NMR (75 MHz, $CDCl_3$) δ 146.34, 144.20, 139.42, 133.67, 129.97, 129.36, 127.88, 124.05, 56.18, 21.29.

(*p*-tolylmethylene)bis(naphthalen-2-ylsulfane) (**3′p**). White solid, 71% yield. ^{1}H NMR (300 MHz, $CDCl_3$) δ 7.86 (s, 2H), 7.78 (dd, *J* = 5.8, 3.5 Hz, 2H), 7.74–7.65 (m, 4H), 7.52–7.34 (m, 8H), 7.12 (d, *J* = 7.9 Hz, 2H), 5.69 (s, 1H), 2.34 (s, 3H). ^{13}C NMR (75 MHz, $CDCl_3$) δ 138.11, 136.69, 133.61, 132.62, 132.25, 131.51, 129.57, 129.42, 128.44, 127.90, 127.77, 127.69, 126.54, 126.43, 60.34, 21.31, 1.16.

(Phenylmethylene)bis(benzylsulfane) (**3′q**) [28]. White solid, 73% yield. ^{1}H NMR (300 MHz, $CDCl_3$) δ 7.32–7.10 (m, 11H), 7.09–7.00 (m, 4H), 4.38 (s, 1H), 3.68 (d, *J* = 13.4 Hz,

2H), 3.45 (d, *J* = 13.4 Hz, 2H). ^{13}C NMR (75 MHz, $CDCl_3$) δ 139.64, 137.81, 129.05, 128.68, 128.57, 128.09, 128.03, 127.07, 50.96, 36.61.

4-Methyl-4-(*p*-tolylthio)pentan-2-one (**4a**) [20]. Oil liquid, 86% yield. ^{1}H NMR (300 MHz, $CDCl_3$) δ 7.39 (d, *J* = 8.0 Hz, 2H), 7.14 (d, *J* = 7.9 Hz, 2H), 2.64 (s, 2H), 2.34 (s, 3H), 2.13 (s, 3H), 1.36 (s, 6H). ^{13}C NMR (75 MHz, $CDCl_3$) δ 206.75, 139.31, 137.62, 129.54, 127.92, 54.41, 46.91, 32.26, 28.13, 21.30. HRMS (ESI-MS) $[M + Na]^+$: found 245.0985; calculated for $C_{13}H_{18}NaOS^+$: 245.0971.

4-Methyl-4-(*m*-tolylthio)pentan-2-one (**4b**) [29]. Oil liquid, 84% yield. ^{1}H NMR (300 MHz, $CDCl_3$) δ 7.38–7.26 (m, 2H), 7.20 (m, *J* = 17.9, 7.3 Hz, 2H), 2.65 (s, 2H), 2.34 (s, 3H), 2.13 (s, 3H), 1.38 (s, 6H). ^{13}C NMR (75 MHz, $CDCl_3$) δ 206.66, 138.39, 138.22, 134.61, 131.12, 129.91, 128.47, 54.43, 46.94, 32.21, 28.23, 21.27. HRMS (ESI-MS) $[M + Na]^+$: found 245.0982; calculated for $C_{13}H_{18}NaOS^+$: 245.0971.

4-Methyl-4-(*o*-tolylthio)pentan-2-one (**4c**) [29]. Oil liquid, 78% yield. ^{1}H NMR (300 MHz, $CDCl_3$) δ 7.54 (d, *J* = 7.6 Hz, 1H), 7.34–7.25 (m, 2H), 7.23–7.13 (m, 1H), 2.75 (s, 2H), 2.54 (s, 3H), 2.17 (s, 3H), 1.42 (s, 6H). ^{13}C NMR (75 MHz, CDCl3) δ 206.66, 144.03, 138.99, 130.99, 130.66, 129.32, 125.97, 54.63, 48.56, 32.27, 28.22, 21.87. HRMS (ESI-MS) $[M + Na]^+$: found 245.0984; calculated for $C_{13}H_{18}NaOS^+$: 245.0971.

4-((4-Isopropylphenyl)thio)-4-methylpentan-2-one (**4d**). Oil liquid, 77% yield. ^{1}H NMR (300 MHz, $CDCl_3$) δ 7.42 (d, *J* = 8.1 Hz, 2H), 7.18 (d, *J* = 8.1 Hz, 2H), 2.89 (dt, *J* = 13.8, 6.9 Hz, 1H), 2.65 (s, 2H), 2.13 (s, 3H), 1.36 (s, 6H), 1.23 (d, *J* = 6.9 Hz, 6H). ^{13}C NMR (75 MHz, $CDCl_3$) δ 206.81, 150.07, 137.69, 128.23, 126.87, 54.45, 46.93, 33.86, 32.26, 28.15, 23.90. HRMS (ESI-MS) $[M + Na]^+$: found 273.1290; calculated for $C_{15}H_{22}NaOS^+$: 273.1284.

4-((2,4-Dimethylphenyl)thio)-4-methylpentan-2-one (**4e**). Oil liquid, 89% yield. ^{1}H NMR (300 MHz, $CDCl_3$) δ 7.39 (d, *J* = 7.8 Hz, 1H), 7.11 (s, 1H), 6.97 (d, *J* = 7.7 Hz, 1H), 2.71 (s, 2H), 2.47 (s, 3H), 2.32 (s, 3H), 2.15 (s, 3H), 1.37 (s, 6H). ^{13}C NMR (75 MHz, $CDCl_3$) δ 206.94, 143.90, 139.50, 139.06, 131.58, 127.59, 126.92, 54.73, 48.46, 32.37, 28.21, 21.84, 21.25. HRMS (ESI-MS) $[M + Na]^+$: found 259.1136; calculated for $C_{14}H_{20}NaOS^+$: 259.1127.

4-((4-Methoxyphenyl)thio)-4-methylpentan-2-one (**4f**) [20]. Oil liquid, 87% yield. ^{1}H NMR (300 MHz, $CDCl_3$) δ 7.38 (d, *J* = 8.6 Hz, 2H), 6.82 (d, *J* = 8.6 Hz, 2H), 3.75 (s, 3H), 2.59 (s, 2H), 2.09 (s,3H), 1.31 (s, 6H). ^{13}C NMR (75 MHz, $CDCl_3$) δ 206.71, 160.47, 139.01, 122.19, 114.15, 55.24, 54.22, 46.74, 32.16, 27.94. HRMS (ESI-MS) $[M + Na]^+$: found 261.0920; calculated for $C_{13}H_{18}NaO_2S^+$: 261.0920.

4-((4-Fluorophenyl)thio)-4-methylpentan-2-one (**4g**) [30]. Oil liquid, 76% yield. ^{1}H NMR (300 MHz, $CDCl_3$) δ 7.67–7.33 (m, 2H), 7.01 (t, *J* = 8.5 Hz, 2H), 2.62 (s, 2H), 2.12 (s, 3H), 1.34 (s, 6H). ^{13}C NMR (75 MHz, $CDCl_3$) δ 206.53 (s), 163.68 (d, *J* = 249.7 Hz), 139.63 (d, *J* = 8.4 Hz), 126.94 (d, *J* = 3.5 Hz), 115.93 (d, *J* = 21.6 Hz), 54.39, 47.20 (d, *J* = 1.3 Hz), 32.28, 28.18. ^{19}F NMR (282 MHz, Chloroform-*d*) δ -111.91 (tt, *J* = 8.6, 5.5 Hz). HRMS (ESI-MS) $[M + Na]^+$: found 249.0726;calculated for $C_{12}H_{15}FNaOS^+$: 249.0720.

4-((4-Chlorophenyl)thio)-4-methylpentan-2-one (**4h**) [20]. Oil liquid, 85% yield. ^{1}H NMR (300 MHz, $CDCl_3$) δ 7.43 (d, *J* = 8.4 Hz, 2H), 7.30 (d, *J* = 8.4 Hz, 2H), 2.63 (s, 2H), 2.12 (s, 3H), 1.36 (s, 6H). ^{13}C NMR (75 MHz, $CDCl_3$) δ 206.35, 138.89, 135.74, 130.09, 128.99, 54.39, 47.45, 32.23, 28.24. HRMS (ESI-MS) $[M + Na]^+$: found 265.0429; calculated for $C_{12}H_{15}ClNaOS^+$: 265.0424.

4-((4-Bromophenyl)thio)-4-methylpentan-2-one (**4i**) [20]. Oil liquid, 79% yield. ^{1}H NMR (300 MHz, $CDCl_3$) δ 7.45 (d, *J* = 8.4 Hz, 2H), 7.35 (d, *J* = 8.4 Hz, 2H), 2.62 (s, 2H), 2.11 (s, 3H), 1.35 (s, 6H). ^{13}C NMR (75 MHz, $CDCl_3$) δ 206.32, 139.13, 131.93, 130.63, 124.02, 54.32, 47.40, 32.21, 28.21. HRMS (ESI-MS) $[M + Na]^+$: found 308.9923; calculated for $C_{12}H_{15}BrNaOS^+$: 308.9919.

4-((4-Iodophenyl)thio)-4-methylpentan-2-one (**4j**). Oil liquid, 76% yield. ^{1}H NMR (300 MHz, $CDCl_3$) δ 7.64 (d, *J* = 8.2 Hz, 2H), 7.21 (d, *J* = 8.2 Hz, 2H), 2.62 (s, 2H), 2.11 (s, 3H), 1.34 (s, 6H). ^{13}C NMR (75 MHz, $CDCl_3$) δ 206.26, 139.25, 137.85, 131.32, 95.94, 54.27, 47.38, 32.20, 28.18. HRMS (ESI-MS) $[M + Na]^+$: found 356.9786; calculated for $C_{12}H_{15}INaOS^+$: 356.9780.

4-Methyl-4-((4-(trifluoromethyl)phenyl)thio)pentan-2-one (**4k**). Oil liquid, 63% yield. ^{1}H NMR (300 MHz, $CDCl_3$) δ 7.62 (q, *J* = 8.4 Hz, 4H), 2.68 (s, 2H), 2.15 (s, 3H), 1.40 (s, 6H). ^{13}C NMR (75 MHz, $CDCl_3$) δ 206.28, 137.77, 136.40, 131.19 (q, *J* = 32.7 Hz), 125.59 (q, *J* = 3.6 Hz). 124.05 (q, J = 270.75 Hz), 54.46, 48.00, 32.25, 28.42. ^{19}F NMR (282 MHz, $CDCl_3$) δ = −62.74. HRMS (ESI-MS) $[M + Na]^+$: found 299.0678; calculated for $C_{13}H_{15}F_3NaOS^+$: 299.0688.

4-Methyl-4-((4-nitrophenyl)thio)pentan-2-one (**4l**) [20]. Oil liquid, 68% yield. ^{1}H NMR (300 MHz, $CDCl_3$) δ 8.18 (d, *J* = 8.7 Hz, 2H), 7.68 (d, *J* = 8.7 Hz, 2H), 2.71 (s,2H), 2.15 (s, 3H), 1.44 (s, 6H). ^{13}C NMR (75 MHz, $CDCl_3$) δ 205.88, 148.19, 140.87, 137.59, 123.61, 54.47, 48.80, 32.21, 28.61. HRMS (ESI-MS) $[M + Na]^+$: found 276.0668; calculated for $C_{12}H_{15}NNaO_3S^+$: 276.0665.

Methyl 4-((2-methyl-4-oxopentan-2-yl)thio)benzoate (**4m**). Oil liquid, 69% yield. ^{1}H NMR (300 MHz, $CDCl_3$) δ 7.92 (d, *J* = 8.1 Hz, 2H), 7.52 (d, *J* = 8.1 Hz, 2H), 3.84 (s, 3H), 2.61 (s, 2H), 2.06 (s, 3H), 1.33 (s, 6H). ^{13}C NMR (75 MHz, $CDCl_3$) δ 206.10, 166.44, 137.51, 137.08, 130.45, 129.56, 54.27, 52.22, 47.84, 32.06, 28.29. HRMS (ESI-MS) $[M + Na]^+$: found 289.0872; calculated for $C_{14}H_{18}NaO_3S^+$: 289.0869.

4-(Benzylthio)-4-methylpentan-2-one (**4n**). Oil liquid, 85% yield. ^{1}H NMR (300 MHz, $CDCl_3$) δ 7.46–7.27 (m, 5H), 3.84 (s, 2H), 2.74 (s, 2H), 2.19 (s, 3H), 1.52 (s, 6H). ^{13}C NMR (75 MHz, $CDCl_3$) δ 206.76, 137.99, 129.03, 128.57, 127.00, 54.56, 44.32, 33.32, 32.28, 28.49. HRMS (ESI-MS) $[M + Na]^+$: found 245.0978; calculated for $C_{13}H_{18}NaOS^+$: 245.0971.

4-Methyl-4-((4-methylbenzyl)thio)pentan-2-one (**4o**). Oil liquid, 79% yield. ^{1}H NMR (300 MHz, $CDCl_3$) δ 7.22 (d, *J* = 7.9 Hz, 2H), 7.10 (d, *J* = 7.8 Hz, 2H), 3.75 (s, 2H), 2.69 (s, 2H), 2.31 (s, 3H), 2.14 (s, 3H), 1.46 (s, 6H). ^{13}C NMR (75 MHz, $CDCl_3$) δ 206.67, 136.47, 134.72, 129.17, 128.83, 54.47, 44.16, 32.89, 32.20, 28.42, 21.04. HRMS (ESI-MS) $[M + Na]^+$: found 259.1130; calculated for $C_{14}H_{20}NaOS^+$: 259.1127.

4-((4-Methoxybenzyl)thio)-4-methylpentan-2-one (**4p**) [31]. Oil liquid, 52% yield. ^{1}H NMR (300 MHz, $CDCl_3$) δ 7.33 (d, *J* = 8.6 Hz, 2H), 6.91 (d, *J* = 8.6 Hz, 2H), 3.86 (s, 3H), 3.82 (s, 2H), 2.77 (s, 2H), 2.24 (s, 3H), 1.53 (s, 6H). ^{13}C NMR (75 MHz, $CDCl_3$) δ 206.95, 158.66, 130.14, 129.79, 114.04, 55.35, 54.68, 44.30, 32.70, 32.40, 28.56. HRMS (ESI-MS) $[M + Na]^+$: found 275.0924; calculated for $C_{14}H_{20}NaO_2S^+$: 275.1076.

4-((4-Chlorobenzyl)thio)-4-methylpentan-2-one (**4q**) [31]. Oil liquid, 85% yield. ^{1}H NMR (300 MHz, $CDCl_3$) δ 7.29 (s,4H), 3.77 (s, 2H), 2.72 (s, 2H), 2.18 (s, 3H), 1.47 (s, 6H). ^{13}C NMR (75 MHz, $CDCl_3$) δ 206.52, 136.56, 132.65, 130.35, 128.63, 54.54, 44.43, 32.58, 32.24, 28.48. HRMS (ESI-MS) $[M + Na]^+$: found 279.0586; calculated for $C_{13}H_{17}ClNaOS^+$: 279.0581.

4-(Butylthio)-4-methylpentan-2-one (**4r**). Oil liquid, 36% yield. ^{1}H NMR (300 MHz, $CDCl_3$) δ 2.66 (s, 2H), 2.51 (t, *J* = 7.3 Hz, 2H), 2.16 (s, 3H), 1.59–1.45 (m, 2H), 1.45–1.29 (m, 8H), 0.89 (t, *J* = 7.2 Hz, 3H). ^{13}C NMR (75 MHz, $CDCl_3$) δ 207.10, 54.73, 43.35, 32.46, 31.64, 28.57, 27.86, 22.43, 13.81. HRMS (ESI-MS) $[M + Na]^+$: found 211.1126; calculated for $C_{10}H_{20}NaOS^+$: 211.1127.

4-(Cyclohexylthio)-4-methylpentan-2-one (**4s**). Oil liquid, 41% yield. ^{1}H NMR (300 MHz, $CDCl_3$) δ 2.67 (s, 2H), 2.66–2.57 (m, 1H), 2.15 (s, 3H), 1.91 (d, *J* = 9.0 Hz, 2H), 1.73–1.64 (m, 2H), 1.56–1.49 (m, 1H), 1.39 (s, 6H), 1.35–1.09 (m, 5H).^{13}C NMR (75 MHz, $CDCl_3$) δ 207.05, 55.50, 44.70, 41.22, 36.24, 32.54, 29.07, 26.44, 25.52. HRMS (ESI-MS) $[M + Na]^+$: found 237.1288; calculated for $C_{12}H_{22}NaOS^+$: 237.1284.

4-Methyl-4-(naphthalen-2-ylthio)pentan-2-one (**4t**) [29]. Oil liquid, 85% yield. ^{1}H NMR (300 MHz, $CDCl_3$) δ 8.08 (s, 1H), 7.89–7.73 (m, 3H), 7.59 (dd, *J* = 8.4, 1.4 Hz, 1H), 7.51 (dd, *J* = 6.2, 3.3 Hz, 2H), 2.72 (s, 2H), 2.14 (s, 3H), 1.46 (s, 6H). ^{13}C NMR (75 MHz, $CDCl_3$) δ 206.52, 137.45, 134.17, 133.32, 133.16, 128.85, 128.05, 127.91, 127.61, 126.94, 126.43, 54.38, 47.51, 32.16, 28.30. HRMS (ESI-MS) $[M + Na]^+$: found 281.0980; calculated for $C_{16}H_{18}NaOS^+$: 281.0971.

4-Methyl-4-(thiophen-2-ylthio)pentan-2-one (**4u**). Oil liquid, 21% yield. ^{1}H NMR (300 MHz, $CDCl_3$) δ 7.45 (dd, *J* = 5.4, 0.9 Hz, 1H), 7.17 (dd, *J* = 3.4, 0.9 Hz, 1H), 7.06 (dd, *J* = 5.3, 3.6 Hz, 1H), 2.70 (s, 2H), 2.16 (s, 3H), 1.41 (s, 6H). ^{13}C NMR (75 MHz, $CDCl_3$) δ

206.68, 137.63, 131.52, 130.63, 127.87, 54.04, 48.05, 32.30, 27.8 HRMS (ESI-MS) $[M + Na]^+$: found 237.0383; calculated for $C_{10}H_{14}NaOS_2^+$: 237.0378.

4. Conclusions

In summary, the synthesis methods of thioacetal, thioacetone, and 4-methyl-4-(arylsulfide)pentane-2-ketone promoted by methanesulfonic anhydride/sulfuric acid were discussed under mild conditions. In this paper, the mechanism of reaction is explored, and it is suggested that two reaction processes may have occurred simultaneously. Notably, using this synthetic approach eliminates the need for the prior preparation of α, β-unsaturated ketones. This strategy is characterized as simple and efficient while demonstrating good substrate compatibility. It is an effective method to prepare thioacetals/thioketals and thio-Michael addition products.

Supplementary Materials: The following supporting information can be downloaded at https://www.mdpi.com/article/10.3390/molecules29204785/s1: NMR spectra of products.

Author Contributions: Conceptualization, X.B.; methodology, H.Y. and J.L.; investigation, H.Y., X.Z. and Y.F.; data curation, H.L. All authors have read and agreed to the published version of the manuscript.

Funding: This research received no external funding.

Institutional Review Board Statement: This study did not involve humans or animals.

Informed Consent Statement: Not applicable.

Data Availability Statement: The original contributions presented in this study are included in the article; further inquiries can be directed to the corresponding author/s.

Conflicts of Interest: There are no conflicts of interest to declare.

References

1. Xiao, X.; Feng, M.; Jiang, X. Transition-metal-free persulfuration to construct unsymmetrical disulfides and mechanistic study of the sulfur redox process. *Chem. Commun.* **2015**, *51*, 4208–4211. [CrossRef] [PubMed]
2. Smith, A.B.; Adams, C.M. Evolution of dithiane-based strategies for the construction of architecturally complex natural products. *Acc. Chem. Res.* **2004**, *37*, 365–377. [CrossRef] [PubMed]
3. Althoff, F.; Benzing, K.; Comba, P.; McRoberts, C.; Boyd, D.R.; Greiner, S.; Keppler, F. Abiotic methanogenesis from organosulphur compounds under ambient conditions. *Nat. Commun.* **2014**, *5*, 4205. [CrossRef]
4. Sasson, R.; Hagooly, A.; Rozen, S. Novel method for incorporating the CHF_2 group into organic molecules using BrF_3. *Org. Lett.* **2003**, *5*, 769–771. [CrossRef] [PubMed]
5. Chiba, K.; Uchiyama, R.; Kim, S.; Kitano, Y.; Tada, M. Benzylic intermolecular carbon-carbon bond formation by selective anodic oxidation of dithioacetals. *Org. Lett.* **2001**, *3*, 1245–1248.
6. Luh, T.Y. New synthetic applications of the dithioacetal functionality. *Acc. Chem. Res.* **1991**, *24*, 257–263. [CrossRef]
7. Guo, J.; Zhang, S. Involvement of Caveolin-1 in probucol-reduced hERG membrane expression. *Biophys. J.* **2011**, *100*, 102a. [CrossRef]
8. Bursell, S.E.; Della Vecchia, K.M.; Clermont, A.C.; Takahashi, J.; Sundell, C.L.; Luchoomun, J.; Aiello, L.P. Early diabetes–induced retinal vascular abnormalities are ameliorated by the antioxidant Agix-4207 in diabetic rats. *Investig. Ophthalmol. Vis. Sci.* **2004**, *45*, 3207.
9. Nicoll-Griffith, D.A.; Gupta, N.; Twa, S.P.; Williams, H.; Trimble, L.A.; Yergey, J.A. Verlukast (MK-0679) conjugation with glutathione by rat liver and kidney cytosols and excretion in the bile. *Drug Metab. Dispos.* **1995**, *23*, 1085–1093.
10. Sharma, G.; Kumar, R.; Chakraborti, A.K. Fluoroboric acid adsorbed on silica-gel (HBF_4-SiO_2) as a new, highly efficient and reusable heterogeneous catalyst for thia-Michael addition to α, β-unsaturated carbonyl compounds. *Tetrahedron Lett.* **2008**, *49*, 4272–4275. [CrossRef]
11. Perni, R.B. Amberlyst-15 as a convenient catalyst for chemoselective thioacetalization. *Synth. Commun.* **1989**, *19*, 2383–2387. [CrossRef]
12. Shinde, P.D.; Borate, H.B.; Wakharkar, R.D. Thioacetalization of the carbonyl function, transthioacetalization of acetals, ketals, oximes and hydrazones catalysed by aqueous hydrobromic acid. *Arkivoc* **2004**, *14*, 110–117. [CrossRef]
13. Miyake, H.; Nakao, Y.; Sasaki, M. Oxalic acid-promoted preparation of dithioacetals from carbonyl compounds or acetals. *Chem. Lett.* **2007**, *36*, 104–105. [CrossRef]
14. Soderstrom, M.; Matt, C.; Odell, L.R. Thioacetalation and multi-component thiomethylative Friedel-Crafts arylation using BF_3SMe_2. *ACS Omega* **2023**, *8*, 4320–4330. [CrossRef] [PubMed]

15. Wu, Y.C.; Zhu, J. Hafnium trifluoromethanesulfonate (hafnium triflate) as a highly efficient catalyst for chemoselective thioacetalization and transthioacetalization of carbonyl compounds. *J. Org. Chem.* **2008**, *73*, 9522–9524. [CrossRef]
16. Madabhushi, S.; Mallu, K.K.R.; Chinthala, N.; Beeram, C.R.; Vangipuram, V.S. Efficient and chemoselective acetalization and thioacetalization of carbonyls and subsequent deprotection using InF_3 as a reusable catalyst. *Tetrahedron Lett.* **2012**, *53*, 697–701. [CrossRef]
17. Xing, Z.; Yang, M.; Sun, H.; Wang, Z.; Chen, P.; Liu, L.; She, X. Visible-light promoted dithioacetalization of aldehydes with thiols under aerobic and photocatalyst-free conditions. *Green Chem.* **2018**, *20*, 5117–5122. [CrossRef]
18. Du, K.; Wang, S.C.; Basha, R.S.; Lee, C.F. Visible-light Photoredox-catalyzed thioacetalization of aldehydes under metal-free and solvent-free conditions. *Adv. Synth. Catal.* **2019**, *361*, 1597–1605. [CrossRef]
19. Khatik, G.L.; Sharma, G.; Kumar, R.; Chakraborti, A.K. Scope and limitations of $HClO_4$-SiO_2 as an extremely efficient, inexpensive, and reusable catalyst for chemoselective carbon-sulfur bond formation. *Tetrahedron Lett.* **2007**, *63*, 1200–1210. [CrossRef]
20. Peng, A.; Rosenblatt, R.; Nolin, K. Conjugate addition of unactivated thiols to α, β-unsaturated ketones catalyzed by a bifunctional rhenium (V)-oxo complex. *Tetrahedron Lett.* **2012**, *53*, 2712–2714. [CrossRef]
21. Lauder, K.; Toscani, A.; Qi, Y.; Lim, J.; Charnock, S.J.; Korah, K.; Castagnolo, D. Photo-biocatalytic one-pot cascades for the enantioselective synthesis of 1, 3-mercaptoalkanol volatile sulfur compounds. *Angew. Chem. Int. Ed.* **2018**, *130*, 5905–5909. [CrossRef]
22. Cong, Z.S.; Li, Y.G.; Du, G.F.; Gu, C.Z.; Dai, B.; He, L. N-Heterocyclic carbene-catalyzed sulfa-Michael addition of enals. *Chem. Commun.* **2017**, *53*, 13129–13132. [CrossRef] [PubMed]
23. Lorette, N.B.; Howard, W.L.; Brown, J.H., Jr. Preparation of ketone acetals from linear ketones and alcohols. *J. Org. Chem.* **1959**, *24*, 1731–1733. [CrossRef]
24. Leitemberger, A.; Böhs, L.M.; Peixoto, M.L.; Rosa, C.H.; Rosa, G.R.; Godoi, M. Sulfamic acid-catalyzed thioacetalization of aldehydes under solvent and metal-free conditions. *Chem. Sel.* **2020**, *5*, 8253–8257. [CrossRef]
25. Hirschbeck, V.; Boldl, M.; Gehrtz, P.H.; Fleischer, I. Tandem acyl substitution/Michael addition of thioesters with vinylmagnesium bromide. *Org. Lett.* **2019**, *21*, 2578–2582. [CrossRef]
26. Silva-Cuevas, C.; Paleo, E.; León-Rayo, D.F.; Lujan-Montelongo, J.A. An expeditious and efficient bromomethylation of thiols: Enabling bromomethyl sulfides as useful building blocks. *RSC Adv.* **2018**, *8*, 24654–24659. [CrossRef]
27. Arunprasath, D.; Sekar, G. A Transition-metal-free and base-mediated carbene insertion into sulfur-sulfur and selenium-selenium bonds: An easy access to thio-and selenoacetals. *Adv. Synth. Catal.* **2017**, *359*, 698–708. [CrossRef]
28. Bognar, S.; van Gemmeren, M. Direct synthesis of unsymmetrical dithioacetals. *Chem. Eur. J.* **2021**, *27*, 4859–4863. [CrossRef]
29. MacNicol, D.D.; McKendrick, J.J. Formation of benzo [*b*] thiophens and related compounds by a rearrangement involving ring contraction. *J. Chem. Soc. Perkin Trans. 1* **1974**, *6*, 2493–2496. [CrossRef]
30. Xu, B.; Ye, Y.; Lin, Y.; Bai, R.; Ye, X.Y.; Xie, T. Cu-catalyzed coupling of unactivated tertiary alkyl alcohols with thiols via C-O bond cleavage. *Tetrahedron Lett.* **2022**, *89*, 153604. [CrossRef]
31. Tian, X.; Cassani, C.; Liu, Y.; Moran, A.; Urakawa, A.; Galzerano, P.; Melchiorre, P. Diastereodivergent asymmetric sulfa-Michael additions of α-branched enones using a single chiral organic catalyst. *J. Am. Chem. Soc.* **2011**, *133*, 17934–17941. [CrossRef] [PubMed]

molecules

MDPI

Review

Towards More Practical Methods for the Chemical Synthesis of Thioamides Using Sulfuration Agents: A Decade Update

Qiang Zhang [1], Laurent Soulère [2] and Yves Queneau [2,*]

1 Hubei Key Laboratory of Purification and Application of Plant Anti-Cancer Active Ingredients, Hubei University of Education, 129 Second Gaoxin Road, Wuhan 430205, China; qiang.zhang@hue.edu.cn

2 Univ Lyon, INSA Lyon, Université Claude Bernard Lyon 1, CNRS, UMR5246, ICBMS, Institut de Chimie et de Biochimie Moléculaires et Supramoléculaires, Bât. E. Lederer, 1 rue Victor Grignard, F-69622 Villeurbanne, France

* Correspondence: yves.queneau@insa-lyon.fr

Abstract: Compounds possessing a thioamide function play a crucial role in organic synthesis, serving as key building blocks. They are also important in the pharmaceutical chemistry and drug design, owing to their ability to mimic the amide function in biomolecules while retaining or developing biological activity. From the synthetic viewpoint, several methods have been developed for preparing thioamides using sulfuration agents. The purpose of this review is to give an update of the last decade of contributions focusing on the formation of thioamides employing different sulfur sources. When appropriate, the cleanness and practicality of the new methods are highlighted.

Keywords: thioamide; sulfur; sulfuration agent

Citation: Zhang, Q.; Soulère, L.; Queneau, Y. Towards More Practical Methods for the Chemical Synthesis of Thioamides Using Sulfuration Agents: A Decade Update. *Molecules* **2023**, *28*, 3527. https://doi.org/10.3390/molecules28083527

Academic Editor: Ming Wang

Received: 24 March 2023
Revised: 12 April 2023
Accepted: 13 April 2023
Published: 17 April 2023

1. Introduction

The thioamide function is one of the vital structural components present in many natural products and pharmaceutical molecules [1–3], such as prothionamide [4], closthioamide [5], cycasthioamide [6], 6-thioguanine [7], and 4-thiouridine [8] (Figure 1). Naturally occurring thioamide-containing peptides being important biomolecules, the chemical pathways enabling the biosynthesis of thiopeptides in different organisms have been investigated [9]. The thioamide functional group, as a bioisostere of the amide bond, has been associated with enhanced chemical stabilities and improved biological activities of pharmaceuticals compared with the corresponding molecule-bearing amide functions [10]. Rationales for these effects have been proposed, discussing notably the hydrogen bonding strength and the structural impact of thioamide backbone modifications [11,12].

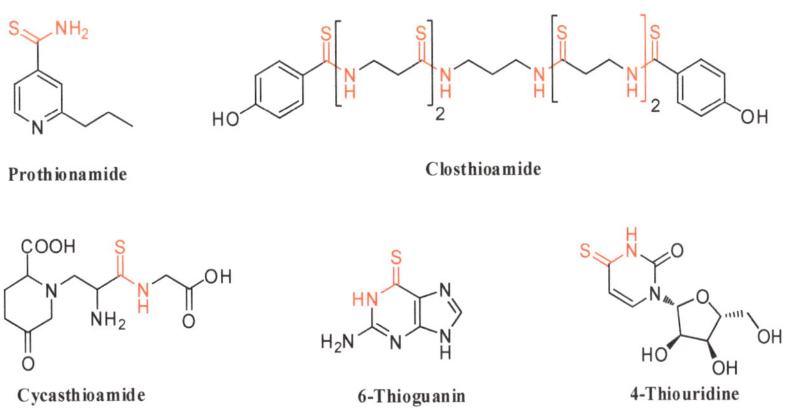

Figure 1. Representative pharmaceuticals or natural products containing a thioamide scaffold.

The usefulness of thioamide-containing molecules relies also on their ability to serve as reactive intermediates towards various heterocyclic compounds by reaction with di-electrophilic agents [13]. Owing to different reactive centers in the thioamide function, their heterocyclization reaction may lead to the formation of thiazoles [14], thiazolines [15], thiazines [16], and benzothiazoles [17,18].

Thioamide compounds have therefore attracted considerable attention from organic chemists, and recent years have been rich in reported novel protocols looking for better practicality, efficiency, or environmentally friendliness. A wide range of starting materials have been employed to construct thioamides, including aldehydes, alkynes, alkenes, amines, benzyl halides, phenyl acetonitrile, cinnamic acids, α-azido ketones, and several others, as depicted in Figure 2.

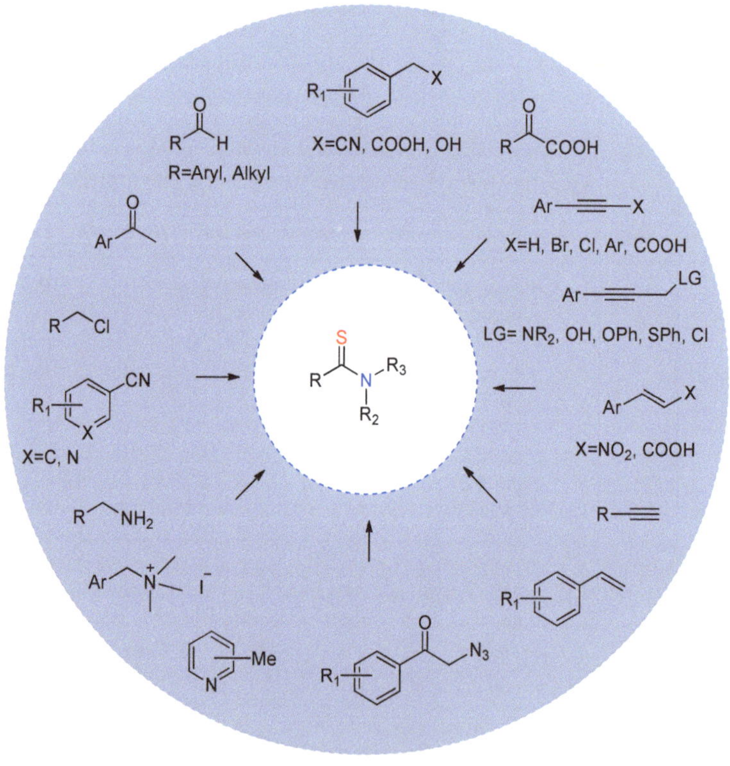

Figure 2. Applicable substrates to synthesize thioamides.

The purpose of this review is to give an overview of the synthetic routes towards thioamides published in the last decade or so, giving to organic chemists a practical tool for their design and their investigation. For Lawesson's reagent or P_4S_{10} as thionation methods, readers can rely on two comprehensive reviews [19,20]. This account will focus mainly on all other methods employing sulfuration agents including elemental sulfur and inorganic sulfide, which often offer several advantages over the classical Lawesson's reagent- or P_4S_{10}-mediated reactions in terms of practicality, selectivity, toxicity, and severe reaction conditions.

2. Elemental Sulfur as a Sulfuration Agent

Elemental sulfur (S_8) has been extensively utilized for O-S exchange reactions or C-S bond formations, which can play the roles of reagent, oxidant, reducing agent, or catalyst depending on the specific process. The typical Willgerodt–Kindler reaction involves the oxidation/rearrangement of a ketone by using elemental sulfur and primary or secondary amines, leading to the thioamide linkage. When aldehydes are used as the carbonyl substrate instead of ketones, the reaction does not involve the rearrangement step (Figure 3).

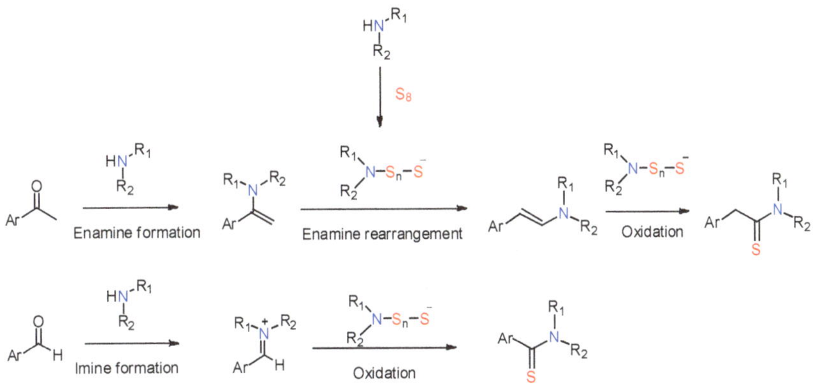

Figure 3. The mechanism of the Willgerodt–Kindler reaction starting from ketone or aldehyde.

Though a comprehensive review on the Willgerodt–Kindler reaction was reported in 2013 [21], important updates over the last decade on this reaction and their variations are worth mentioning. Indeed, the classical reaction conditions, often performed in organic solvents and at high temperature, can however exhibit moderate yields and lead to complex mixtures. In order to be efficiently applied to the synthesis of complex compounds of pharmaceutical interest, the challenge for this reaction is therefore to find practical conditions that allow reaching higher yields, while exhibiting high functional group tolerance.

2.1. Characteristics and Practical Issues

Elemental sulfur is nontoxic to humans, naturally abundant, easily available with high purity, stable under ambient conditions, insoluble in water, and easy to handle. It is therefore a reagent of choice which provides a pathway to develop new protocols in the synthesis of thioamide. The reactivity of elemental sulfur relies normally on the additional presence of a chemical activating agent [22]. Reactions using elemental sulfur generally require very easy and classical work-up, involving washing the crude organic layer with water or a $NaHCO_3$ solution, followed by column chromatography purification.

2.2. Thionylation of Aldehydes, Ketones, and Acids

A three-component synthesis of aromatic thioamides **3** was established in a one-pot procedure, which involved substituted benzaldehydes **1**, primary amines **2**, and elemental sulfur (Scheme 1) [23]. The reaction is performed under catalyst-free and no-organic-solvent conditions, providing thus a range of desired aryl-substituted thioamides in a very practical way.

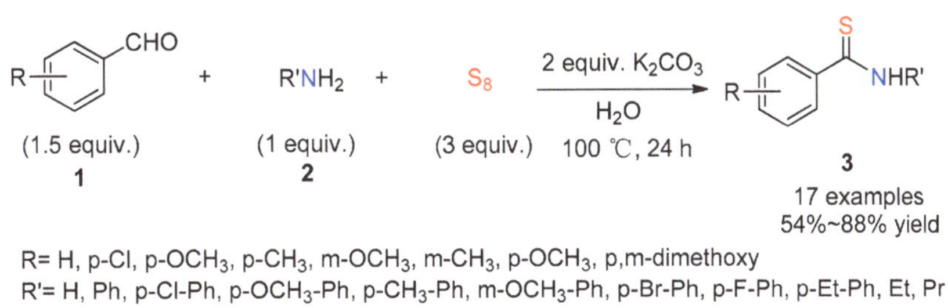

Scheme 1. K_2CO_3-mediated synthesis of aromatic thioamides in water by Zhou and co-workers [23].

DMF can act as both solvent and reactant in this transformation, as exemplified for the Willgerodt–Kindler reaction involving either variously substituted benzaldehydes or acetophenones **4** (Scheme 2). In this conversion, the key intermediate dimethylamine originated from a base-mediated cleavage of DMF. Compared with the established methods, this route provides a straightforward and rapid protocol to access *N*,*N*-dimethyl thioamides **6** [24].

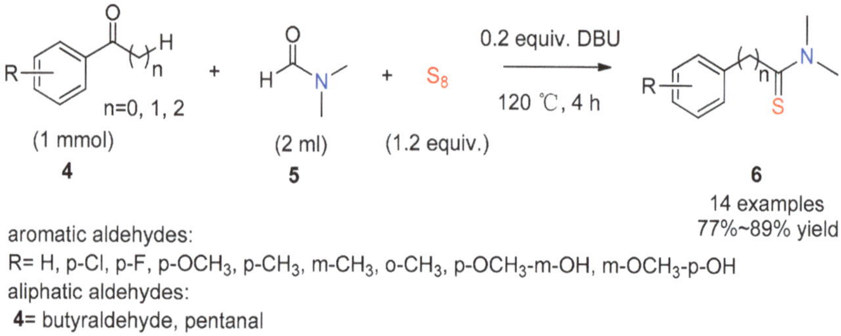

Scheme 2. DBU-mediated synthesis of *N*,*N*-dimethyl thioamides in DMF by Liu and co-workers [24].

This protocol is extendable to other amides. For example, Yuan's group [25] recently reported a thioamidation reaction using various amides as amine sources in water (Scheme 3). Screening different organic bases, inorganic bases, or their mixtures, they found the combination of sodium carbonate and triethylamine afforded the desired aryl thioamides **9** in a high yield. Their study of the scope of formamides and aryl aldehydes demonstrated that the method was compatible with many functional groups, including halides and heteroaromatic rings. Interestingly, this protocol avoided the use of an excess of *N*-substituted formamide or acetamide.

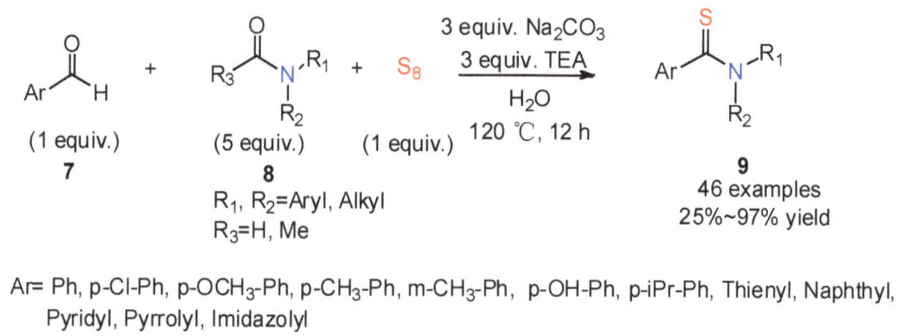

Scheme 3. Mixed bases promoted synthesis of aryl thioamides in water by Yuan and co-workers [25].

A catalyst-free and solvent-free Willgerodt–Kindler reaction was developed by Dalal's team to prepare aryl thioamides **11** in good yields (Scheme 4). The reaction was performed at 100 °C using cyclic secondary amines including pyrrolidine, piperidine, and morpholine [26]. Some acyclic amines were also explored, such as diethylamine, which led to the corresponding aryl thioamides in good-to-excellent yields at room temperature. This method is remarkable for its clean reaction conditions which avoid the use of organic solvent and catalyst.

A catalyst-free protocol using alkyl or aryl aldehydes toward thioamides **14** was reported by Gururaja's group (Scheme 5) [27]. The reaction can proceed in water with small amounts of THF, in the absence of any catalysts, additives, or metal oxides, and with excellent functional group compatibility. Control experiments revealed that water plays an essential role by bringing the reactive substrates into the aqueous medium where the intermediate polysulfide can form. Several examples of biologically active molecules such as thionicotinamide exhibiting anticancer activity have been obtained using this mild protocol.

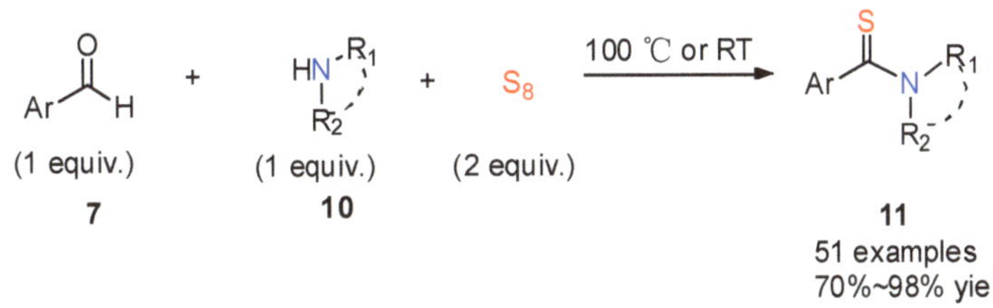

Ar= Ph, p-Cl-Ph, p-OCH$_3$-Ph, p-CH$_3$-Ph, m-CH$_3$-Ph, p,m-dimethoxy-Ph, p-Br-Ph, p-CN-Ph, p-NO$_2$-Ph, m-NO$_2$-Ph, p-(dimethyl amino)-Ph, Thienyl

Amine= morpholine, piperidine, pyrrolidine, 1-phenylpiperazine, 1-ethylpiperazine, 1-methylpiperazine, diethylamine, benzylamine,butylamine.

Scheme 4. Catalyst- and solvent-free synthesis of aryl thioamides by Dalal and co-workers [26].

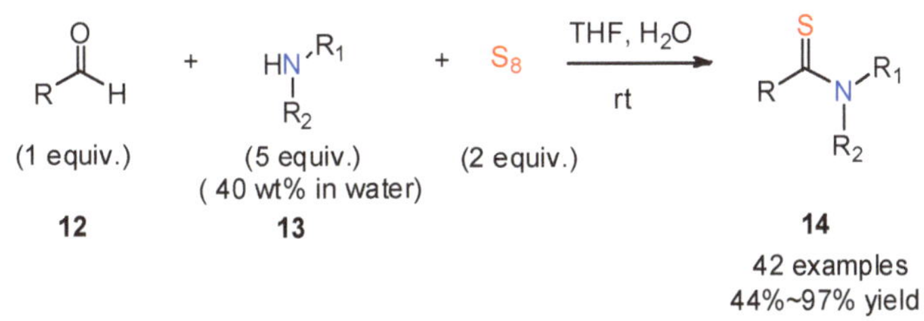

R= Ph, p-Cl-Ph, p-F-Ph, m-Cl-Ph, m-F-Ph, o-F-Ph, p-OCH$_3$-Ph, p-CH$_3$-Ph, p-OH-Ph, o-OH-Ph, o,m-dimethoxy-Ph, trimethoxy-Ph, p-Br-Ph, p-NO$_2$-Ph, p-(dimethyl amino)-Ph, p-(diethyl amino)-Ph, Thienyl, Naphthyl, Pyridyl, Furyl, Pyrrolyl, Pyrenyl, Pentyl, Hexyl, Cyclohexyl.

Amine= methylamine, dimethylamine, ethylamine, morpholine, thiomorpholine, piperidine, pyrrolidine, diethylamine, propylamine, ammonium hydroxide.

Scheme 5. Water-mediated synthesis of aryl/alkyl thioamides by Gururaja and co-workers [27].

The formation of thioamide peptides **17** by reaction of amino acids with amino aldehydes in the presence of S_8 and sodium sulfide has been systematically explored by Jiang and co-workers (Scheme 6) [28]. The presence of copper II chloride appeared essential to the reaction by forming a N-Cu-N chelate fixing the imine, thus preventing racemization. A family of thioamide-containing analogs of dipeptides were prepared with high enantiomeric purity. A 15 mol% amount of $CuCl_2$ was identified as the optimal catalyst amount to provide the desired products in good yields and with high enantioselectivity. This method provides a direct strategy for the synthesis of thioamide-containing peptides with chirality retention, and was efficiently applied to targets of pharmaceutical interest. Here, the presence of the catalyst is essential for preserving the chirality of the substrate into the product.

Decarboxylative reactions being a powerful strategy to construct the C-N bond, Singh's group [29] reported a novel route towards benzothioamides **19** or 2-phenylethanethioamides **21** (Scheme 7). The reaction of aryl acetic or cinnamic acids with amines and elemental sulfur led to the formation of thioamides in the absence of any catalyst or solvent. The structural scope included various substituted arylacetic and cinnamic acids found to efficiently provide the target thioamides in good yields. In the case of nitro substituted substrates, the nitro group was reduced into an NH_2 group that remained in the final product. This method, which proceeds without the use of catalyst or external oxidants, extended the scope of the Willgerodt–Kindler reaction to carboxylic acids, which are readily available starting materials.

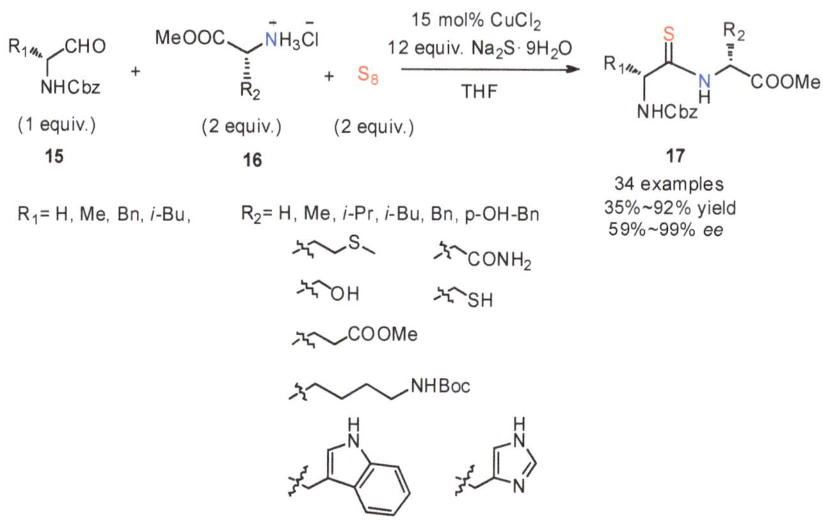

Scheme 6. Cu-catalyzed synthesis of thioamide peptide by Jiang and co-workers [28].

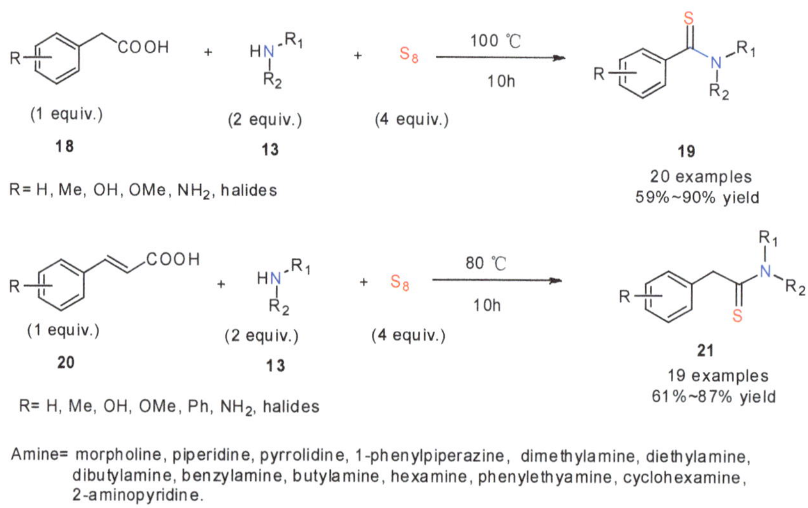

Scheme 7. Decarboxylative reaction for the synthesis of thioamides by Singh and co-workers [29].

In 2020, Takemoto's group [30] described a facile and site-selective approach to access thioamides using a α-keto carboxylic acid, amine, and sulfur via the nucleophilic addition of thiols to elemental sulfur (Scheme 8). The decarboxylative thioamidation between α-keto carboxylic acids **22** and amines **13** furnished the aryl/alkyl thioamides **23** in good yields. The reaction exhibited broad functional group tolerance, including unprotected alcohols, carboxylic acids, phenols, and unsaturated bonds. Examples of applications of the method included biologically active compounds such as thiomyristoyl lysine, possessing anticancer properties. The key to this approach is the activation of elemental sulfur by thiols.

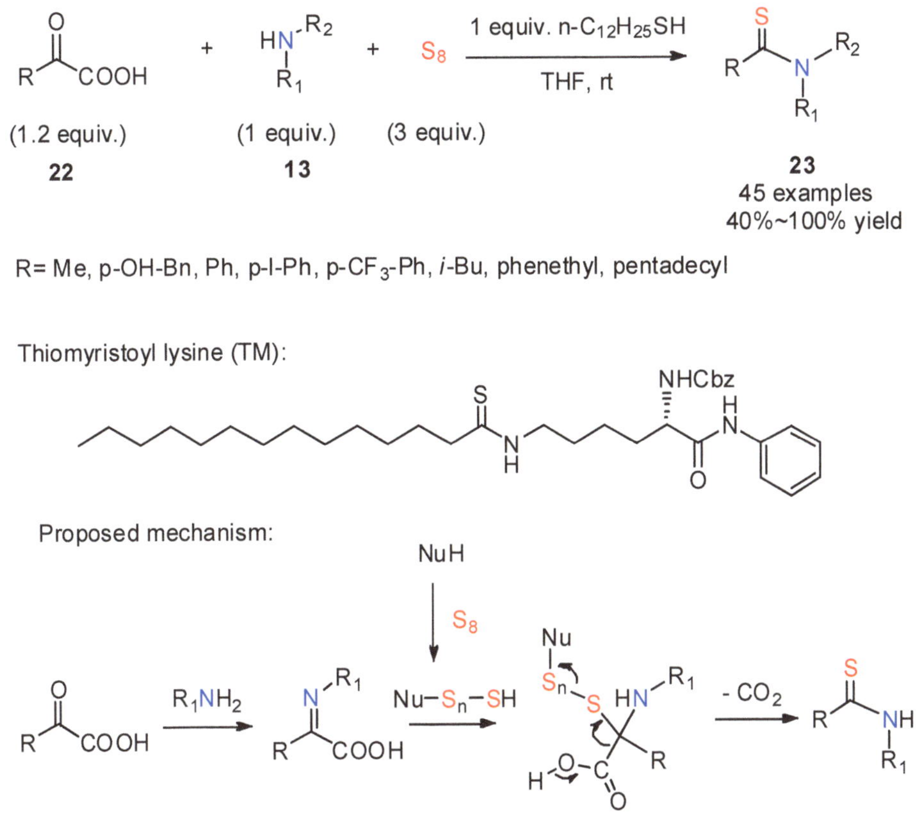

Scheme 8. Thiol-mediated synthesis of aryl or alkyl thioamides by Takemoto's group [30].

2.3. Thionylation of Cyanides, Halides, Azides, α-Nitroketones, Alcohols, Sulfoxonium Ylides, and Their Derivatives

Copper-mediated selective thionylation of phenylacetonitrile with sulfur and dimethylformamide (DMF) was realized by Zhou's group [31] to provide *N*,*N*-dimethylthiobenzamides **25** (Scheme 9). The exploration of the mechanism showed that the reaction proceeded via the formation of a benzene radical intermediate generated from Cu-catalyzed C-C bond cleavage of phenylacetonitrile and *N*,*N*-dimethyl sulfide amide generated from the reaction of elemental sulfur and DMF. When the reaction was performed in the presence of *n*-pentanal, *N*,*N*-dimethyl-2-phenylethanethioamides **26** were obtained (Scheme 9). Though the role of the aldehyde remained unclear, its ability to prevent the formation of benzene radicals was hypothesized. This method requires the presence of the copper catalyst to transform the initial phenylacetonitrile into the radical species.

Benzyl chlorides can also lead to benzothioamides, thanks to a three-component reaction with amines and elemental sulfur, such as in the synthetic route established by Han and co-workers [32]. Under the optimized conditions, the reaction generated the desired benzothioamides **28** in yields up to 88% (Scheme 10). The scope of the reaction was investigated using a variety of amines and substituents on the aromatic ring of benzyl chloride. This is a very practical transformation which exhibits good functional group tolerance, and proceeds in a one-pot procedure without the utilization of transition metals or oxidants.

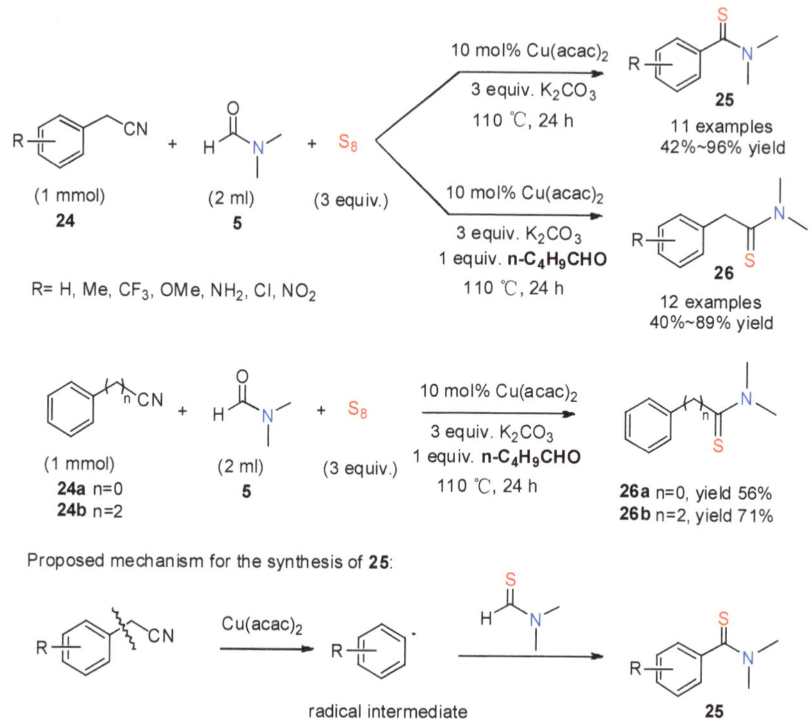

Scheme 9. Cu-catalyzed synthesis of *N*,*N*-dimethylthioamides by Zhou and co-workers [31].

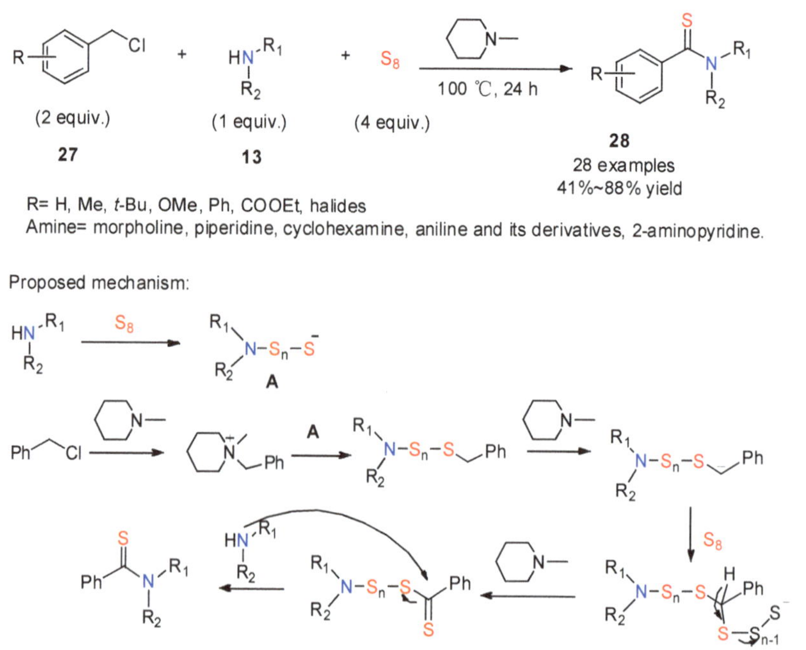

Scheme 10. *N*-methylpiperidine-mediated synthesis of benzothioamides by Han and co-workers [32].

Recently, another three-component thionylation involving aryl or alkyl chlorides, dimethylformamide, and elemental sulfur in a one-pot procedure was reported (Scheme 11) [33]. In this protocol, the starting dimethylformamide was not basic enough to activate the elemental sulfur, requiring addition of NaOH as a basic additive. Interestingly, this strategy extends the scope to alkyl chloride, while most other studies are limited to benzylic substrates.

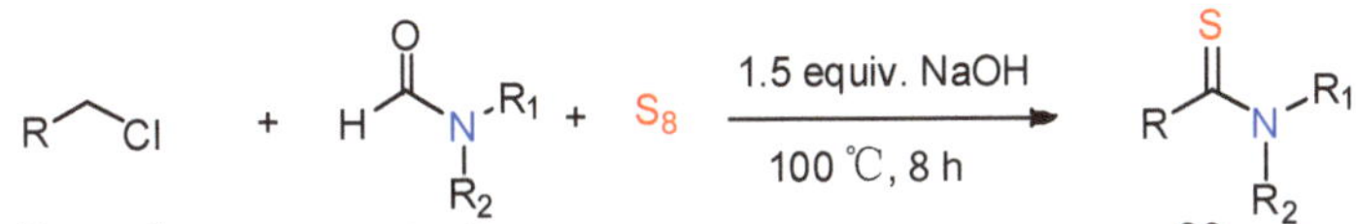

(2 mmol) **29** (2 ml) **8** (1.5 equiv.) **30**

39 examples
38%~91% yield

R= Ph, X-substituted phenyl (X=halogen, Me, MeO, OH, NH_2, CN, NO_2, CF_3), Thienyl, Naphthyl, Pyridyl, Benzofuryl, Benzothienyl, Bn, Cyclohexyl, Phenethyl, Et, *i*-Pr, Heptyl, Pr, Diethoxymethyl.

Scheme 11. NaOH-mediated synthesis of thioamides by Zhou and co-workers [33].

Phosphinic chlorides are interesting substrates towards phosphoryl thioamides and are precious intermediates in organophosphorus chemistry. Volkova's group [34] developed an efficient, catalyst-free three-component preparation of phosphoryl thioamides **32** from phosphinic chlorides, amines, and elemental sulfur in water or under neat condition (Scheme 12). An excess of amines and sulfur was essential to provide the desired product in good yields, but the easy work-up allowed the complete recovery of excesses of substrates and reagents. Furthermore, the investigation of the substrate scope demonstrated the generality of the strategy.

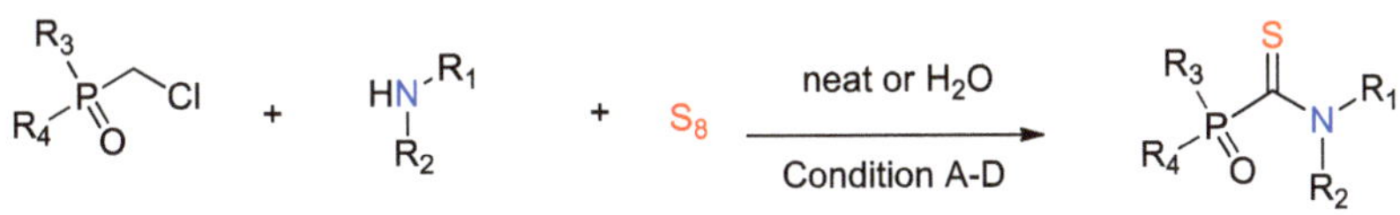

31 **13** **32**

R_3, R_4=Ar, NHAlk, NHAr, $NAlk_2$

31 examples
19%~99% yield

Condition A: neat, **31** (1 equiv.), S_8 (3 equiv.), **13** (20 equiv.), 40~100 °C, 8 h
Condition B: neat, **31** (1 equiv.), S_8 (3 equiv.), **13** (3 equiv.), DIPEA (3 equiv.), 100 °C, 8 h
Condition C: H_2O, **31** (1 equiv.), S_8 (3 equiv.), **13** (1.5 equiv.), DIPEA (3 equiv.), reflux, 8 h
Condition D: H_2O, **31** (1 equiv.), S_8 (3 equiv.), **13** (4.5 equiv.), reflux, 5 h

Scheme 12. Synthesis of phosphoryl thioamides by Volkova and co-workers [34].

A three-component reaction to access *N*-phenyl-benzothioamides **35** involving benzyl alcohol or phenylacetic acid, nitroarenes, and elemental sulfur was developed in a one-pot procedure [35]. This method employed 1,4-diazabicyclo[2.2.2]octane (DABCO) as the base which activated the elemental sulfur (Scheme 13). The authors investigated the scope of substitution on the aryl acetic acid and nitroarene moieties, observing a good tolerance with respect to a wide range of functional groups. The extension of the reaction to benzyl alcohols is currently explored. Compared with other methods, the benefit of this approach is the use of nitroarenes as stable and readily available substrates.

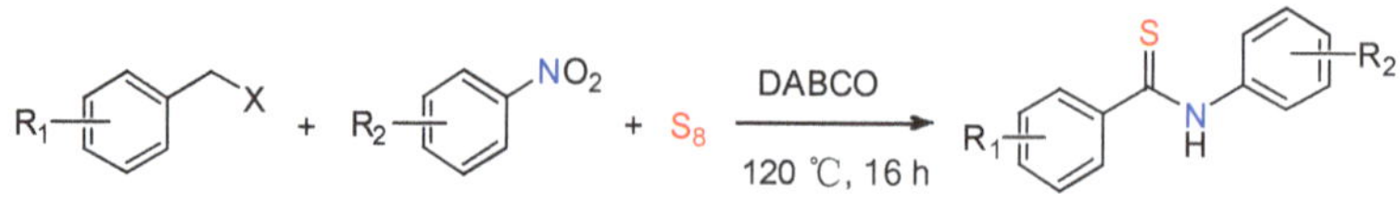

33 34 35

17 examples
28%~86% yield

X=COOH: **33** (1.5 equiv.), **34** (1 equiv.), S_8 (2 equiv.), DABCO (1 equiv.)
X=OH: **33** (1 equiv.), **34** (2 equiv.), S_8 (3 equiv.), DABCO (2 equiv.)

R_1= H, Me, Cl, OMe, CF_3
R_2= H, Me, Br, Ph

Scheme 13. DABCO-mediated synthesis of benzothioamides by Phan and co-workers [35].

More recently, a general strategy was developed for the one-pot synthesis of aryl or alkyl thioamide from variously substituted aryl methanol or their formate esters, in the presence of amides and elemental sulfur (Scheme 14) [36]. In this transformation, ethylene diamine tetraacetic acid (EDTA) served as a hydrogen delivery agent which transferred a proton from the benzylic carbon to a sulfur atom in the C-S coupling concerted C-H bond activation step. Alkyl alcohols were not able to undergo the same reaction because the hydroxyl group was preferably deprotonated in the initial C-H bond activation step. Benzyl formate being an intermediate in the formation of thioamides from benzylic alcohols, the authors found that alkyl formates were also able to provide the desired alkylthioamides. Formates thus appear as general substrates for thioamides synthesis.

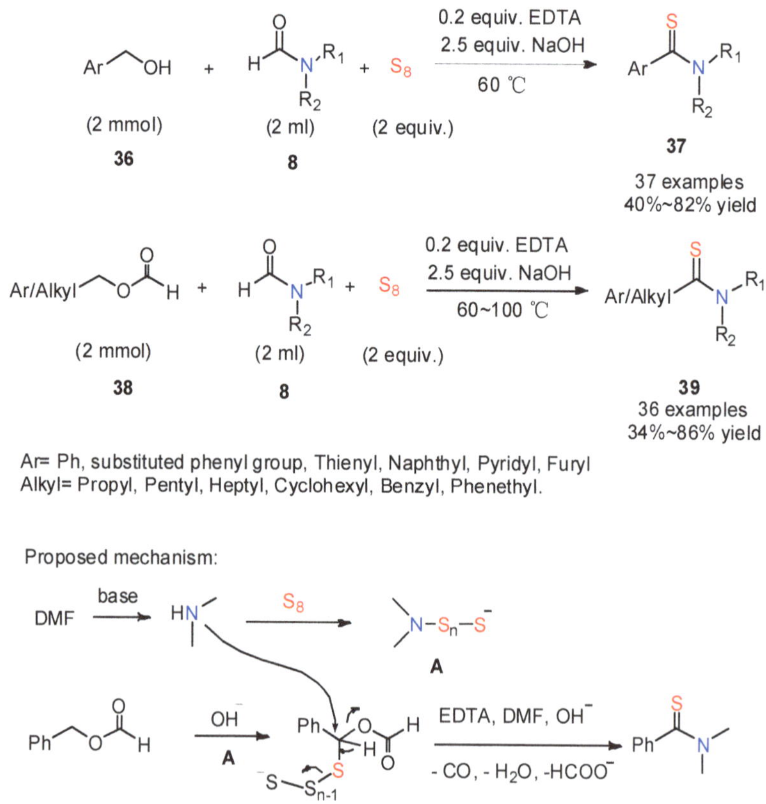

Scheme 14. EDTA-mediated synthesis of aryl or alkyl thioamides by Zhou and co-workers [36].

Benefiting from the ability of azides to behave as leaving groups, Chen and his colleagues [37] described a metal-free synthetic approach towards α-ketothioamides **41** from α-azido ketones, amines, and elemental sulfur (Scheme 15). This reaction involving the cleavages of C-N_3 displayed a broad substrate scope and high yields under mild conditions. On the basis of a series of control experiments, the authors proposed a plausible reaction mechanism via enol tautomerization, desulfhydrylation, and the departure of the azido group. This work provides a new development of α-azido ketone chemistry, and offers an alternative protocol for the synthesis of α-ketothioamide.

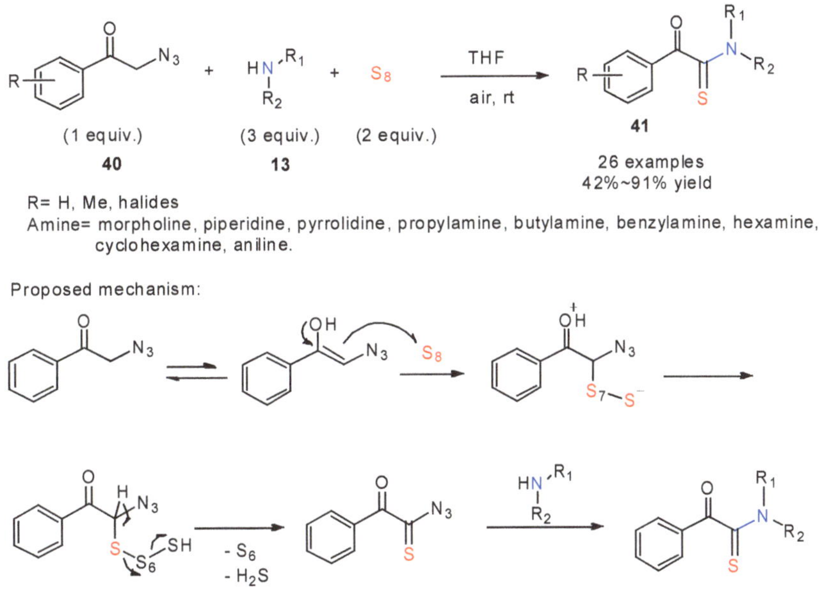

Scheme 15. Oxidative amidation for the synthesis of α-keto thioamides by Chen and his colleagues [37].

α-Nitroketones are also useful substrates for the preparation of α-keto thioamides. In 2021, Sheng and his colleagues [38] described an efficient protocol for the synthesis of diverse α-keto thioamides **43** from α-nitroketones and amines in the presence of sulfur (Scheme 16). This transformation, which proceeds under mild condition, tolerates a broad range of substitutions, especially for the alkyl α-keto thioamides, which are rarely reported in the literature.

In 2022, Chen's group reported an efficient synthesis of α-keto thioamides via thiocarbonylation of C(sp^3)-H of α-bromo ketones (Scheme 17) [39]. A wide range of α-bromo ketones and amines was well-tolerated in this transformation, leading to the formation of phenyl α-keto thioamides in good yields.

The same year, another access to α-keto thioamides was reported by Pandey's group, who used sulfoxonium ylides as substrates with primary or secondary amines (Scheme 18) [40]. The reaction, driven by the release of DMSO, provided α-keto thioamides **47** at room temperature in good-to-excellent yields. This work provides an original development of the chemistry of sulfoxonium ylides, which are easy to synthesize and stock usually as stable crystalline solids. Using a ready operational procedure and mild conditions, this reaction leads to the formation of α-keto thioamides without the need of any catalysts or additional base.

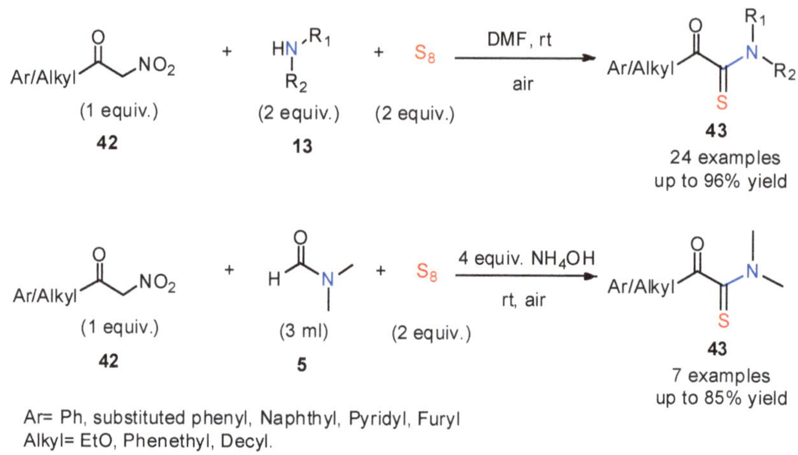

Scheme 16. Oxidant-free synthesis of α-keto thioamides by Sheng and his colleagues [38].

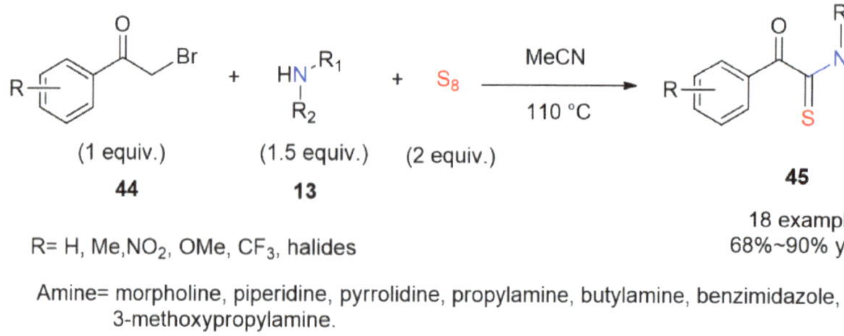

Scheme 17. Synthesis of α-keto thioamides by Chen and co-workers [39].

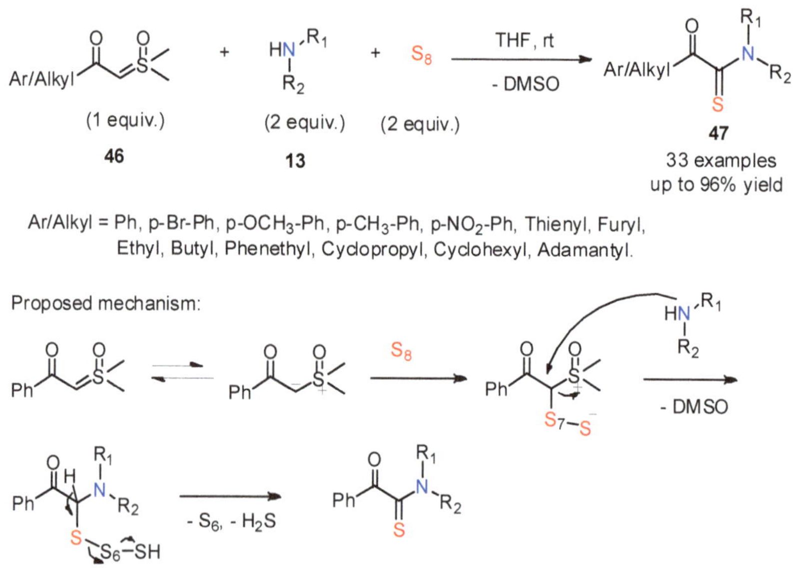

Scheme 18. Synthesis of α-keto thioamides by Pandey and co-workers [40].

2.4. Thionylation of Alkynes and Alkenes

Alkenes or alkynes can serve as substrates in thionylation reactions. For example, Nguyen and his colleagues [41] developed a straightforward approach to alkyl thioamides **49** through a three-component reaction from alkynes, aliphatic amines, and elemental sulfur (Scheme 19). The conversion of substrates was accelerated by the presence of pyridine making the reaction mixture more homogeneous. The reaction exhibits a remarkable tolerance for a wide range of functional groups and susbtituents on the alkyne and the alkylamine including cyclic secondary amines. Only aniline remained untransformed under these conditions. Mechanistic investigations led the authors to suggest that the reaction proceeded via the generation of a polysulfide **A**, its subsequent addition reaction onto the triple bond, then an addition reaction of the amino group onto the double bond, before final elimination of polysulfide. This work shows a remarkable example of atom- and step-economical reaction.

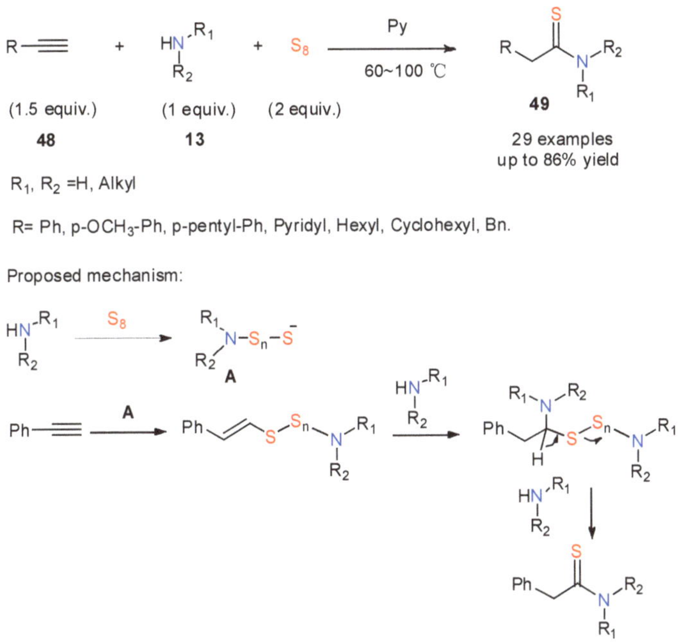

Scheme 19. Synthesis of alkyl thioamides from alkynes by Nguyen and his colleagues [41].

When diynes are used, the strategy can be applied to the preparation of polythioamides, which are interesting sulfur-containing polymers. Tang and his colleagues [42] employed aromatic diynes, elemental sulfur, and aliphatic diamines as monomers to synthesize the polythioamide **52** in high yield through multicomponent polymerization (Scheme 20). The resulting polymer exhibited high molecular weight and well-defined structure.

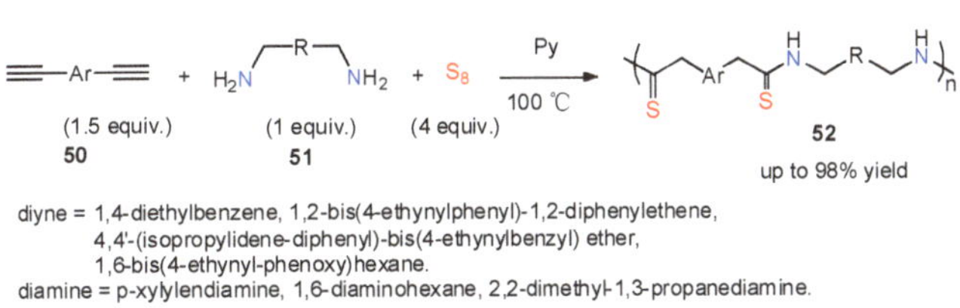

Scheme 20. Synthesis of polythioamides by Tang and co-workers [42].

In 2018, Liu and co-workers [43] reported the efficient three-component reaction of alkynes, amides, and elemental sulfur, which provided aryl thioamides such as **54** in moderate-to-good yields (Scheme 21). A first advantage of this protocol is to proceed in transition-metal-free conditions. Moreover, the cleavage of the C-C triple bond of aromatic alkynes can be promoted by elemental sulfur without any additional oxidants. This unexpected alkyne cleavage reaction was found to be of high tolerance with respect to the functional group substitution on the aryl alkynes.

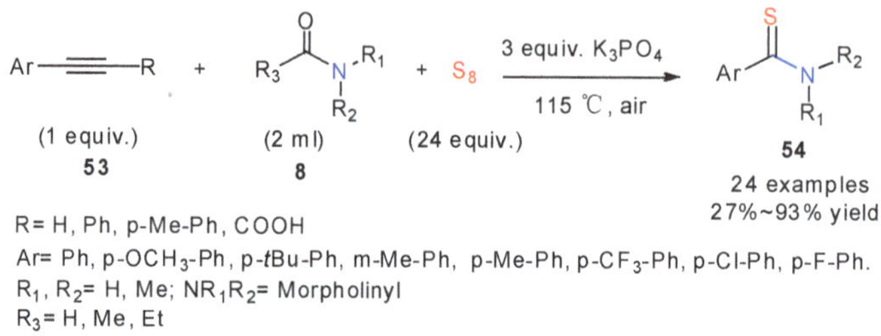

Scheme 21. K_3PO_4-mediated synthesis of thioamides by Liu and co-workers [43].

A close example using alkenes was later described by Wu and his colleagues [44], who performed their three-component reaction with amines and sulfur in the presence of KF or K_3PO_4. Surprisingly, the two bases led chemoselectively to two different kinds of thioamides, the K_3PO_4-mediated reaction giving the 2-phenylethanethioamides **57** while the KF-mediated reaction provided benzothioamides **58**. Though still under investigation, the authors proposed a mechanism supported by control experiments involving the formation of the polysulfide $RNHS_5K$ (Scheme 22).

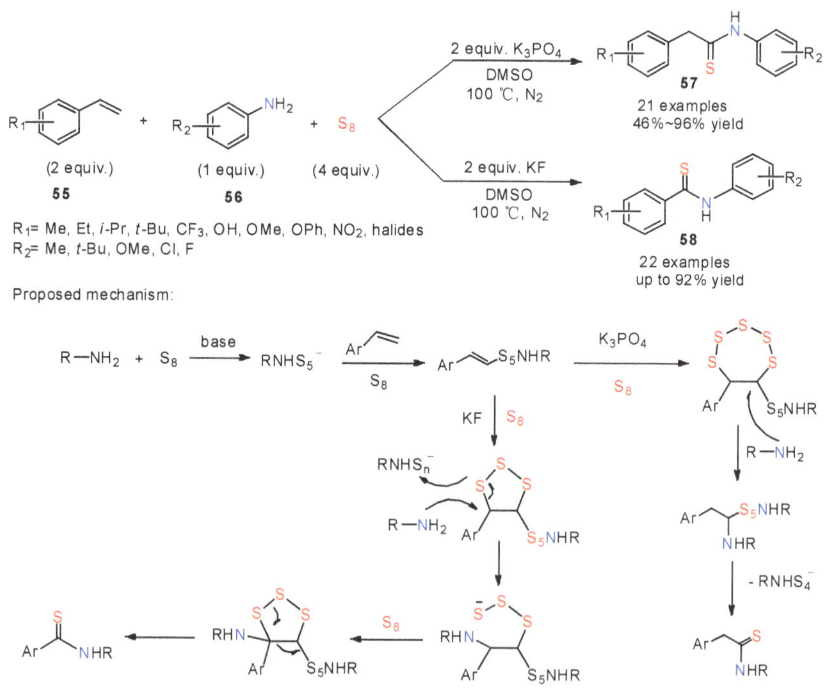

Scheme 22. K_3PO_4-/KF-mediated synthesis of thioamides by Wu and co-workers [44].

The C=C bond of enaminoketones can also be involved in such a cleavage reaction. This was used by Wan and colleagues, who applied it to the synthesis of α-keto thioamides **60** in moderate-to-good yields by transformation of enaminoketones **59** in the presence of sulfur and DMAP under air (Scheme 23) [45]. This protocol provides an elegant and practical alternative for the construction of α-keto thioamides with broad substrate tolerance from easily available tertiary enaminones. The reaction failed to provide the desired product with *N*-methyl-*N*-phenyl tertiary enaminones, giving instead *N,N*-dimethyl α-keto thioamide, suggesting that DMF competed as an amido/amino source. Control experiments indicated that the cleavage of the C=C bond produced formaldehyde, which was further oxidized to formic acid.

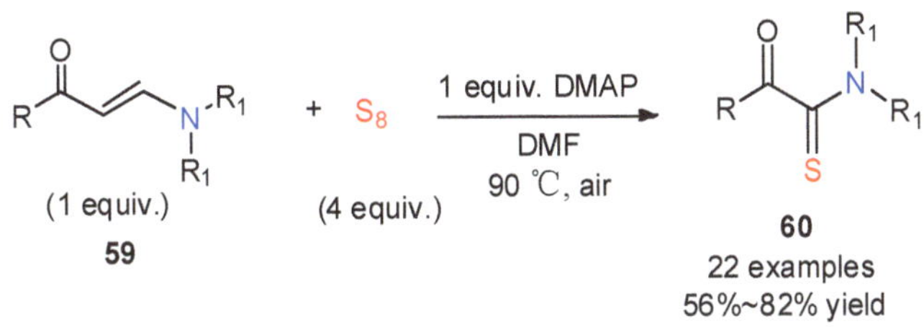

R= Ph, p-Cl-Ph, m,p-dichloro-Ph, o-CH_3-Ph, p-OCH_3-Ph, p-CH_3-Ph, p-Br-Ph, p-NO_2-Ph, o-NO_2-Ph, m-OCH_3-Ph, Thienyl, Naphthyl, Pyridyl, Furyl, Me, vinyl.
R_1=R_2= Me; NR_1R_2= Morpholinyl, Piperidinyl, Pyrrolidinyl

Scheme 23. Synthesis of α-keto thioamides by Wan and co-workers [45].

Formamides can also be used in place of amines, such as in the 2020 work reported by Zhang's group [46] on the three-component reaction of arylpropynes **61**, formamides, and elemental sulfur (Scheme 24). The KF-mediated reaction led efficiently to the targeted aryl propanethioamides **62** with wide structural scope. The reaction mechanism was suggested to proceed first via hydrolysis of the C-C triple bond, to form the ketone, and concomitant hydrolysis of formamides, forming the secondary amines, and subsequent formation of the C-N bond and C-S double bond. Aryl propanethioamides **62** could be obtained from internal alkynes by this transformation, indicating that this process shared some similarities with the Willgerodt rearrangement.

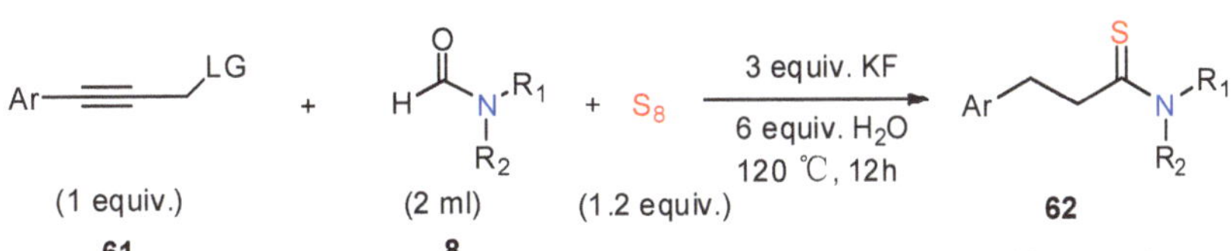

LG= $N(CH_3)_2$, NHPh, NMePh, Morpholinyl, Piperidinyl, OH, OPh, SPh, Cl

27 examples
50%~86% yield

Ar= Ph, o-CH_3-Ph, m-CH_3-Ph, p-CH_3-Ph, p-Et-Ph, p-*t*Bu-Ph, m-Cl-Ph, m-F-Ph, m-Br-Ph, p-OCH_3-Ph, m-OCH_3-Ph, p-CF_3-Ph, Thienyl, Naphthyl, Biphenyl

R_1=R_2= Me; R_1=H, R_2= Et; NR_1R_2= Morpholinyl, Piperidinyl

Scheme 24. KF-mediated synthesis of aryl propanethioamides by Zhang and co-workers [46].

Alternatively, nitrostyrene can undergo a similar C=C bond cleavage process. In 2021, a three-component strategy towards aryl thioamides **64**/**65** was developed in Zheng's group through a base-mediated C=C bond cleavage of β-nitrostyrene in the presence of sulfur and formamides or amines (Scheme 25) [47]. KO-*t*-Bu was found to be the most effective among organic or inorganic basic promoters of the reaction, leading to the corresponding thioamides in moderate-to-high yields.

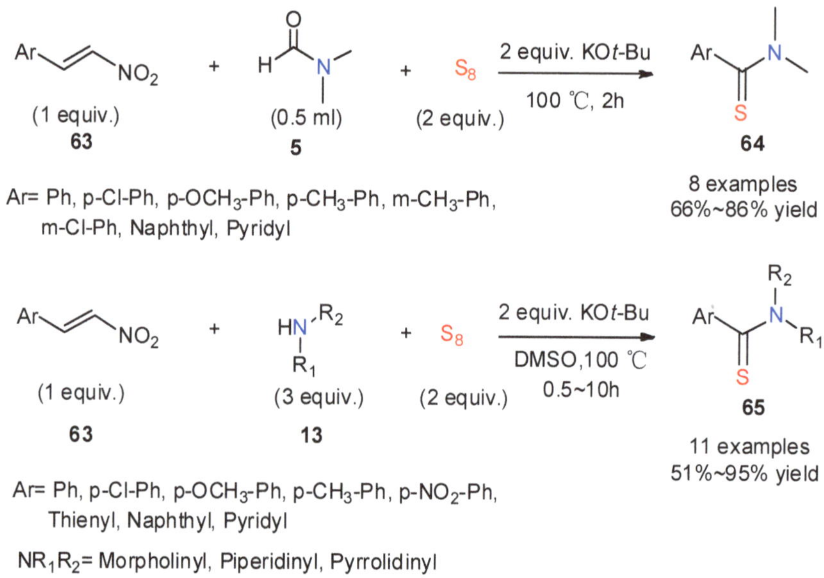

Scheme 25. KO-*t*-Bu-mediated synthesis of aryl thioamides by Zeng and co-workers [47].

Singh and co-workers reported in 2021 a very straightforward approach to construct α-keto thioamides, starting from readily available cinnamic acids and secondary amines (Scheme 26) relying on a copper-catalyzed decarboxylative thioamidation in the presence of sulfur under neat conditions [48]. α-Keto thioamides **67** were obtained as the final oxidative products in moderate yields. The presence of radical scavengers fully inhibited the formation of final oxidative products, suggesting that the reaction proceeded through a radical mechanism. Different from the one shown in Scheme 7, this reaction of cinnamic acids led to the formation α-keto thioamides. The presence of the copper catalyst is, however, indispensable, as it promotes the oxygenation step.

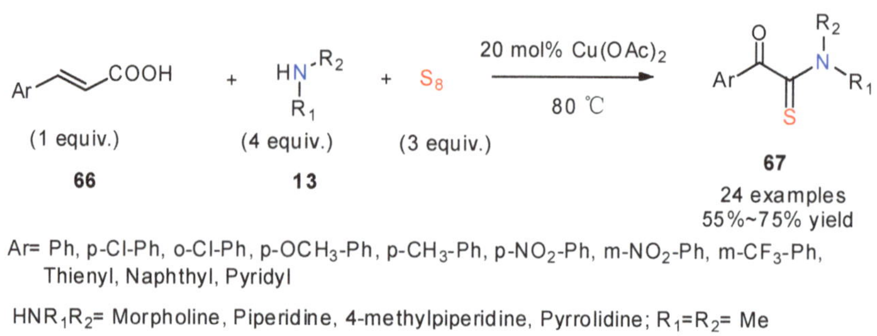

Scheme 26. Cu-catalyzed synthesis of α-keto thioamides by Singh and co-workers [48].

In 2021, Gururaja and co-workers reported the use of *gem*-dibromostyrenes as substrates in a catalyst-free, three-component reaction in aqueous medium leading to 2-arylethanethioamides **69** (Scheme 27) [49]. As in the case of aldehydes as substrates mentioned earlier (Scheme 5), water is essential for the thioamidation, which could be applied to a variety of aliphatic amines.

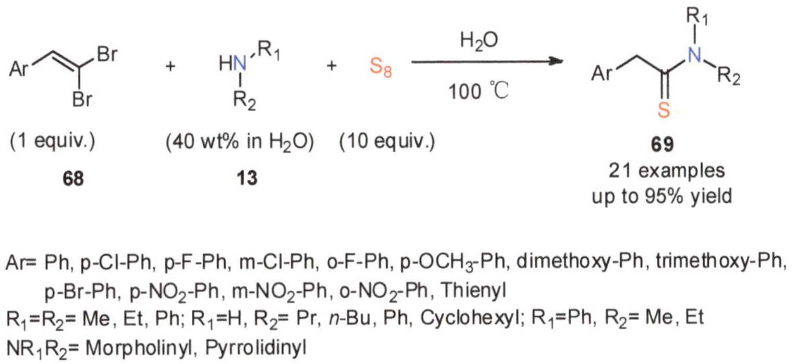

Scheme 27. Synthesis of 2-arylethanethioamides in an aqueous medium by Gururaja and co-workers [49].

2.5. Thionylation of Methylheteroarenes

Nguyen's group [50] developed a thioamidation method via DMSO-promoted oxidative coupling of methylhetarenes with amines (Scheme 28). In their procedure, the loading of sulfur could be decreased as low as 1.25 equivalent, and a wide scope of substrates were tolerated to provide a variety of thioamides in high yield. Both DMSO and S_8 were considered to act as oxidizing species. Addition of acetic acid was required for the reaction of aliphatic amines, which gave the expected thioamides **72**, and aromatic amines could be transformed to the desired products **74** in high yields even without any additive. Compared with the protocols without DMSO, this protocol is advantageous by decreasing the reaction temperature and the load of sulfur.

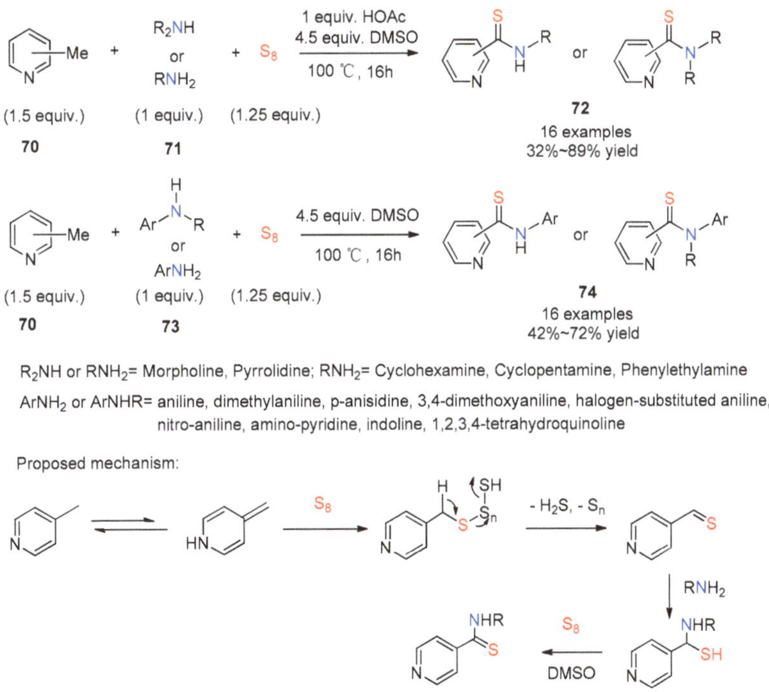

Scheme 28. DMSO-promoted synthesis of aryl thioamides by Nguyen and co-workers [50].

2.6. Thionylation Involving Amines Only

In the absence of any other partner, primary or secondary amines can be transformed to thioamides by reaction with S_8. For example, under microwave irradiation and in solvent-free conditions, variously substituted benzylamines **75** or **77** could be converted in moderate-to-good yields to the aromatic thioamides **76** or **78**, respectively (Scheme 29) [51]. The main benefit of this method is the by far shorter reaction time compared to previously reported methods. The generality of the reaction needs further investigation for its extension to secondary amines, as only low yields were observed for some examples such as *N*-tert-butyl and *N*-phenylbenzylamines.

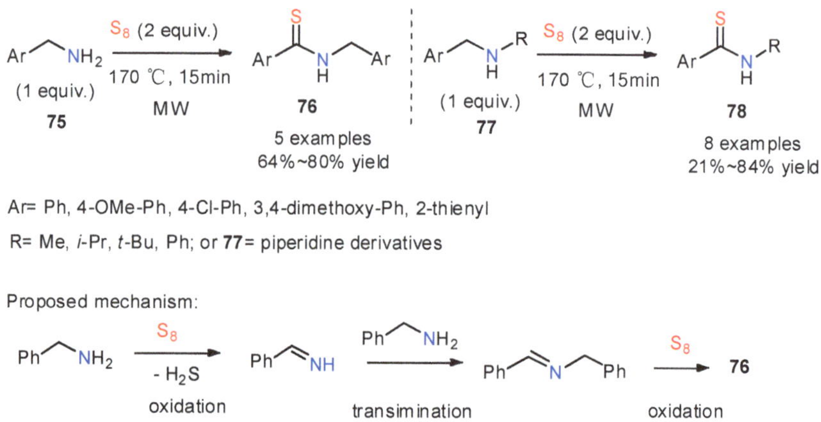

Scheme 29. Microwave-assisted synthesis of aromatic thioamides by Keglevich and co-workers [51].

If two different primary amines are mixed together in the presence of elemental sulfur, four possible thioamides can generally be formed in the reaction. Nguyen and his colleagues found, however, that cross-coupling products could be obtained with significant selectivity because benzyl amines are more prone to oxidation [52]. They reported the preparation of a series of cross-coupled thioamides, indicating that only one heterocoupled thioamide **80a** was the major product, while only small amounts of homocoupled thioamides **80b** and **80c** were observed as by-products (Scheme 30).

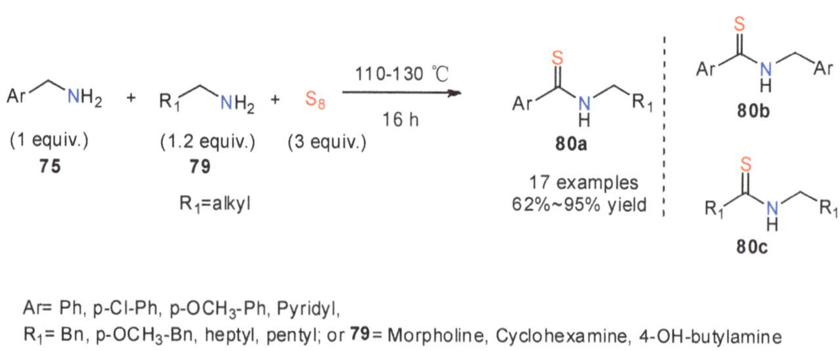

Scheme 30. Oxidative coupling of two amines into aromatic thioamides by Nguyen and his colleagues [52].

The formation of thioamides from amines and elemental sulfur can be photocatalyzed. A recent example is the work of Savateev's group in 2018 [53]. They reported the synthesis of aryl or alkyl thioamides **82** under visible light irradiation in the presence of potassium poly(heptazine imide) (K-PHI) as a photocatalyst (Scheme 31). In terms of scope, a wide range of substituted benzylamines and heterocyclic or aliphatic methylamines were transformed to thioamides in high isolated yields, except for 2-methyl benzylamine, which was unreactive under these conditions. The authors suggested that steric hindrance of the methyl group near the reactive site was responsible for this lack of reactivity.

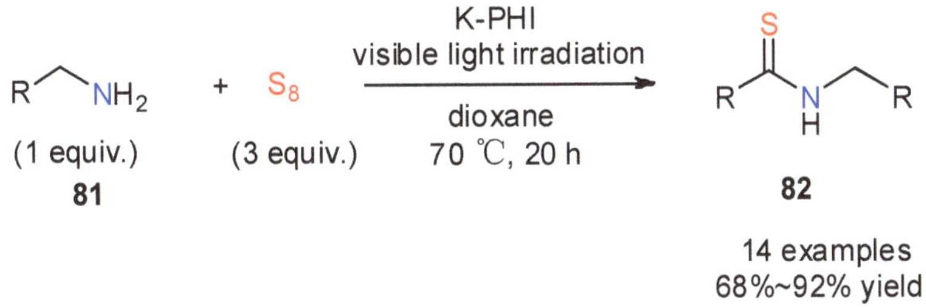

R= Ph, p-OCH_3-Ph, p-CH_3-Ph, m-CH_3-Ph, p-NH_2-Ph, Pyridyl, Furyl, Et, Bu.

Scheme 31. Photocatalytic synthesis of aryl/alkyl thioamides by Savateev and co-workers [53].

Catalysis of the thioamidation can also come from the addition of tetrabutylammonium hydroxide (TBAOH), as reported by Adimurthy and his colleagues, who described the transformation of benzyl amines to *N*-benzylbenzothioamides **83** with elemental sulfur under aerobic conditions (Scheme 32) [54]. The corresponding thioamides were provided in good yields using different benzyl amines with electron-withdrawing or electron-donating groups. Interestingly, 2-chlorobenzylamine could be transformed to the corresponding thioamide under these conditions, despite the steric hindrance of the chloride, which normally limits its reactivity. In this formation, ionic liquid TBAOH acted as catalyst and aerobic oxygen as a sole oxidant under solvent- and base-free conditions. The authors further expanded the scope of their study to heterothioamides.

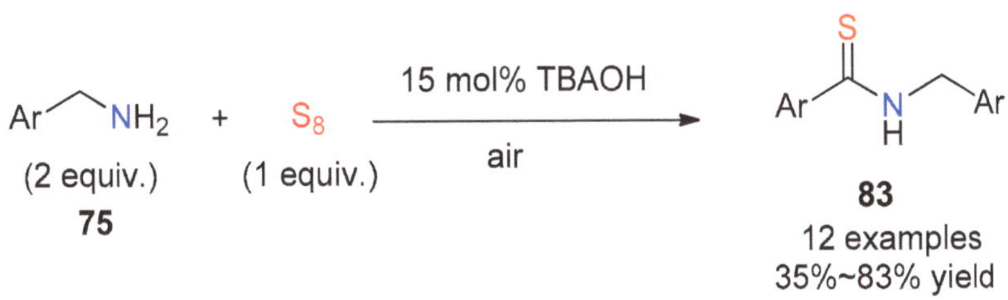

Ar= Ph, p-Cl-Ph, o-Cl-Ph, m-Cl-Ph, o-OCH_3-Ph, p-OCH_3-Ph, p-CH_3-Ph, p-F-Ph, o-F-Ph, p-CF_3-Ph, Thienyl, Pyridyl

Scheme 32. Ionic liquid TBAOH catalyzed synthesis of *N*-benzylbenzothioamides by Adimurthy and co-workers [54].

A solvent-driven C(sp^3)-H thiocarbonylation of bicyclic benzylic amines such as **84** into 2-phenyl-3,4-dihydroisoquinoline-1(2*H*)-thione derivatives **85** was developed by Wei and co-workers (Scheme 33) [55]. Conducted under catalyst-free conditions, the reaction was specific to dipolar aprotic solvents such as DMF, DMA, and DMPU. Mechanistic calculations of the complete reaction free energy profiles for the different benzylamines in DMPU as solvent led to the proposal that the reaction proceeded via the formation of $S_3^{\bullet-}$ species and the solvent radical anion (DMPU$^{\bullet-}$).

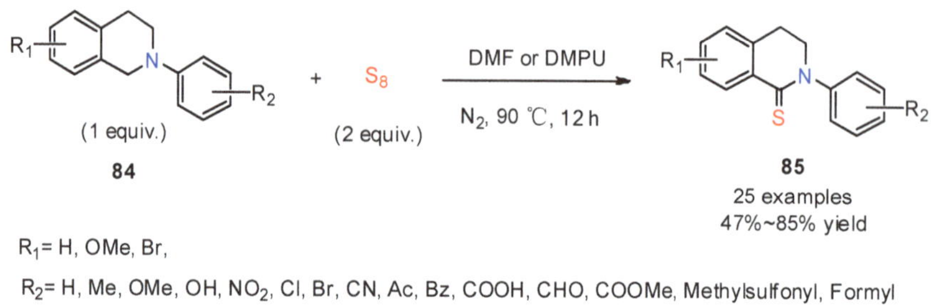

Scheme 33. C(sp^3)-H thiocarbonylation for thioamides by Wei and his co-workers [55].

3. Inorganic Sulfides as a Sulfuration Agent

In this section, we review how inorganic sulfides such as sodium hydrosulfide (NaSH), sodium sulfide (Na_2S), and sodium disulfide (Na_2S_2) have been recently used as sulfur sources in strategies targeting thioamides. Using inorganic sulfide is indeed an attractive source of sulfur for organic reactions with numerous occurrences and applications in industrial processes and material science.

Though not as convenient as compared to elemental sulfur because some hydrogen sulfide is generally released, inorganic sulfides are, however, useful reagents, being still preferable compared to organosulfur reagents or Lawesson and P_4S_{10} reagents in terms of cleanness and practicality, while covering a wide range of possible substrates. These practical aspects have confirmed them as useful reagents in organic synthesis. Some recent efforts to improve the cleanness and practicality of these latter methods are briefly discussed below. Mechanistically, in the case of aldehydes, the reaction process involves the formation of an imine, which undergoes sulfur anion or HS- addition before the final oxidation step (Figure 4).

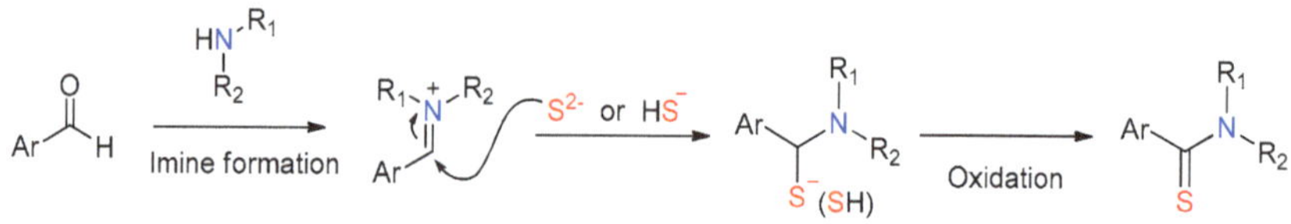

Figure 4. The mechanism of the inorganic sulfur-mediated thioamide synthesis.

3.1. *Sodium Hydrosulfide as a Sulfuration Agent*

Sodium hydrosulfide is a water-soluble and stable salt. When dissolved in water, a highly alkaline sodium hydrosulfide (pH range of 11.5 to 12.5) is obtained. NaSH is not very cheap, but it can be readily prepared by reacting hydrogen sulfide with sodium ethoxide in ethanol.

A recent study by Wu's group focused on the use of sodium hydrosulfide as a sulfur source in a three-component reaction starting from amines and aryl methyl ketones promoted by iodine (Scheme 34) [56]. This method allowed the direct and efficient conversion of aryl methyl ketones to α-keto thioamides **87**, which were not able to be directly prepared through the original Willgerodt–Kindler reaction from acetophenone. The optimal reaction conditions (1.5 equiv I_2 at 120 °C in DMSO) were applied to the reaction of acetophenone with a variety of secondary amines, leading smoothly to the corresponding α-ketothioamides in good yields. In this process, iodine played the role of an oxidant which could be regenerated from the Kornblum oxidation of the byproduct HI in the presence of DMSO.

NaSH served as the sulfur source in another approach converting carboxylic acids to thioamides. In this method, the carboxylic acids were first transformed into corresponding thiocarboxylic acids in the presence of NaSH. A second step is the addition of the thiocarboxylic acids on the ynamide (*N*-methylynetoluenesulfonamide MYTsA), forming an intermediate thioester **89** (Scheme 35). The thioamidation then occurs by reaction of this latter with amines, leading to aryl or alkyl thioamides **90** in excellent yields and wide functional group tolerance [57]. Thanks to the mildness of the conditions avoiding any epimerization or racemization of chiral substrates, this method has found remarkable applications in the synthesis of thiopeptides [58].

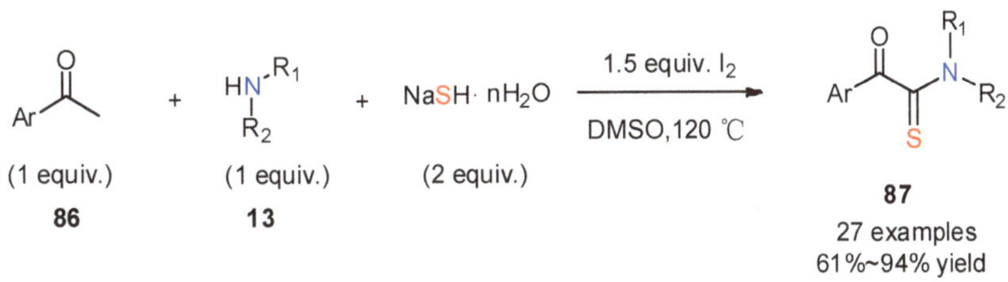

Ar= Ph, p-Cl-Ph, p-F-Ph, m-OCH_3-Ph, p-OCH_3-Ph, p-OEt-Ph, p-CH_3-Ph, p-NO_2-Ph, Thienyl, Naphthyl, Furyl, Biphenyl, Benzofuryl
HNR_1R_2= Morpholine, Piperidine, Cyclohexamine, 4-Methylpiperazine, 4-Ethylpiperazine

Scheme 34. I_2-mediated synthesis of α-keto thioamides by Wu and co-workers [56].

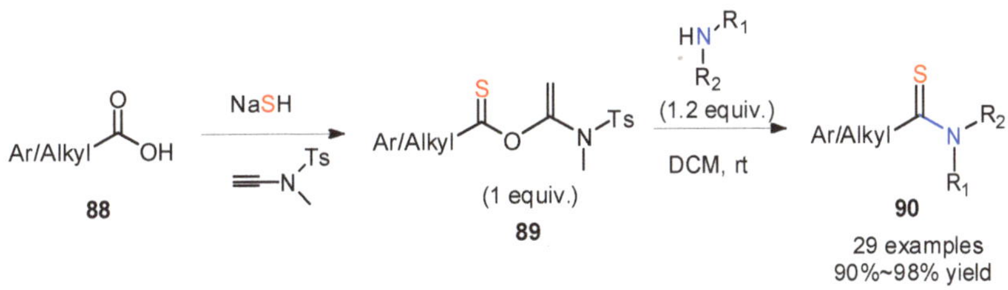

Ar= Ph, p-Cl-Ph, p-F-Ph, p-OCH_3-Ph, p-CN-Ph, p-CF_3-Ph, 2-OMe-4-NH_2-5-SO_2Et-Ph, Furyl 5-Br-furyl, 4,5-Dibromo-thienyl
Alkyl= Me

Scheme 35. Ynamide-mediated synthesis of thioamides by Zhao and co-workers [57].

3.2. Sodium Sulfide as a Sulfuration Agent

Sodium sulfide, often used as its hydrate, is a water-soluble salt that is readily available and very cheap. Sodium sulfide is stable in basic conditions.

Jiang's group [59] reported an efficient and practical three-component preparation of aryl/alkyl thioamides **91** in water involving aryl or alkyl aldehydes, *N*-substituted formamides, and sodium sulfide (Scheme 36). *N*-substituted formamides are essential for this transformation. Comparatively, the corresponding amines led to a trace amount of the targeted thioamides, owing to the hydrolysis of formamide by sodium sulfide to release hydrogen sulfide, acting as a promoter of the reaction. This protocol proved to tolerate several functional groups, notably for aryl aldehydes and unsaturated aliphatic aldehydes. This is an important advantage of this method, which could be applied to late-stage modifications of bioactive drugs containing carbon-carbon double bonds.

It is worth mentioning another example of sodium sulfide-mediated thionylation which benefited from the use of an ionic liquid as medium. This work concerned the formation of thiobenzamide derivatives **93** from aryl nitriles promoted by [DBUH][OAc] ionic liquid at room temperature (Scheme 37) [60]. This was well-applicable to a wide range of functional aryl nitriles. The ionic liquid allowed easy purification from the reaction mixture of the target products and could be reused several times (same catalytic ability even five rounds).

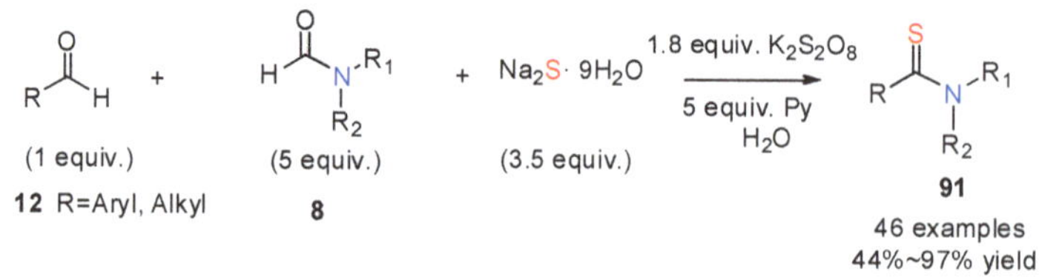

Ar= Ph, X-substituted phenyl (X=halogen, Me, MeO, OH, CF_3), Thienyl, Naphthyl, Pyridyl, Benzofuryl, Benzothienyl, N-benzylindolyl
R=Alkyl, **12**= isovaleraldehyde, phenylpropanal, citronellal, lilyaldehyde, melonal, cholesterol derivative

Scheme 36. Na_2S-mediated synthesis of aryl or alkyl thioamides in water by Jiang and co-workers [59].

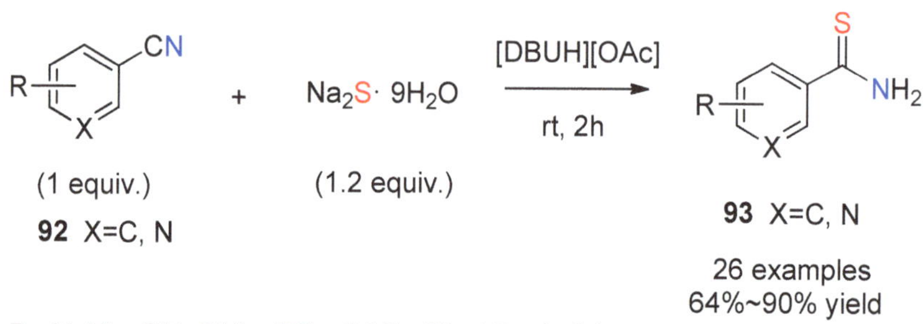

R= H, Me, CN, OMe, CF_3, OCF_3, Ph, *t*-Bu, halides

Scheme 37. Ionic liquid [DBUH][OAc]-promoted synthesis of thioamides by Zhang and co-workers [60].

A recent example of thionylation using sodium sulfide as a source of sulfur was reported by Sun and co-workers for the transformation of 2-aryl haloalkynes into 2-arylethanethioamides (Scheme 38) [61]. They established an efficient three-component reaction under rather mild conditions using DMF as the best solvent among several others, such as 1,4-dioxane, and DME. A wide range of amines and different functional haloalkynes were examined to afford the expected thioamides **95** in good yield. The proposed mechanism based on isotopic exchange control experiments showed that, among the two hydrogen atoms of the methylene group in **95**, one arose from the amine, and one from water, indicating that the presence of the water molecules (arising from $Na_2S{\cdot}9H_2O$) is essential.

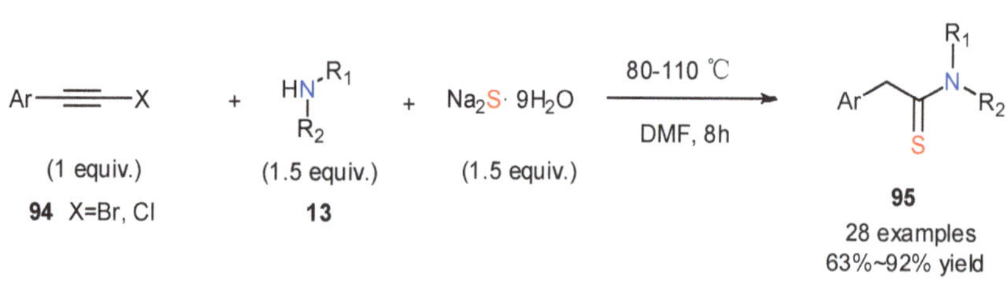

Ar= Ph, X-substituted phenyl (X=halogen, Me, MeO, Et, EtO, OH, Ac, $CONH_2$, ethylcyclohexyl), Thienyl, Pyridyl
R_1= R_2= Me, Et, Pr, Bu, Bn; R_1=Bn, R_2= Me, Et; R_1=Pr, R_2= Et; R_1= H, R_2= Pr, Bu;
HNR_1R_2= morpholine, thiomorpholine, pyrrolidine, tetrahydroisoquinoline

Scheme 38. Na_2S-mediated synthesis of 2-arylethanethioamides in DMF by Jiang and co-workers [61].

Doddi and co-workers used sodium sulfide and *N*-substituted formamide in a three-component transformation of *gem*-dibromostyrenes (Scheme 39) [62]. A series of 2-arylethanethioamides **96** were obtained in fair-to-excellent yields in the absence of any catalyst. The authors indicated that the transformation starts by the formation of an intermediate alkyne.

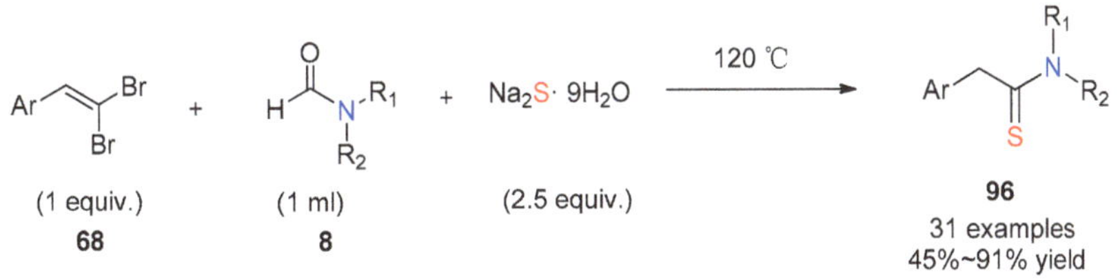

Ar= X-substituted phenyl (X=halogen, Me, MeO, Pr, EtO, OH, CN, NO_2), Thienyl, Furyl, Naphthyl, anthryl, pyrenyl
R_1= R_2= Me; R_1= H, R_2= Ph

Scheme 39. Na_2S-mediated synthesis of 2-arylethanethioamides by Doddi and co-workers [62].

3.3. Sodium Disulfide as a Sulfuration Agent

Sodium disulfide is a water-soluble yellow salt. Sodium disulfide is not stable and causes the release of H_2S in the air and light. Though unstable, it can be a useful sulfur source for thionylation reactions, such as in the transformation of quaternary ammonium salts and amides reported recently by Cheng and his colleagues [63]. They developed a catalyst-free protocol towards aryl thioamides **98** involving aryl trimethyl ammonium iodide, *N*-substituted formamides, and aqueous sodium disulfide (Scheme 40). This transformation gave the expected thioamides in moderate-to-good yields with high functional group tolerance. In their study, several inorganic sulfur sources were compared, including elemental sulfur, sodium sulfide, potassium sulfide, and sodium disulfide. The advantage of sodium disulfide was to provide good results without additional oxidant required.

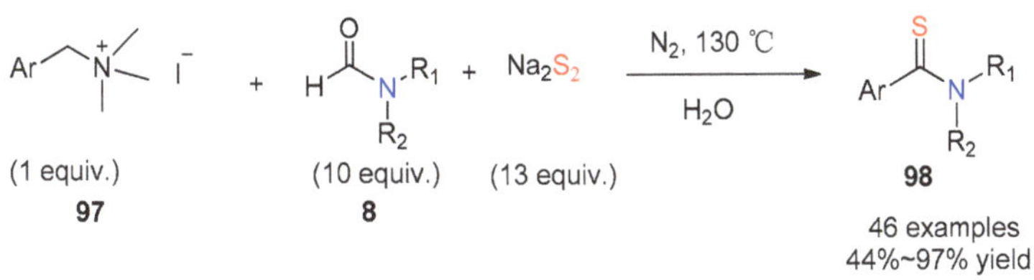

Ar= Ph, X-substituted phenyl (X=Br, Me, MeO, NH_2, CF_3), Thienyl, Pyridyl, Pyrrolyl
R_1= R_2= Me; R_1= H, R_2= Ph, Me, Cyclohexyl, Cyclopentyl, Bn, Et; NR_1R_2= Morpholinyl, Piperidinyl

Scheme 40. Na_2S_2-mediated synthesis of aryl thioamides by Cheng and co-workers [63].

4. Recent Practical Improvements of Other Thionylation Methods

Though less attractive with respect to cleanness and toxicity, we want to mention some efforts which have been made for improving other traditional methods, namely, the use of P_4S_{10} or Lawesson's reagent, and reactions of organosulfur compounds, as these methods offer in some cases a shorter option with respect to the access to the substrate itself. For example, as the LR is inefficient at relatively high temperatures due to its rapid decomposition, Svensson and co-workers [64] used P_4S_{10} in pyridine as a thionating agent in DMSO at high temperatures. In the workup procedure, the remaining thionating reagent can be transferred to solvents by addition of water and the target products can be easily obtained with high purity by recrystallization. Organosulfur is also a well-documented source of sulfur to construct thioamides, especially thiols [65], disulfides [66–68], and thiourea [69], primary or secondary thioamides [70,71], and 1,2,3-thiadiazoles [72]. However, organosulfur compounds are featured as catalyst poisons for transition metals due to their ability to coordinate metallic species [73], which sometimes limits their application. Overcoming these drawbacks has thus been an aim, as in some protocols based on transition-metal-free transamidation reactions, operationally simple and broadly functional group-compatible [70,71].

The transamidation reaction of thioamides has also been recently considered as a sensible solution for the synthesis of substituted thioamides. In 2021, Zeng and co-workers [70] applied NaOTs as promoter to activate unsubstituted thioamides towards new C-N bond formation (Scheme 41), which provides a simple pathway to access substituted thioamides.

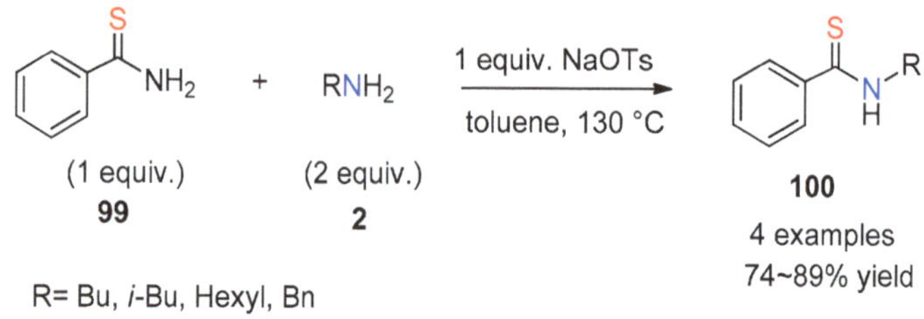

Scheme 41. Transamidation of aryl thioamides by Zeng and co-workers [70].

In the same trend, Szostak and co-workers reported in 2022 a simple and direct method for transamidation of *N*-tert-butoxycarbonyl aryl thioamides **101** based on the uncommon cleavage of the C(S)-N bond under mild conditions [71]. This process occurred well in the presence of NaHMDS, whereas other bases (NaOH, KO-*t*-Bu, *n*-BuLi) were found ineffective in this reaction (Scheme 42). This protocol is broadly compatible with different functional groups, thus offering an interesting alternative for the construction of substituted thioamides.

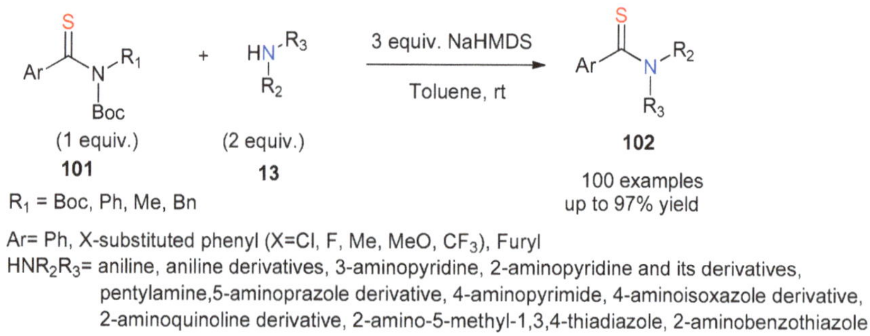

Scheme 42. Transamidation of aryl thioamides by Szostak and co-workers [71].

Regarding P_4S_{10} and Lawesson's reagent, which often produce toxic phosphor-containing aqueous chemical wastes and require tedious procedures, some improved work-ups have been proposed. Recently, Hu and co-workers [74] developed an efficient work-up for their thionation reactions with LR by a treatment of ethylene glycol (Scheme 43) which avoided purification by column chromatography and the generation of phosphor-containing aqueous waste.

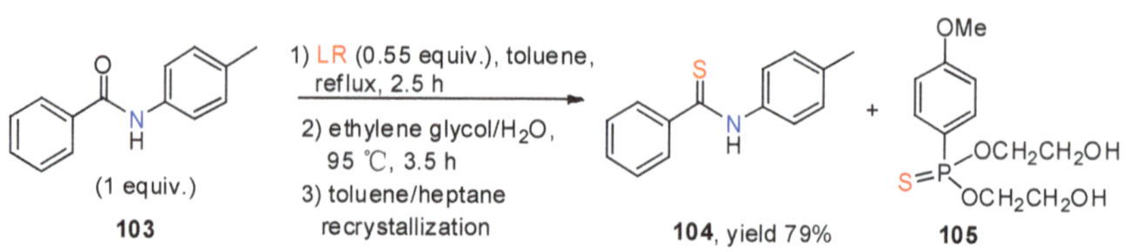

Scheme 43. Thionation reactions with LR by Hu and co-workers [74].

5. Conclusions

While advanced chemical synthesis techniques for directly preparing thioamide moieties have become common, there is an increasing interest in efficient and more practical methods applied to thioamide-containing natural products. However, sulfur incorporation in organic molecules remains a challenge from the practical viewpoint. In this review, we have tried to highlight some recent studies which all try to improve the feasibility of the preparation of thioamides, which are very important targets in organic and medicinal chemistry. Focusing first on elemental sulfur, which is by far the most practical and clean reagent, we have highlighted numerous very useful methods which continue to widen the scope of substrates, solvents, catalysts, and the remarkably wide scope of possible targets.

Many protocols performed in water or solvent-free conditions provide various types of thioamides in high yields, which makes their process clean and significantly improved in the context of more environmentally friendly chemical transformations. The transamidation of thioamides appears also as a sensible alternative as it is operationally easy, user-friendly, and conducted under rather mild conditions, though necessitating the addition of base. The ability of a transformation to be performed on a very complex substrate is of high importance. Thus, the processes which exhibit wide functional group tolerance are the most attractive ones as they can be applied to the late-stage functionalization of drug candidates. Important also is the ability to preserve chirality of chiral substrates, and in this regard, the presence of catalysts remains often essential.

These methods, more practical and already quite general, can though still be improved with respect to the amine substrate scope, some of them such as aliphatic primary amines remaining more challenging. Other methods, yet not as ideal as compared to elemental sulfur, have also been mentioned when some kind of improvement is proposed. Overall, these new protocols and methods open up new chances to discover unconventional reactions for thioamides and their derivatives.

Author Contributions: Q.Z. conducted the initial writing of the manuscript, which was then completed by L.S. and Y.Q. Y.Q. supervised all the work. All authors have read and agreed to the published version of the manuscript.

Funding: This work was supported by grants from the Scientific Research Project of Education Department of Hubei Province (B2021259).

Institutional Review Board Statement: Not applicable.

Informed Consent Statement: Not applicable.

Data Availability Statement: See bibliographic data cited in the References section below.

Acknowledgments: This work was also supported by the Scientific Research Foundation of Hubei University of Education, CNRS and MESRI.

Conflicts of Interest: The authors declare no conflict of interest.

References

1. Mueller, E.G. Trafficking in persulfides: Delivering sulfur in biosynthetic pathways. *Nat. Chem. Biol.* **2006**, *2*, 185–194. [CrossRef]
2. Schwalen, C.J.; Hudson, G.A.; Kille, B.; Mitchell, D.A. Bioinformatic expansion and discovery of thiopeptide antibiotics. *J. Am. Chem. Soc.* **2018**, *140*, 9494–9501. [CrossRef]
3. Kenney, G.E.; Dassama, L.M.; Pandelia, M.-E.; Gizzi, A.S.; Martinie, R.J.; Gao, P.; DeHart, C.J.; Schachner, L.F.; Skinner, O.S.; Ro, S.Y. The biosynthesis of methanobactin. *Science* **2018**, *359*, 1411–1416. [CrossRef]
4. Hooper, M. The medicinal chemistry of anti-leprosy drugs. *Chem. Soc. Rev.* **1987**, *16*, 437–465. [CrossRef]
5. Lincke, T.; Behnken, S.; Ishida, K.; Roth, M.; Hertweck, C. Closthioamide: An unprecedented polythioamide antibiotic from the strictly anaerobic bacterium *Clostridium cellulolyticum*. *Angew. Chem. Int. Ed.* **2010**, *49*, 2011–2013. [CrossRef] [PubMed]
6. Banala, S.; Süssmuth, R.D. Thioamides in nature: In search of secondary metabolites in anaerobic microorganisms. *ChemBioChem* **2010**, *11*, 1335–1337. [CrossRef]
7. Mahanta, N.; Szantai-Kis, D.M.; Petersson, E.J.; Mitchell, D.A. Biosynthesis and chemical applications of thioamides. *ACS Chem. Biol.* **2019**, *14*, 142–163. [CrossRef]

8. Litomska, A.; Ishida, K.; Dunbar, K.L.; Boettger, M.; Coyne, S.; Hertweck, C. Enzymatic thioamide formation in a bacterial antimetabolite pathway. *Angew. Chem. Int. Ed.* **2018**, *57*, 11574–11578. [CrossRef] [PubMed]
9. Chatterjee, S.; Hausinger, R.P. Sulfur incorporation into biomolecules: Recent advances. *Crit. Rev. Biochem. Mol. Biol.* **2022**, 1–16. [CrossRef]
10. Kumari, S.; Carmona, A.V.; Tiwari, A.K.; Trippier, P.C. Amide bond bioisosteres: Strategies, synthesis, and successes. *J. Med. Chem.* **2020**, *63*, 12290–12358. [CrossRef] [PubMed]
11. Lampkin, B.J.; VanVeller, B. Hydrogen bond and geometry effects of thioamide backbone modifications. *J. Org. Chem.* **2021**, *86*, 18287–18291. [CrossRef]
12. Fiore, K.E.; Patist, M.J.; Giannakoulias, S.; Huang, C.H.; Verma, H.; Khatri, B.; Cheng, R.P.; Chatterjee, J.; Petersson, E.J. Structural impact of thioamide incorporation into a β-hairpin. *RSC Chem. Biol.* **2022**, *3*, 582–591. [CrossRef]
13. Jagodziński, T.S. Thioamides as useful synthons in the synthesis of heterocycles. *Chem. Rev.* **2003**, *103*, 197–228. [CrossRef]
14. Gasteiger, J.; Herzig, C. Synthetic applications of 2-chlorooxiranes: Preparation of thiazoles, dihydrothiazoles and selenazoles. *Tetrahedron* **1981**, *37*, 2607–2611. [CrossRef]
15. Singh, H.; Sarin, R. Carbon transfer reactions of Δ^2-oxazolinium and thiazolinium cations. *Tetrahedron* **1986**, *42*, 1449–1460. [CrossRef]
16. Engman, L. Organoselenium- and proton-mediated cyclization reactions of allylic amides and thioamides. Syntheses of 2-oxazolines and 2-thiazolines. *J. Org. Chem.* **1991**, *56*, 3425–3430. [CrossRef]
17. Joyce, L.L.; Evindar, G.; Batey, R.A. Copper- and palladium-catalyzed intramolecular C–S bond formation: A convenient synthesis of 2-aminobenzothiazoles. *Chem. Commun.* **2004**, 446–447. [CrossRef]
18. Cheng, Y.; Peng, Q.; Fan, W.; Li, P. Room-temperature ligand-free Pd/C-catalyzed C–S bond formation: Synthesis of 2-substituted benzothiazoles. *J. Org. Chem.* **2014**, *79*, 5812–5819. [CrossRef]
19. Ozturk, T.; Ertas, E.; Mert, O. A berzelius reagent, phosphorus decasulfide (P_4S_{10}) in organic syntheses. *Chem. Rev.* **2010**, *110*, 3419–3478. [CrossRef]
20. Ozturk, T.; Ertas, E.; Mert, O. Use of Lawesson's reagent in organic syntheses. *Chem. Rev.* **2007**, *107*, 5210–5278. [CrossRef]
21. Priebbenow, D.L.; Bolm, C. Recent advances in the Willgerodt-Kindler reaction. *Chem. Soc. Rev.* **2013**, *42*, 7870–7880. [CrossRef] [PubMed]
22. Nguyen, T.B. Recent advances in organic reactions involving elemental sulfur. *Adv. Synth. Catal.* **2017**, *359*, 1066–1130. [CrossRef]
23. Xu, H.; Deng, H.; Li, Z.; Xiang, H.; Zhou, X. Synthesis of thioamides by catalyst-free three-component reactions in water. *Eur. J. Org. Chem.* **2013**, *2013*, 7054–7057. [CrossRef]
24. Liu, W.; Chen, C.; Liu, H. Dimethylamine as the key intermediate generated in situ from dimethylformamide (DMF) for the synthesis of thioamides. *Beilstein J. Org. Chem.* **2015**, *11*, 1721–1726. [CrossRef]
25. Li, J.; Ren, X.; Li, G.; Liang, H.; Zhao, Y.; Wang, Z.; Li, H.; Yuan, B. Mixed bases mediated synthesis of thioamides in water. *J. Sulfur. Chem.* **2020**, *41*, 229–237. [CrossRef]
26. Kale, A.D.; Dalal, D.S. Catalyst- and solvent-free thioamidation of aromatic aldehydes through a Willgerodt-Kindler reaction. *ChemistrySelect* **2022**, *7*, e202203497. [CrossRef]
27. Gupta, A.; Vankar, J.K.; Jadav, J.P.; Gururaja, G.N. Water mediated direct thioamidation of aldehydes at room temperature. *J. Org. Chem.* **2022**, *87*, 2410–2420. [CrossRef]
28. Liao, Y.; Jiang, X. Construction of thioamide peptide via sulfur-involved amino acids/amino aldehydes coupling. *Org. Lett.* **2021**, *23*, 8862–8866. [CrossRef]
29. Guntreddi, T.; Vanjari, R.; Singh, K.N. Decarboxylative thioamidation of arylacetic and cinnamic acids: A new approach to thioamides. *Org. Lett.* **2014**, *16*, 3624–3627. [CrossRef]
30. Saito, M.; Murakami, S.; Nanjo, T.; Kobayashi, Y.; Takemoto, Y. Mild and chemoselective thioacylation of amines enabled by the nucleophilic activation of elemental sulfur. *J. Am. Chem. Soc.* **2020**, *142*, 8130–8135. [CrossRef]
31. Qu, Y.; Li, Z.; Xiang, H.; Zhou, X. Copper (II)-catalyzed reactions of dimethylformamide with phenylacetonitrile and sulfur to form *N,N*-dimethylthioamides. *Adv. Synth. Catal.* **2013**, *355*, 3141–3146. [CrossRef]
32. Li, X.; Pan, Q.; Hu, R.; Wang, X.; Yang, Z.; Han, S. S_8-mediated one-pot synthesis of thioamides from benzyl chlorides and amines. *Asian J. Org. Chem.* **2016**, *5*, 1353–1358. [CrossRef]
33. Jin, H.; Chen, X.; Qian, C.; Ge, X.; Zhou, S. Transition-metal-free, general construction of thioamides from chlorohydrocarbon, amide and elemental sulfur. *Eur. J. Org. Chem.* **2021**, *2021*, 3403–3406. [CrossRef]
34. Kozlov, M.; Kozlov, A.; Komkov, A.; Lyssenko, K.; Zavarzin, I.; Volkova, Y. Synthesis of phosphoryl thioamides via three-component reaction of phosphinic chlorides with amines and sulfur. *Adv. Synth. Catal.* **2019**, *361*, 2904–2915. [CrossRef]
35. Do, N.T.; Tran, K.M.; Phan, H.T.; To, T.A.; Nguyen, T.T.; Phan, N.T. Functionalization of activated methylene C–H bonds with nitroarenes and sulfur for the synthesis of thioamides. *Org. Biomol. Chem.* **2019**, *17*, 8987–8991. [CrossRef] [PubMed]
36. Jin, H.; Ge, X.; Zhou, S. General construction of thioamides under mild conditions: A stepwise proton transfer process mediated by EDTA. *Eur. J. Org. Chem.* **2021**, *2021*, 6015–6021. [CrossRef]
37. Yu, P.; Wang, Y.; Zeng, Z.; Chen, Y. Metal-free C–N or C–C bond cleavages of α-azido ketones: An oxidative-amidation strategy for the synthesis of α-ketothioamides and amides. *J. Org. Chem.* **2019**, *84*, 14883–14891. [CrossRef] [PubMed]
38. Zhang, Z.; Yang, J.; Yu, R.; Wu, K.; Bu, J.; Li, S.; Qian, P.; Sheng, L. Efficient synthesis of α-ketothioamides from α-nitroketones, amines or DMF and elemental sulfur under oxidant-free conditions. *Eur. J. Org. Chem.* **2021**, *37*, 5209–5212. [CrossRef]

39. Tian, H.; Guo, F.; Chen, X. C_{sp3}-H bond functionalization of α-bromo ketones for the synthesis of α-keto thioamides using elemental sulfur. *Russ. J. Org. Chem.* **2022**, *58*, 1260–1266. [CrossRef]
40. Chaubey, T.N.; Borpatra, P.J.; Sharma, A.; Pandey, S.K. Metal-free syntheses of α-keto thioamide and α-ketoamide derivatives from sulfoxonium ylides. *Org. Lett.* **2022**, *24*, 8062–8066. [CrossRef]
41. Nguyen, T.B.; Tran, M.Q.; Ermolenko, L.; Al-Mourabit, A. Three-component reaction between alkynes, elemental sulfur, and aliphatic amines: A general, straightforward, and atom economical approach to thioamides. *Org. Lett.* **2014**, *16*, 310–313. [CrossRef]
42. Li, W.; Wu, X.; Zhao, Z.; Qin, A.; Hu, R.; Tang, B.Z. Catalyst-free, atom-economic, multicomponent polymerizations of aromatic diynes, elemental sulfur, and aliphatic diamines toward luminescent polythioamides. *Biomacromolecules* **2015**, *48*, 7747–7754. [CrossRef]
43. Xu, K.; Li, Z.; Cheng, F.; Zuo, Z.; Wang, T.; Wang, M.; Liu, L. Transition-metal-free cleavage of C–C triple bonds in aromatic alkynes with S_8 and amides leading to aryl thioamides. *Org. Lett.* **2018**, *20*, 2228–2231. [CrossRef]
44. Zhang, P.; Chen, W.; Liu, M.; Wu, H. Base-controlled three component reactions of amines, elemental sulfur, and styrenes: Synthesis of thioamides under metal-free conditions. *J. Org. Chem.* **2018**, *83*, 14269–14276. [CrossRef]
45. Gan, L.; Gao, Y.; Wei, L.; Wan, J.-P. Synthesis of α-keto thioamides by metal-free C=C bond cleavage in enaminones using elemental sulfur. *J. Org. Chem.* **2019**, *84*, 1064–1069. [CrossRef]
46. Xu, H.H.; Zhang, X.G.; Zhang, X.H. Thioamidation of arylpropyne derivatives with sulfur and formamides for the synthesis of aryl propanethioamides. *Asian J. Org. Chem.* **2020**, *9*, 111–115. [CrossRef]
47. Peng, L.; Ma, L.; Ran, Y.; Chen, Y.; Zeng, Z. Metal-free three-component synthesis of thioamides from β-nitrostyrenes, amines and elemental sulfur. *Tetrahedron Lett.* **2021**, *74*, 153092. [CrossRef]
48. Jaiswal, A.; Sharma, A.K.; Singh, K.N. Copper-catalyzed decarboxylative synthesis of α-ketothioamides using α,α-unsaturated arylcarboxylic acids, alicyclic secondary amines and elemental sulfur. *Asian J. Org. Chem.* **2021**, *10*, 1748–1751. [CrossRef]
49. Vankar, J.K.; Gupta, A.; Jadav, J.P.; Nanjegowda, S.H.; Gururaja, G.N. The thioamidation of *gem*-dibromoalkenes in an aqueous medium. *Org. Biomol. Chem.* **2021**, *19*, 2473–2480. [CrossRef]
50. Nguyen, T.T.T.; Ngo, Q.A.; Koleski, M.; Nguyen, T.B. The catalytic role of elemental sulfur in the DMSO-promoted oxidative coupling of methylhetarenes with amines: Synthesis of thioamides and bis-aza-heterocycles. *Org. Chem. Front.* **2021**, *8*, 1593–1598. [CrossRef]
51. Milen, M.; Ábrányi-Balogh, P.; Dancsó, A.; Keglevich, G. Microwave-assisted synthesis of thioamides with elemental sulfur. *J. Sulfur. Chem.* **2012**, *33*, 33–41. [CrossRef]
52. Nguyen, T.B.; Ermolenko, L.; Al-Mourabit, A. Efficient and selective multicomponent oxidative coupling of two different aliphatic primary amines into thioamides by elemental sulfur. *Org. Lett.* **2012**, *14*, 4274–4277. [CrossRef]
53. Kurpil, B.; Kumru, B.; Heil, T.; Antonietti, M.; Savateev, A. Carbon nitride creates thioamides in high yields by the photocatalytic Kindler reaction. *Green Chem.* **2018**, *20*, 838–842. [CrossRef]
54. Joshi, A.; Kumar, R.; Semwal, R.; Rawat, D.; Adimurthy, S. Ionic liquid catalysed aerobic oxidative amidation and thioamidation of benzylic amines under neat conditions. *Green Chem.* **2019**, *21*, 962–967. [CrossRef]
55. Zhou, J.; Wang, S.; Lu, Y.; Li, L.; Duan, W.; Wang, Q.; Wang, H.; Wei, W. Solvent-driven C (sp^3)–H thiocarbonylation of benzylamine derivatives under catalyst-free conditions. *Green Chem.* **2021**, *23*, 767–773. [CrossRef]
56. Li, H.-Z.; Xue, W.-J.; Wu, A.-X. Direct synthesis of α-ketothioamides from aryl methyl ketones and amines via I_2-promoted sp^3 C–H functionalization. *Tetrahedron* **2014**, *70*, 4645–4651. [CrossRef]
57. Wang, C.; Han, C.; Yang, J.; Zhang, Z.; Zhao, Y.; Zhao, J. Ynamide-mediated thioamide and primary thioamide syntheses. *J. Org. Chem.* **2022**, *87*, 5617–5629. [CrossRef]
58. Yang, J.; Wang, C.; Xu, S.; Zhao, J. Ynamide-mediated thiopeptide synthesis. *Angew. Chem. Int. Ed.* **2019**, *58*, 1382–1386. [CrossRef]
59. Wei, J.; Li, Y.; Jiang, X. Aqueous compatible protocol to both alkyl and aryl thioamide synthesis. *Org. Lett.* **2016**, *18*, 340–343. [CrossRef]
60. Cao, X.-T.; Qiao, L.; Zheng, H.; Yang, H.-Y.; Zhang, P.-F. A efficient protocol for the synthesis of thioamides in [DBUH][OAc] at room temperature. *RSC Adv.* **2018**, *8*, 170–175. [CrossRef]
61. Sun, Y.; Jiang, H.; Wu, W.; Zeng, W.; Li, J. Synthesis of thioamides via one-pot A3-coupling of alkynyl bromides, amines, and sodium sulfide. *Org. Biomol. Chem.* **2014**, *12*, 700–707. [CrossRef]
62. Morri, A.K.; Thummala, Y.; Adepu, R.; Sharma, G.V.; Ghosh, S.; Doddi, V.R. Synthesis of substituted thioamides from *gem*-dibromoalkenes and sodiumsulfide. *Eur. J. Org. Chem.* **2019**, *42*, 7159–7163. [CrossRef]
63. Zhou, Z.; Yu, J.-T.; Zhou, Y.; Jiang, Y.; Cheng, J. Aqueous MCRs of quaternary ammoniums, *N*-substituted formamides and sodium disulfide towards aryl thioamides. *Org. Chem. Front.* **2017**, *4*, 413–416. [CrossRef]
64. Bergman, J.; Pettersson, B.; Hasimbegovic, V.; Svensson, P.H. Thionations using a P_4S_{10}-pyridine complex in solvents such as acetonitrile and dimethyl sulfone. *J. Org. Chem.* **2011**, *76*, 1546–1553. [CrossRef]
65. Wang, X.; Ji, M.; Lim, S.; Jang, H.-Y. Thiol as a synthon for preparing thiocarbonyl: Aerobic oxidation of thiols for the synthesis of thioamides. *J. Org. Chem.* **2014**, *79*, 7256–7260. [CrossRef]
66. Chen, S.; Li, Y.; Chen, J.; Xu, X.; Su, L.; Tang, Z.; Au, C.-T.; Qiu, R. Iodine-promoted synthesis of thioamides from 1,2-dibenzylsulfane and difurfuryl disulfide. *Synlett* **2016**, *27*, 2339–2344.

67. Chen, J.; Mei, L.; Liu, J.; Zhong, C.; Yuan, B.; Li, Q. Microwave-assisted iodine-catalyzed oxidative coupling of dibenzyl (difurfuryl) disulfides with amines: A rapid and efficient protocol for thioamides. *RSC Adv.* **2019**, *9*, 28576–28580. [CrossRef]
68. Rezapour, M.; Abbasi, M. Oxidative coupling of dibenzyl disulfide with amines catalyzed by quinoline-bromine complex: Access to thioamides. *Mol. Divers.* **2023**, *27*, 159–165. [CrossRef]
69. Bian, Y.; Qu, X.; Chen, Y.; Li, J.; Liu, L. $K_2S_2O_8$-promoted aryl thioamides synthesis from aryl aldehydes using thiourea as the sulfur source. *Molecules* **2018**, *23*, 2225. [CrossRef]
70. Zhang, Y.; Ye, X.; Liu, S.; Chen, W.; Majeed, I.; Liu, T.; Zhu, Y.; Zeng, Z. NaOTs-promoted transition metal-free C-N bond cleavage to form C-X (X= N, O, S) bonds. *Org. Biomol. Chem.* **2021**, *19*, 8566–8571. [CrossRef]
71. Li, G.; Xing, Y.; Zhao, H.; Zhang, J.; Hong, X.; Szostak, M. Chemoselective transamidation of thioamides by transition-metal-free N-C(S) transacylation. *Angew. Chem. Int. Ed.* **2022**, *61*, e202200144.
72. Lu, C.; Li, X.; Chang, S.; Zhang, Y.; Xing, D.; Wang, S.; Lin, Y.; Jiang, H.; Huang, L. Thioamide synthesis via copper-catalyzed C-H activation of 1,2,3-thiadiazoles enabled by slow release and capture of thioketenes. *Org. Chem. Front.* **2022**, *9*, 2382–2389. [CrossRef]
73. Hegedus, L.L.; McCabe, R.W. *Catalyst Poisoning*; Marcel Dekker Incorporated: New York, NY, USA, 1984.
74. Wu, K.; Ling, Y.; Ding, A.; Jin, L.; Sun, N.; Hu, B.; Shen, Z.; Hu, X. A chromatography-free and aqueous waste-free process for thioamide preparation with Lawesson's reagent. *Beilstein J. Org. Chem.* **2021**, *17*, 805–812. [CrossRef]

molecules

MDPI

Review

Synthesis of Thionated Perylenediimides: State of the Art and First Investigations of an Alternative to Lawesson's Reagent †

Oksana Kharchenko [1], Anna Hryniuk [2], Oksana Krupka [1,*] and Piétrick Hudhomme [2,*]

1 Univ Angers, Inserm, CNRS, MINT, SFR ICAT, F-49000 Angers, France; oksana.kharchenko@univ-angers.fr
2 Univ Angers, CNRS, MOLTECH-Anjou, SFR MATRIX, F-49000 Angers, France; annagrynyuk20@gmail.com
* Correspondence: oksana.krupka@univ-angers.fr (O.K.); pietrick.hudhomme@univ-angers.fr (P.H.); Tel.: +33-2-41-73-85-59 (O.K.); +33-2-41-73-50-94 (P.H.)

† Dedicated in memory to Professor Guy Duguay and Professor Hervé Quiniou.

Abstract: Perylenediimides (PDIs) are composed of a central perylene ring, on which are grafted two imide groups at the peri positions. Thionated PDIs are characterized by the substitution of one or more oxygen atoms of these imide functions with sulfur atoms. This structural modification alters the electronic properties with a redshift of the optical absorption accompanied by modification of the charge transport characteristics compared to their non-thionated counterparts. These properties make them suitable candidates for applications in optoelectronic devices, such as organic light-emitting diodes and organic photovoltaics. Moreover, the presence of sulfur atom(s) can favor the promotion of reactive oxygen species production for photodynamic and photothermal therapies. These thionated PDIs can be synthesized through the post-functionalization of PDIs by using a sulfurizing reagent. Nevertheless, the main drawbacks remain the difficulties in adjusting the degree of thionation and obtaining tri- and tetrathionated PDIs. Up to now, this thionation reaction has been described almost exclusively using Lawesson's reagent. In the current study, we present our first investigations into an alternative reagent to enhance selectivity and achieve a greater degree of thionation. The association of phosphorus pentasulfide with hexamethyldisiloxane (Curphey's reagent) clearly demonstrated higher reactivity compared with Lawesson's reagent to attain multi-thionated PDIs.

Keywords: perylenediimide; thionation; phosphorus pentasulfide

Citation: Kharchenko, O.; Hryniuk, A.; Krupka, O.; Hudhomme, P. Synthesis of Thionated Perylenediimides: State of the Art and First Investigations of an Alternative to Lawesson's Reagent. *Molecules* **2024**, *29*, 2538. https://doi.org/10.3390/molecules29112538

Academic Editor: Ming Wang

Received: 30 April 2024
Revised: 16 May 2024
Accepted: 24 May 2024
Published: 28 May 2024

1. Introduction

Perylenediimides (PDIs) are among the most interesting polycyclic aromatic hydrocarbon structures for chemists, physicists and materials scientists [1,2]. These molecules are composed of a central perylene ring on which are grafted two imide groups at the 3,4 and 9,10 positions (Figure 1). Interest in them continues to grow because they combine thermal and photostability, as well as remarkable optical properties, with a high absorption coefficient and fluorescence quantum yield close to unity. They exhibit strong electron-accepting character with their two electron-withdrawing imide moieties on each side of the perylene backbone and, consequently, they are now considered as one of the best n-type semiconductors, making them ideally suited to applications in organic electronics [3,4], in particular for their use in organic field-effect transistors (OFETs) [5], organic light-emitting diodes (OLEDs) [6] and organic photovoltaic (OPV) [7–10] devices. However, academic interest has increased in recent years toward the development of new PDI derivatives focusing on biological applications [11–13], and their photochemical properties are now widely exploited for developing novel systems for applications in bioimaging, photodynamic therapy (PDT) and photothermal therapy (PTT) [14,15]. Certainly, organic chemistry has played a pivotal role in the development of synthetic strategies for the advancement of such applications [16]. Indeed, it is well-established that optoelectrical properties can be modified significantly by the introduction of substituents in the bay (1, 6, 7 and 12) and ortho (2, 5, 8 and 11) positions. Whereas PDI derivatives were firstly reported in 1913 [17], transformation in the thionated analogs by the substitution of the oxygen atoms of both imide groups with sulfur atoms was only reported and patented almost a century later [18]. It has been demonstrated that the optoelectronic properties depend directly on the degree of thionation and the high potential of these thionated PDIs for specific applications in materials science. In the first part, we present an overview of the methods to synthesize thionated PDIs, showing that this post-functionalization of PDIs has been almost exclusively limited to the use of Lawesson's reagent to transform imide groups into thioimide groups. In the second part, we describe our initial research into the development of new thionation methods with the aim of obtaining greater selectivity, given that a mixture of monosubstituted **PDI-1S**, disubstituted **PDI-2S-cis** and **PDI-2S-trans**, trisubstituted **PDI-3S** and tetrasubstituted **PDI-4S** is conventionally obtained.

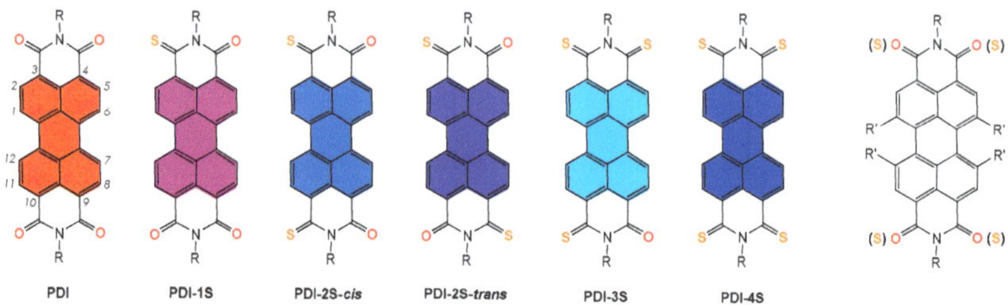

Figure 1. Representation of perylenediimide (PDI) and its thionated derivatives.

2. Overview of the Synthesis of Thionated Perylenediimides

Thionation is a suitable and efficient method for the substitution of an oxygen atom of the carbonyl group with a sulfur atom, using a wide range of thionating reagents such as elemental sulfur (S_8) [19], hydrogen sulfide [20], phosphorus pentasulfide (P_2S_5 or its dimer phosphorus decasulfide P_4S_{10}) [21,22], Lawesson's reagent (LR) [23–26], Davy's reagent [27], Heimgartner's reagent [28], Curphey's reagent (P_4S_{10} with hexamethyldisiloxane HMDSO) [29–31], Bergman's reagent (P_4S_{10}/pyridine) [32,33], Kaushik's reagent (P_4S_{10}/Al_2O_3) [34], Bernthsen's reagent (S_8/I_2) [35], and bis(trimethylsilyl)sulfide or hexamethyldisilathiane (HMDST) (Figure 2) [36].

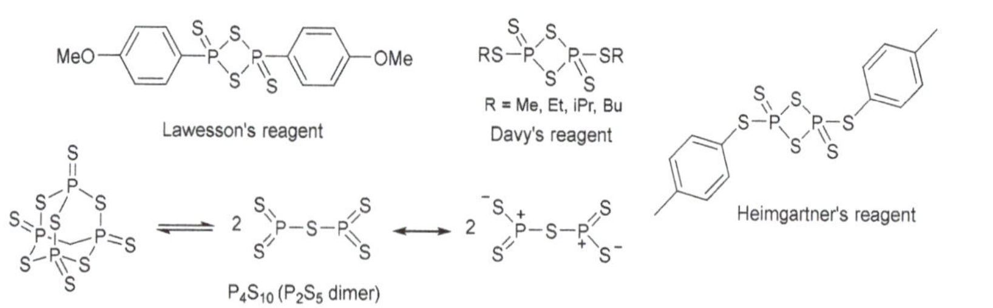

Figure 2. Usual reagents for the thionation of carbonyl groups.

Among these methods, firstly reported by Lecher et al. in 1956 [37], LR as a phosphorus–sulfur compound has been the most popular for several decades. The LR-mediated thionation reaction is widely used due to its reliability, efficiency, and compatibility with various functional groups. It was successfully applied for the conversion of alcohols, carboxylic acids, ketones, esters and amides. Typically, the thionation reaction involving LR proceeds under mild conditions, usually in an appropriate solvent, often toluene, xylene or dichloromethane. However, LR can release toxic hydrogen sulfide gas upon exposure to water or moisture and is unstable in solution at temperatures above 110 °C with slow decomposition [37,38]. The mechanism of the thionation reaction using LR involves the formation of a highly reactive dithiophosphine ylide that can react with the carbonyl function to form a thiaoxaphosphetane intermediate that evolves into a Wittig-like reaction to give the corresponding thione derivative. This mechanism clearly indicates that each equivalent of LR is capable of delivering two sulfur atoms per reaction (Scheme 1).

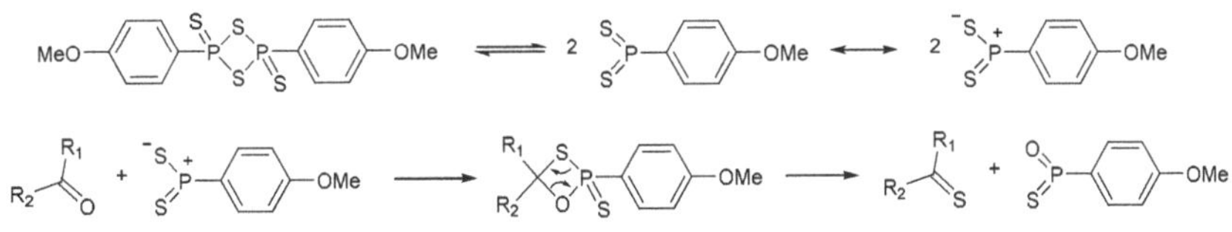

Scheme 1. Mechanism of thionation using Lawesson's reagent.

As a preamble to the development of the thionation of PDI derivatives, it is important to point out that the thionation reaction of imides has only been described in rare cases [25]. Nevertheless, the first study of the thionation reaction of cyclic imides should be highlighted [39]. Thionation in the phthalimides and naphthtalimides series using LR led to mono- and dithioimides in good yields thanks to the high polarity of carbonyl groups. But, on the other hand, steric hindrance on the nitrogen atom of the imide group was shown to strongly inhibit the replacement of the oxygen atom with a sulfur atom.

The first synthesis of thionated PDI derivatives was patented by A. Fachetti and coll [18]. Thionation was carried out starting from a PDI derivative substituted with (*S*)-1-methylhexyl chains at the imide positions and using LR in 1-methylnaphthalene at 180 °C for 30 min (Table 1, entry 1). Thionated PDI derivatives were separated by silica gel chromatography using toluene as the eluent, from the less polar **PDI-4S** (Rf = 0.92), then **PDI-3S** (Rf = 0.83), **PDI-2S-trans** (Rf = 0.67), **PDI-2S-cis** (Rf = 0.50) to the more polar **PDI-1S** (Rf = 0.23). Optical properties were determined in chloroform, showing a bathochromic shift of the λ_{onset} which increases with the number of sulfur atoms, **PDI-4S** (λ = 765 nm, E_g = 1.62 eV), **PDI-3S** (λ = 710 nm, E_g = 1.75 eV), **PDI-2S-trans** (λ = 645 nm, E_g = 1.92 eV), **PDI-2S-cis** (λ = 645 nm, E_g = 1.92 eV), and **PDI-1S** (λ = 605 nm, E_g = 2.05 eV), to be compared with PDI starting material (λ = 540 nm, E_g = 2.30 eV). Then, this study focused on the production of the **PDI-2S-trans** isomer from PDI derivatives diversely substituted on the imide position using (*R*)-1-methylheptyl (25% yield), racemic 1-methylheptyl (22% yield), 1,3-dimethylbutyl (17% yield), 2-octyldodecyl (25% yield) groups. It should be noted that Davy's reagent was investigated here as an alternative to the LR yielding **PDI-2S-cis** and **PDI-2S-trans** compounds in 26% and 27% yield, respectively (Table 1, entry 2), or to **PDI-2S-trans** in 22% yield for the 1,6 and 1,7 mixture of dicyanoPDI derivative (Table 1, entry 3) [18].

A few years later, in 2014, D. S. Seferos and coll. nicely synthesized a series of thionated PDIs using LR in refluxing toluene giving from **PDI-1S** to **PDI-3S** in 10–24% yield (Table 1, entry 4). A larger excess of LR and extended reaction time was required to attain **PDI-4S** compound in 29% yield (Table 1, entry 5) [40]. It was noted that the recrystallization of LR from toluene improved yields of the thionation reaction. Additionally, the effect of stoichiometry and the reaction time were studied by C.F.J. Faul and coll. who carried out the synthesis of thionated PDIs in order to investigate the effects of heteroatom substitution in supramolecular polymer systems (Table 1, entry 6) [41]. The best results were obtained using LR in large excess (8 equivalents) and refluxing in toluene for 48 h.

The degree of thionation can be easily determined by ^{1}H NMR spectra (Figure 3), with an exception for **PDI-2S-cis** and **PDI-2S-trans,** for which 2D NMR was required for the assignment of the two isomers [40].

Table 1. Reported experimental conditions and yields in thionated PDI derivatives (Tol.: Toluene; 1-MeNaphth.: 1-Methylnaphthalene; Xyl.: Xylene; *o*-$C_6H_4Cl_2$: *o*-dichlorobenzene; LR: Lawesson's reagent; DR: Davy's reagent; MW = Microwave).

Entry	R	R′	Experimental Conditions	Yield PDI 1S to 4S	Ref
1	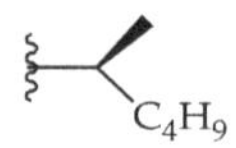	H	LR (2.7 eq.) 1-MeNaphth., 180 °C, 30 min	PDI-1S: < 5% PDI-2S cis: 20–30% PDI-2S trans: 29% PDI-3S: 1% PDI-4S: traces	[18]
2	-$CH(CH_2CH_3)_2$	H	DR (2 eq.) *o*-$C_6H_4Cl_2$, 180 °C, 7 min	PDI-2S cis: 26% PDI-2S trans: 27%	[18]
3	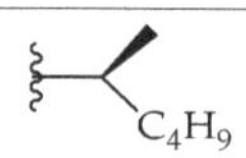	CN (1,6 and 1,7 isomers)	DR (2.2 eq.) *o*-$C_6H_4Cl_2$, 180 °C, 15 min	PDI-2S trans: 22%	[18]
4	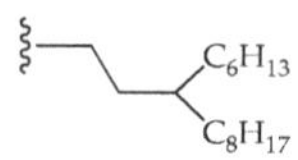	H	LR (2 eq.) Tol., 110 °C, 18 h	PDI-1S: 10% PDI-2S cis: 13% PDI-2S trans: 17% PDI-3S: 24% PDI-4S: traces	[40]
5	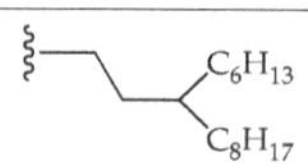	H	LR (5 eq.) Tol., 110 °C, 50 h	PDI-3S: 13% PDI-4S: 29%	[40]
6	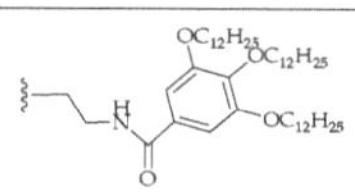	H	LR (8 eq.) Tol., 110 °C, 48 h	PDI-1S: 21% PDI-2S cis: 10% PDI-2S trans: 9%	[41]
7	-C_4H_9	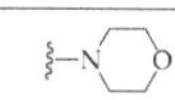 (1,6 and 1,7 isomers)	LR (6 eq.) Tol., 110 °C, 48 h	1,7-PDI-4S: 16% 1,6-PDI-4S: 17%	[42]
8	-C_6H_{11}	Br 1,7 isomer	LR (4 eq.) Tol., 85 °C, 36 h	PDI-1S: 13% and 9% PDI-2S cis: 11% PDI-2S trans: 15%	[43]
9	-$CH(C_6H_{13})_2$	H	LR (10 eq.) Tol., 110 °C, 18 h	PDI-2S cis: 5.4% PDI-2S trans: 10.9%	[44]
10	-C_8H_{17}	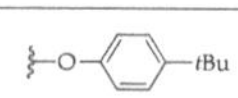 (1,6,7,12 tetra substituted)	LR (4 eq.) Xyl., MW 150 W, 103 °C, 20 min	PDI-1S: 10% PDI-2S trans: 15% PDI-3S: 19% PDI-4S: 24%	[45]
11	-$CH(C_2H_5)_2$	H	LR (6 eq.) Tol., 110 °C, overnight	PDI-1S: 20% PDI-2S cis: 30% PDI-2S trans: 35%	[46]
12	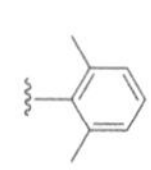	H	LR (6 eq.) Tol., 110 °C, overnight	PDI-1S: 10% PDI-2S cis: 18% PDI-2S trans: 19% PDI-3S: 15% PDI-4S: 9%	[46]
13	-C_6H_{11}	-NH-C_6H_{11} 1,7 isomer	LR (6 eq.) Tol., 110 °C, 3 days	PDI-1S: 20% PDI-2S-trans: 35% PDI-3S: 12%	[47]

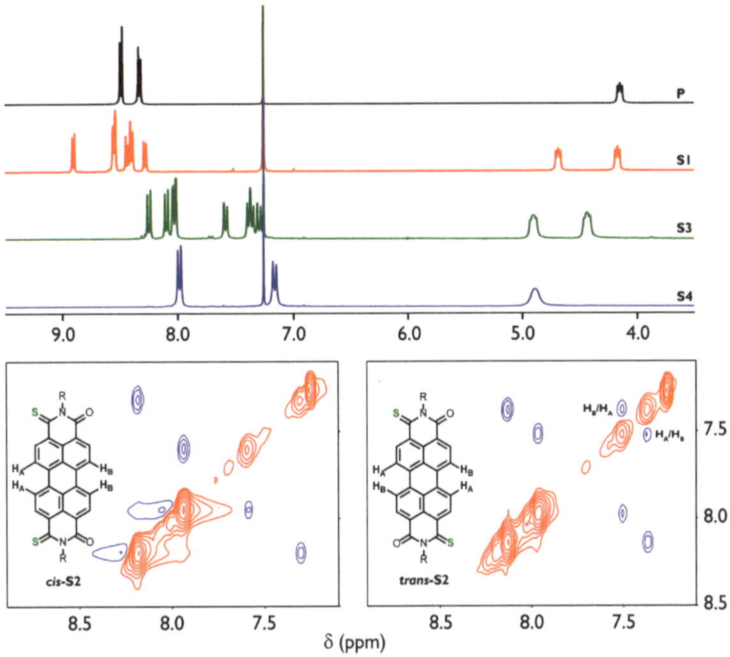

Figure 3. ^{1}H NMR spectra of the aromatic part of **PDI** (**P**) (R = 3-hexylundecyl chain) and its corresponding thionated derivatives **PDI-1S** (**S1**), **PDI-3S** (**S3**), **PDI-4S** (**S4**), and partial ROESY spectra of **PDI-2S-cis** (**cis-S2**) and **PDI-2S-trans** (**trans-S2**), recorded in $CDCl_3$ at 25 °C. Reproduced with permission from reference [40]. Copyright 2014 American Chemical Society.

All thionated PDIs are characterized by an absorption maximum ranging from 574 nm for **PDI-1S** to 706 nm for **PDI-4S** (Figure 4, Table 2), but none of them exhibit fluorescence. The presence of a single C=S functional group is sufficient to completely quench PDI emission due to a rapid and highly efficient intersystem crossing (ISC) to a triplet state, this phenomenon being independent of the degree of thionation and attributed to a reordering of the molecular electronic structure.

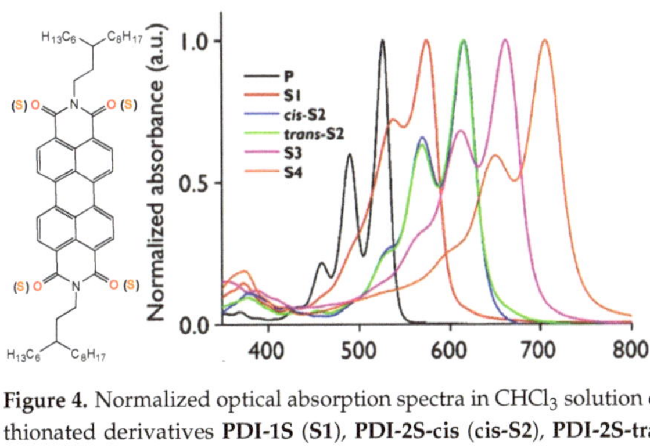

Figure 4. Normalized optical absorption spectra in $CHCl_3$ solution of **PDI** (**P**) and its corresponding thionated derivatives **PDI-1S** (**S1**), **PDI-2S-cis** (**cis-S2**), **PDI-2S-trans** (**trans-S2**), **PDI-3S** (**S3**) and **PDI-4S** (**S4**). Reproduced with permission from reference [40]. Copyright 2014 American Chemical Society.

Table 2. Optical and electrochemical data of PDI and its thionated derivatives from reference [40].

Compound	λ_{max} (nm)	Optical Band Gap (eV)	$E^{1/2}_{red1}$ (V)	$E^{1/2}_{red2}$ (V)
PDI (P)	526	2.25	−0.68	−0.91
PDI-1S (S1)	574	2.06	−0.55	−0.72
PDI-2S-cis (S2-cis)	616	1.91	−0.48	−0.57
PDI-2S-trans (S2-trans)	615	1.90	−0.51	−0.61
PDI-3S (S3)	663	1.78	−0.36	−0.45
PDI-4S (S4)	706	1.64	−0.23	−0.33

Both calculated and experimental HOMO/LUMO energies confirmed the slight increase in HOMO energies and the sharp decrease in LUMO energies with increasing sulfur atoms, which justifies the redshift of the maximum absorption (Table 3). Moreover, the influence of the sulfur atoms on the electronic structure and their significant contribution in comparison to the oxygen of diimide groups were achieved.

Table 3. Experimental and calculated (in parentheses) HOMO/LUMO energies of thionated derivatives from reference [40]. Copyright 2014 American Chemical Society.

Compound	HOMO (eV)		LUMO (eV)	
PDI (P)	−5.92 (−6.23)		−3.67 (−3.76)	
PDI-1S (S1)	−5.85 (−6.15)		−3.80 (−3.88)	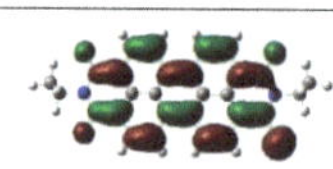
PDI-2S-cis (S2-cis)	−5.78 (−6.09)		−3.87 (−3.99)	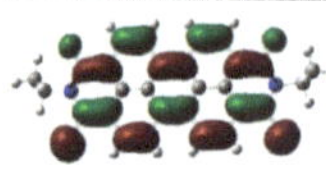
PDI-2S-trans (S2-trans)	−5.74 (−6.08)	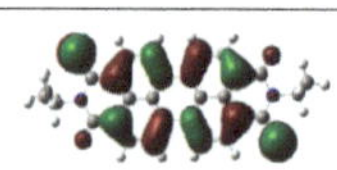	−3.84 (−3.97)	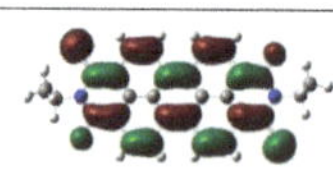
PDI-3S (S3)	−5.77 (−6.04)	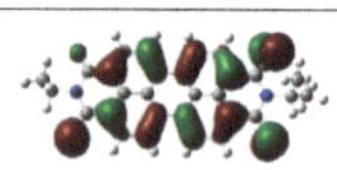	−3.99 (−4.07)	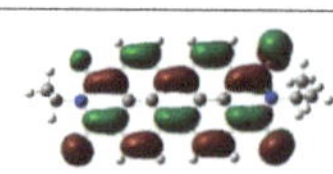
PDI-4S (S4)	−5.76 (−5.98)	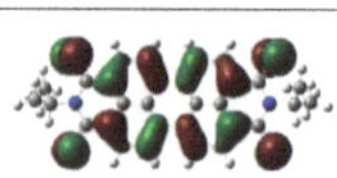	−4.12 (−4.15)	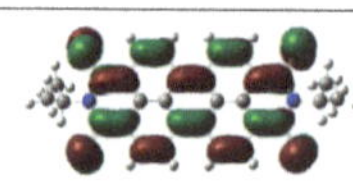

Furthermore, D.S. Seferos and coll. demonstrated that thionation led to an increase in thin-film transistor electron mobility by two orders of magnitude from **PDI-4O** to **PDI-4S** (0.16 $cm^2 \cdot V^{-1} \cdot s^{-1}$) [48]. This synthetic strategy was later expanded in the naphthalenediimide (NDI) series to attain S1 to S4 compounds, replacing the branched 3-hexylundecyl chain by a linear dodecyl chain for studying the influence on solid-state packing [49]. Interestingly, the rate and extent of thionation was increased by heating the reaction mixture more efficiently and at higher temperatures using microwave irradiation to reach thionated NDI derivatives [50].

While the introduction of electron-withdrawing groups into the PDI bay region decreases the energy level of the lowest unoccupied molecular orbital (LUMO), thus increasing the n-type semiconducting character [51,52], the substitution with strong electron-donating groups induces a significant redshift in absorption combined with the fluorescence quenching arising from the electron transfer between the donor groups and the PDI framework. In further developments of thionated PDIs chemistry, N.R. Champness and coll. combined functionalization with electron-donating morpholino groups at the 1,6 and 1,7 bay positions (Table 1, entry 7) and the full thionation of the imide functions to extend absorption in the near infrared (NIR) region (864 nm for the **1,7-PDI-4S** isomer and 838 nm for the **1,6-PDI-4S** isomer in dichloromethane solution (Figure 5) [42]. Moreover, spectroelectrochemical experiments recorded in *o*-dichlorobenzene showed absorption bands at 1568 nm and 1491 nm for the anion-radical species of 1,7-isomer and 1,6-isomer, respectively. These radical anions were more extensively studied by P. Mukhopadhyay and coll. who prepared **PDI-1S** isolated as two isomers and **PDI-2S-cis** and **PDI-2S-trans** (Figure 5) starting from 1,7-dibromoPDI material (Table 1, entry 8) [43]. Remarkably, it was shown that the stability of the radical anion increases with the degree of thionation (the highest 18.8 h for **PDI-2S-trans**), with the vacant *d* orbital of the sulfur atom playing a crucial role in the delocalization of the unpaired electron, thus stabilizing the reduced species.

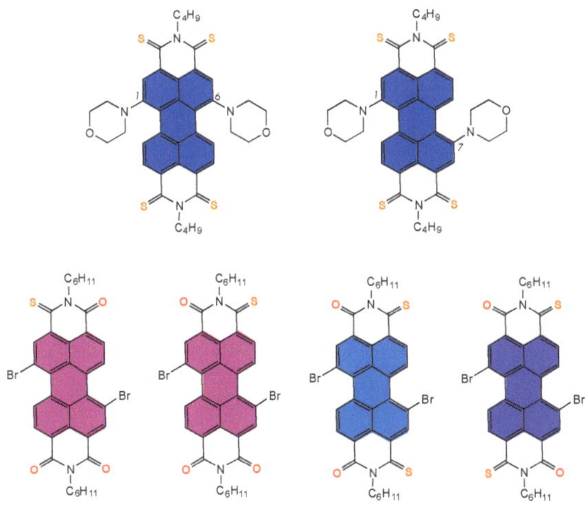

Figure 5. Syntheses of 1,6- and 1,7-isomers of dimorpholino PDI-4S [42], and PDI-1S (two isomers), PDI-2S-cis and PDI-2S-trans from 1,7-dibromoPDI [43].

While these thionated PDI derivatives were designed for use in organic electronics, their synthesis presents notable challenges, resulting in relatively average yields and low selectivity. These factors currently inhibit their large-scale synthesis. More recently, opportunities for their application in PDT and PTT have emerged and aroused growing interest. This is particularly relevant with the very recent progress made in the development of heavy-atom-free photosensitizers (HAF-PSs) because of their potential biocompatibility and prospective applications in PDT [53–58]. Indeed, these PDT materials, incorporating sulfur instead of oxygen atoms, are likely to promote ISC, leading to the generation of triplet excitons for reactive oxygen species (ROS) generation. Dithionated PDIs with 1-hexylheptyl imide chains were synthesized by Y. Huang and coll. using LR in refluxing toluene, giving **PDI-2S-cis** and **PDI-2S-trans** in 10.9% and 5.4% yields, respectively (Table 1, entry 9) [44]. Polyethylene glycol-based nanoparticles incorporating these dithionated PDIs were prepared to target tumor tissues via the enhanced permeability retention (EPR) effect. Corresponding **PDI-2S-trans**-based nanoparticles were shown to induce photothermal depression on A549 cells under 660 nm light irradiation both in vitro and in vivo, with a higher photothermal conversion efficiency (PCE) of 58.4% compared to 41.6% for **PDI-2S-cis** based nanoparticles. Furthermore, **PDI-2S-trans** based nanoparticles were shown to generate ROS upon 660 nm laser irradiation, demonstrating an inhibitory effect on tumor growth.

The thionation of the PDI backbone substituted by four 4-*tert*-butylphenoxy groups on the bay region was investigated by M. Yin and coll. (Figure 6) [45]. The synthesis was carried out using LR in xylene under microwave irradiation (Table 1, entry 10).

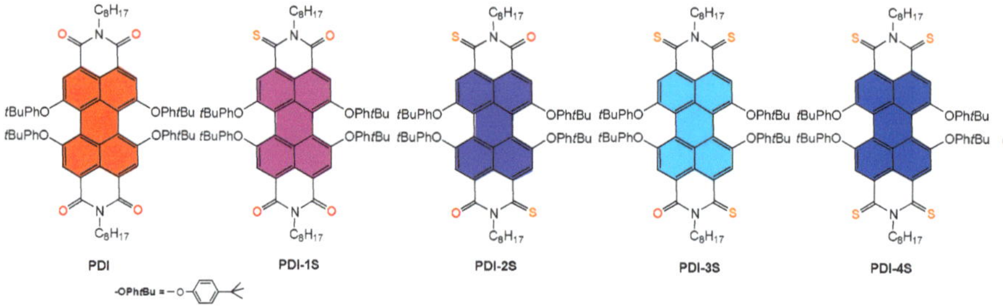

Figure 6. 1,6,7,12-tetra-*p*-*tert*butylphenoxy PDI and corresponding thionated derivatives.

Using 1,3-diphenylisobenzofuran as a probe to measure singlet oxygen ($^1O_2$) generation and comparing with the photosensitizer methylene blue standard, the $^1O_2$ quantum yields in these 4-*tert*-butylphenoxy bay-substituted PDIs were shown to gradually decrease with an increase in the thionation degree (Table 4).

Table 4. Optical data and photosensitizing properties of PDI and its thionated derivatives from reference [45]. Measurements were obtained in dichloromethane (DCM), and $^1O_2$ quantum yields were determined according to methylene blue (Φ_Δ = 57% in DCM).

Compound	λ_{max} (nm)	Optical Band Gap (eV)	Φ_{PL} (%)	Φ_Δ (%)
PDI	575	2.31	0.92	1
PDI-1S	623	2.13	-	95.6
PDI-2S	666	1.99	-	45.8
PDI-3S	719	1.81	-	11.1
PDI-4S	769	1.68	-	0.5

This study demonstrated the remarkable capabilities of thionated PDIs in various biomedical applications. By highlighting the influence of the degree of sulfur substitution on $^1O_2$ generation and photothermal conversion efficiency, as well as the link with the ISC rate constant, the in vitro experiments showed that **PDI-1S**, with its enhanced photodynamic capacity, could be used in tumor phototherapy, while **PDI-4S** might be more suitable as a photothermal and photoacoustic agent in tumor theranostics (Figure 7).

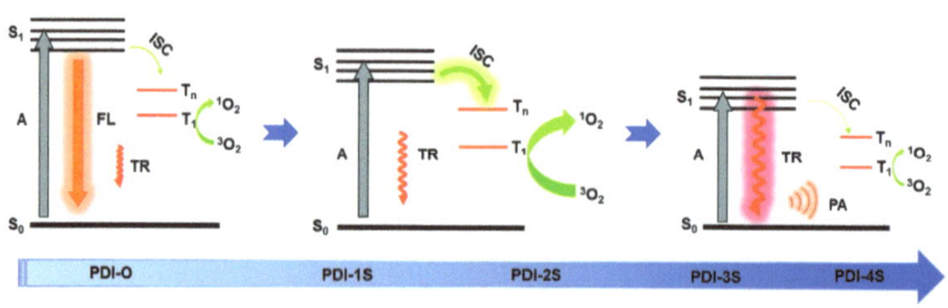

Figure 7. Representation of phototheranostic effects through the degree of PDI thionation (A: absorption, FL: fluorescence, TR: thermal radiation, ISC: intersystem crossing, PA: photoacoustic). Reproduced with permission from reference [45]. Copyright 2021 Royal Society of Chemistry.

The thionation of PDI substituted with 2-ethylpropyl groups (Table 1, entry 11) and 2,6-dimethylphenyl groups (Table 1, entry 12) on the imide positions was carried out using LR in refluxing toluene by P.T. Chou and coll [46]. It was noted from the last series that there was an increase in the extinction coefficient with the number of sulfur atoms (**PDI-1S**: λ_{max} = 575 nm, ε = 29,200 $M^{-1} \cdot cm^{-1}$ in toluene; **PDI-2S-cis**: λ_{max} = 610 nm, ε = 64,300 $M^{-1} \cdot cm^{-1}$; **PDI-2S-trans**: λ_{max} = 610 nm, ε = 47,700 $M^{-1} \cdot cm^{-1}$; **PDI-3S**: λ_{max} = 660 nm, ε = 66,200 $M^{-1} \cdot cm^{-1}$; **PDI-4S**: λ_{max} = 700 nm, ε = 97,700 $M^{-1} \cdot cm^{-1}$). In addition, the ability of **PDI-1S** to act as a photosensitiser was exploited with its coupling with FC131 and Cy5 peptides. The key synthetic step was using an LR-mediated thionation in refluxing toluene giving a **PDI-1S** intermediate in 34% yield, which was post-functionalized with the FC131 peptide grafted on each imide position affording the FC131-PDI-1S-FC131 triad (Figure 8). This thionated PDI was also linked, on one imide side, with peptide FC131, and on the other side, with cyanine5 dye, yielding an FC131-PDI-1S-Cy5 assembly. In vitro and in vivo evaluations confirmed the selectivity of these assemblies as active materials in PDT by exhibiting strong two-photon absorption and imaging capabilities of notable anticancer effects, with evidence of exceptional in vivo antitumor efficacy in A549 xenografted tumor mice.

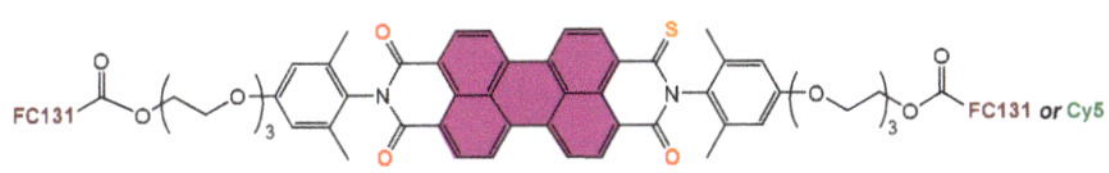

FC131-PDI-1S-FC131 or FC131-PDI-1S-Cy5

Figure 8. Structures of FC131-PDI-1S-FC131 and FC131-PDI-1S-Cy5 assemblies.

As previously mentioned, the introduction of the less electronegative sulfur atoms on the imide groups leads to a significant extension of the absorption in the long wavelength region. The combination of this phenomenon with an intramolecular charge transfer (ICT) resulting from the introduction of amino groups in the bay position should induce a synergistic shift towards the NIR spectrum. Density Functional Theory (DFT) calculations of the tetrathionated PDI-bearing cyclohexylamino groups in the 1,7 positions showed the high impact of thionation on the LUMO energy level, with a maximum absorption band calculated at 1085 nm (LUMO: −3.72 eV) to be compared with the tetraoxygenated PDI analog presenting an absorption band at 732 nm (LUMO: −3.10 eV), the latter presenting a maximum absorption at 654 nm in dichloromethane solution [59]. Sun and coll. demonstrated this phenomenon with the synthesis of corresponding **PDI-1S** (λ_{max} = 746 nm in DCM), **PDI-2S-trans** (λ_{max} = 795 nm) and **PDI-3S** compounds, the latter exhibiting a remarkable λ_{max} = 854 nm (Table 1, entry 13) [47]. The authors showed $^1O_2$ generation by these PDIs under 650 nm laser irradiation. In order to solve the problem of solubility in water, silica nanocapsules (SNCs) with encapsulated PDIs were formulated. These **PDI-3S@SNC** exhibited a remarkable power conversion efficiency (PCE) reaching 88% under 808 nm laser irradiation. Additionally, an exceptional photothermal effect under 1064 nm laser irradiation was observed, highlighting its potential as an NIR photothermal agent.

3. Investigations of Novel Reagents in Thionated Perylenediimides Synthesis

Lawesson's reagent (LR) is a widely used tool for the synthesis of thionated compounds, and especially thionated PDIs. Our primary objective was to find an effective alternative for enhancing selectivity and yields for the less available **PDI-3S** and **PDI-4S** compounds. At that point, we decided to investigate this thionation reaction, focusing our research on the phosphorus pentasulfide (P_2S_5 or its dimer phosphorus decasulfide P_4S_{10}) reagent. LR and P_4S_{10} are both commonly used reagents in organic synthesis for converting carbonyl compounds to thiocarbonyl analogs. While both reagents serve a similar purpose, some advantages of using LR over P_4S_{10} are commonly described. It is often noted that LR proceeds in milder reaction conditions and in a shorter reaction time compared to P_4S_{10}. Moreover, LR is described to provide cleaner reactions with fewer side products compared to P_4S_{10}, leading to higher yields and easier purification of the desired product. Concerning the functional group compatibility, LR is generally more compatible with a wider range of functional groups present in the substrate molecule compared to P_4S_{10}. For instance, LR reagent is less likely to react with sensitive functional groups such as esters and amides. Finally, LR has a relatively milder odor compared to P_4S_{10}, making it more "pleasant" to work with in the laboratory.

The ancestor thionation reagent P_4S_{10} was first used in 1869 by Henry [60] and Wislicenus [61]. Then, A.W. Hofmann described, in 1878, the transformation of carboxamides into thionoamides, exemplified by the conversion of formanilide into thioformanilide [62]. Due to its low solubility, the reaction is normally carried out with an excess of P_4S_{10} in refluxing solvent which includes toluene, xylene, dioxan, dimethoxyethane, pyridine and dichloromethane. Furthermore, it was demonstrated that reaction times and reaction temperatures can be reduced significantly when using ultrasound for thionation reaction. Under ultrasound conditions, the use of P_4S_{10} has been reported to be more selective than LR, producing no reaction side-products [63].

In this initial investigation of a novel thionation method, several PDI derivatives were selected as starting materials (Figure 9). PDI derivatives **A** and **B** bearing 2-ethylpropyl and 2,6-dimethylphenyl groups as imide substituents were chosen for comparison with thionated compounds obtained using LR and described in the literature. These compounds were prepared by reacting perylene 3,4,9,10-tetracarboxylic dianhydride (PTCDA) with the corresponding amine in imidazole at 150 °C, then purified by column chromatography and precipitated in a mixture of dichloromethane and methanol [64,65]. Compound **C**, prepared as reported in the literature [66], was studied to verify the feasibility of the new procedure, with a hindered PDI tetrasubstituted in the bay region. In addition, this compound can be used to demonstrate the electronic impact of electron-donating tetraphenoxy groups by the mesomeric effect on the reactivity of the thionation reaction. Finally, we proposed PDI **D**, synthesized according to the described procedure [65], bearing a strong electron-withdrawing nitro group in the bay region that is also extremely sensitive to substitution.

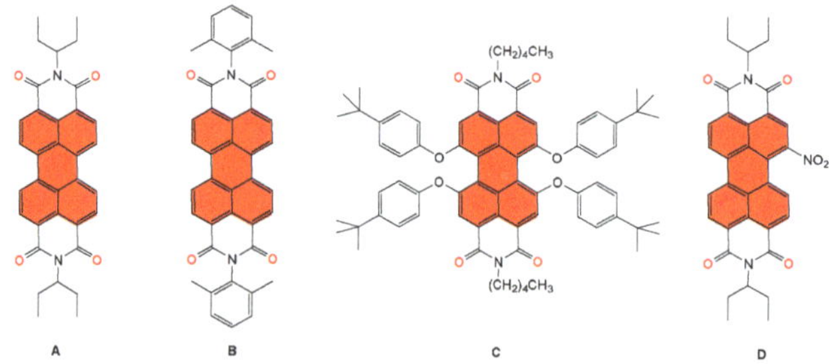

Figure 9. PDI derivatives diversely substituted in the imide and bay positions used in this study.

Initial attempts using P_4S_{10} as the sole reagent in refluxing toluene or xylene quickly proved unsuccessful when the reaction was carried out on compound **A**. So, we naturally turned to Curphey's reagent (CR), which combines P_4S_{10} and hexamethyldisiloxane (HMDSO). The combination of P_4S_{10} and HMDSO is described to efficiently convert esters, lactones, lactams and ketones to their corresponding thionated derivatives in yields comparable or superior to those obtained with LR [30]. As our main goal was to optimize the thionation reaction in order to obtain the **PDI-4S** derivative as efficiently as possible, we considered the following stoichiometry presented below (Scheme 2). In this multi-variable optimization study, 0.5 mmol of PDI was employed in anhydrous toluene or xylene (40 mL) at 110 °C or 150 °C, respectively. Consequently, the stoichiometric conditions required 0.33 mmol of P_4S_{10} and 1.66 mmol of HMDSO.

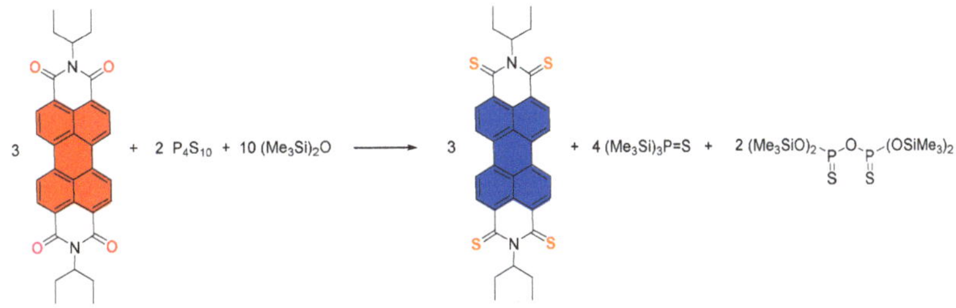

Scheme 2. Stoichiometry of the reaction converting PDI A into its corresponding thionated PDI-4S.

The first objective aimed to compare reactions carried out with LR (Table 5, entry 1) or CR (Table 5, entry 2). Under the same reaction conditions, i.e., for 24 h at reflux in toluene, the reaction carried out with LR (6 equiv.) led mainly to the formation of compound **PDI-1S** (35%) alongside small quantities of **PDI-2S-cis** (9%) and **PDI-2S-trans** (8%). The multi-thionated compounds **PDI-3S** and **PDI-4S** were not detected, in agreement with results described for this same compound **A** (Table 1, entry 11) [46]. The first investigation using P_4S_{10}/HMDSO (CR) in the ratio defined above immediately showed that CR significantly accelerated the thionation reaction. The characteristic purple color of the **PDI-1S** compound appeared after about 30 min in refluxing toluene, followed rapidly by a blue color, indicating the formation of multi-thionated compounds. This greater reactivity was confirmed by the absence of starting product **PDI A** at the end of the reaction. Regarding work-up, phosphorus-containing by-products were removed by a mild hydrolysis using a 5.3 M K_2CO_3 aqueous solution, according to the literature [31], followed by extraction with chloroform before purification by silica gel chromatography. Elution was carried out using toluene as the eluent affording firstly **PDI-4S** (Rf = 0.96), then **PDI-3S** (Rf = 0.75), **PDI-2S-trans** (Rf = 0.54) and **PDI-2S-cis** (Rf = 0.22), the **PDI-2S-trans** (C_{2h}) isomer being less polar than the **PDI-2S-cis** (C_{2v}) isomer. The **PDI-1S** derivative (Rf = 0.10 in toluene) was obtained after elution using toluene/EtOAc (95:5) as a mixture of solvents. Whereas **PDI-1S** was isolated in only 4% yield, compounds **PDI-2S-cis** and **PDI-2S-trans** were each obtained in around 40% yield, with at their side **PDI-3S** in 8% yield and **PDI-4S** as traces (Table 5, entry 2). It should be noted that **PDI-2S-gem** was detected by thin-layer chromatography (TLC) but could not be isolated (see TLC in Supporting Information).

After obtaining this initial positive result demonstrating the potential of CR as reagent, the influence of the solvent and temperature reaction was studied. Using xylene at 150 °C under identical stoichiometric conditions, but in a Schlenk flask to take account of the lower boiling point of HMDSO compared to xylene, a significant improvement was observed in the production of the **PDI-3S** compound (17% yield) (Table 5, entry 3). This improvement could also be observed when studying the effect of stoichiometry on the reaction, using a small excess of P_4S_{10} and HMDSO (Table 5, entry 4) or, more importantly, increasing the quantity of HMDSO (Table 5, entry 5). In the last case, a higher yield (27%) in **PDI-3S** was obtained. However, further increasing the quantity of reagents (Table 5, entry 6) and the reaction time (Table 5, entry 7) quickly revealed certain limitations with a reduction of the overall yield, probably resulting from the degradation process. The successive additions of excess reagent did not significantly improve the reaction efficiency (Table 5, entry 8). Bis(trimethylsilyl)sulfide or hexamethyldisilathiane (HMDST) has been reported as a versatile reagent in the transformation of carbonyl compounds into their thioxo analogs [36]. To our knowledge, reagents P_4S_{10} and HMDST have never been associated with carrying out a thionation reaction. This reagent was tested under experimental conditions (Table 5, entry 9) allowing for comparison with results obtained using CR (Table 5, entry 4); however, the yields obtained were inconclusive. This initial investigation involving **PDI A** clearly shows the enhanced reactivity of CR compared with LR. It enabled the synthesis of a **PDI-3S** derivative in significant yields, a compound which had not been described until now. Nevertheless, compound **PDI-4S** could only be isolated in trace amounts.

Table 5. Experimental results from reactions carried out using 0.5 mmol of PDI **A**, **B**, **C** or **D** in toluene (Tol.) at 110 °C or xylene (Xyl.) at 150 °C. (nd = not detected).

<table>
<tr><th>Entry</th><th>PDI</th><th>Reagent</th><th>Solvent, Time</th><th colspan="6">Yields (%)</th></tr>
<tr><th></th><th></th><th></th><th></th><th>PDI</th><th>PDI-1S</th><th>PDI-2S-cis</th><th>PDI-2S-trans</th><th>PDI-3S</th><th>PDI-4S</th></tr>
<tr><td>1</td><td>A</td><td>LR (3 mmol)</td><td>Δ Tol., 24 h</td><td>21</td><td>35</td><td>9</td><td>8</td><td>nd</td><td>nd</td></tr>
<tr><td>2</td><td>A</td><td>P_4S_{10} (0.33 mmol)
HMDSO (1.66 mmol)</td><td>Δ Tol., 24 h</td><td>trace</td><td>4</td><td>40</td><td>42</td><td>8</td><td>trace</td></tr>
<tr><td>3</td><td>A</td><td>P_4S_{10} (0.33 mmol)
HMDSO (1.66 mmol)</td><td>Δ Xyl., 24 h</td><td>nd</td><td>5</td><td>33</td><td>31</td><td>17</td><td>trace</td></tr>
<tr><td>4</td><td>A</td><td>P_4S_{10} (0.75 mmol)
HMDSO (3.75 mmol)</td><td>Δ Tol., 24 h</td><td>nd</td><td>trace</td><td>28</td><td>29</td><td>20</td><td>trace</td></tr>
<tr><td>5</td><td>A</td><td>P_4S_{10} (0.75 mmol)
HMDSO (7.5 mmol)</td><td>Δ Tol., 24 h</td><td>nd</td><td>1</td><td>24</td><td>23</td><td>27</td><td>trace</td></tr>
<tr><td>6</td><td>A</td><td>P_4S_{10} (1.5 mmol)
HMDSO (15 mmol)</td><td>Δ Tol., 24 h</td><td>nd</td><td>nd</td><td>12</td><td>15</td><td>7</td><td>trace</td></tr>
<tr><td>7</td><td>A</td><td>P_4S_{10} (1.5 mmol)
HMDSO (15 mmol)</td><td>Δ Tol., 120 h</td><td>nd</td><td>nd</td><td>11</td><td>18</td><td>9</td><td>trace</td></tr>
<tr><td>8</td><td>A</td><td>P_4S_{10} (3 × 0.33 mmol)
HMDSO (3 × 1.66 mmol)</td><td>Δ Tol., 3 × 24 h</td><td>nd</td><td>2</td><td>27</td><td>28</td><td>16</td><td>trace</td></tr>
<tr><td>9</td><td>A</td><td>P_4S_{10} (0.75 mmol)
HMDST (3.75 mmol))</td><td>Δ Tol., 24 h</td><td>2</td><td>13</td><td>18</td><td>23</td><td>1</td><td>trace</td></tr>
<tr><td>10</td><td>B</td><td>LR (3 mmol)</td><td>Δ Tol., 24 h</td><td>trace</td><td>13</td><td>19</td><td>15</td><td>8</td><td>5</td></tr>
<tr><td>11</td><td>B</td><td>P_4S_{10} (0.33 mmol)
HMDSO (1.66 mmol)</td><td>Δ Tol., 24 h</td><td>nd</td><td>11</td><td>14</td><td>17</td><td>20</td><td>12</td></tr>
<tr><td>12</td><td>B</td><td>P_4S_{10} (0.75 mmol)
HMDSO (7.5 mmol)</td><td>Δ Tol., 24 h</td><td>nd</td><td>trace</td><td>2</td><td>2</td><td>18</td><td>32</td></tr>
<tr><td>13</td><td>C</td><td>P_4S_{10} (0.75 mmol)
HMDSO (7.5 mmol)</td><td>Δ Tol., 24 h</td><td>nd</td><td>trace</td><td>trace</td><td>trace</td><td>trace</td><td>89</td></tr>
<tr><td>14</td><td>C</td><td>P_4S_{10} (1 mmol)
HMDSO (10 mmol)</td><td>Δ Tol., 24 h</td><td>nd</td><td>trace</td><td>trace</td><td>trace</td><td>trace</td><td>84</td></tr>
<tr><td>15</td><td>D</td><td>P_4S_{10} (0.75 mmol)
HMDSO (7.5 mmol)</td><td>Δ Tol., 24 h</td><td>nd</td><td>13</td><td colspan="2">16</td><td>trace</td><td>nd</td></tr>
</table>

The optical properties of these thionated PDI derivatives were determined in dichloromethane solution, showing the redshift of the maximum absorption with an increase in the number of sulfur atoms (Figure 10a). Corresponding absorption maxima λ_{max} were determined for **PDI A** (525 nm), **PDI-1S** (572 nm), **PDI-2S-cis** (612 nm), **PDI-2S-trans** (612 nm), **PDI-3S** (653 nm), and **PDI-4S** (697 nm).

Moreover, we observed that the purple and blue spots on the TLC plate were transformed into orange-red spots, partially regaining their fluorescent properties under light and ambient air conditions. Then, solutions were prepared from starting material **PDI A** and corresponding thionated derivatives from **PDI-1S** to **PDI-4S** in dichloromethane and those saturated with oxygen before sunlight irradiation (Figure 11). We observed the rapid disappearance of the magenta color for **PDI-1S**, along with the blue colors of **PDI-2S-cis** and **PDI-2S-trans** isomers, and also **PDI-3S**. The **PDI-4S** derivative seemed to be much less sensitive to these conditions, resulting in photochemical degradation. This naked eye observation could be paralleled by the measurements of $^1O_2$ quantum yields described above (Table 4). Mass spectra and UV–visible spectra (Figure 10b) of these samples confirmed the instability in the proposed conditions and the formation of new products, including a return to the **PDI A** starting material resulting from an exchange between the sulfur and oxygen atoms.

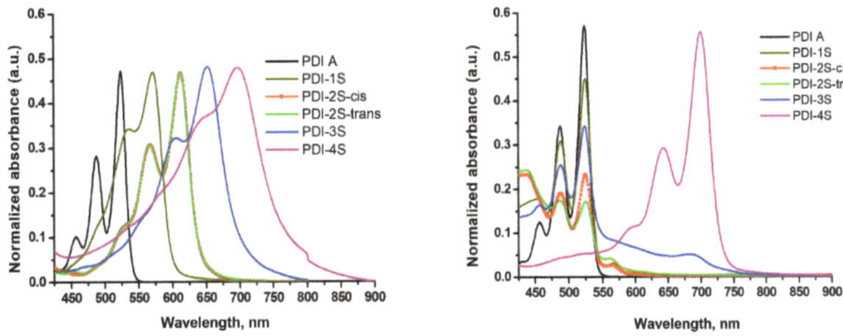

(a) (b)

Figure 10. Normalized absorption spectra of PDI A and its thionated PDI derivatives recorded in dichloromethane solution at (**a**) t = 0; (**b**) after 90 h under sunlight in a saturated oxygen atmosphere.

(a) (b)

(c) (d)

Figure 11. Behavior of PDI A and its thionated derivatives (from left to right: PDI A, PDI-1S, PDI-2S-cis, PDI-2S-trans, PDI-3S, PDI-4S) in dichloromethane under sunlight in a saturated oxygen atmosphere: (**a**) t = 0; (**b**) t = 2 h; (**c**) t = 24 h; (**d**) t = 90 h.

The comparison of LR and CR reagents was continued using a PDI **B** compound. The first observation was that the reaction of **PDI B** with LR (Table 5, entry 10) was more efficient than the similar reaction with **PDI A** (Table 5, entry 1). Starting material was quasi-completely converted into thionated PDIs with moderate yields in **PDI-3S** (8%) and **PDI-4S** (5%). These yields were significantly increased as soon as the CR reagent was used, with yields more than doubling in **PDI-3S** (20%) and **PDI-4S** (12%) (Table 5, entry 11). The best results in **PDI-4S** (32%) were obtained using an excess of P_4S_{10} and HMDSO, under conditions to be compared with those obtained with **PDI A** (Table 5, entry 5). Microwave-assisted thionation of carbonyl compounds using CR was previously described, giving desired products in higher yield and shorter reaction time compared to conventional methods [67,68]. The experiment carried out on **PDI B** under MW irradiation confirmed the importance of solubility (5×10^{-2} M for MW experiment instead of 1.25×10^{-2} M for experiment reported in Table 5), since only the **PDI-1S** derivative accompanied by unreacted starting material in a large proportion was obtained.

The experimental conditions giving the highest yield in **PDI-4S** for **PDI B** (Table 5, entry 12) were applied to **PDI C** (Table 5, entry 13). Thanks to the improved solubility resulting from the torsion of the PDI backbone, which limits aggregation, and despite the steric hindrance provided by the four para *tert*-butylphenoxy groups in the bay positions, an exceptional reactivity was achieved, leading to **PDI-4S** in an 89% yield (Scheme 3). A slight increase in the quantities of reagents did not improve the yield of the reaction (Table 5, entry 14). As the **PDI-4S** derivative was the only product isolated, the other **PDI-1S**, **PDI-2S** and **PDI-3S** compounds were only detected by mass spectrometry. Once again, this result clearly demonstrates the superiority of CR over LR for obtaining multi-thionated derivatives. The single **PDI-4S** derivative obtained in the example described here should be compared with the mixture containing 24% of a similar derivative obtained using LR in xylene under MW irradiation at 103 °C (Table 1, entry 10).

Scheme 3. Synthesis of thionated PDI-4S derivative-bearing para *tert*-butylphenoxy groups in the bay positions.

The optical properties of PDI **C** and its **PDI-4S** derivative were determined by UV–Vis absorption spectroscopy in dichloromethane solution, confirming the NIR absorption of the **PDI-4S** compound (λ_{max} = 770 nm) compared with tetraoxygenated PDI **C** (λ_{max} = 577 nm). (Figure 12). It can be estimated that each replacement of an oxygen atom of the diimide moieties by a sulfur atom results in a bathochromic shift of approximately 50 nm, whatever the substitution or not on bay region. Solutions of PDI **C** and its **PDI-4S** derivatives in dichloromethane were saturated with oxygen and irradiated with sunlight. We could confirm the good photochemical stability of the **PDI-4S** derivative, as previously noted, for tetrathionated PDI unsubstituted in the bay region.

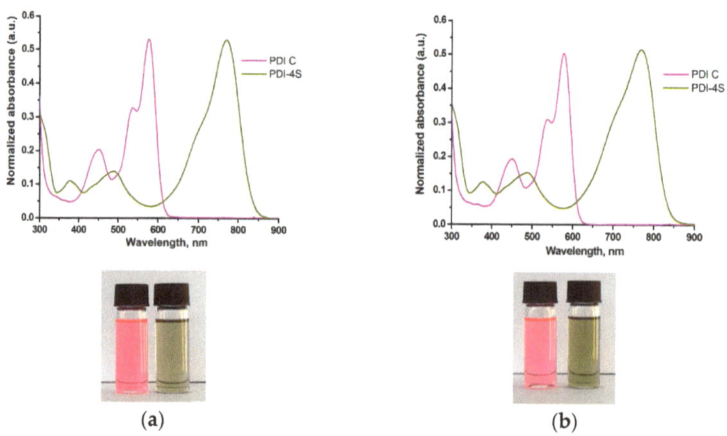

Figure 12. Normalized absorption spectra of PDI C and its PDI-4S derivative recorded in dichloromethane solution. Behavior of PDI C (pink color) and its thionated PDI-4S derivative (grey-green color) at (**a**) t = 0; (**b**) after 90 h under sunlight in a saturated oxygen atmosphere.

NitroPDIs are currently attracting significant attention due to their ease of access and higher selectivity of mononitration compared with monobromination [69]. In addition, many types of reactions have been successfully applied to nitroPDIs, ranging from nucleophilic substitution to palladium catalyzed cross-couplings [70–72]. The nitro group can also be reduced to an amino group, and further transformation in the corresponding imine using an aldehyde allows azabenzannulated PDI-based materials to be obtained by subsequent photocyclization [73,74]. Consequently, access to thionated PDI derivatives bearing a nitro group in the bay position would open the way to a whole series of post-functionalization processes that are extremely interesting from a synthetic and application point of view. The thionation reaction using CR in refluxing toluene led to a mixture of products, complexed for each compound **PDI-1S**, **PDI-2S cis**, **PDI-2S trans** and **PDI-3S** by the presence of isomers due to the nitro group in the bay position (Scheme 4). The major products isolated were **PDI-1S** and **PDI-2S** compounds, which were characterized by HRMS (Table 5, entry 15). Compound **PDI-3S** was also characterized as a trace by HRMS.

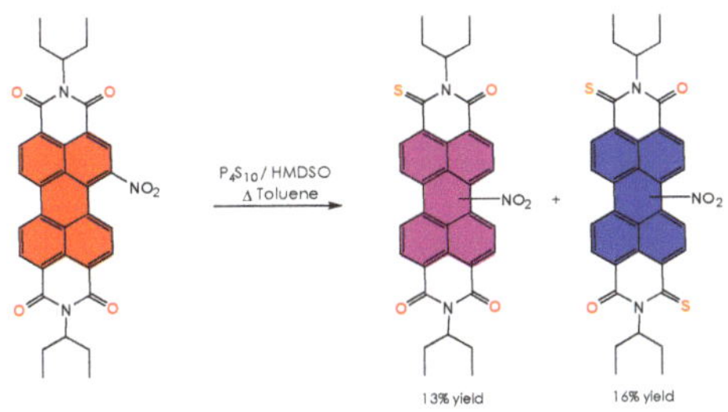

Scheme 4. Synthesis of thionated PDI-1S- and PDI-2S-bearing nitro group in bay position.

This initial study demonstrates the potential of CR in accessing thionated-PDI derivatives. The impact of the group present at the imide position has an effect on solubility and aggregation [75], and for this thionation reaction, an aromatic substituent seems more favorable than the introduction of a short alkyl chain. On the other hand, the addition of four electron-donating groups in the bay positions, despite the high steric hindrance inherent in this tetrasubstitution, provides a remarkable example of accessing the **PDI-4S** derivative in a practically quantitative yield. Other studies are currently underway to validate the influence of the electronic effect on this thionation reaction. Finally, the presence of a nitro group sensitive to substitution seems compatible with such a reaction. However, achieving selectivity becomes difficult in the case of a monosubstituted derivative in the bay position due to the presence of many possible isomers. This preliminary result seems to indicate that electron-withdrawing groups on the PDI backbone do not favor the thionation reaction. Finally, it is important to highlight that the combination of the P_4S_{10} reagent with HMDSO would make it possible, for the first time, to synthesize selenated analogues **PDI-1Se** to **PDI-4Se** using the analog P_4Se_{10} reagent [76]. Whereas remarkable intersystem crossing rates with a complete fluorescence quenching have been observed for thionated PDIs, the theoretical calculations recently reported on selenated PDIs suggest five-order larger ISC rates, signifying high efficiency for photosensitization applications [77].

4. Materials and Methods

Experimental procedure for synthesis from PDI derivatives A and B: The reaction and workup should be carried out under an efficient laboratory fume hood. To a solution of PDI (0.5 mmol) in anhydrous toluene or xylene, P_4S_{10} and HMDSO were added in stoichiometry, as presented in Table 5. The reaction mixture was heated at 110 °C (toluene) or at 150 °C (xylene) under argon atmosphere. After cooling in an ice bath, an aqueous solution of K_2CO_3 5.3 M (1 mL/0.75 mmol P_4S_{10}), then acetone (5 mL), was added. The solution was stirred for 15 min at 0 °C and poured in a separating funnel. Water (100 mL) was added, and the aqueous phase was extracted with chloroform (2 × 100 mL). The organic layer was washed with brine (100 mL), dried over $MgSO_4$ and concentrated under vacuum. The crude product was purified by silica gel column chromatography using toluene as the eluent for PDI-4S, PDI-3S, PDI-2S-trans, PDI-2S-cis, and toluene/ethyl acetate (95:5) for PDI-1S.

Experimental procedure for synthesis from PDI-derivative C: To a solution of PDI (561 mg, 0.5 mmol) in anhydrous toluene (40 mL), P_4S_{10} (334 mg, 0.75 mmol) and HMDSO 1.6 mL (7.5 mmol) were added. The reaction mixture was heated at 110 °C under argon atmosphere for 24 h. After cooling in an ice bath, an aqueous solution of K_2CO_3 5.3 M (1 mL) was added, then acetone (5 mL). The solution was stirred for 15 min at 0 °C and poured in a separating funnel. Water (100 mL) was added, and the aqueous phase was extracted with chloroform (2 × 100 mL). The organic layer was washed with brine (100 mL), dried over $MgSO_4$ and concentrated under vacuum. The crude product was purified by silica gel column chromatography using CH_2Cl_2/petroleum ether (1/1) as the mixture of eluents. Compound PDI-4S was dissolved in a minimum of CH_2Cl_2 and precipitated using MeOH before filtration, giving a green-dark powder (525 mg, 89% yield).

5. Conclusions

In conclusion, the current state of the art in the synthesis of thionated perylenediimides (PDIs) clearly shows that the methods used rely almost exclusively on Lawesson's reagent. However, the reactivity of this reagent seems limited, particularly in terms of access to the multi-thionated derivatives PDI-3S and PDI-4S. Here, we describe a possible alternative using the Curphey's reagent which combines the P_2S_5 reagent (in dimeric form P_4S_{10}) with hexamethyldisiloxane (HMDSO). This initial study clearly demonstrated higher reactivity than Lawesson's reagent. While direct access to fully thionated PDI-4S appears challenging for PDI derivatives unsubstituted on the perylene backbone, the incorporation of four para *tert*-butylphenoxy substituents in the bay region showed remarkable efficiency with this new synthetic method. In addition, this thionation method seems compatible with the presence of the substitution-sensitive nitro group grafted in the bay position. This preliminary work clearly paves the way for an effective alternative to Lawesson's reagent in the synthesis of thionated PDIs with possible new developments to access these electron acceptors on a larger scale for applications in organic electronics and biomedicine, especially as metal-free photosensitizers in photodynamic therapy or photothermal therapy.

Supplementary Materials: The following supporting information can be downloaded at: https://www.mdpi.com/article/10.3390/molecules29112538/s1. Figures S1–S22: ^{1}H, ^{13}C and HRMS spectra of thionated perylenediimides; Figures S23–S24: Photochemical degradation study.

Author Contributions: All co-authors (O.K. (Oksana Kharchenko), A.H., O.K. (Oksana Krupka), P.H.) carried out the experimental work described in the manuscript. O.K. (Oksana Krupka) and P.H. participated equally to the redaction of the manuscript and the project coordination. All authors have read and agreed to the published version of the manuscript.

Funding: Authors are grateful to the University of Angers for financial support. Oksana Krupka thanks the Agence Nationale de la Recherche (ANR) for the Chair Professor Junior support towards the project ANR-22-CPJ1-0026-01. This work received financial support from the European Union through the MSCA4Ukraine (ID 1233281) project for Oksana Kharchenko and also the EUR LUMOMAT project with the Investments for the Future program ANR-18-EURE-0012 and the Erasmus+ 2022-KA171 program for Anna Hryniuk as Master LUMOMAT student.

Institutional Review Board Statement: Not applicable.

Informed Consent Statement: Not applicable.

Data Availability Statement: Not applicable.

Acknowledgments: The authors acknowledge the University of Angers and SFR MATRIX for the access to the CARMA platform, Ingrid Freuze for Mass Spectrometry measurements, Benjamin Siegler for NMR experiments.

Conflicts of Interest: The authors declare no conflicts of interest.

References

1. Stępień, M.; Gońka, E.; Żyła, M.; Sprutta, N. Heterocyclic Nanographenes and Other Polycyclic Heteroaromatic Compounds: Synthetic Routes, Properties, and Applications. *Chem. Rev.* **2017**, *117*, 3479–3716. [CrossRef]
2. Borissov, A.; Maurya, Y.K.; Moshniaha, L.; Wong, W.-S.; Żyła-Karwowska, M.; Stępień, M. Recent Advances in Heterocyclic Nanographenes and Other Polycyclic Heteroaromatic Compounds. *Chem. Rev.* **2022**, *122*, 565–788. [CrossRef] [PubMed]
3. Zang, L.; Che, Y.; Moore, J.S. One-Dimensional Self-Assembly of Planar π-Conjugated Molecules: Adaptable Building Blocks for Organic Nanodevices. *Acc. Chem. Res.* **2008**, *41*, 1596–1608. [CrossRef] [PubMed]
4. Nowak-Król, A.; Shoyama, K.; Stolte, M.; Würthner, F. Naphthalene and perylene diimides–better alternatives to fullerenes for organic electronics? *Chem. Commun.* **2018**, *54*, 13763–13772. [CrossRef] [PubMed]
5. Quinn, J.T.E.; Zhu, J.; Li, X.; Wang, J.; Li, Y. Recent progress in the development of n-type organic semiconductors for organic field effect transistors. *J. Mater. Chem. C* **2017**, *5*, 8654–8681. [CrossRef]
6. Qin, Y.; Li, G.; Qi, T.; Huang, H. Aromatic imide/amide-based organic small-molecule emitters for organic light-emitting diodes. *Mater. Chem. Front.* **2020**, *4*, 1554–1568. [CrossRef]
7. Zink-Lorre, N.; Font-Sanchis, E.; Sastre-Santos, Á.; Fernández-Lázaro, F. Perylenediimides as more than just non-fullerene acceptors: Versatile components in organic, hybrid and perovskite solar cells. *Chem. Commun.* **2020**, *56*, 3824–3838. [CrossRef] [PubMed]
8. Sharma, V.; Koenig, J.D.B.; Welch, G.C. Perylene diimide based non-fullerene acceptors: Top performers and an emerging class featuring N-annulation. *J. Mater. Chem. A* **2021**, *9*, 6775–6789. [CrossRef]
9. Murugan, P.; Ravindran, E.; Sangeetha, V.; Liu, S.-Y.; Jung, J.W. Perylene-diimide for organic solar cells: Current scenario and prospects in molecular geometric, functionalization, and optoelectronic properties. *J. Mater. Chem. A* **2023**, *11*, 26393–26425. [CrossRef]
10. Akash; Tiwari, J.P. Recent advancements in perylene diimide as an electron acceptor in organic solar cells. *J. Mater. Chem. C* **2024**, *12*, 838–853. [CrossRef]
11. Görl, D.; Zhang, X.; Würthner, F. Molecular Assemblies of Perylene Bisimide Dyes in Water. *Angew. Chem. Int. Ed.* **2012**, *51*, 6328–6348. [CrossRef]
12. Sun, M.; Müllen, K.; Yin, M. Water-soluble perylenediimides: Design concepts and biological applications. *Chem. Soc. Rev.* **2016**, *45*, 1513–1528. [CrossRef]
13. Zhang, X.; Rehm, S.; Safont-Sempere, M.M.; Würthner, F. Vesicular perylene dye nanocapsules as supramolecular fluorescent pH sensor systems. *Nat. Chem.* **2009**, *1*, 623–629. [CrossRef]
14. Krupka, O.; Hudhomme, P. Recent Advances in Applications of Fluorescent Perylenediimide and Perylenemonoimide Dyes in Bioimaging, Photothermal and Photodynamic Therapy. *Int. J. Mol. Sci.* **2023**, *24*, 6308. [CrossRef]
15. Sun, H.; Zhang, Q. Recent Advances in Perylene Diimides (PDI)-based Small Molecules Used for Emission and Photothermal Conversion. *ChemPhotoChem* **2024**, *8*, e202300213. [CrossRef]
16. Nowak-Król, A.; Würthner, F. Progress in the synthesis of perylene bisimide dyes. *Org. Chem. Front.* **2019**, *6*, 1272–1318. [CrossRef]
17. Kardos, M. Über einige Aceanthrenchinon- und 1.9-Anthracen-Derivate. *Berichte Dtsch. Chem. Ges.* **1913**, *46*, 2086–2091. [CrossRef]
18. Quinn, J.; Zheng, Y.; Chen, Z.; Usta, H.; Newman, C.; Yan, H.; Facchetti, A. Organic Semiconductors and Devices Incorporating Same. U.S. Patent 8.440,828 B2, 29 December 2010.
19. Shibahara, F.; Sugiura, R.; Murai, T. Direct Thionation and Selenation of Amides Using Elemental Sulfur and Selenium and Hydrochlorosilanes in the Presence of Amines. *Org. Lett.* **2009**, *11*, 3064–3067. [CrossRef]
20. McGregor, W.M.; Sherrington, D.C. Some recent synthetic routes to thioketones and thioaldehydes. *Chem. Soc. Rev.* **1993**, *22*, 199–204. [CrossRef]
21. Ozturk, T.; Ertas, E.; Mert, O. A Berzelius Reagent, Phosphorus Decasulfide (P4S10), in Organic Syntheses. *Chem. Rev.* **2010**, *110*, 3419–3478. [CrossRef] [PubMed]
22. Kaur, N. Phosphorus Pentasulfide in Heterocycle Synthesis. In Lawesson's Reagent in Heterocycle Synthesis, Springer: Singapore, 2021; pp. 245–306.
23. Cava, M.P.; Levinson, M.I. Thionation reactions of lawesson's reagents. *Tetrahedron* **1985**, *41*, 5061–5087. [CrossRef]
24. Jesberger, M.; Davis, T.P.; Barner, L. Applications of Lawesson's Reagent in Organic and Organometallic Syntheses. *Synthesis* **2003**, *2003*, 1929–1958. [CrossRef]
25. Ozturk, T.; Ertas, E.; Mert, O. Use of Lawesson's Reagent in Organic Syntheses. *Chem. Rev.* **2007**, *107*, 5210–5278. [CrossRef] [PubMed]
26. Khatoon, H.; Abdulmalek, E. A Focused Review of Synthetic Applications of Lawesson's Reagent in Organic Synthesis. *Molecules* **2021**, *26*, 6937. [CrossRef] [PubMed]
27. Jackson, Y.A.; Rajagopal, D.; Bendolph, J.; Guillory, M.; Lakshmikantham, M.V.; Yang, J.; Cava, M.P. Thiophene Isosteres of 9,10-Dithioanthraquinone. *Org. Lett.* **2003**, *5*, 1883–1885. [CrossRef] [PubMed]
28. Wipf, P.; Jenny, C.; Heimgartner, H. 2,4-Bis(4-methylpheylthio)-1,3,2λ5,4λ5-dithiadiphosphetan-2,4-dithion: Ein neues Reagens zur Schwefelung von N,N-disubstituierten Amiden. *Helv. Chim. Acta* **1987**, *70*, 1001–1011. [CrossRef]
29. Curphey, T.J. A superior procedure for the conversion of 3-oxoesters to 3H-1,2-dithiole-3-thiones. *Tetrahedron Lett.* **2000**, *41*, 9963–9966. [CrossRef]

30. Curphey, T.J. Thionation of esters and lactones with the reagent combination of phosphorus pentasulfide and hexamethyldisiloxane. *Tetrahedron Lett.* **2002**, *43*, 371–373. [CrossRef]
31. Curphey, T.J. Thionation with the Reagent Combination of Phosphorus Pentasulfide and Hexamethyldisiloxane. *J. Org. Chem.* **2002**, *67*, 6461–6473. [CrossRef]
32. Bergman, J.; Pettersson, B.; Hasimbegovic, V.; Svensson, P.H. Thionations using a P4S10-pyridine complex in solvents such as acetonitrile and dimethyl sulfone. *J. Org. Chem.* **2011**, *76*, 1546–1553. [CrossRef]
33. Kingi, N.; Bergman, J. Thionation of Tryptanthrin, Rutaecarpine, and Related Molecules with a Reagent Prepared from P4S10 and Pyridine. *J. Org. Chem.* **2016**, *81*, 7711–7716. [CrossRef] [PubMed]
34. Polshettiwar, V.; Kaushik, M.P. A new, efficient and simple method for the thionation of ketones to thioketones using P_4S_{10}/Al_2O_3. *Tetrahedron Lett.* **2004**, *45*, 6255–6257. [CrossRef]
35. Morel, S.; Chatel, F.; Boyer, G.; Galy, J.-P. Synthesis of New Cyclopenta-acridinone and -phenothiazine Derivatives. *J. Chem. Res. (S)* **1998**, *1*, 4–5. [CrossRef]
36. Degl'Innocenti, A.; Capperucci, A.; Nocentini, T.; Castagnoli, G.; Malesci, I.; Cerreti, A. HMDST as Useful Tool in Organic Synthesis: A Further Step in the Delivery of Sulfur Functionalities. *Phosphorus Sulfur Silicon Relat. Elem.* **2005**, *180*, 1247–1251. [CrossRef]
37. Lecher, H.Z.; Greenwood, R.A.; Whitehouse, K.C.; Chao, T.H. The Phosphonation of Aromatic Compounds with Phosphorus Pentasulfide. *J. Am. Chem. Soc.* **1956**, *78*, 5018–5022. [CrossRef]
38. Perregaard, J.; Scheibye, S.; Meyer, H.J.; Thomsen, I.; Lawesson, S.-O. Studies on Organophosphorus Compounds XVIII. Oxidation of Tertiary Alicyclic Amines with Elemental Sulfur in Hexamethylphosphoric Triamide (HMPA). Oxidative Rearrangements of Hexahydroazepines and Octahydroazocines to Bis(3-Pyrrolyl)Polysulfides. *Bull. Soc. Chim. Belg.* **1977**, *86*, 679–691. [CrossRef]
39. Orzeszko, A.; Maurin, J.K.; Melon-Ksyta, D. Investigation of the Thionation Reaction of Cyclic Imides. *Z. Naturforschung B* **2001**, *56*, 1035–1040. [CrossRef]
40. Tilley, A.J.; Pensack, R.D.; Lee, T.S.; Djukic, B.; Scholes, G.D.; Seferos, D.S. Ultrafast Triplet Formation in Thionated Perylene Diimides. *J. Phys. Chem. C* **2014**, *118*, 9996–10004. [CrossRef]
41. Symons, H.E.; Hagemann, M.J.L.; Harniman, R.L.; Faul, C.F.J. Thionated PDI supramolecular polymers: Controlling aggregation mechanisms, morphology and function. *J. Mater. Chem. C* **2022**, *10*, 2828–2837. [CrossRef]
42. Llewellyn, B.A.; Davies, E.S.; Pfeiffer, C.R.; Cooper, M.; Lewis, W.; Champness, N.R. Thionated perylene diimides with intense absorbance in the near-IR. *Chem. Commun.* **2016**, *52*, 2099–2102. [CrossRef]
43. Mandal, K.; Yadav, D.; Saini, P.; Mukhopadhyay, P. Synthesis, optical and redox attributes of core-/bay-substituted thionated NDIs, PDIs and their diverse radical anions. *J. Mater. Chem. C* **2023**, *11*, 12543–12549. [CrossRef]
44. Liu, Z.; Gao, Y.; Jin, X.; Deng, Q.; Yin, Z.; Tong, S.; Qing, W.; Huang, Y. Regioisomer-manipulating thio-perylenediimide nanoagents for photothermal/photodynamic theranostics. *J. Mater. Chem. B* **2020**, *8*, 5535–5544. [CrossRef] [PubMed]
45. Liu, C.; Ji, C.; Fan, Z.; Ma, R.; Yin, M. A facile design of thio-perylenediimides with controllable fluorescent, photodynamic and photothermal effects towards cancer theranostics. *Chem. Commun.* **2021**, *57*, 13126–13129. [CrossRef] [PubMed]
46. Lee, Y.-L.; Chou, Y.-T.; Su, B.-K.; Wu, C.-c.; Wang, C.-H.; Chang, K.-H.; Ho, J.-a.A.; Chou, P.-T. Comprehensive Thione-Derived Perylene Diimides and Their Bio-Conjugation for Simultaneous Imaging, Tracking, and Targeted Photodynamic Therapy. *J. Am. Chem. Soc.* **2022**, *144*, 17249–17260. [CrossRef]
47. An, F.; Zhao, Y.; Li, H.; Meng, J.; Jiao, L.; Zhang, Z.; Li, X.; Sun, X. Intramolecular charge transfer versus intersystem crossing: The way toward super-high photothermal efficiency by thionation. *Dyes Pigment.* **2023**, *217*, 111411. [CrossRef]
48. Tilley, A.J.; Guo, C.; Miltenburg, M.B.; Schon, T.B.; Yan, H.; Li, Y.; Seferos, D.S. Thionation Enhances the Electron Mobility of Perylene Diimide for High Performance n-Channel Organic Field Effect Transistors. *Adv. Funct. Mater.* **2015**, *25*, 3321–3329. [CrossRef]
49. Kozycz, L.M.; Guo, C.; Manion, J.G.; Tilley, A.J.; Lough, A.J.; Li, Y.; Seferos, D.S. Enhanced electron mobility in crystalline thionated naphthalene diimides. *J. Mater. Chem. C* **2015**, *3*, 11505–11515. [CrossRef]
50. Pahlavanlu, P.; Tilley, A.J.; McAllister, B.T.; Seferos, D.S. Microwave Synthesis of Thionated Naphthalene Diimide-Based Small Molecules and Polymers. *J. Org. Chem.* **2017**, *82*, 12337–12345. [CrossRef]
51. Weitz, R.T.; Amsharov, K.; Zschieschang, U.; Villas, E.B.; Goswami, D.K.; Burghard, M.; Dosch, H.; Jansen, M.; Kern, K.; Klauk, H. Organic n-Channel Transistors Based on Core-Cyanated Perylene Carboxylic Diimide Derivatives. *J. Am. Chem. Soc.* **2008**, *130*, 4637–4645. [CrossRef]
52. Gsänger, M.; Oh, J.H.; Könemann, M.; Höffken, H.W.; Krause, A.-M.; Bao, Z.; Würthner, F. A Crystal-Engineered Hydrogen-Bonded Octachloroperylene Diimide with a Twisted Core: An n-Channel Organic Semiconductor. *Angew. Chem. Int. Ed.* **2010**, *49*, 740–743. [CrossRef]
53. Ortiz-Rodríguez, L.A.; Crespo-Hernández, C.E. Thionated organic compounds as emerging heavy-atom-free photodynamic therapy agents. *Chem. Sci.* **2020**, *11*, 11113–11123. [CrossRef] [PubMed]
54. Tang, J.; Wang, L.; Loredo, A.; Cole, C.; Xiao, H. Single-atom replacement as a general approach towards visible-light/near-infrared heavy-atom-free photosensitizers for photodynamic therapy. *Chem. Sci.* **2020**, *11*, 6701–6708. [CrossRef] [PubMed]
55. Nguyen, V.-N.; Park, S.J.; Qi, S.; Ha, J.; Heo, S.; Yim, Y.; Baek, G.; Lim, C.S.; Lee, D.J.; Kim, H.M.; et al. Design and synthesis of efficient heavy-atom-free photosensitizers for photodynamic therapy of cancer. *Chem. Commun.* **2020**, *56*, 11489–11492. [CrossRef] [PubMed]

56. Nguyen, V.-N.; Qi, S.; Kim, S.; Kwon, N.; Kim, G.; Yim, Y.; Park, S.; Yoon, J. An Emerging Molecular Design Approach to Heavy-Atom-Free Photosensitizers for Enhanced Photodynamic Therapy under Hypoxia. *J. Am. Chem. Soc.* **2019**, *141*, 16243–16248. [CrossRef] [PubMed]
57. Ortiz-Rodríguez, L.A.; Hoehn, S.J.; Loredo, A.; Wang, L.; Xiao, H.; Crespo-Hernández, C.E. Electronic Relaxation Pathways in Heavy-Atom-Free Photosensitizers Absorbing Near-Infrared Radiation and Exhibiting High Yields of Singlet Oxygen Generation. *J. Am. Chem. Soc.* **2021**, *143*, 2676–2681. [CrossRef] [PubMed]
58. Deiana, M.; Josse, P.; Dalinot, C.; Osmolovskyi, A.; Marqués, P.S.; Castán, J.M.A.; Abad Galán, L.; Allain, M.; Khrouz, L.; Maury, O.; et al. Site-selected thionated benzothioxanthene chromophores as heavy-atom-free small-molecule photosensitizers for photodynamic therapy. *Commun. Chem.* **2022**, *5*, 142. [CrossRef] [PubMed]
59. Li, H.; Meng, J.; Sun, X. Substitution effects on NIR-absorbing Perylene Diimide based on DFT calculation. *Inorg. Chem. Commun.* **2019**, *105*, 194–198. [CrossRef]
60. Henry, L. Ueber eine neue Bildungs- und Darstellungsweise der Nitrile. *Ann. Chem. Pharm.* **1869**, *152*, 148–152. [CrossRef]
61. Wislicenus, J. Vorläufige Mittheilungen. *Z. Chem.* **1869**, *12*, 324–326.
62. Hofmann, A.W. Ueber die Daratellung der geachwefelten Amide. *Berichte Dtsch. Chem. Ges.* **1878**, *11*, 338–340. [CrossRef]
63. Sureshbabu, V.V.; Nagendra, G.; Venkataramanarao, R. Ultrasound accelerated conversion of Nα-urethane protected peptide esters to their thiopeptides using P_2S_5. *Ultrason. Sonochem.* **2008**, *15*, 927–929. [CrossRef] [PubMed]
64. Demmig, S.; Langhals, H. Leichtlösliche, lichtechte Perylen-Fluoreszenzfarbstoffe. *Chem. Berichte* **1988**, *121*, 225–230. [CrossRef]
65. Hendsbee, A.D.; Sun, J.-P.; Law, W.K.; Yan, H.; Hill, I.G.; Spasyuk, D.M.; Welch, G.C. Synthesis, Self-Assembly, and Solar Cell Performance of N-Annulated Perylene Diimide Non-Fullerene Acceptors. *Chem. Mater.* **2016**, *28*, 7098–7109. [CrossRef]
66. Perrin, L.; Hudhomme, P. Synthesis, Electrochemical and Optical Absorption Properties of New Perylene-3,4:9,10-bis(dicarboximide) and Perylene-3,4:9,10-bis(benzimidazole) Derivatives. *Eur. J. Org. Chem.* **2011**, *2011*, 5427–5440. [CrossRef]
67. Polshettiwar, V.; Nivsarkar, M.; Paradashani, D.; Kaushik, M.P. Thionation of carbonyl compounds using phosphorus pentasulfide and hexamethyldisiloxane under microwave irradiations. *J. Chem. Res.* **2004**, *2004*, 474–476. [CrossRef]
68. Krstić, N.M.; Bjelaković, M.S.; Dabović, M.M.; Pavlović, V.D. Thionation of some alpha,beta-unsaturated steroidal ketones. *Molecules* **2010**, *15*, 3462–3477. [CrossRef] [PubMed]
69. Rocard, L.; Goujon, A.; Hudhomme, P. Nitro-Perylenediimide: An Emerging Building Block for the Synthesis of Functional Organic Materials. *Molecules* **2020**, *25*, 1402. [CrossRef] [PubMed]
70. El-Berjawi, R.; Hudhomme, P. Synthesis of a perylenediimide-fullerene C60 dyad: A simple use of a nitro leaving group for a Suzuki-Miyaura coupling reaction. *Dyes Pigment.* **2018**, *159*, 551–556. [CrossRef]
71. Rocard, L.; Hatych, D.; Chartier, T.; Cauchy, T.; Hudhomme, P. Original Suzuki–Miyaura Coupling Using Nitro Derivatives for the Synthesis of Perylenediimide-Based Multimers. *Eur. J. Org. Chem.* **2019**, *2019*, 7635–7643. [CrossRef]
72. Rocard, L.; Hudhomme, P. Recent Developments in the Suzuki–Miyaura Reaction Using Nitroarenes as Electrophilic Coupling Reagents. *Catalysts* **2019**, *9*, 213. [CrossRef]
73. Goujon, A.; Rocard, L.; Cauchy, T.; Hudhomme, P. An Imine Photocyclization as an Alternative to the Pictet–Spengler Reaction for the Synthesis of AzaBenzannulated Perylenediimide Dyes. *J. Org. Chem.* **2020**, *85*, 7218–7224. [CrossRef] [PubMed]
74. El-Berjawi, R.; Rocard, L.; Goujon, A.; Cauchy, T.; Hudhomme, P. Visible-Light-Mediated Synthesis of AzaBenzannulated Perylenediimide-Based Light-Harvesting Dyads. *J. Org. Chem.* **2020**, *85*, 12252–12261. [CrossRef] [PubMed]
75. Sun, J.-P.; Hendsbee, A.D.; Dobson, A.J.; Welch, G.C.; Hill, I.G. Perylene diimide based all small-molecule organic solar cells: Impact of branched-alkyl side chains on solubility, photophysics, self-assembly, and photovoltaic parameters. *Org. Electron.* **2016**, *35*, 151–157. [CrossRef]
76. Kudchadker, M.V.; Zingaro, R.A.; Irgolic, K.J. Chemistry of phosphorus pentaselenide. I. Its reaction with alcohols. *Can. J. Chem.* **2011**, *46*, 1415–1424. [CrossRef]
77. Tedy, A.M.; Manna, A.K. Nature and energetics of low-lying excited singlets/triplets and intersystem crossing rates in selone analogs of perylenediimide: A theoretical perspective. *J. Chem. Phys.* **2024**, *160*, 114306. [CrossRef]

molecules

MDPI

Review

Desulfonylative Functionalization of Organosulfones via Inert (Hetero)Aryl C(sp^2)–SO_2 Bond Cleavage

Rui Huang [1,†], Boning Gu [2,†], Ming Wang [2], Yinsong Zhao [1,*] and Xuefeng Jiang [1,2]

1 State Key Laboratory of Estuarine and Coastal Research, Hainan Institute of East China Normal University, East China Normal University, Shanghai 200241, China
2 State Key Laboratory of Petroleum Molecular & Process Engineering, Shanghai Key Laboratory of Green Chemistry and Chemical Processes, School of Chemistry and Molecular Engineering, East China Normal University, Shanghai 200062, China
* Correspondence: yszhao@sklec.ecnu.edu.cn
† These authors contributed equally to this work.

Abstract: As "chemical chameleons," organosulfones have been widely applied in various desulfonylative functionalization reactions. However, the desulfonylative functionalization of (hetero)arylsulfones through the cleavage of inert C(sp^2)–SO_2 bonds remains a challenging and underexplored task. Over the past twenty years, the use of (hetero)arylsulfones as arylation reagents has gradually gained attention in diverse cross-coupling reactions under specific catalytic conditions, especially in transition metal-catalysis and photocatalysis chemistry. In this review, we discuss the representative accomplishments and mechanistic insights achieved in desulfonylative reactions of inactive C(sp^2)–SO_2 bonds in (hetero)arylsulfones, including: (i) transition-metal-catalyzed desulfonylative cross-coupling reactions and (ii) photo-/electrocatalytic radical desulfonylative coupling reactions. We anticipate that this review will provide an overall perspective in this area to a general audience of researchers and stimulate further innovative strategies for desulfonylative functionalization of inert arylsulfones.

Keywords: (hetero)arylsulfones; desulfonylative functionalization; C(sp^2)–SO_2 bond cleavage; transition-metal catalysis; photocatalysis

Citation: Huang, R.; Gu, B.; Wang, M.; Zhao, Y.; Jiang, X. Desulfonylative Functionalization of Organosulfones via Inert (Hetero)Aryl C(sp^2)–SO_2 Bond Cleavage. *Molecules* **2024**, *29*, 4137. https://doi.org/10.3390/molecules29174137

Academic Editor: Diego A. Alonso

Received: 20 July 2024
Revised: 23 August 2024
Accepted: 30 August 2024
Published: 31 August 2024

1. Introduction

Organosulfones are highly versatile building blocks widely utilized in pharmaceutical science, functional materials, and polymer science [1–3]. Known as "chemical chameleons", organosulfones have participated in a diverse array of reactions, enabling the valuable formations of new chemical bonds [1,4]. The desulfonylative functionalization of organosulfones—substituting the sulfonyl group with other functional groups—is an attractive alternative that enhances molecular complexity [5]. Owing to its strong electron-withdrawing ability and inert reactivity, the sulfonyl functionality typically serves as a leaving group in elimination transformations or as an auxiliary moiety to stabilize carbanions in traditional organic synthesis. Over the past decades, studies on organosulfones have garnered significant attention, focusing on utilization as organic (pseudo)halides in cross-coupling reactions [6–11]. By integrating novel catalytic methods and customizing substituents on organosulfones, numerous new methodologies have emerged to activate C–SO_2 bonds, leading to remarkable advancements in desulfonylative functionalizations, even in innovative transformations that are challenging to achieve with conventional electrophiles or nucleophiles. Despite several excellent reviews on C–S bond functionalization [9] and radical desulfonylative functionalization [6,7,10,11] through transition-metal catalysis or photocatalysis, which highlight the cleavage of aryl, alkyl, vinylic, and alkynyl C–SO_2

bonds in thioethers, sulfamides, sulfonates, sulfones, and their derivatives, there is a limited number of reviews that focus specifically on the desulfonylative functionalization of organosulfones via inert (hetero)aryl C(sp^2)–SO_2 bond cleavage [8].

In this review, we will succinctly recount the history of desulfonylative functionalization of inactive (hetero)arylsulfones, focusing on representative accomplishments and progress as arylation reagents to access functionalized aromatics. The review is divided into two main sections: (i) transition-metal-catalyzed desulfonylative cross-coupling reactions and (ii) photo-/electrocatalytic desulfonylative coupling reactions (Figure 1).

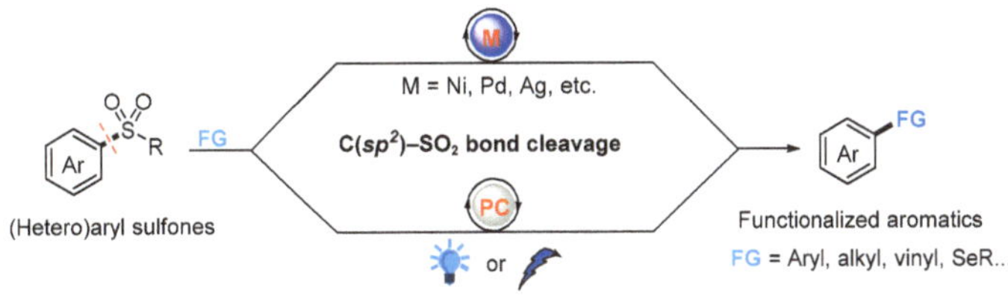

Figure 1. Desulfonylative functionalization of (hetero)arylsulfones via C(sp^2)–SO_2 bond cleavage.

2. Transition Metal-Catalyzed Desulfonylative Cross-Coupling Reaction

2.1. Kumada-Type Cross-Coupling Reaction

In 1979, Wenkert and co-workers reported the first Ni-catalyzed Kumada-type cross-coupling reaction of phenyl sulfones with methyl/*p*-tolyl magnesium bromides, giving desulfonylative methylation/*p*-tolylation products in medium to high yields (Scheme 1) [12]. When methyl phenyl sulfone was tested under the optimized conditions, the insertion of Ni(0) complex into the aryl C(sp^2)–SO_2 bond rather than the methyl C(sp^3)–SO_2 bond occurred selectively. This seminal achievement indicated the possibility of substituted arylsulfones as (pseudo)halide reagents in cross-coupling reactions.

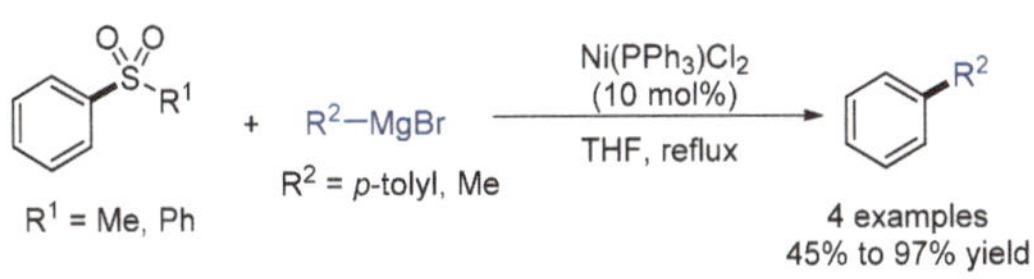

Scheme 1. Seminal desulfonylative cross-coupling of arylsulfones with Grignard reagents.

In the 1990s, Clayden and Julia developed a modified Ni-catalyzed cross-coupling reaction of aryl *tert*-butylsulfones with aryl magnesium halides (Scheme 2) [13,14]. This protocol established expanded scopes of arylsulfones and Grignard reagents, yielding various *ortho*-substituted unsymmetrical biaryls in satisfactory yields. However, mixtures of cross-coupling, homocoupling, and reduction products were also afforded due to the steric effect of *ortho*-substituents.

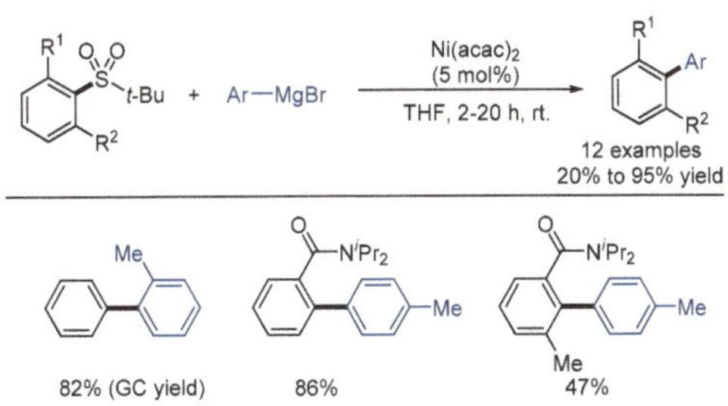

Scheme 2. Ni-catalyzed cross-coupling of aryl sulfones with Ar-MgBr.

In 2008, García and co-workers developed the desulfonylative functionalization of dibenzothiophene dioxides with methyl magnesium bromide (MeMgBr) catalyzed by nickel-complexes (Scheme 3) [15]. A series of nickel complexes including $[Ni(dippe)H]_2$, $[Ni(dcype)H]_2$, $[Ni(dtbpe)H]_2$, or $[Ni(dippe)(Me)_2]$ in mixed solvents of toluene and THF could yield methyl biaryl or dimethyl biaryls in moderate to good yields. Dibenzothiophene dioxides reacted with 6.0 equivalents of MeMgBr to give dimethyl biaryls in quantitative yields in the presence of low nickel catalyst loading (1.0 mol%). As shown in the proposed catalytic cycle, the reaction of $[Ni(dippe)H]_2$ with dibenzothiophene dioxide yields a nickel-complex **I**. Further oxidative addition of C(sp^2)–SO_2 bond followed by extrusion of SO_2 with the aid of MeMgBr furnishes cyclic nickel intermediate **III**. The intermediate **III** undergoes metathesis with MeMgBr followed by reductive elimination to give dimethyl-substituted biaryls.

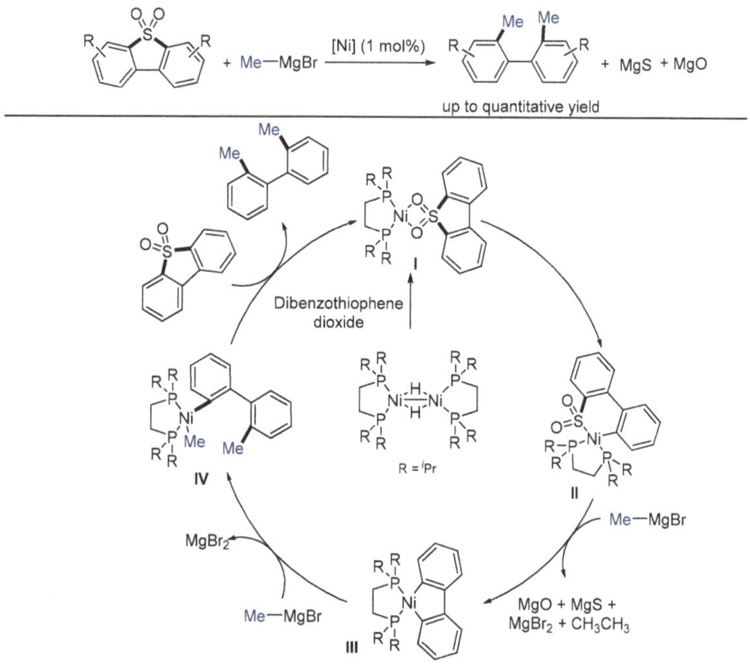

Scheme 3. Ni-catalyzed desulfonylative cross-coupling of dibenzothiophene dioxides.

Very recently, Xu, Zhou, and Shen's group established a Ni-catalyzed cross-coupling of inactivated (hetero)aryl sulfones with aryl bromides at ambient temperature with the assistance of excess magnesium powder (Scheme 4) [16]. Control experiments indicated that $Ni(PPh_3)_2Cl_2$, 2,2′-bipyridine ligand (bpy), magnesium powder, and THF are all indispensable for the successful realization of the reaction. A variety of (hetero)aryl sulfones and aryl bromides were well compatible under the optimized conditions with/without LiCl, enabling the facile synthesis of biaryl compounds with moderate to high yields. It was noteworthy that the late-stage desulfonylative coupling reactions with complex molecules derived from *L*-borneol and *L*-menthol also proceeded smoothly under the established conditions, affording the corresponding product in moderate yields. In addition, the practical utility of this cross-coupling was demonstrated by performing a 5 mmol scale reaction, delivering the final products in 66% yield. For unsymmetrical (hetero)aryl sulfones, the desulfonylative cross-coupling reaction occurred selectively through the cleavage of the more electron-poor aryl C(sp^2)–SO_2 bond. Initially, the reduction of $Ni(PPh_3)_2Cl_2$ by

Mg powder generates active Ni(0) species, which undergo oxidative addition with aryl sulfones ($ArSO_2R^2$) to produce an Ar–Ni(II)–SO_2R^2 complex **I**. Meanwhile, the reaction of aryl bromides with Mg powder with/without the aid of lithium chloride produces the active aryl magnesium compounds. The subsequent transmetallation between intermediate **I** and Grignard reagents affords the biaryl-nickel intermediates **II**, which finally undergo reductive elimination to generate the desired diaryls.

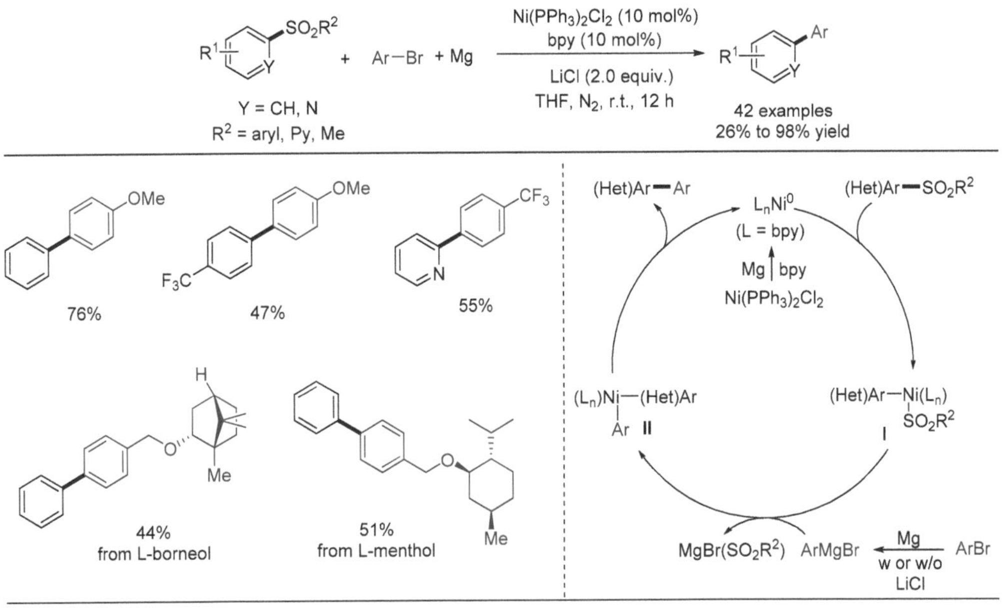

Scheme 4. Ni-catalyzed desulfonylative cross-coupling with in situ-generated Ar-MgBr.

2.2. Suzuki–Miyaura-Type Cross-Coupling Reaction

In 2005, the Robins group reported one example of Pd-catalyzed Suzuki–Miyaura-type cross-coupling of 6-[(3-methylbutyl)sulfonyl]-9-[2,3,5-tri-O-(2,4,6-trimethylbenzoyl)-β-D-ribofuranosyl]purine with 4-methoxyphenylboronic acid (*PMP*$B(OH)_2$) [17]. In the presence of 10 mol% of Pd(OAc)2 and 10 mol% of 1,3-bis(2,6-diisopropylphenyl)imidazolin-2-ylidene (IPr), the corresponding desulfonylative arylation product was isolated in 81% yield. The cleavage of heteroaryl C(sp^2)–SO_2 bond preferred to occur (Scheme 5).

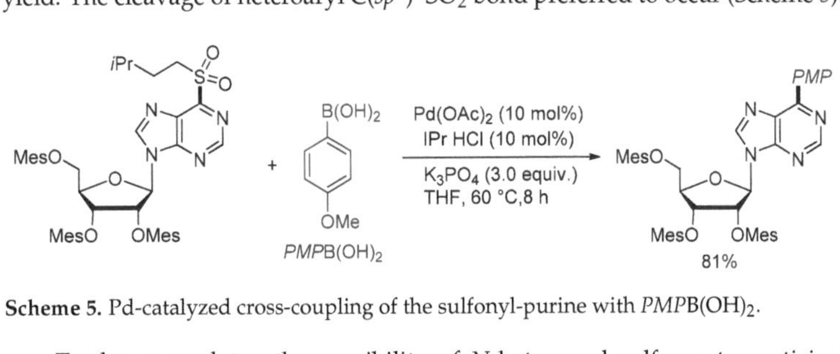

Scheme 5. Pd-catalyzed cross-coupling of the sulfonyl-purine with *PMP*$B(OH)_2$.

Twelve years later, the possibility of *N*-heteroaryl sulfones to participate in Pd-catalyzed Suzuki–Miyaura-type cross-coupling reaction was further investigated by Hennessy and co-workers (Scheme 6) [18]. The synthesis of 1,5-disubstituted tetrazoles was successfully realized through desulfonylative cross-coupling of readily prepared 5-*p*-toluenesulfonyltetrazoles with boron-based nucleophiles including boronic acids, pinacolboronate esters, and trifluoroborate salts. The use of biaryl phosphine ligands such as RuPhos was essential for achieving higher yields. The protocol is compatible with a variety of 5-*p*-toluenesulfonyltetrazoles and organoboron compounds with diverse functional groups. It is noteworthy that a 5-*p*-tolylsulfonyltetrazole substrate containing the bromoaryl group underwent Pd-catalyzed Suzuki–Miyaura cross-coupling reaction selectively through the cleavage of aryl C–Br bond, thus providing new opportunities for step-wise construction of polyaryls.

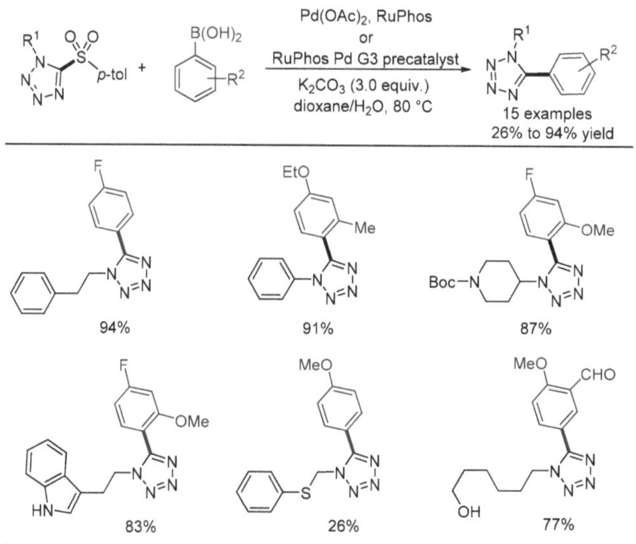

Scheme 6. Pd-catalyzed cross-coupling of *N*-heteroarylsulfones with boronic acids.

Very recently, Niu and co-workers reported a Ni-catalyzed Suzuki–Miyaura-type cross-coupling reaction of readily available and bench-stable phenyl pyrimidinyl sulfones with (hetero)arylboronic acids (Scheme 7) [19]. This desulfonylative cross-coupling reaction involving the pyrimidinyl $C(sp^2)–SO_2$ bond cleavage offered a facile route for the rapid synthesis of synthetically challenging 2,4-diarylated pyrimidines with moderate to high yields. The presence of an *ortho*-nitrogen atom in sulfone substrates is critical in this method. In the proposed catalytic cycle, the ligand exchange of Ni(0) species with phenyl pyrimidinyl sulfones first generates the intermediate **I**. Then, a chemoselective $C(sp^2)–SO_2$ bond insertion affords the oxidative additive Ni(II) complex **II**. Transmetallation with arylboronic acids furnishes the pyrimidinyl-Ni(II)-aryl complex **III**, which undergoes reductive elimination to give the final product.

Compared to *N*-heteroaryl $C(sp^2)–SO_2$ bond, the activation of inert aryl $C(sp^2)–SO_2$ bond in arylsulfones is much more challenging due to its weak coordinating ability and higher bond dissociation energy. In 2019, the Moran group described the Suzuki–Miyaura cross-coupling using the trifluoromethyl arylsulfones as the substrates (Scheme 8) [20]. Control experiments indicated that the desulfonylative reactions of diphenyl sulfone, phenyl methyl sulfone and its mono- and difluorinated analogs were inert or much less reactive, suggesting the important role of the trifluoromethyl group in promoting aryl $C(sp^2)–SO_2$ bond activation. In addition, the presence of a small amount of DMSO (1%, v/v) was critical to obtain high yields by solubilizing the inorganic base. This protocol could provide various functionalized di(hetero)aryls from readily available trifluoromethyl arylsulfones and arylboronic acids. Furthermore, the reactivity of different electrophiles such as aryl chlorides, nitroarenes, and aryl tosylates was investigated in one pot. The reactivity of aryl sulfones showed at least two orders of magnitude slower than aryl chlorides, but at least two orders faster than nitroarenes and aryl tosylates under standard conditions, which provided new opportunities for multiple functionalization reactions in a step-wise manner. Mechanistic experiments and DFT calculations demonstrated that the relatively rare oxidative addition of palladium species into the aryl C(sp^2)–SO_2 bond was the turnover-limiting step.

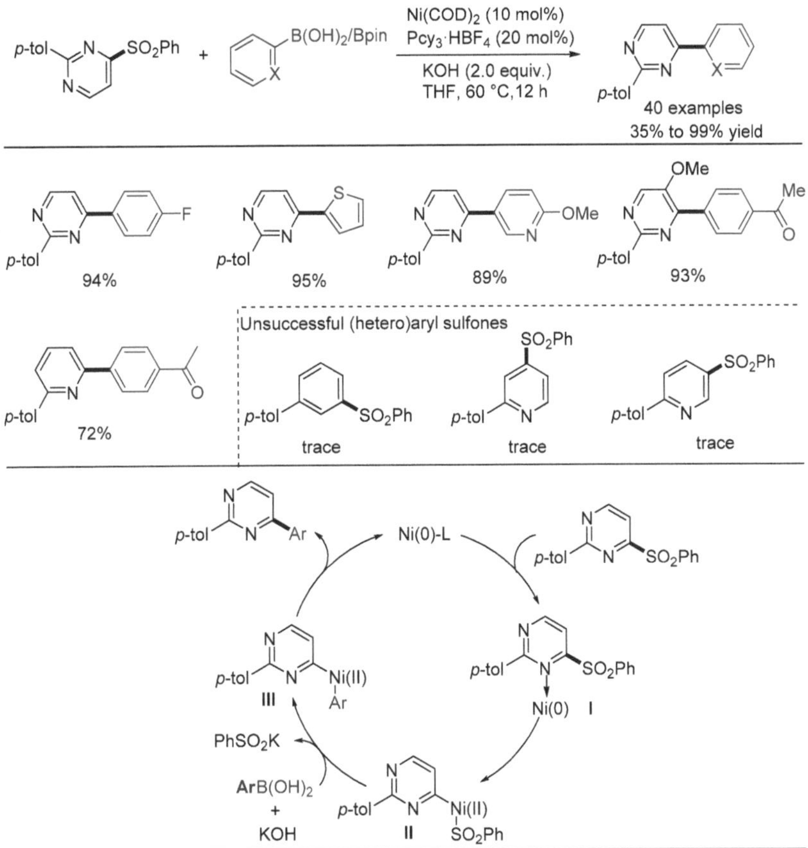

Scheme 7. Ni-catalyzed cross-coupling of phenyl pyrimidinyl sulfones with boronic acids.

In the same year, the Yorimitsu group developed similar desulfonylative transformations enabled by cooperative Pd/Rh catalysis [21]. Under the optimized condition, the model reaction could obtain 91% yield. In the absence of $[Rh(cod)_2]Cl_2$ catalyst, the reaction showed much poor efficiency (<40% yield) and higher loadings of Pd catalyst was required. Kinetic experiments indicated that the rhodium catalyst probably promoted the transfer of the aryl ring from arylboronate to palladium species, thus accelerating the transmetalation process on the palladium center, and the reductive elimination step to form C–C bond assisted by IPr^{NiPr2} ligand might be the turnover-limiting step. Despite the acyclic arylsulfones, the Suzuki–Miyaura type cross-coupling reaction of dibenzothiophene sulfones with arylboronic acids were achieved by García and co-workers. However, stoichiometric amounts of nickel salts were needed in this reaction, yielding complex desulfonylative arylation products [22].

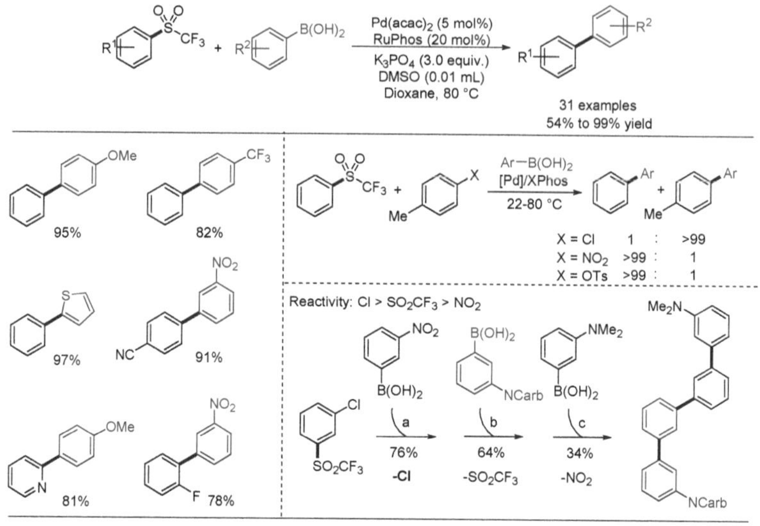

Scheme 8. Pd-catalyzed cross-coupling of trifluoromethyl arylsulfones with boronic acids.

2.3. *Desulfonylative Cross-Coupling Reaction with (Hetero)Aryl Halides*

As a class of organic (pseudo)halides, the desulfonylative cross-coupling of sulfones with electrophiles such as aryl halides has remained underexplored. In 2018, the Willis group demonstrated that *N*-heteroaryl allylsulfones could act as precursors of sulfinate reagents under palladium(0) catalysis, leading to efficient Pd-catalyzed cross-coupling reactions with (hetero)aryl halides in one pot (Scheme 9) [23]. Under the optimized conditions, the constructions of synthetically challenging *N*-heteroaryl-(hetero)aryl products were realized efficiently. Notably, the desulfonylative cross-coupling reaction showed high regioselectivity. In the proposed catalytic cycle, the oxidative addition reaction of active Pd(0) species with allylsulfones would generate the π-allyl-Pd(II) intermediate with the release of *N*-heteroaryl sulfinate. The π-allyl-Pd(II) intermediate was intercepted by a nucleophile to regenerate active Pd(0) species, thus allowing the subsequent cross-coupling of in situ-generated *N*-heteroaryl sulfinate with aryl halides to proceed. Later, the same group developed β-nitrile/ester alkyl (hetero)aryl sulfones as new latent sulfinate reagents [24]. In the presence of base, β-nitrile/ester alkyl (hetero)aryl sulfones could release the corresponding sulfinates and vinylic nitrile/ester by-product. These latent sulfinate reagents are stable and readily available, facilitating multistep elaborations that are challenging for traditional sulfinates with poor solubility and reactivity.

Very recently, the Ni-catalyzed reductive cross-coupling of arylsulfones with aryl bromides was realized by Ma and co-workers (Scheme 10) [25]. In this protocol, trifluoromethyl arylsulfones showed the best result, while other arylsulfones containing CF_2H, CH_3, Ph, and F instead of the CF_3 group resulted in poor or no efficiency. By using 10 mol% of $NiCl_2$ as the catalyst, the combination of 6 mol% of dmbpy and 6 mol% of dcype as ligands, and Zn powder as reductant, a variety of functionalized biaryl compounds were obtained with moderate to good isolated yields. However, aryl (trifluoromethyl) sulfones containing strong electron-withdrawing groups such as ketone, ester, and cyano groups failed to afford the desired product. In addition, the applicability of this method was then investigated using several aryl bromides derived from natural products and drug molecules such as ibuprofen and cholesterol, affording the desired products in moderate yields.

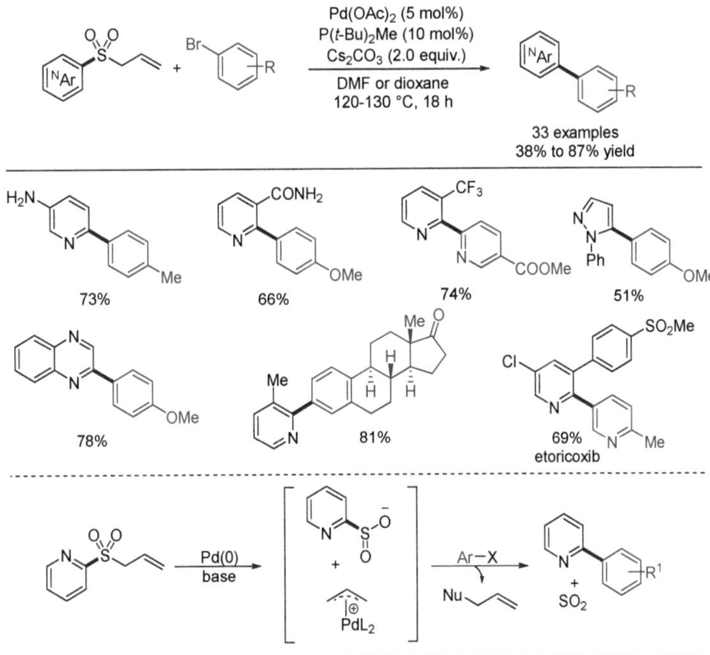

Scheme 9. Pd-catalyzed cross-coupling of allyl heteroaryl sulfones with aryl halides.

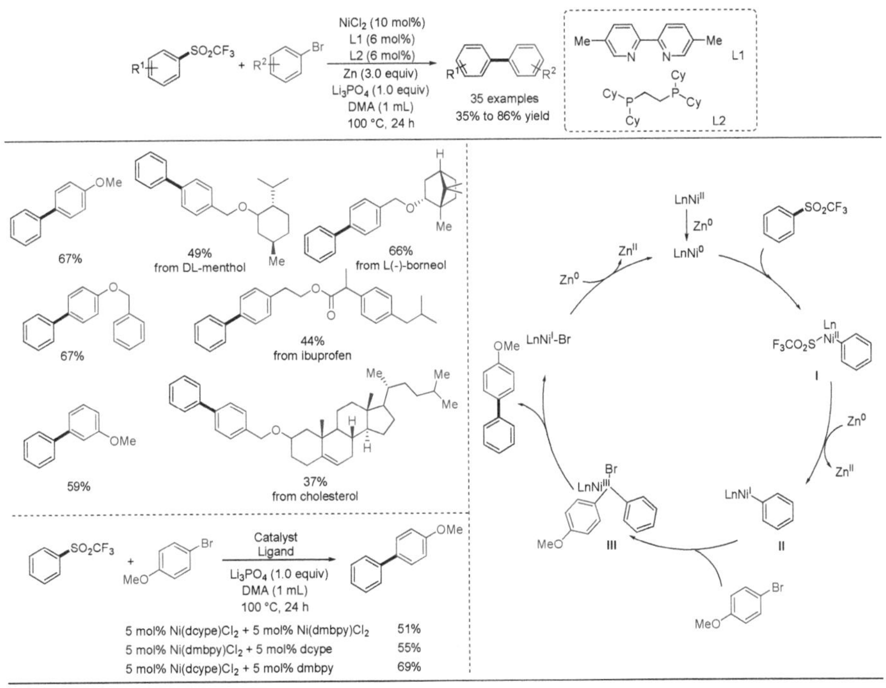

Scheme 10. Ni-catalyzed desulfonylative cross-coupling of arylsulfones with aryl halides.

To gain insight into the mechanism, a series of control experiments were then conducted. Under the optimized conditions, the cross-coupling of phenyl (trifluoromethyl) sulfone with preformed phenylzinc bromide did not occur without Zn powder, excluding the possible formation of organozinc reagents in situ. The reaction of Ni-complex PMP-(dmbpy)Ni(II)-Br with arylsulfones gave no desired product, while the reaction of Ni-complex Ph-(dcype)Ni(II)–SO_2CF_3, generated from oxidative addition of aryl sulfone to Ni(0) species and confirmed by an X-ray analysis, could give the final product in 49% GC yield. However, the reaction of Ni-complex PMP-(dcype)Ni(II)-Br with arylsulfones also gave the desired product in 20% GC yield, thus the oxidative addition of aryl halides to nickel species could not be excluded. A possible catalytic cycle was also proposed. The key intermediate **II**, generated from intermediate **I** in the presence of Zn, inserts C–Br bond of aryl halides to give intermediate **III**. Reductive elimination of **III** gives the desired product.

2.4. Intramolecular Desulfonylative Cross-Coupling Reaction

Without the addition of coupling reagents, transition metal-catalyzed intramolecular desulfonylative coupling reactions occurred in some cases. In 2018, Yorimitsu and coworkers developed a Ni-catalyzed intramolecular desulfonylative coupling of π-extended *N*-hetero-2-aryl arylsulfones (Scheme 11) [26]. In the presence of 20 mol% of Ni(cod)$_2$, 40 mol% of ligand, 60 mol% of *t*-BuONa, and one equivalent of Mg, sulfones with π-extended *N*-aromatic rings and arylsulfonyl group at *ortho*-nitrogen carbon atom were well compatible to the corresponding unsymmetric biaryls with moderate to high yield. However, sulfones containing 2-pyridyl, 2-pyrazinyl, and 3-/4-quinolyl rings showed much lower efficiency or were completely inert under the optimized conditions.

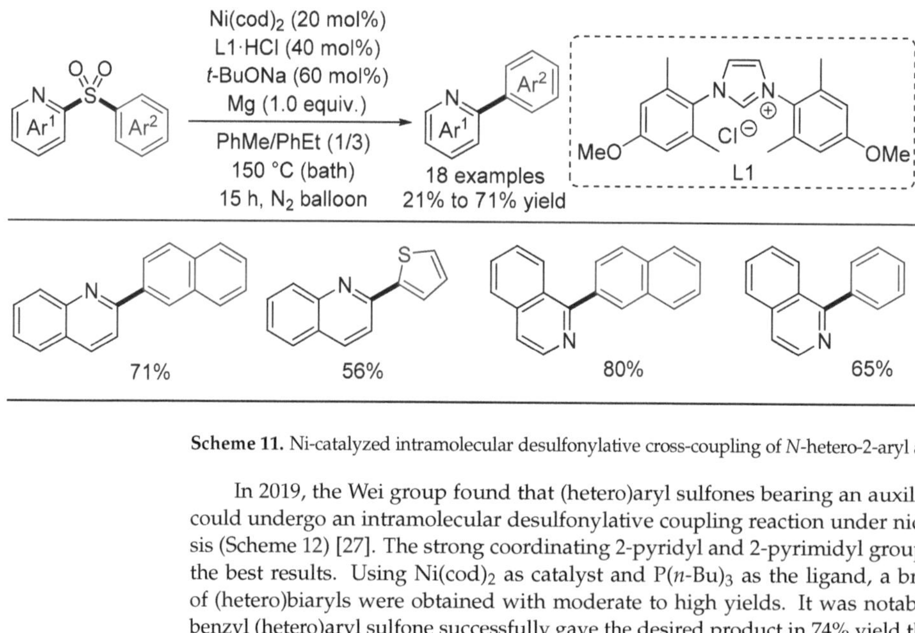

Scheme 11. Ni-catalyzed intramolecular desulfonylative cross-coupling of *N*-hetero-2-aryl arylsulfones.

In 2019, the Wei group found that (hetero)aryl sulfones bearing an auxiliary group could undergo an intramolecular desulfonylative coupling reaction under nickel catalysis (Scheme 12) [27]. The strong coordinating 2-pyridyl and 2-pyrimidyl groups showed the best results. Using Ni(cod)$_2$ as catalyst and P(n-Bu)$_3$ as the ligand, a broad range of (hetero)biaryls were obtained with moderate to high yields. It was notable that the benzyl (hetero)aryl sulfone successfully gave the desired product in 74% yield through the cleavages of both C(sp^2)–SO_2 and C(sp^2)–SO_2 bonds. Without a directing group, the intramolecular desulfonylative coupling reaction also occurred, but required a stoichiometric amount of Ni(cod)$_2$.

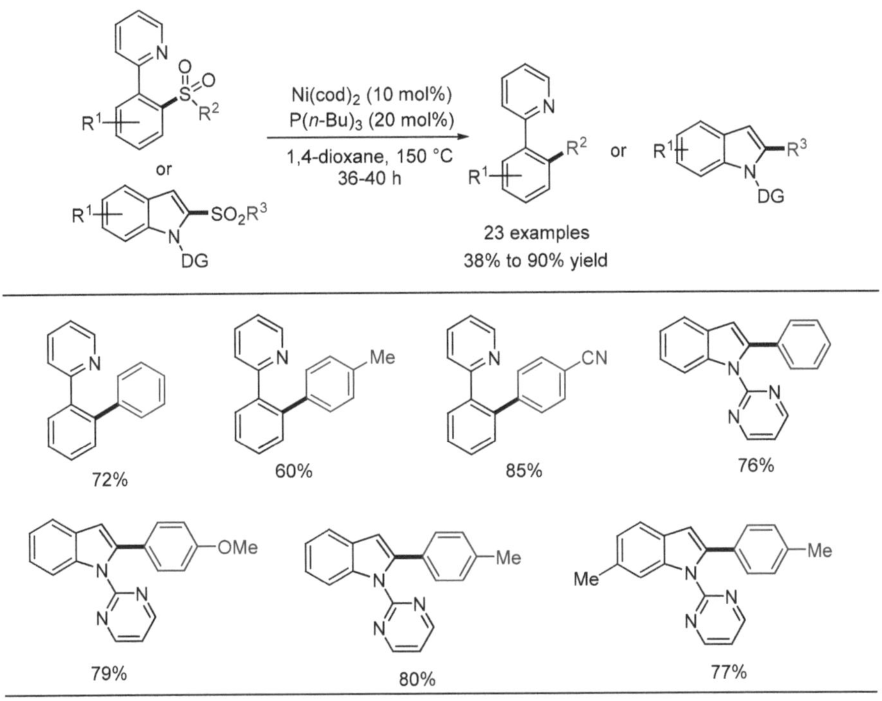

Scheme 12. Ni-catalyzed intramolecular desulfonylative cross-coupling of *N*-hetero-2-aryl arylsulfones.

2.5. Desulfonylative Cross-Coupling Reaction with Other Reagents

In 2006, Ruano and co-workers reported a Pd-catalyzed Mizoroki–Heck-type reaction of alkenyl or alkynyl aryl sulfones with acrylate esters (Scheme 13) [28]. By using $Pd(OAc)_2/PCy_3$ as the catalyst and Ag_2CO_3 as an additive, the cleavage of aryl C(sp^2)–SO_2 bond within alkenyl or alkynyl aryl sulfones proceeded primarily, leading to further arylation of acrylate esters to form cinnamyl esters. Diphenyl sulfone or methyl phenyl sulfone were unreactive under the optimal conditions, suggesting that coordination of the π-bond of the alkene and alkyne moiety to palladium center is likely involved. As shown in the proposed catalytic cycle, the coordination of the π-bonds followed by oxidative addition to Pd-species affords the intermediate **II**. Then, coordination and insertion of π-bond in the acrylate ester generates an alkyl-Pd species **III**, which undergoes the β-elimination process to give the desired product.

Transition metal-catalyzed desulfonylative cross-coupling reactions of (hetero)aryl sulfones with radical precursors have also been achieved. In 2020, Matsubara and co-workers developed the Ag-catalyzed radical cross-coupling of arylsulfonyl-substituted furoxans with aliphatic carboxylic acids (Scheme 14) [29]. A variety of primary, secondary, and tertiary alkyl carboxylic acids successfully reacted with arylsulfonyl-substituted furoxans to give the corresponding alkyl-substituted furoxans with moderate to good yields. Various functional groups such as ketone, ether, ester, bromo, amine, and aromatic rings, as well as several bioactive and drug skeletons, were well tolerated in this method. Notably, this protocol showed exclusive site-selectivity favoring the cleavage of sulfonyl group at the 3-position of disulfonyl furoxan. The obtained 3-alkyl-4-sulfonyl furoxan products could be further converted into dual-functionalized furoxans, enabling the modular synthesis of diverse functionalities through "build-and-scrap" strategy.

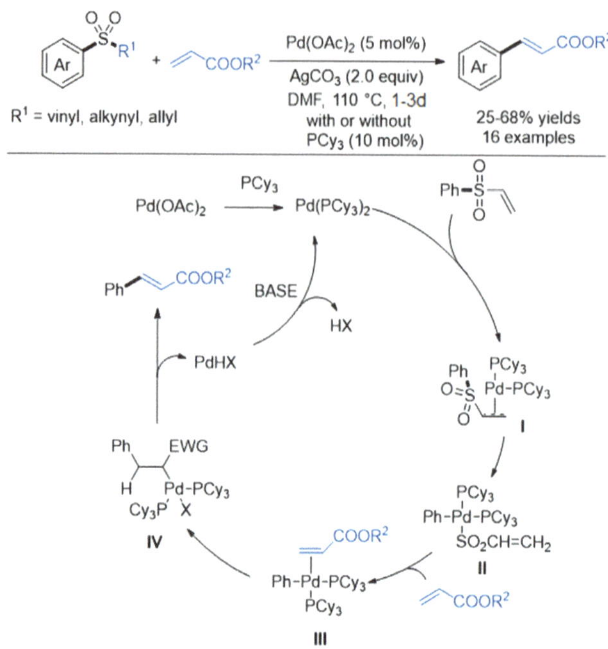

Scheme 13. Pd-catalyzed Heck-type desulfonylative cross-coupling.

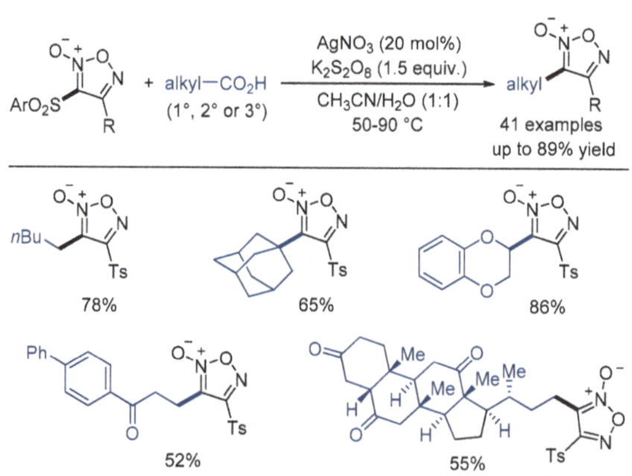

Scheme 14. Ag-catalyzed desulfonylative cross-coupling with alkyl carboxylic acids.

In recent years, the transition metal-catalyzed desulfonylative functionalization reactions to form carbon–heteroatom bonds have also been realized. In 2016, Chen and co-workers developed an efficient Ni-catalyzed desulfonylative cross-coupling of arylsulfones with phosphine oxides (Scheme 15) [30]. This protocol proceeded with both aryl and alkyl secondary phosphine oxides to offer functionalized *tertiary*-phosphine oxides in good to high yields. The desulfonylative reaction could be performed at 10 mmol scale in high yield with the addition of Ni(cod)$_2$ at a loading of 0.1 mol%. Interestingly, the desulfonylative reaction of arylsulfones bearing ester and cyano groups could work well without nickel catalyst. This protocol provides an efficient method for the construction of C–P bonds. Later, the formation of C–B bond was achieved by the Yorimitsu group through Pd-catalyzed coupling of diphenyl sulfone with diboronates, yet with only one example [31].

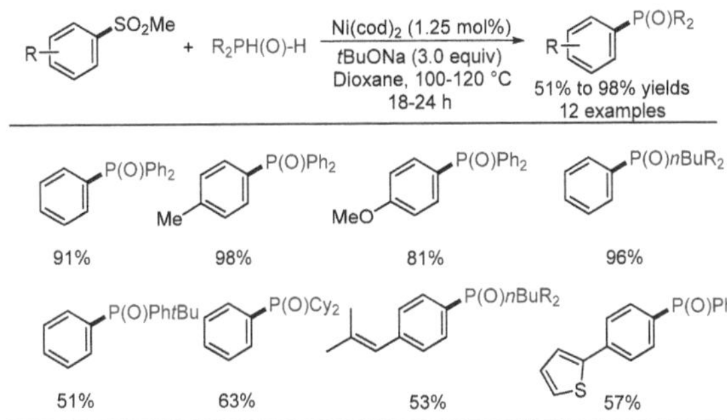

Scheme 15. Ni-catalyzed intramolecular desulfonylative cross-coupling of *N*-hetero-2-aryl arylsulfones.

Organic selenium compounds have attracted a lot of attention due to their unique biological activity. Recently, the Wang group developed a Ni-catalyzed cross-coupling reaction of heteroaryl sulfones with diselenides (Scheme 16) [32]. Using 5 mol% of $Ni(PPh_3)Cl_2$ as catalyst, excess amounts of Mn powder as reductant, and MeOH or DMF as solvent, various heteroaryl sulfones and diselenides were well compatible to afford the anticipated heteroaryl selenides with moderate to high yields. In this protocol, the nickel catalyst undergoes oxidative addition with diselenide ether to furnish Ni(II)-selenide species **I**. Then, Ni(II)-selenide species can generate the selenium radical **II** and Ni(I)-selenide species **III**, which regenerates active Ni(0) species and the selenium negative ion **IV**. Finally, radical/nucleophilic addition into C=N double bonds within *N*-heteroarylsulfones followed by rearomatization releases the sulfite radical/negative ion and the target heteroaryl selenides.

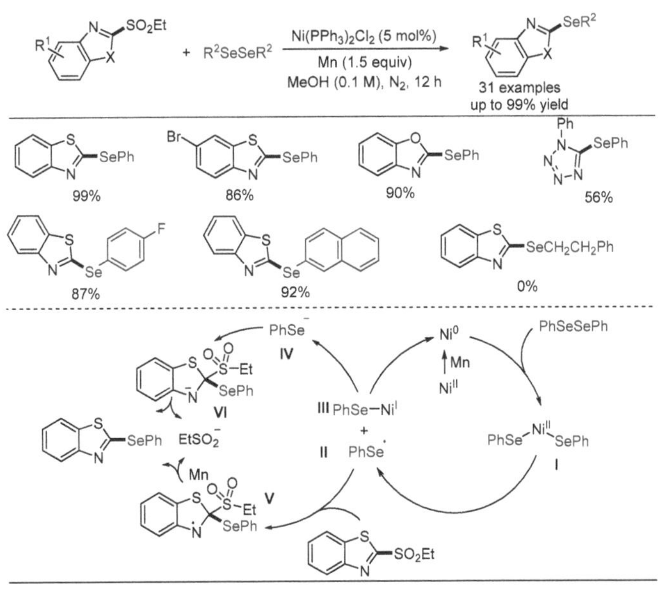

Scheme 16. Ni-catalyzed desulfonylative cross-coupling with diselenides.

3. Photo-/Electrocatalytic Desulfonylative Cross-Coupling

Over the past decades, photoredox catalysis has emerged as a powerful and promising strategy in organic synthesis [33,34]. Photocatalysis enables reactivity up-conversions by harnessing the energy of visible light and has been recently applied in a series of radical desulfonylative functionalization of sulfones. Recently, photoinduced desulfonylative functionalization that involves the cleavage of (hetero)aryl C(sp^2)–SO_2 bond of sulfones has also been achieved, providing a powerful and efficient route to access functional aromatics.

3.1. Desulfonylative Cross-Coupling with Alkylating Agents

In 2017, the Kamijo group reported a Ph_2CO-mediated photoinduced desulfonylative alkylation of sulfonylpyrimidines (Scheme 17) [35]. In the presence of 0.1 or 1.0 equivalent of Ph_2CO, the reactions could give various alkylated pyrimidines with moderate to good yields under the irradiation of medium-pressure Hg-lamp at room temperature. In this case, only electron-deficient substrates showed relatively higher yields. A proposed catalytic cycle was illustrated in Scheme 17b. Hydrogen atom abstraction (HAA) of the C(sp^3)–H assisted by photoexcited diaryl ketone with the release of the ketyl radical occurs to generate a carbon radical **I**. The addition of the carbon radical to sulfonylpyrimidines furnishes the radical intermediate **II**, which undergoes rearomatization in the presence of ketyl radical to provide the desired product.

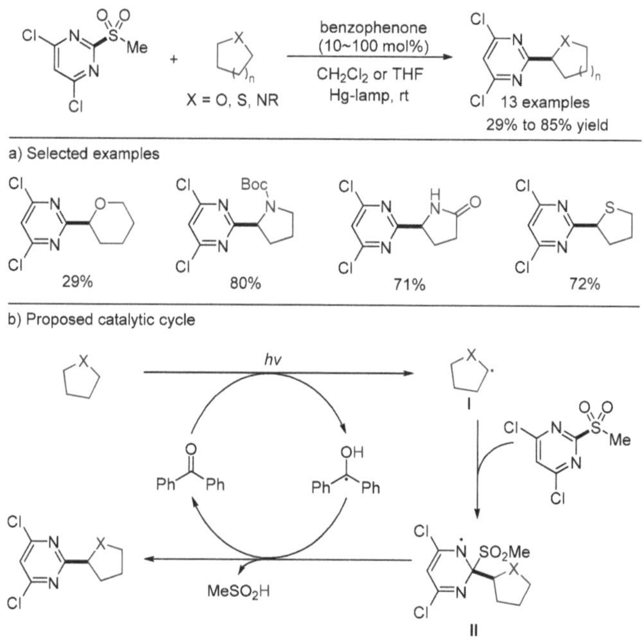

Scheme 17. Ph_2CO-mediated photoinduced desulfonylative alkylation.

In 2019, Molander and co-workers reported a photoredox-catalyzed desulfonylative alkylation of *N*-heteroaryl sulfones with alkyl bis(catecholato)silicates or 4-alkyl-1,4-dihydropyridines (DHPs) (Scheme 18) [36]. When using alkyl bis(catecholato)silicates with low oxidation potentials as substrates, the desulfonylative reactions gave the best results in the presence of $[Ru(bpy)_3][PF_6]_2$. Meanwhile, a variety of 4-alkyl-1,4-dihydropyridines also successfully reacted with the *N*-heteroaryl sulfones to yield comparable yields using 4CzIPN as the photocatalyst. Primary and secondary alkyl substrates bearing potentially reactive functional groups such as alkenes, esters, alkyl chlorides, amines, and pyrroles were well tolerated. To check the utility of this method, the desulfonylative reaction of a saccharide-derived DHP was subjected to the optimized conditions, yielding the final product in 54% yield with excellent diastereoselectivity. In the proposed cycle, excited photocatalyst by light absorption, followed by SET to the alkyl radical precursor. The newly formed alkyl radical inserts into the heteroaryl sulfone at the α-position relative to the sulfone with the concomitant elimination of a sulfinate anion by SET between the transient aryl radical and the reduced photocatalyst.

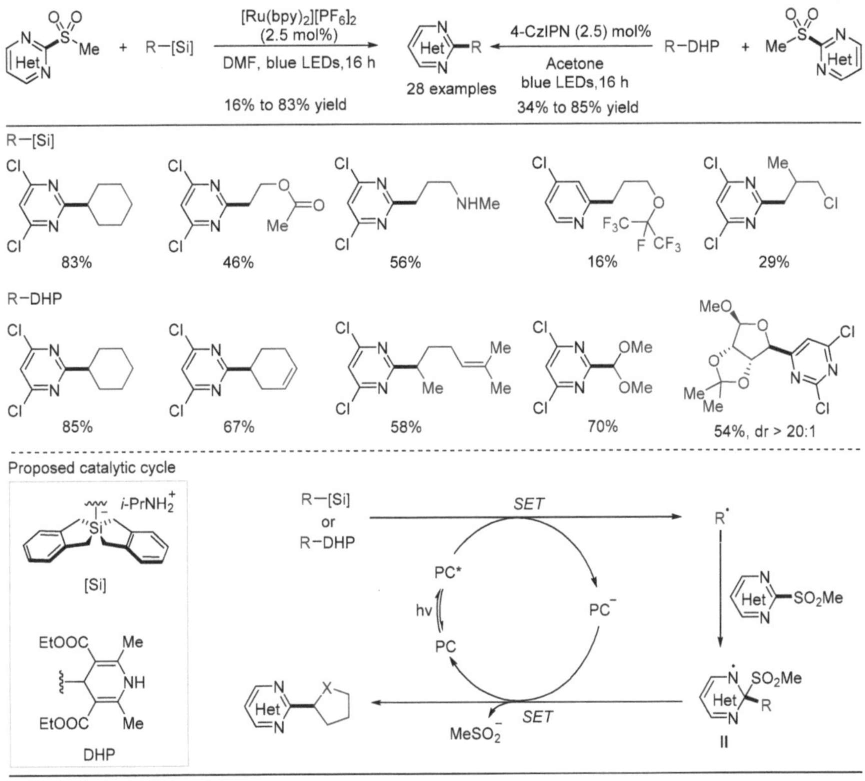

Scheme 18. Photoinduced desulfonylative alkylation using functionalized reagents.

In 2019, the Kamijo group developed similar photoinduced desulfonylative alkylation of sulfone-functionalized benzazoles mediated by 4-benzoylpyridine (4-BzPy) (Scheme 19) [37]. Sulfone-functionalized benzothiazoles, benzoxazole, and benzimidazoles could react successfully to give the desired products in 24–92% yield. However, this protocol required stoichiometric amounts of 4-BzPy as the photocatalyst due to its decomposition during the reaction. Recently, the Xia and Yang group achieved similar transformation through a photoinduced Fe-catalyzed ligand-to-metal charge transfer strategy [38]. In this method, the generated chlorine radical served as the hydrogen atom absorbing reagent.

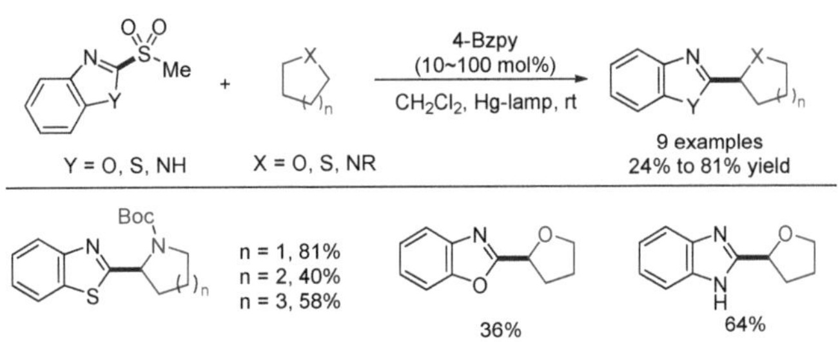

Scheme 19. 4-BzPy-promoted photoinduced desulfonylative alkylation.

In 2023, Shirakawa and co-workers developed an electrochemical direct α-arylation of trialkylamines with (hetero)arylsulfones (Scheme 20) [39]. By using a glassy carbon (GC) anode and a platinum (Pt) cathode in an undivided cell and lithium perchlorate (LiClO4) as a supporting electrolyte, the reactions successfully yielded model product in 97% yield under a constant cell voltage of 2.0 V (corresponding to an anodic potential of +0.86 V vs $Ag/AgNO_3$) in a MeCN/MeOH (3:1) solution at 25 °C for 2 h. This protocol was compatible with diverse alkylamines and (hetero)arylsulfones containing various functional groups and aromatic rings. When using aryl(dimethyl)amines as substrates, one equivalent of 2,2,6,6-tetramethylpiperidine (TMP) was employed as the base to obtain a higher yield. The reaction is initiated by the direct oxidation of amines on a glassy carbon (GC) anode to give a cation radical followed by deprotonation process to form α-aminoalkyl radical **I**. Generally, a sterically compact carbon radical is kinetically favored to form. Then, the reaction of intermediate **I** with sulfones through homolytic aromatic substitution (HAS) yields the final product with the release of $PhSO_2^{\bullet}$, which reacts with amines to form $PhSO_2^{-}$ and a cation radical. Very recently, a similar aminomethylation of sulfonylthiazoles enabled by the photoredox strategy has been achieved by the Fan and Chen group [40].

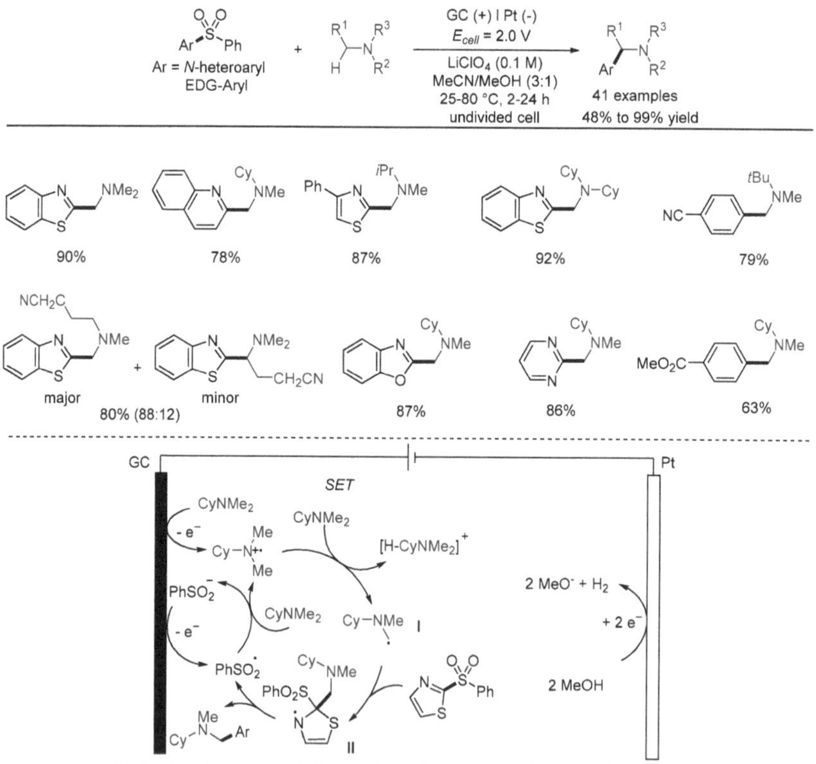

Scheme 20. Electroinduced direct desulfonylative alkylation with amines.

3.2. *Desulfonylative Cross-Coupling with Alkenes*

Alkenes are ubiquitous in natural products and pharmaceuticals and represent useful feedstocks in synthetic chemistry. The dual difunctionalization of alkenes provides a powerful route for olefin utilization. In 2018, Zhu and co-workers developed a conceptually new docking–migration strategy for radical difunctionalization of alkenes using bromodifluoromethyl *N*-heteroaryl sulfones (Scheme 21) [41]. This protocol employed *fac*-Ir(ppy)$_3$ as photocatalyst (2 mol%) and blue LEDs as light source (14 W). Under the standard conditions, both aryl alkenes and inactivated aliphatic alkenes were suitable substrates for this reaction, leading to formation of diverse heteroaryl- and difluoromethyl-difunctionalized alkanes. It is noteworthy that the final products could be further converted into synthetically valuable fluorine-containing molecules. As shown in Scheme 21, the bromodifluoromethyl *N*-heteroaryl sulfone undergoes SET process with excited Ir(III)* to generate difluoromethyl radical species **I** and Ir(IV) complex. Then, addition of **I** to an alkene gives the active alkyl radical intermediate **II**, which simultaneously undergoes intramolecular cyclization to form a five-membered spiro-*N*-radical **III**. Next, aromatization and extrusion of SO_2 results in the difluoroalkyl radical **IV**. Finally, the in situ formed iodine from the reaction of TBAI with Ir(IV) complex reacts with intermediate **IV** to yield the final difunctionalized product and regenerate the Ir(III) catalyst.

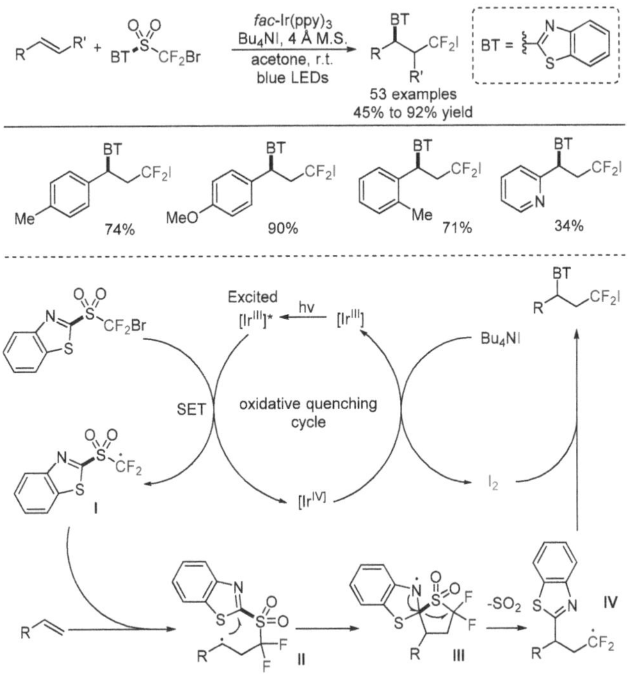

Scheme 21. Photocatalytic desulfonylative dual functionalization of alkenes.

Later, the same group utilized easily accessed, bromoalkyl *N*-heteroarylsulfones as the dual-functional reagent, and developed a new "polarity umpolung" strategy for radical alkylation/*N*-heteroarylation of alkenes [42]. The *tert*-dodecylthiol was crucial for this transformation reaction as it served as both the hydrogen source and the reductant. Notably, conjugated dienes and enynes were well compatible with the protocol with high regioselectivity. Importantly, radical trapping agents such as $PhSO_2SPh$, $PhSO_2SePh$, $PhSO_2SCF_3$, CH_2Br_2, allyl sulfone, and even nitrogen-containing heteroarene, could also give dual functionalized products in satisfactory yields.

Very recently, Hong and co-workers reported a photocatalytic desulfonylative carboarylation of alkenes using arylsulfonyl acetates as both arylating and carbonylating reagents (Scheme 22) [43]. Both activated and nonactivated alkenes are proved to be suitable substrates to access the corresponding aryl esters. This reaction is proposed to initiate with a single-electron oxidation between the excited Ir(III)* species and the deprotonated arylsulfonyl acetate **I**, yielding the carbonyl radical **II**. Then, addition of intermediate **II** to an alkene affords an intermediate **III**, which undergoes intramolecular radical addition at the *ipso*-position regioselectively followed by homolytic cleavage of the C–SO_2 bond to give the intermediate **IV**. Finally, the SO_2 extrusion and single-electron oxidation in the presence of Ir(II) yield the desired product through proton transfer. This method offers an effective route for rapidly increasing molecular complexity with two valuable groups.

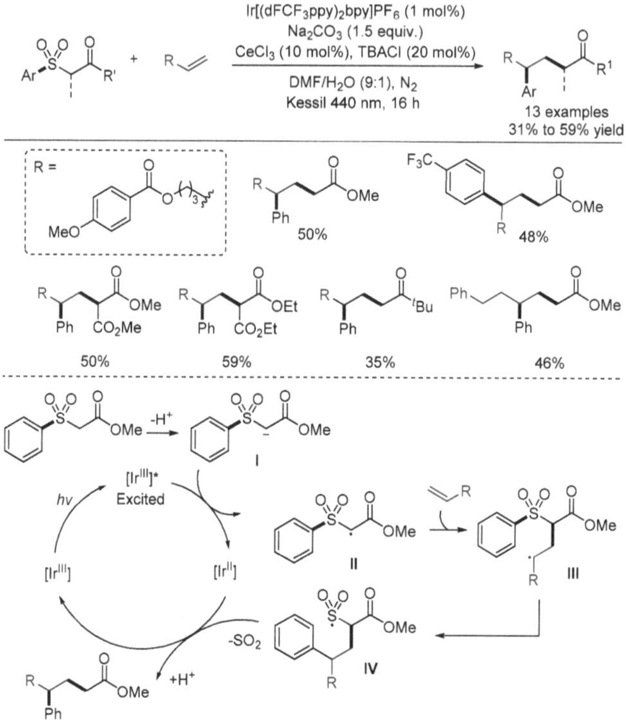

Scheme 22. Photocatalytic desulfonylative dual functionalization of alkenes.

4. Conclusion

In summary, we have demonstrated desulfonylative functionalization reactions of organosulfones through the selective cleavages of inert (hetero)aryl C(sp^2)−SO_2 bonds. Since the seminal achievement in the 1970s, various desulfonylative cross-coupling reactions have been developed to form valuable carbon–carbon and carbon–heteroatom bonds, along with the discoveries of novel catalytic protocols including transition metal catalysis, photocatalysis, and others. Mechanism studies showed the involvement of (hetero)aryl-metal-SO_2 species or radical processes. Despite the excellent progress in this field, several aspects should be stressed in developing new desulfonylative functionalization reactions in the future. For example, the cleavage of aromatic C(sp^2)−SO_2 bonds in unsymmetrical diarylsulfones generally tended to occur at the one linked to more electron-withdrawing substituents. For the desulfonylative functionalization of aryl sulfones, specific directing groups or alkyl substituents or harsh conditions such as the use of excess amounts of organometallic reagents or reductants, high catalyst loadings, and high temperature are usually required. Due to the inert reactivity of (hetero)aryl C(sp^2)−SO_2 bonds, examples of desulfonylative cross-coupling reactions to afford carbon–heteroatom bonds are much limited. Therefore, expansions of the sulfone substrate and reaction scopes, employments of milder reaction conditions with the developments of innovative catalytic protocols will be crucial in the future.

Author Contributions: Writing—original draft preparation, R.H. and B.G.; writing—review and discussion, M.W.; writing—original draft preparation, review and editing, Y.Z.; and supervision, Y.Z. and X.J. All authors have read and agreed to the published version of the manuscript.

Funding: This research was funded by NSFC (22125103 and 22301077) and Shanghai Pujiang Program (22PJ1403200).

Conflicts of Interest: The authors declare no conflicts of interest.

References

1. Trost, B.M. Chemical Chameleons. Organosulfones as Synthetic Building Blocks. *Bull. Chem. Soc. Jpn.* **2006**, *61*, 107–124. [CrossRef]
2. Elena, N.P. Sulfones and Sulfoxides in the Total Synthesis of Biologically Active Natural Compounds. *Russ. Chem. Rev.* **2000**, *69*, 367.
3. Wang, X.; Chen, L.; Chong, S.Y.; Little, M.A.; Wu, Y.; Zhu, W.-H.; Clowes, R.; Yan, Y.; Zwijnenburg, M.A.; Sprick, R.S.; et al. Sulfone-containing covalent organic frameworks for photocatalytic hydrogen evolution from water. *Nat. Chem.* **2018**, *10*, 1180–1189. [CrossRef] [PubMed]
4. Trost, B.M.; Kalnmals, C.A. Sulfones as Chemical Chameleons: Versatile Synthetic Equivalents of Small-Molecule Synthons. *Chem. Eur. J.* **2019**, *25*, 11193–11213. [CrossRef] [PubMed]
5. Nájera, C.; Yus, M. Desulfonylation Reactions: Recent Developments. *Tetrahedron* **1999**, *55*, 10547–10658. [CrossRef]
6. Chu, X.-Q.; Ge, D.; Cui, Y.-Y.; Shen, Z.-L.; Li, C.-J. Desulfonylation via Radical Process: Recent Developments in Organic Synthesis. *Chem. Rev.* **2021**, *121*, 12548–12680. [CrossRef]
7. Nambo, M.; Crudden, C.M. Transition Metal-Catalyzed Cross-Couplings of Benzylic Sulfone Derivatives. *Chem. Rec.* **2021**, *21*, 3978–3989. [CrossRef]
8. Corpas, J.; Kim-Lee, S.-H.; Mauleón, P.; Arrayás, R.G.; Carretero, J.C. Beyond Classical Sulfone Chemistry: Metal- and Photocatalytic Approaches for C–S Bond Functionalization of Sulfones. *Chem. Soc. Rev.* **2022**, *51*, 6774–6823. [CrossRef]
9. Huang, S.; Wang, M.; Jiang, X. Ni-catalyzed C–S bond construction and cleavage. *Chem. Soc. Rev.* **2022**, *51*, 8351–8377. [CrossRef]
10. Nambo, M.; Maekawa, Y.; Crudden, C.M. Desulfonylative Transformations of Sulfones by Transition-Metal Catalysis, Photocatalysis, and Organocatalysis. *ACS Cat.* **2022**, *12*, 3013–3032. [CrossRef]
11. Paul, B.; Paul, H.; Chatterjee, I. Photoredox-Mediated Desulfonylative Radical Reactions: An Excellent Approach Towards C–C and C–Heteroatom Bond Formation. *Synthesis* **2022**, *54*, 5409–5422. [CrossRef]
12. Wenkert, E.; Ferreira, T.W.; Michelotti, E.L. Nickel-Induced Conversion of Carbon–Sulphur into Carbon–Carbon Bonds. One-Step Transformations of Enol Sulphides into Olefins and Benzenethiol Derivatives into Alkylarenes and Biaryls. *J. Chem. Soc. Chem. Commun.* **1979**, *14*, 637–638. [CrossRef]
13. Clayden, J.; Julia, M. *ortho*-Substituted Unsymmetrical Biaryls from Aryl *tert*-Butyl Sulfones. *J. Chem. Soc. Chem. Commun.* **1993**, 1682–1683. [CrossRef]
14. Clayden, J.; Cooney, J.J.A.; Julia, M. Nickel-Catalysed Substitutions of Aryl *tert*-Butyl Sulfones with Organometallic Reagents: Synthesis of *ortho*-Substituted Unsymmetrical Biaryls. *J. Chem. Soc. Perkin Trans.* **1995**, *1*, 7–14. [CrossRef]
15. Oviedo, A.; Torres-Nieto, J.; Arévalo, A.; García, J.J. Deoxydesulfurization of Sulfones Derived from Dibenzothiophene Using Nickel Compounds. *J. Mol. Catal. A Chem.* **2008**, *293*, 65–71. [CrossRef]
16. Li, W.-X.; Huo, B.-J.; Huang, J.-Y.; Rao, W.; Xu, H.; Zhou, X.; Shen, Z.-L. Nickel-Catalyzed Cross-electrophile Coupling of Unactivated (hetero)aryl Sulfone with Aryl Bromide. *J. Catal.* **2024**, *430*, 115359. [CrossRef]
17. Liu, J.; Robins, M.J. Fluoro, Alkylsulfanyl, and Alkylsulfonyl Leaving Groups in Suzuki Cross-Coupling Reactions of Purine 2′-Deoxynucleosides and Nucleosides. *Org. Lett.* **2005**, *7*, 1149–1151. [CrossRef]
18. Hennessy, E.J.; Cornebise, M.; Gingipalli, L.; Grebe, T.; Hande, S.; Hoesch, V.; Huynh, H.; Throner, S.; Varnes, J.; Wu, Y. Preparation of Highly Functionalized 1,5-Disubstituted Tetrazoles via Palladium-Catalyzed Suzuki Coupling. *Tetrahedron Lett.* **2017**, *58*, 1709–1713. [CrossRef]
19. Zhang, Q.; Zhang, W.; Cui, R.; Zhang, Y.; Niu, D. Synthesis of 2,4-Diarylated Pyrimidines Enabled by Ni-Catalyzed C–Sulfone Bond Activation. *Org. Chem. Front.* **2023**, *10*, 645–650. [CrossRef]
20. Chatelain, P.; Sau, A.; Rowley, C.N.; Moran, J. Suzuki–Miyaura Coupling of (Hetero)Aryl Sulfones: Complementary Reactivity Enables Iterative Polyaryl Synthesis. *Angew. Chem. Int. Ed.* **2019**, *58*, 14959–14963. [CrossRef]
21. Fukuda, J.-I.; Nogi, K.; Yorimitsu, H. Cross-Coupling of Aryl Trifluoromethyl Sulfones with Arylboronates by Cooperative Palladium/Rhodium Catalysis. *Org. Lett.* **2019**, *21*, 8987–8991. [CrossRef] [PubMed]
22. Gutiérrez-Ordaz, R.; García, J.J. Desulfurization of Dibenzothiophene and Dibenzothiophene Sulfone via Suzuki–Miyaura Type Reaction: Direct Access to *o*-Terphenyls and Polyphenyl Derivatives. *Polyhedron* **2018**, *154*, 373–381. [CrossRef]
23. Markovic, T.; Murray, P.R.D.; Rocke, B.N.; Shavnya, A.; Blakemore, D.C.; Willis, M.C. Heterocyclic Allylsulfones as Latent Heteroaryl Nucleophiles in Palladium-Catalyzed Cross-Coupling Reactions. *J. Am. Chem. Soc.* **2018**, *140*, 15916–15923. [CrossRef] [PubMed]

24. Cook, X.A.F.; Pantaine, L.R.E.; Blakemore, D.C.; Moses, I.B.; Sach, N.W.; Shavnya, A.; Willis, M.C. Base-Activated Latent Heteroaromatic Sulfinates as Nucleophilic Coupling Partners in Palladium-Catalyzed Cross-Coupling Reactions. *Angew. Chem. Int. Ed.* **2021**, *60*, 22461–22468. [CrossRef]
25. Huang, X.; Tang, L.; Song, Z.; Jiang, S.; Liu, X.; Ma, M.; Chen, B.; Ma, Y. Nickel-Catalyzed Desulfonylative Reductive Cross-Coupling of Aryl Sulfones with Aryl Bromides. *Org. Lett.* **2023**, *25*, 1198–1203. [CrossRef] [PubMed]
26. Takahashi, F.; Nogi, K.; Yorimitsu, H. Intramolecular Desulfitative Coupling: Nickel-Catalyzed Transformation of Diaryl Sulfones into Biaryls via Extrusion of SO_2. *Org. Lett.* **2018**, *20*, 6601–6605. [CrossRef]
27. Yu, T.-Y.; Zheng, Z.-J.; Bai, J.-H.; Fang, H.; Wei, H. Nickel-Catalyzed Intramolecular Coupling of Sulfones via the Extrusion of Sulfur Dioxide. *Adv. Synth. Catal.* **2019**, *361*, 2020–2024. [CrossRef]
28. García Ruano, J.L.; Alemán, J.; Paredes, C.G. Oxidative Addition of Pd(0) to $Ar{-}SO_2R$ Bonds: Heck-Type Reactions of Sulfones. *Org. Lett.* **2006**, *8*, 2683–2686. [CrossRef]
29. Matsubara, R.; Kim, H.; Sakaguchi, T.; Xie, W.; Zhao, X.; Nagoshi, Y.; Wang, C.; Tateiwa, M.; Ando, A.; Hayashi, M.; et al. Modular Synthesis of Carbon-Substituted Furoxans via Radical Addition Pathway. Useful Tool for Transformation of Aliphatic Carboxylic Acids Based on "Build-and-Scrap" Strategy. *Org. Lett.* **2020**, *22*, 1182–1187. [CrossRef]
30. Yang, J.; Xiao, J.; Chen, T.; Yin, S.-F.; Han, L.-B. Efficient nickel-catalyzed phosphinylation of C–S bonds forming C–P bonds. *Chem. Commun.* **2016**, *52*, 12233–12236. [CrossRef]
31. Saito, H.; Nogi, K.; Yorimitsu, H. Palladium-Catalyzed Double Borylation of Diaryl Sulfoxides with Diboron. *Synthesis* **2017**, *49*, 4769–4774.
32. Liu, X.-Y.; Dou, Y.-X.; Hasan, M.; Rao, W.; Shen, D.; Shen, S.; Wang, S.-Y. Ni-catalyzed cross-coupling of heteroaryl sulfones and diselenides via deheteroaromatization and heteroaromatization: Synthesis of heteroaryl selenides. *Org. Chem. Front.* **2024**, *11*, 1469–1472. [CrossRef]
33. Romero, N.A.; Nicewicz, D.A. Organic Photoredox Catalysis. *Chem. Rev.* **2016**, *116*, 10075–10166. [CrossRef]
34. Crisenza, G.E.M.; Melchiorre, P. Chemistry Glows Green with Photoredox Catalysis. *Nat. Commun.* **2020**, *11*, 803. [CrossRef] [PubMed]
35. Kamijo, S.; Kamijo, K.; Murafuji, T. Synthesis of Alkylated Pyrimidines via Photoinduced Coupling Using Benzophenone as a Mediator. *J. Org. Chem.* **2017**, *82*, 2664–2671. [CrossRef] [PubMed]
36. Wang, Z.-J.; Zheng, S.; Matsui, J.K.; Lu, Z.; Molander, G.A. Desulfonative Photoredox Alkylation of *N*-Heteroaryl Sulfones—An Acid-Free Approach for Substituted Heteroarene Synthesis. *Chem. Sci.* **2019**, *10*, 4389–4393. [CrossRef]
37. Kamijo, S.; Kamijo, K.; Murafuji, T. Aryl Ketone Mediated Photoinduced Radical Coupling for the Alkylation of Benzazoles Employing Saturated Heterocyclic Compounds. *Synthesis* **2019**, *51*, 3859–3864.
38. Li, P.; Tu, J.-L.; Gao, H.; Shi, C.; Zhu, Y.; Guo, L.; Yang, C.; Xia, W. $FeCl_3$-Catalyzed $C(sp^3){-}H$ Heteroarylation Enabled by Photoinduced Ligand-to-Metal Charge Transfer. *Adv. Synth. Catal.* **2024**, *366*, 220–224. [CrossRef]
39. Yonekura, K.; Murooka, M.; Aoki, K.; Shirakawa, E. Electrochemical Direct α-Arylation of Alkylamines with Sulfonylarenes. *Org. Lett.* **2023**, *25*, 6682–6687. [CrossRef]
40. Meng, L.; Dong, J.; Tang, Y.; Yang, H.; Sun, L.; Chen, J.; Fan, B. Photoredox Catalytic Aminomethylation of Sulfonylthiazoles. *Green Chem.* **2024**, *26*, 6063–6067. [CrossRef]
41. Yu, J.; Wu, Z.; Zhu, C. Efficient Docking–Migration Strategy for Selective Radical Difluoromethylation of Alkenes. *Angew. Chem. Int. Ed.* **2018**, *57*, 17156–17160. [CrossRef] [PubMed]
42. Liu, J.; Wu, S.; Yu, J.; Lu, C.; Wu, Z.; Wu, X.; Xue, X.-S.; Zhu, C. Polarity Umpolung Strategy for the Radical Alkylation of Alkenes. *Angew. Chem. Int. Ed.* **2020**, *59*, 8195–8202. [CrossRef] [PubMed]
43. Kim, M.; Lee, K.; Kim, S.; Hong, S. Photoinduced Carboarylation of Alkenes by Using Bifunctional Reagents. *Synlett* **2023**, *34*, 1437–1441.

molecules

MDPI

Article

Selenium Biofortification Effect on Glucosinolate Content of *Brassica oleracea* var. *italic* and *Eruca vesicaria*

Azra Đulović [1], Katarina Usanović [1], Lea Kukoč Modun [2] and Ivica Blažević [1,*]

[1] Department of Organic Chemistry, Faculty of Chemistry and Technology, University of Split, Ruđera Boškovića 35, 21000 Split, Croatia; azra@ktf-split.hr (A.Đ.); katarina.usanovic@ktf-split.hr (K.U.)

[2] Department of Analytical Chemistry, Faculty of Chemistry and Technology, University of Split, Ruđera Boškovića 35, 21000 Split, Croatia; lea.kukoc-modun@ktf-split.hr

* Correspondence: blazevic@ktf-split.hr

Abstract: Glucosinolates (GSLs) in different plant parts of broccoli (*Brassica oleracea* var. *italic*) and rocket (*Eruca vesicaria*) were analyzed qualitatively and quantitatively before and after treatment with sodium selenate (2 and 5 mM), by their desulfo-counterparts using the UHPLC-DAD-MS/MS technique. Twelve GSLs were detected in broccoli (five aliphatic, one arylaliphatic, and six indolic), where 4-(methylsulfanyl)butyl GSL (glucoerucin) was the main one in the roots (4.88–9.89 μmol/g DW), 4-(methylsulfinyl)butyl GSL (glucoraphanin) in stems (0.44–1.11 μmol/g DW), and 4-hydroxyindol-3-ylmethyl GSL (4-hydroxyglucobrassicin) in leaves (0.51–0.60 μmol/g DW). No GSL containing selenium was detected in the treated broccoli. Ten GSLs were detected in rocket (seven aliphatic and three indolic), where 4-(methylsulfanyl)butyl GSL (glucoerucin) was the main one in the roots (4.50–20.59 μmol/g DW) and 4-methoxyindol-3-ylmethyl GSL (4-methoxyglucobrassicin) in the aerial part (0.57–5.69 μmol/g DW). As a result of induced stress by selenium fertilization, the total GSL content generally increased in both plants. In contrast to broccoli, the roots and the aerial part of the rocket treated with a high concentration of sodium selenate contained 4-(methylseleno)butyl GSL (glucoselenoerucin) (0.36–4.48 μmol/g DW). Although methionine-derived GSLs are the most abundant in both plants, the plants' ability to tolerate selenate and its regulation by selenoglucosinolate production is species- and growth-stage-dependent.

Keywords: glucosinolates; broccoli; rocket; 4-(methylseleno)butyl glucosinolate (glucoselenoerucin); UHPLC-MS/MS

Citation: Đulović, A.; Usanović, K.; Kukoč Modun, L.; Blažević, I. Selenium Biofortification Effect on Glucosinolate Content of *Brassica oleracea* var. *italic* and *Eruca vesicaria*. *Molecules* **2023**, *28*, 7203. https://doi.org/10.3390/molecules28207203

Academic Editor: Jalal Hawari

Received: 29 August 2023
Revised: 18 October 2023
Accepted: 19 October 2023
Published: 21 October 2023

1. Introduction

Selenium (Se) is an important micronutrient required by most living organisms. The difference between deficiency and toxicity is incredibly narrow compared to other micronutrients, with toxic dosages of >400 μg/day and a dietary deficiency of 40 μg/day, which is an emerging global problem [1–3]. Plants are important sources of organic Se, as they have the capacity to accumulate inorganic Se or metabolites and store them as organic Se forms. Selenates (SeO_4^{2-}) and selenites (SeO_3^{2-}) are the most significant inorganic forms of Se that are easily taken by plants [4]. Selenate is the most common and fastest-assimilating form of bioavailable Se in alkaline soils with high oxygen content, whereas selenite is more prevalent in anoxic settings with higher acidity and humidity [4]. Plants usually intake selenate, which is reduced, incorporated into organic compounds (selenomethionine, selenocysteine) and further transported through the food chain, and finally decomposed and excreted from the organism, thus completing the biogeochemical cycle of Se [5]. Se levels in soil generally reflect its presence in foods and, therefore, its availability to humans. With an estimated mean Se content in human soft tissues of 110 μg/kg, it was found that the Se intake status in Europe was low, which means that the Se levels in European soils are not sufficient [1,3]. Therefore, biofortification is an important strategy to increase Se in the edible parts of plants.

Based on their ability to accumulate Se, plants can be categorized into three main groups: non-accumulators (accumulate less than 100 mg Se/kg dry weight, DW), secondary accumulators or accumulators (accumulate up to 1000 mg Se/kg DW), and hyperaccumulators (accumulate over 1000 mg Se/kg DW) [6]. The ability to hyperaccumulate Se appears to have evolved within the Asteraceae, Brassicaceae, and Fabaceae families [6]. Broccoli (*Brassica oleracea* var. *italica*) and rocket (*Eruca vesicaria*) are members of Brassicaceae family, which is the largest family in the order Brassicales. Some known Se secondary accumulators include several plants of the *Brassica* genus, such as *Brassica juncea*, *Brassica napus*, and *Brassica oleracea* var. *italic*, whereas Se-hyperaccumulators include plants of the *Cardamine* genus (*Cardamine hupingshanensis*, *Cardamine violofolia*) and *Stanleya* genus (*Stanleya pinnata*, *Stanleya bipinnata*) [6,7].

Glucosinolates (GSLs) are well-defined specialized plant metabolites containing sulfur and nitrogen which represent the molecular tags of plants from the Brassicaceae family. Generally, they are found in 16 families of the order Brassicales, as well as several plants outside of this order, such as the genus *Drypetes* (family Putranjivaceae, order Malpighiales) and *Rinorea* genus (family Violaceae, order Malpighiales) [8,9]. GSLs can be divided into three classes based on the structure of various amino acid precursors: aliphatic GSLs derived from methionine, isoleucine, leucine, or valine; arylaliphatic GSLs derived from phenylalanine or tyrosine; and indole GSLs derived from tryptophan. Only 90 of the 139 GSLs discovered in the plant kingdom have been thoroughly described by relevant spectroscopy techniques (MS, NMR) to date [10–12]. Se-biofortification influences GSL production with possible exchange positions of sulfur (S) by selenium (Se) in three places in the side chain, in thioglucose, and in the sulfate group (Figure 1) [7].

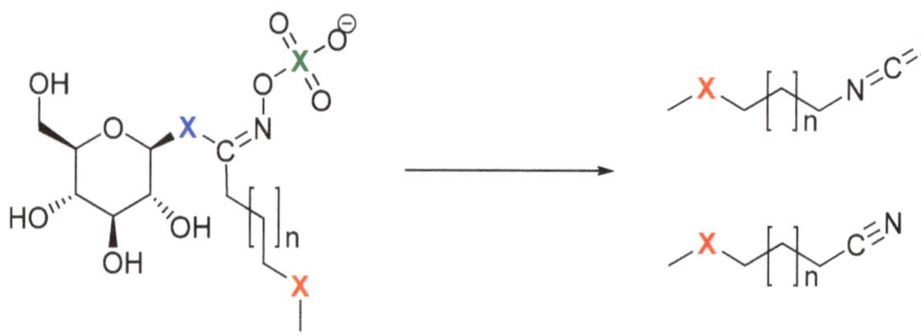

Figure 1. Possible sites (X) for selenium to replace sulfur: green—incorporated in sulfate group; blue—incorporated in thioglucose, and then it is selenoGSL, which produces isoselenocyanides; red—incorporated in side chain, and then it is selenoMet-derived GSL, as breakdown results in isothiocyanates.

Several crop species have been biofortified with Se in field or greenhouse experiments where various Se sources and techniques were used. According to Stewart et al. (1974), sinigrin, which was isolated from horseradish (*Armoracia rusticana*) after receiving $Na_2{}^{75}SeO_4$, was shown to contain Se in its structure [13]. After three weeks at high Se concentrations (100 ppm) in water-cultured specimens of the Se-tolerant species desert princes' plume (*Stanleya pinnata*), Bertelsen et al. (1988) detected trace quantities of GSL containing Se named as but-3-enylselenoglucosinolate, whereas those grown in environments with lower, more typical Se amounts did not contain this compound [14]. Additionally, no such selenoglucosinolates were found in garden cress (*Lepidium sativum*) and horseradish (*Armoracia rusticana*) [14]. This selenoglucosinolate, originally defined as a compound in which the Se atom is replaced by an S atom in the anomeric position in the GSL structure, is converted into isoselenocyanates by enzymatic hydrolysis [14]. Since the formation of isoselenocyanates rather than isothiocyanates significantly departs from the characteristics of GSLs, they are, by definition, not GSLs (Figure 1).

Matich et al. (2012) treated broccoli (*B. oleracea* var. *italica*), cauliflower (*B. oleracea* L.), and forage rape (*B. napus*) with elevated sodium selenate and examined them for the presence of organoselenides. Based on the analysis of broccoli and cauliflower florets, and the roots of forage rape, the discovery of putative GSL hydrolysis products using synthesized standards confirmed by MS and NMR, as well as by LC-MS analysis, three Se-Met artificially derived GSLs of the general type (methylseleno)alkyl GSL are accepted: 3-(methylseleno)propyl GSL (glucoselenoibervirin **[144]**), 4-(methylseleno)butyl GSL (glucoselenoerucin, **[145]**), and 5-(methylseleno)pentyl GSL (glucoselenoberteroin, **[146]**) [15]. This analysis showed that Se is preferentially incorporated into the methylselenyl moiety rather than the sulfate or β-thioglucose groups. A further Se-biofortification study by Matich et al. (2015) using HPLC-MS/MS tentatively indicated glucoselenoraphanin in broccoli (*B. oleracea* var. *italica*) florets with Se envisioned in the side chain and glucoselenonasturtiin in forage rape (*B. napus*) roots, but with Se, that could not be envisioned to be on the side chain, suggesting a selenoGSL sensu Bertelsen et al. [14,16].

Black mustard (*Brassica nigra*) seeds grown on naturally Se-rich soils in the Punjab region of India, where the Se soil content ranges from 2 to 7 mg/kg, were studied by Ouerdane et al. (2013) [17]. Using HPLC ESI Orbitrap MS(/MS) and GC APCI TQ MS/MS for the extract analysis, among over 30 Se species, they detected the presence of "selenoglucosinolates", namely glucoselenoibervirin and glucoselenoerucin, as well as their corresponding degradation products 3-(methylseleno)propylisothiocyanate and 4-(methylseleno)butanenitrile, i.e., 4-(methylseleno)butylisothiocyanate and 5-(methylseleno) pentanenitrile, respectively. To avoid confusion with selenoglucosinolates according to Bertelsen et al. [14], GSLs with Se in the side chain are suggested to be referred to as selenoMet-derived GSLs [8]. The presence of Se in the side chains is fully compatible with the definition of GSLs because they form isothiocyanates and not isoselenocyanates as hydrolysis products (Figure 1).

The effects of Se-fertilization on GSL production in a radish (*Raphanus sativus*) have been studied by McKenzie et al. (2019) throughout a period of five developmental phases (from seed to fully developed salad greens). With the double bond geometry still unclear, they used tandem mass spectrometry to determine the existence of a novel Se-containing GSL, 4-(methylseleno)but-3-enyl GSL (also named selenoglucoraphenin) [18]. By using GC-MS, two similar isothiocyanates of 4-(methylseleno)but-3-enyl isothiocyanate were tentatively identified as (*E*/*Z*?) isomers. Se-biofertilization of mature radish led to the presence of selenoglucosinolates in the seed [18].

The aim of this study is to investigate the influence of Se intake and its metabolism, with special emphasis on the identification and quantification of GSLs and selenoMet-derived GSLs. For this purpose, cultivated broccoli (*Brassica olaracea* var. *italic*) and rocket (*Eruca vesicaria*) were watered with sodium selenate salt solutions (Na_2SeO_4) of different concentrations for 35 days and GSLs were analyzed by UHPLC-DAD-MS/MS according to the ISO 9167-1 official method [19].

2. Results and Discussion

2.1. Effect of Selenium Treatment on Broccoli

Brassica oleracea var. *italica*, a young plant which, after 18 days, had no developed flowers or florets, was exposed to Na_2SeO_4 concentrations of 2 mM and 5 mM over a period of 35 days. During that time period, the plants did not exhibit any signs of decay of the existing plant parts throughout cultivation. The detrimental influence of selenate solution on the aerial part's growth was notable when compared to the reference plant's height, which was 33 cm, and in comparison to 25 cm of the plant exposed to 5 mM selenate solution (Figure S1, Table S1). Se-biofertilization had a detrimental effect on root growth, as evidenced by the fact that the reference plant's measured roots were 24 cm long compared to 6 cm length of roots watered with the highest Na_2SeO_4 concentration (Figure S1, Table S1). The symptom of Se exposure is a reduction in root elongation. Selenate treatment was shown to decrease cytokinin oxidase transcript levels. Cytokinin oxidases in

Arabidopsis are known to catalyze the irreversible degradation of the hormone cytokinin, in which accumulation in the root tips may partly be responsible for the root meristem shortening [20–22]. A previous report by Adamopoulou et al. included the study of Se toxicity on broccoli florets grown hydroponically in a greenhouse for 12 weeks [23,24]. The plants were fortified with two different concentrations of sodium selenate (1.5 mM and 3.0 mM) between the 5th and 10th week. Although broccoli florets grown in the presence and absence of sulfur had the same weight, the absence of sulfur also led to increased Se toxicity, which reduced weight by up to 65%. Also, the leaves' optical characteristics revealed no visible signs of the biofortification [23,24]. Tian et al. reported the influence of selenium treatment on broccoli growth and showed that when S nutrition was low, Se was particularly harmful to plants and dramatically reduced plant sizes. They proposed that the Se toxicity could be counteracted by increasing sulfate supplementation, which would likely occur via the decreasing non-specific integration of Se into proteins and altering the redox system [25].

GSLs were qualitatively and quantitatively analyzed via desulfated forms using UHPLC-DAD-MS/MS, and the results are presented in Table 1 and Figure 2, as well as in the Supplementary Materials (Figures S3–S6).

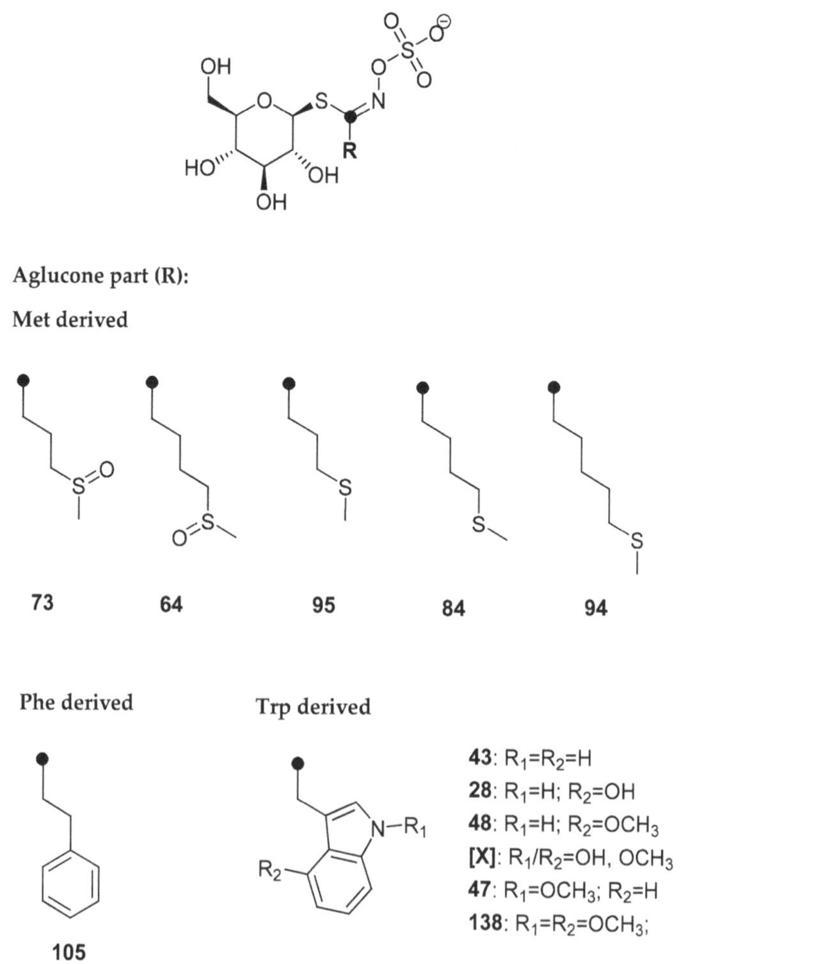

Figure 2. Structures of the GSLs identified in *Brassica oleracea* var. *italica* (*cf.* Table 1).

Table 1. Glucosinolate content (μmol/g DW) in *Brassica oleracea* var. *italica*.

No.	Glucosinolate	Formula (desulfo)	Desulfo M	Desulfo $[M + Na]^+$	Reference Root	Reference Stem	Reference Leaf	Na_2SeO_4 Solution 2 mM Root	2 mM Stem	2 mM Leaf	5 mM Root	5 mM Stem	5 mM Leaf
	Met derived												
73	3-(Methylsulfinyl)propyl GSL (glucoiberin) [a]	$C_{11}H_{21}NO_7S_2$	343	366	tr	tr	n.d.	tr	tr	n.d.	tr	tr	n.d.
64	4-(Methylsulfinyl)butyl GSL (glucoraphanin) [a]	$C_{12}H_{23}NO_7S_2$	357	380	2.09 ± 0.12 ^{A}	1.11 ± 0.06 ^{D}	tr	1.40 ± 0.13 ^{E}	0.58 ± 0.04 ^{B}	tr	3.68 ± 0.87 ^{C}	0.44 ± 0.12 ^{E}	tr
95	3-(Methylsulfanyl)propyl GSL (glucoibervirin) [b]	$C_{11}H_{21}NO_6S_2$	327	350	n.d.	n.d.	n.d.	n.d.	n.d.	n.d.	1.03 ± 0.12	n.d.	n.d.
84	4-(Methylsulfanyl)butyl GSL (glucoerucin) [a]	$C_{12}H_{23}NO_6S_2$	341	364	4.88 ± 0.72 ^{A}	0.13 ± 0.00 ^{D}	n.d.	9.05 ± 0.12 ^{B}	0.11 ± 0.00 ^{E}	n.d.	9.89 ± 0.57 ^{B}	tr	n.d.
94	5-(Methylsulfanyl)pentyl GSL (glucoberteroin) [a]	$C_{15}H_{21}NO_6S$	355	378	tr	n.d.	tr	n.d.	n.d.	n.d.	tr	n.d.	n.d.
	Phe derived												
105	2-Phenylethyl GSL (gluconasturtiin) [a]	$C_{11}H_{21}NO_7S_2$	343	366	tr	n.d.	tr	tr	n.d.	n.d.	0.31 ± 0.00	n.d.	n.d.
	Trp derived												
43	Indol-3-ylmethyl GSL (glucobrassicin) [a]	$C_{16}H_{20}N_2O_6S$	368	391	0.68 ± 0.14 ^{A}	0.07 ± 0.00 ^{D}	0.31 ± 0.04 ^{G}	2.34 ± 0.13 ^{B}	0.55 ± 0.12 ^{E}	0.09 ± 0.00 ^{H}	1.20 ± 0.18 ^{C}	0.25 ± 0.00 ^{F}	0.58 ± 0.12 ^{I}
28	4-Hydroxyindol-3-ylmethyl GSL (4-hydroxyglucobrassicin) [a]	$C_{16}H_{20}N_2O_7S$	384	407	0.62 ± 0.10 ^{A}	tr	0.60 ± 0.20 ^{G}	0.78 ± 0.19 ^{A}	0.22 ± 0.09	0.51 ± 0.15 ^{G}	tr	tr	0.57 ± 0.00 ^{G}
48	4-Methoxyindol-3-ylmethyl GSL (4-methoxyglucobrassicin) [a]	$C_{17}H_{22}N_2O_7S$	398	421	3.23 ± 0.43 ^{A}	0.15 ± 0.00 ^{D}	0.09 ± 0.00	4.23 ± 0.12 ^{B}	0.15 ± 0.00 ^{D}	0.10 ± 0.00	3.80 ± 0.43 ^{A,B}	0.53 ± 0.12 ^{E}	0.45 ± 0.00
[X]	Hydroxymethoxy-glucobrassicin	$C_{17}H_{22}N_2O_8S$	414	437	tr	tr	0.24 ± 0.08 ^{G}	0.04 ± 0.00	0.04 ± 0.00	0.02 ± 0.00 ^{H}	n.d.	n.d.	0.03 ± 0.00 ^{H}
47	*N*-Methoxyindol-3-ylmethyl GSL (neoglucobrassicin) [a]	$C_{17}H_{22}N_2O_7S$	398	421	1.97 ± 0.31 ^{A}	0.20 ± 0.05 ^{D}	0.04 ± 0.00 ^{G}	2.73 ± 0.33 ^{B}	0.56 ± 0.12 ^{E}	tr	2.43 ± 0.12 ^{A,B}	0.20 ± 0.00 ^{D}	0.18 ± 0.00 ^{G}
138	1,4-Dimethoxyindol-3-ylmethyl GSL (1,4-dimethoxyglucobrassicin) [b]	$C_{18}H_{24}N_2O_8S$	428	451	n.d.	n.d.	n.d.	n.d.	n.d.	n.d.	tr	n.d.	n.d.
	Total (μmol/g DW)				13.48 ± 1.82 ^{J}	1.66 ± 0.11 ^{M,N}	1.28 ± 0.32 ^{P,R}	20.57 ± 1.02 ^{K}	2.21 ± 0.37 ^{M,N}	0.72 ± 0.15 ^{P,R}	22.34 ± 2.29 ^{K}	1.42 ± 0.24 ^{M}	1.81 ± 0.12 ^{P}

No.—numbers are related to the glucosinolate number given in review paper by Blažević et al. [10]; Desulfo $[M + Na]^+$, sodium adduct of desulfoglucosinolate used for GSL identification. [a] Compound identified by MS^2 spectra and t_R comparison with standard. [b] Compound identified by MS^2 spectra and t_R comparison with the literature. The structures are shown in Figure 2. All chromatograms are given in Figures S3–S5, and MS^2 spectra of all identified desulfoGSLs are given in Figure S10, except of hypothesized dGSL designated as **[X]**, the MS^2 spectra of which are given in Figure S6. GSL—glucosinolate; tr—traces (<0.01 μmol/g DW); n.d.—not detected; DW—dry weight of plant material. Data are expressed as the mean value ± standard error (n = 3). Statistical analyses are performed by one-way ANOVA followed by Tukey's honest significant test. Mean values with a different letter differ significantly ($p \leq 0.05$). Statistical significance: root—letters $^{A–C}$, stem—letters $^{D–F}$, leaf—letters $^{G–I}$, root total—letters ^{J,K} stem total—letters ^{M,N}, leaf total ^{P,R}.

Five Met-derived desulfoGSLs (dGSLs) with sulfur in the side chain, one Phe-derived dGSL, and six Trp-derived dGSLs were identified. Based on the obtained $[M+Na]^+$, no GSLs with Se in the structure were detected. At t_R = 8.30 min, $[M + Na]^+$ = 437 was observed, which was assumed to be a hitherto unidentified dGSL, and was labeled **[X]**. The UV spectrum of dGSL indicated that it is an indole GSL, and the MS^2 spectrum was recorded in order to identify the characteristic fragments (Figure S6). Fragments m/z 185 ("a", Na^+ adduct of $C_6H_{10}O_5$) and m/z 274 ("c", loss of $C_6H_{10}O_5$) are observed, which suggested that it is a GSL. The fragment m/z 389 indicated the loss of the methoxy and hydroxyl groups. Based on this information, hydroxymethoxyglucobrassicin was proposed, since it is known that hydroxylation and the subsequent methylation of the indole ring takes place at positions 1 and 4 (R_1 and R_2, respectively, Figure 2) [26]. Namely, the key enzymes are cytochrome P450 monooxygenases of the CYP81F subfamily, which carry out hydroxylation reactions at position 4 or 1 of the indole ring. IGMT enzymes, which belong to the family of plant *O*-methyltransferases, convert hydroxy intermediates into methoxylated products, and the enzyme IGMT5 specific for position 1 of the indole ring was found, i.e., IGMT1-4 enzymes specific for position 4 [27].

The aliphatic and indole GSL type represent the major GSL found in the analyzed plant parts, with the roots having the highest GSL content. The total content of GSLs in the roots increased for both solution concentrations added (2 mM and 5 mM) when compared to the control and differed significantly ($p \leq 0.05$). Glucoerucin (**84**), as the main GSL in roots, significantly increased from 4.88 µmol/g DW to 9.05 and 9.89 µmol/g DW, respectively.

Several studies have reported the existence of Se-GSLs in certain plant parts, such as florets and roots, after the Se-biofertilization of GSL-containing plants. Matich et al. (2012, 2015) showed the presence of Se-GSLs in biofortified broccoli florets [15,16]. However, the same plant in our study did not reach this stage of development (no florets) and no Se-GSLs were detected. According to Tian et al., the total GSL contents of broccoli sprouts determined by glucose from GSL hydrolysis were not significantly changed by the impacts of Se treatments (100 µmol/L selenite and selenate) [25]. However, since the analysis was not conducted in accordance with the ISO 9167 standard method [19], this should only be accepted cautiously.

2.2. *Effect of Selenium Treatment on Rocket*

Eruca vesicaria (rocket), a 24-day old plant, was also exposed to Na_2SeO_4 concentrations of 2 mM and 5 mM over a period of 35 days. A similar detrimental influence of Se-biofortification on morphological characteristics was observed for rocket as in broccoli. When sodium selenate (5 mM) was applied, the plant's height decreased from ca. 23 cm to 8 cm and its root length from ca. 15 cm to 10 cm. In this case, the aerial part biomass decreased, and the beginning of decay was observed for the highest concentration (Figure S2, Table S2).

GSLs were qualitatively and quantitatively analyzed via desulfated forms using UHPLC-DAD-MS/MS, and the results are presented in Table 2 and Figures 3 and 4, as well as in the Supplementary Materials (Figures S7–S12).

Six methionine-derived GSLs and three indole ones were identified in the rocket. By comparing the chromatograms obtained for the root and aerial component, it was discovered that Se-treated samples exhibit a signal that is not present in the reference plant, particularly in the root. Its $[M+Na]^+$ was 412 at t_R = 7.6 min, which did not correspond to previously known GSL of that mass. It was assumed that it is selenomethionine-derived selenoglucosinolate, namely glucoselenoerucin ([**145**]), whose desulfo form structure is given in Figure 4, since the obtained mass corresponds to this dGSL when sulfur is replaced by selenium.

Table 2. Glucosinolate content (µmol/g DW) in cultivated *Eruca vesicaria*.

No.	Glucosinolate (GSL)	Formula (desulfo)	Desulfo M	Desulfo $[M + Na]^+$	Reference		Na_2SeO_4 solution			
							2 mM		5 mM	
					Root	Aerial Part	Root	Aerial Part	Root	Aerial Part
	Met derived									
64	4-(Methylsulfinyl)butyl GSL (glucoraphanin) [a]	$C_{12}H_{23}NO_7S_2$	357	380	3.23 ± 0.63 [A]	0.87 ± 0.22 [D]	1.69 ± 0.34 [B]	0.95 ± 0.11 [D]	0.75 ± 0.15 [C]	0.74 ± 0.12 [D]
135	4-(β-D-Glucopyranosyldisulfanyl)butyl GSL (diglucothiobeinin) [b]	$C_{17}H_{31}NO_{11}S_3$	521	544	tr	1.14 ± 0.24	n.d.	n.d.	n.d.	n.d.
95	3-(Methylsulfanyl)propyl GSL (glucoibervirin) [b]	$C_{11}H_{21}NO_6S_2$	327	350	1.88 ± 0.33	1.88 ± 0.17	n.d.	n.d.	n.d.	n.d.
84	4-(Methylsulfanyl)butyl GSL (glucoerucin) [a]	$C_{12}H_{23}NO_6S_2$	341	364	20.59 ± 1.32 [A]	2.78 ± 0.33 [D]	17.27 ± 1.12 [A]	1.30 ± 0.08 [E]	4.50 ± 0.55 [B]	0.21 ± 0.05 [F]
134	Dimeric 4-mercaptobutyl GSL [b]	$C_{22}H_{40}N_2O_{12}S_4$	652	675	4.33 ± 0.62 [A]	0.20 ± 0.00 [D]	17.14 ± 0.93 [B]	1.49 ± 0.22 [E]	6.78 ± 0.77 [C]	0.55 ± 0.12 [F]
65	10-(Methylsulfinyl)decyl GSL (glucocamelinin) [a]	$C_{18}H_{35}NO_7S_2$	441	464	tr	0.21 ± 0.12	tr	tr	n.d.	tr
	Se-Met derived									
[145]	4-(Methylseleno)butyl GSL (glucoselenoerucin)	$C_{12}H_{23}NO_6SSe$	389	412	n.d.	n.d.	4.48 ± 0.55 [A]	1.24 ± 0.33 [D]	0.78 ± 0.29 [B]	0.36 ± 0.10 [E]
	Trp derived									
28	4-Hydroxyindol-3-ylmethyl GSL (4-Hydroxyglucobrassicin) [a]	$C_{16}H_{20}N_2O_7S$	384	407	0.23 ± 0.00	n.d.	tr	n.d.	n.d.	n.d.
48	4-Methoxyindol-3-ylmethyl GSL (4-Methoxyglucobrassicin) [a]	$C_{17}H_{22}N_2O_7S$	398	421	1.16 ± 0.22 [A]	5.96 ± 1.03 [D]	0.80 ± 0.11 [B]	1.64 ± 0.36 [E]	0.07 ± 0.00 [C]	0.57 ± 0.12 [F]
138	1,4-Dimethoxyindol-3-ylmethyl GSL (1,4-dimethoxyglucobrassicin) [b]	$C_{18}H_{24}N_2O_8S$	428	451	0.92 ± 0.12 [A]	0.20 ± 0.00	2.06 ± 0.18 [B]	0.15 ± 0.00	1.33 ± 0.31 [A]	0.04 ±0.00
	Total (µmol/g DW)				32.34 ± 3.24 [G]	13.24 ± 2.11 [J]	43.44 ± 3.23 [H]	6.77 ± 1.10 [K]	14.21 ± 2.07 [I]	2.47 ± 0.51 [L]

No.—numbers are related to the glucosinolate number given in review paper by Blažević et al. [10]; Desulfo $[M + Na]^+$, sodium adduct of desulfoglucosinolate used for GSL identification. [a] Compound identified by MS^2 spectra and t_R comparison with standard. [b] Compound identified by MS^2 spectra and t_R comparison with the literature. The structures are shown in Figure 3. All chromatograms are given in Figures S7 and S8, whereas MS^2 spectra of all identified sodium adduct of desulfoGSLs are given in Figure S10, except for 4-(methylseleno)butyl desulfoglucosinolate, designated as d**[145]**, for which the UV, MS, and MS^2 of are given in Figures S11 and S12. GSL—glucosinolate; tr—traces < 0.01 µmol/g DW; n.d.—not detected; DW—dry weight of plant material. Data are expressed as the mean value ± standard error (n = 3). Statistical analyses are performed by one-way ANOVA followed by Tukey's honest significant test. Mean values with a different letter differ significantly ($p \leq 0.05$). Statistical significance: root—letters [A–C], aerial part—letters [D–F], root total—letters [G–I], aerial part total—letters [J–L].

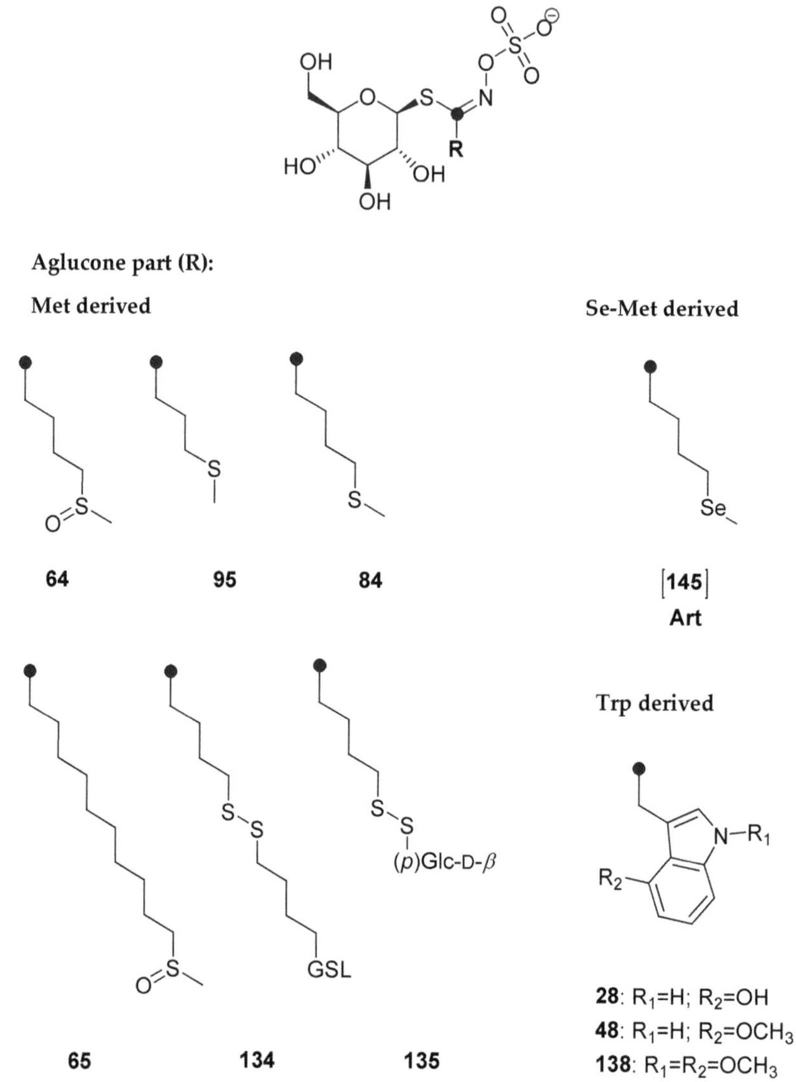

Figure 3. Structures of the GSLs identified in *Eruca vesicaria* (*cf.* Table 2).

In the MS^2 spectrum, fragments marked "a", "b", "c", and "d" are present, indicating that Se was not found in the thioglucose part, but in the side chain derived from selenomethionine (Se-Met). Right next to 412, 410 appears, which is assumed to be the isotope of selenium 78, so MS^2 was obtained for this $[M + Na]^+$ and fragment m/z 185 (a) was also observed (Figure S12). According to previous literature, this selenoglucosinolate was identified for the first time through its desulfated form and its MS^2 was shown for the first time.

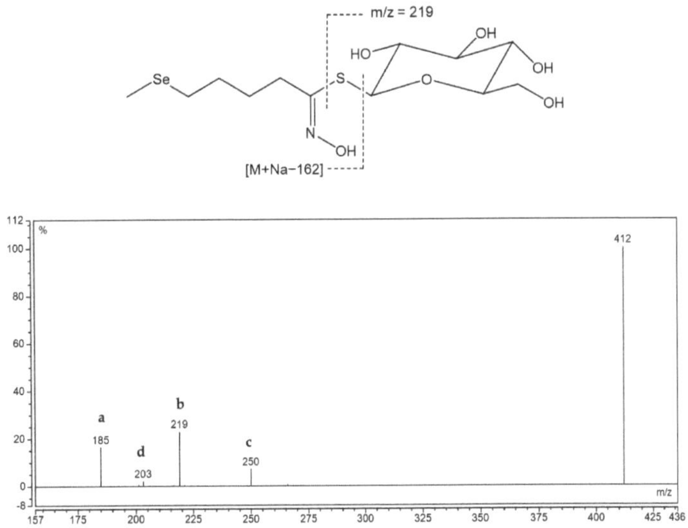

Figure 4. Structure of desulfoglucoselenoerucin **d[145]** and MS^2 spectra of its sodium adduct. Fragment types observed, alone or in combination, in MS^2 spectra desulfoglucosinolates in positive mode: **a**—Na^+ adduct of anhydroglucose, $C_6H_{10}O_5$ (at *m*/*z* 185) or an acyl derivative; **b**—Na^+ adduct of thioglucose, $C_6H_{11}O_5SH$ (at *m*/*z* 219) or an acyl derivative; **c**—Loss of anhydroglucose (*m*/*z* 162) or an acyl derivative; **d**—Na^+ adduct of glucose, $C_6H_{12}O_6$ (at *m*/*z* 203) [28].

The total GSL content is related mostly to the content of aliphatic GSLs, with the highest found in the roots. The occurrence of this GSL derived from Se-Met can be related to the high content of its sulfur analogue glucoerucin (**84**), and the highest content was obtained in the roots of the plant treated with 2 mM solution, 4.48 µmol/g DW. With the addition of 5 mM sodium solution, rocket showed the signs of decay and the total GSL content decreased, having 14.21 and 2.47 µmol/g DW in the roots and aerial parts, in comparison to the control having 32.34 and 13.24 µmol/g DW, respectively, and differed significantly ($p \leq 0.05$). Consequently, the content of GSL **[145]** decreased. Dall'Acqua et al. studied the Se-biofortification of two rocket species (*Eruca sativa* = *E. vesicaria* and *Diplotaxis tenuifolia*) grown at 0–40 µM Na_2SeO_4 in hydroponics and also showed a GSL decrease [29]. Contrary to our investigation, selenoglucosinolates have not been reported in these species. Also, they showed that selenium treatment reduced cysteine and methionine content, and glutathione (GSH), whose precursor is cysteine, also showed a trend toward reduction; thus, a decrease in GSLs, also observed in this study, may be explained by the negative effects of selenate on aminoacids biosynthesis, during which selenoamino acids can be formed and be incorporated into proteins and/or form selenoGSLs [30].

3. Materials and Methods

3.1. Materials and Reagents

4-(Methylsulfinyl)butyl GSL (glucoraphanin, **64**), 4-(methylsulfanyl)butyl GSL (glucoerucin, **84**), 5-(methylsulfanyl)pentyl GSL (glucoberteroin, **94**), 9-(methylsulfinyl)nonyl GSL (glucoarabin, **68**), 10-(methylsulfinyl)decyl GSL (glucocamelinin, **65**), 2-phenylethyl GSL (gluconasturtiin, **105**), indol-3-ylmethyl GSL (glucobrassicin, **43**), 4-hydroxyindol-3-ylmethyl GSL (4-hydroxyglucobrassicin, **28**), 4-methoxyindol-3-ylmethyl GSL (4-methoxyglucobrassicin, **48**), and *N*-methoxyindol-3-ylmethyl GSL (neoglucobrassicin, **47**) were purchased from Phytoplan Diehm & Neuberger GmbH (Heidelberg, Germany). 4- Hydroxybenzyl GSL (glucosinalbin, **23**) was isolated from *Sinapis alba* seeds and 3-(methylsulfinyl) propyl GSL (glucoiberin, **73**) was isolated from *Anastatica hierochuntica*, both confirmed by NMR. All other chemicals and reagents were of analytical grade.

3.2. Plant Growth and Harvesting

Young plants of *Brassica oleracea* var. *italica* (broccoli), one month old and with a total height of 18 cm (root 4 cm), were transplanted into pots (one plant per pot) (dimensions, (w × d × h) 12 × 12 × 10 cm) with soil Potgrond H (Potgrond H 70, Klasmann-Deilmann GmbH, Geeste, Germany) (pH = 6.0) and watered with tap water during the next fifteen days in order to adapt. On the fifteenth day after transplantation, the height of the above-ground part was measured, which was 18 cm, and the height of the stem was 10 cm. The reference plant was watered only with water, whereas the remaining two plants were watered with prepared solutions of sodium selenate with a concentration of 2 mM and 5 mM (three repetition per experiment). During cultivation, the plants were watered with tap water for 35 days, and they were biofortified twice a week with a 20 mL selenate solution. We estimate this selenium application to be equivalent to ca. 22 and 55 kg/ha, respectively, which is considerably higher than the 3–76 g/ha applied to agricultural and forage crops [15,16]. The growth and development of the above-ground part were recorded throughout the period, and on the 35th day from the beginning of watering, the length of the roots and the height of the above-ground part were measured.

Eruca vesicaria (rocket) was grown from the seeds of the Italian distributor Impex s.r.l. cement, obtained from the Sjeme store, Split, Croatia. Seeds (100 mg) were placed in each cultivation pot (n = 3) with soil Potgrond H and germination began on the 3rd day after watering. On the 10th day, the plants were transplanted into three larger pots (dimensions, (w × d × h) 12 × 12 × 10 cm) and on the 24th day after sowing, watering with sodium selenate solutions was started. The height of the aerial part of the plants before watering was 7.5 cm, and the length of the roots was 3 cm.

After collection, all plants were washed with tap water and lyophilized. They were divided into the root and aerial part for the purpose of the qualitative and quantitative analysis of GSLs. The lengths of plant parts were measured by a ruler with an accuracy of 0.1 cm, for the control and each treatment (n = 3).

3.3. Isolation and Chemical Analysis

3.3.1. Isolation of Desulfoglucosinolates

GSLs were extracted as previously reported [31]. Freeze-dried plant parts were ground to a fine powder, from which 100 mg was extracted for 5 min at 80 °C in 2 × 1 mL MeOH/H_2O (70:30 v/v) to inactivate the endogenous myrosinase. Each extract (1 mL) was loaded onto a mini-column filled with 0.5 mL of DEAE-Sephadex A-25 anion-exchange resin (GE Healthcare) conditioned with 25 mM acetate buffer (pH 5.6). After washing the column with 70% MeOH and 1 mL of ultrapure water, the optimal conditions for desulfation were set by adding a buffer solution. Each mini-column was loaded with 20 µL (0.35 U/mL) of purified sulfatase and left to stand for 18 h at room temperature. The desulfoGSLs were then eluted with 1.5 mL of ultra-pure H_2O, lyophilized, and diluted to 1 mL. The samples were stored at −20 °C until further analysis by UHPLC-DAD-MS/MS.

3.3.2. UHPLC-MS/MS Analysis

Analysis was performed on UHPLC-MS/MS (Ultimate 3000RS with TSQ Quantis MS/MS detector, Thermo Fisher Scientific, Waltham, MA, USA) using a Hypersil GOLD column (3.0 µm, 3.0 × 100 mm, Thermo Fisher Scientific, Waltham, MA, USA). A gradient consisting of solvent A (50 µM NaCl in H_2O) and solvent B (acetonitrile:H_2O 30:70 v/v) was applied at a flow rate of 0.5 mL/min as follows: 0.14 min 96% A; 7.84 min 14% A; 8.96 min 14% A; 9.52 min 5% A; 13.16 min 5% A; 13.44 min 96% A; 15.68 min 96% A. The column temperature was held at 15 °C for *E. vesicaria* dGSLs and 25 °C for *B. oleracea* var. *italica* dGSLs and the injection volume was 5 µL. The electrospray interface was an H-ESI source operating with a capillary voltage of 3.5 kV at 350 °C. The system was operated in the positive ion electrospray mode.

Qualitative analysis was performed by the manual examination of chromatograms and MS^2 spectra of each peak of desulfoglucosinolate as sodium adduct ($[M+Na]^+$). The amount of GSLs was quantified using a calibration curve of pure desulfosinigrin solution (range from 0.14 to 1.4 mM) and RPFs for each individual desulfoGSL [31]. The RPF values for quantification of desulfoGSLs were as follows: RPF 1.07 for 3-(methylsulfinyl)propyl GSL (glucoiberin, **73**) and 4-(methylsulfinyl)butyl GSL (glucoraphanin, **64**); 0.80 for 3-(methylsulfanyl)propyl GSL (glucoibervirin, **95**); 1.04 for 4-(methylsulfanyl)butyl GSL (glucoerucin, **84**); 0.50 for glucosinalbin (**23**); 0.95 for gluconasturtiin (**105**); 0.29 for glucobrassicin (**43**); 0.28 for 4-hydroxyglucobrassicin (**28**); 0.25 for 4-methoxyglucobrassicin (**48**); 0.20 for N-methoxyglucobrassicin (**47**) and 1,4-dimethoxyindol-3-ylmethyl GSL (**138**) [32,33]; arbitrary RPF 0.25 for hydroxymethoxyglucobrassicin (**[X]**), and 1.0 for 5-(methylsulfanyl) pentyl GSL (glucoberteroin, **94**), 9-(methylsulfinyl)nonyl GSL (glucoarabin, **68**), 9-(methylsulfonyl)nonyl GSL (**79**), 10-(methylsulfinyl)decyl GSL (**65**), 4-(methylseleno)butyl GSL **[145]**, 4-(β-D-glucopyranosyldisulfanyl)butyl GSL **(135)**, and dimeric 4-mercaptobutyl GSL (**134**). The bold numbers are related to the GSL number given in the review paper by Blažević et al. [10].

3.4. Statistical Analysis

One-way analysis of variance (ANOVA) was performed with SPSS software, version 25.0 (IBM Corporation, New York, NY, USA), in order to determine the difference between treatments. First, the data were tested for normal distribution and log-transformed for further analysis. One-way ANOVA was performed if the requirement of homogeneity of variance was fulfilled; otherwise, Welch correction was performed first. If there was a statistically significant effect between different treatments, Tukey's honest significant test was performed with a significance level of $p \leq 0.05$.

4. Conclusions

The morphological changes and changes in GSL content in broccoli and rocket were used to track the stress caused by selenate. Morphological alterations were detected in the form of decreased root development, which can be explained by the competing action of sulfate and selenate. Both plants' principal GSLs were produced from methionine biosynthesis, with the major GSL being 4-(methylsulfanyl)butyl GSL (glucoerucin, **84**). However, only rocket biosynthesized the Se analogue of **84**, 4-(methylseleno)butyl GSL (glucoselenoerucin, **[145]**). Se was determined in the side chain based on distinctive fragmentations in the MS^2 spectrum, which was recorded and analyzed for the first time for this desulfoselenoglucosinolate. Although other Met-derived GSLs with sulfur in the side chain have been identified, no Se analogues have been found. This study emphasizes the importance of plant tolerance to selenate and the growth stage, such as in broccoli, where biofortification should begin as soon as the florets begin to emerge from the meristerm and not before.

Supplementary Materials: The following are available online at https://www.mdpi.com/article/10.3390/molecules28207203/s1, Figure S1. Broccoli (*Brassica oleracea* var. *italic*) watered with water (reference) and different concentrations of sodium selenate solution; Table S1. The effect of selenate treatments on *B. oleracea* var. *italic*; Figure S2. Rocket (*Eruca vesicaria*) watered with water (reference) and different concentrations of sodium selenate solution.; Table S2. The effect of selenate treatments on *E. vesicaria*; Figure S3. Chromatograms of desulfoglucosinolates (column temperature 25 °C) obtained from the roots of *Brassicca oleracea* var. *italica* before and after the treatment with Na_2SeO_4; Figure S4. Chromatograms of desulfoglucosinolates (column temperature 25 °C) obtained from the stem of *Brassicca oleracea* var. *italica* before and after the treatment with Na_2SeO_4; Figure S5. Chromatograms of desulfoglucosinolates (column temperature 25 °C) obtained from the leaf of *Brassicca oleracea* var. *italica* before and after the treatment with Na_2SeO_4; Figure S6. MS^2 spectra at 15V and 20V of desulfo-4-hydroxy-*N*-methoxyglucobrassicin; Figure S7. Chromatograms of desulfoglucosinolates (column temperature 15 °C) obtained from the roots of *Eruca vesicaria* before and after the treatment with Na_2SeO_4; Figure S8. Chromatogram of desulfoglucosinolates (column temperature 15 °C) obtained from the aerial part of *Eruca vesicaria* before and after the treatment with Na_2SeO_4; Figure S9. Chromatograms of desulfoglucosinolate (dGSL) standards.; Figure S10. MS^2 spectra of desulfoglucosinolates.; Figure S11. MS and UV spectra of detected desulfoglucoselenoerucin. Figure S12. MS^2 spectra at 15V of desulfoglucoselenoerucin formed during biofortification (two isotopes of Se—m/z 410 and 412).

Author Contributions: Conceptualization, I.B.; Methodology, A.Đ., L.K.M. and I.B.; Formal analysis, A.Đ., K.U. and I.B.; Investigation, A.Đ., L.K.M. and I.B.; Data curation, A.Đ., K.U. and I.B.; Writing—original draft, I.B.; Writing—review & editing, A.Đ., K.U., L.K.M. and I.B.; Supervision, I.B.; Project administration, I.B.; Funding acquisition, I.B. All authors have read and agreed to the published version of the manuscript.

Funding: This research has been fully supported by the Croatian Science Foundation (Grant IP-2016-06-1316).

Institutional Review Board Statement: Not applicable.

Informed Consent Statement: Not applicable.

Data Availability Statement: Not applicable.

Acknowledgments: We are also thankful for the scientific research equipment financed by EU grant "Functional integration of the University of Split, PMF-ST, PFST and KTFST through the development of the scientific and research infrastructure" (KK.01.1.1.02.0018).

Conflicts of Interest: The authors declare that they have no conflict of interest.

Sample Availability: Samples of the compounds are available from the authors.

References

1. Kabata-Pendias, A.; Mukherjee, A.B. *Trace Elements from Soil to Human*; Springer: Berlin/Heidelberg, Germany; New York, NY, USA, 2007. [CrossRef]
2. Fordyce, F.M. Selenium deficiency and toxicity in the environment. In *Essentials of Medical Geology*; Revised Edition; Selinus, O., Ed.; British Geological Survey: Nottingham, UK, 2013; pp. 375–419. [CrossRef]
3. dos Reis, A.R.; El-Ramady, H.; Santos, E.F.; Gratão, P.L.; Schomburg, L. Overview of Selenium Deficiency and Toxicity Worldwide: Affected Areas, Selenium-Related Health Issues, and Case Studies. In *Selenium in Plants*; Plant Ecophysiology; Pilon-Smits, E., Winkel, L., Lin, Z.Q., Eds.; Springer: Cham, Switzerland, 2017; Volume 11. [CrossRef]
4. Li, H.F.; McGrath, S.P.; Zhao, F.J. Selenium uptake, translocation and speciation in wheat supplied with selenate or selenite. *New Phytol.* **2008**, *178*, 92–102. [CrossRef] [PubMed]
5. Ajwa, H.A.; Banuelos, G.S.; Mayland, H.F. Selenium uptake by plants from soils amended with inorganic and organic materials. *J. Environ. Qual.* **1998**, *27*, 1218–1227. [CrossRef]
6. White, P.J. Selenium accumulation by plants. *Ann. Bot.* **2016**, *117*, 217–235. [CrossRef]
7. Wiesner-Reinhold, M.; Schreiner, M.; Baldermann, S.; Schwarz, D.; Hanschen, F.S.; Kipp, A.P.; Rowan, D.D.; Bentley-Hewitt, K.L.; McKenzie, M.J. Mechanisms of selenium enrichment and measurement in Brassicaceous vegetables, and their application to human health. *Front. Plant Sci.* **2017**, *8*, 1365. [CrossRef]
8. Blažević, I.; Montaut, S.; Burčul, F.; Rollin, P. Glucosinolates: Novel sources and biological potential. In *Glucosinolates, Reference Series in Phytochemistry*; Mérillon, J.-M., Ramawat, K.G., Eds.; Springer: Cham, Switzerland, 2017; pp. 3–60. [CrossRef]
9. Montaut, S.; De Nicola, G.R.; Agnaniet, H.; Issembe, Y.; Rollin, P.; Menut, C. Probing for the presence of glucosinolates in three *Drypetes* spp. (*Drypetes euryodes* (Hiern) Hutch., *Drypetes gossweileri* S. Moore, *Drypetes laciniata* Hutch.) and two *Rinorea* spp. (*Rinorea subintegrifolia* O. Ktze and *Rinorea woermanniana* (Büttner) Engl.) from Gabon. *Nat. Prod. Res.* **2017**, *31*, 308–313. [CrossRef]
10. Blažević, I.; Montaut, S.; Burčul, F.; Olsen, C.E.; Burow, M.; Rollin, P.; Agerbirk, N. Glucosinolate structural diversity, identification, chemical synthesis and metabolism in plants. *Phytochemistry* **2020**, *169*, 112100. [CrossRef]
11. Montaut, S.; Read, S.; Blažević, I.; Nuzillard, J.-M.; Roje, M.; Harakat, D.; Rollin, P. Investigation of the glucosinolates in *Hesperis matronalis* L. and *Hesperis laciniata* All.: Unveiling 4′-*O*-β-D-apiofuranosylglucomatronalin. *Carbohydr. Res.* **2020**, *488*, 107898. [CrossRef] [PubMed]
12. Trabelcy, B.; Chinkov, N.; Samuni-Blank, M.; Merav, M.; Izhaki, I.; Carmeli, S.; Gerchman, Y. Investigation of glucosinolates in the desert plant *Ochradenus baccatus* (Brassicales: Resedaceae). Unveiling glucoochradenin, a new arabinosylated glucosinolate. *Phytochemistry* **2021**, *187*, 112760. [CrossRef]
13. Stewart, J.M.; Nigiam, S.N.; McConnel, W.B. Metabolism of $Na_2{}^{75}SeO_4$ in horseradish: Formation of selenosinigrin. *Can. J. Biochem.* **1974**, *52*, 144–145. [CrossRef]
14. Bertelsen, F.; Gissel-Nielsen, G.; Kjær, A.; Skrydstrup, T. Selenoglucosinolates in nature: Fact or myth? *Phytochemistry* **1988**, *27*, 3743–3749. [CrossRef]
15. Matich, A.J.; McKenzie, M.J.; Lill, R.E.; Brummell, D.A.; McGhie, T.K.; Chen, R.K.-Y.; Rowan, D.D. Selenoglucosinolates and their metabolites produced in *Brassica* spp. fertilised with sodium selenate. *Phytochemistry* **2012**, *75*, 140–152. [CrossRef] [PubMed]

16. Matich, A.J.; McKenzie, M.J.; Lill, R.E.; McGhie, T.K.; Chen, R.K.; Rowan, D.D. Distribution of selenoglucosinolates and their metabolites in *Brassica* treated with sodium selenate. *J. Agric. Food Chem.* **2015**, *63*, 1896–1905. [CrossRef]
17. Ouerdane, L.; Aureli, F.; Flis, P.; Bierla, K.; Preud'homme, H.; Cubadda, F.; Szpunar, J. Comprehensive speciation of low-molecular weight selenium metabolites in mustard seeds using HPLC-electrospray linear trap/Orbitrap tandem mass spectrometry. *Metallomics* **2013**, *5*, 1294–1304. [CrossRef] [PubMed]
18. McKenzie, M.; Matich, A.; Hunter, D.; Esfandiari, A.; Trolove, S.; Chen, R.; Lill, R. Selenium application during radish (*Raphanus sativus*) plant development alters glucosinolate metabolic gene expression and results in the production of 4-(methylseleno)but-3-enyl glucosinolate. *Plants* **2019**, *8*, 427. [CrossRef] [PubMed]
19. European Economic Community, Commisssion Regulation. Oilseeds—determination of glucosinolates high performance liquid chromatography. *Off. J. Eur. Comm.* **1990**, *L170*, 27–34.
20. Van Hoewyk, D.; Takahashi, H.; Inoue, E.; Hess, A.; Tamaoki, M.; Pilon-Smits, E.A.H. Transcriptome analyses give insights into selenium-stress responses and selenium tolerance mechanisms in *Arabidopsis*. *Physiol. Plant.* **2008**, *132*, 236–253. [CrossRef]
21. Lehotai, N.; Feigl, G.; Koos, A.; Molnar, A.; Ördög, A.; Peto, A.; Erdei, L.; Kolbert, Z.S. Nitric oxide-cytokinin interplay influences selenite sensitivity in *Arabidopsis*. *Plant Cell Rep.* **2016**, *35*, 2181–2195. [CrossRef]
22. Kolbert, Z.; Lehotai, N.; Molnár, Á.; Feigl, G. "The roots" of selenium toxicity: A new concept. *Plant Signal. Behav.* **2016**, *11*, e1241935. [CrossRef] [PubMed]
23. Adamopoulou, M.; Bouzas, E.A.; Siyiannis, V.; Perouli, M.; Kokotou, M.; Chorianopoulou, S.N.; Constantinou-Kokotou, V.; Bouranis, D.L. Selenium assimilation by broccoli: Effect of Se inputs on the biosynthesis of secondary metabolites under normal or reduced S inputs. In *Proceedings of the 28th International Symposium of CIEC, Fertilization and Nutrient Use Efficiency in Mediterranean Environments*; Bouranis, D.L., Haneklaus, S.H., Chorianopoulou, S.N., Li, J., De Kok, L.J., Schnug, E., Ji, L., Eds.; Utopia Publishing Ltd.: Athens, Greece, 2020; pp. 169–173.
24. Bouranis, D.L.; Stylianidis, G.P.; Manta, V.; Karousis, E.N.; Tzanaki, A.; Dimitriadi, D.; Bouzas, E.A.; Siyiannis, V.F.; Constantinou-Kokotou, V.; Chorianopoulou, S.N.; et al. Floret biofortification of broccoli using amino acids coupled with selenium under different surfactants: A case study of cultivating functional foods. *Plants* **2023**, *12*, 1272. [CrossRef]
25. Tian, M.; Hui, M.; Thannhauser, T.W.; Pan, S.; Li, L. Selenium-induced toxicity is counteracted by sulfur in broccoli (*Brassica oleracea* L. var. *italica*). *Front. Plant Sci.* **2017**, *8*, 1425. [CrossRef] [PubMed]
26. Olsen, C.E.; Huang, X.C.; Hansen, C.I.C.; Cipollini, D.; Ørgaard, M.; Matthes, A.; Geu-Flores, F.; Koch, M.A.; Agerbirk, N. Glucosinolate diversity within a phylogenetic framework of the tribe Cardamineae (Brassicaceae) unraveled with HPLC-MS/MS and NMR-based analytical distinction of 70 desulfoglucosinolates. *Phytochemistry* **2016**, *132*, 33–56. [CrossRef] [PubMed]
27. Pfalz, M.; Mukhaimar, M.; Perreau, F.; Kirk, J.; Hansen, C.I.C.; Olsen, C.E.; Agerbirk, N.; Kroymann, J. Methyl transfer in glucosinolate biosynthesis mediated by indole glucosinolate *O*-Methyltransferase 5. *Plant Physiol.* **2016**, *172*, 2190–2203. [CrossRef]
28. Agerbirk, N.; Hansen, C.C.; Olsen, C.E.; Kiefer, C.; Hauser, T.P.; Christensen, S.; Jensen, K.R.; Ørgaard, M.; Pattison, D.I.; Lange, C.B.A.; et al. Glucosinolate profiles and phylogeny in Barbarea compared to other tribe Cardamineae (Brassicaceae) and Reseda (Resedaceae), based on a library of ion trap HPLC-MS/MS data of reference desulfoglucosinolates. *Phytochemistry* **2021**, *185*, 112658. [CrossRef] [PubMed]
29. Dall'Acqua, S.; Ertani, A.; Pilon-Smits, E.A.H.; Fabrega-Prats, M.; Schiavon, M. Selenium biofortification differentially affects sulfur metabolism and accumulation of phytochemicals in two rocket species (*Eruca sativa* Mill. and *Diplotaxis tenuifolia*) grown in hydroponics. *Plants* **2019**, *8*, 68. [CrossRef] [PubMed]
30. Wang, J.; Mao, S.; Xu, H.; Wu, Q.; Liang, M.; Yuan, Y.; Liu, M.; Huang, K.; Wu, Q. Effects of sulfur and selenium on glucosinolate biosynthesis in cabbage. *Plant Mol. Biol. Rep.* **2020**, *38*, 62–74. [CrossRef]
31. Blažević, I.; Đulović, A.; Čikeš Čulić, V.; Popović, M.; Guillot, X.; Burčul, F.; Rollin, P. Microwave-assisted versus conventional isolation of glucosinolate degradation products from *Lunaria annua* L. and their cytotoxic activity. *Biomolecules* **2020**, *10*, 215. [CrossRef]
32. Wathelet, J.-P.; Iori, R.; Leoni, O.; Quinsac, A.; Palmieri, S.; Rollin, P. Guidelines for glucosinolate analysis in green tissues used for biofumigation. *Agroindustria* **2004**, *3*, 257–344.
33. Brown, P.D.; Tokuhisa, J.G.; Reichelt, M.; Gershenzon, J. Variation of glucosinolate accumulation among different organs and developmental stages of *Arabidopsis thaliana*. *Phytochemistry* **2003**, *62*, 471–481. [CrossRef]

molecules

MDPI

Article

Upper Rim-Bridged Calix[4]arenes via Cyclization of *meta* Alkynyl Intermediates with Diphenyl Diselenide

Anastasia Surina [1], Karolína Salvadori [2,3], Matěj Poupě [1], Jan Čejka [4], Ludmila Šimková [2] and Pavel Lhoták [1,*]

1 Department of Organic Chemistry, University of Chemistry and Technology Prague, Technická 5, 166 28 Prague, Czech Republic
2 J. Heyrovský Institute of Physical Chemistry, Czech Academy of Sciences v.v.i., Dolejškova 2155/3, 182 23 Prague, Czech Republic
3 Department of Physical Chemistry, University of Chemistry and Technology Prague, Technická 5, 166 28 Prague, Czech Republic
4 Department of Solid State Chemistry, University of Chemistry and Technology Prague, Technická 5, 166 28 Prague, Czech Republic; cejkaj@vscht.cz
* Correspondence: lhotakp@vscht.cz; Tel.: +420-440-225-055

Abstract: A Sonogashira coupling of *meta*-iodocalix[4]arene with various terminal acetylenes confirmed that the *meta* position of calixarene is well addressable, and that both thermal and microwave protocols led to good yields of alkynylcalixarenes. Alkynes thus obtained were subjected to the ferric chloride and diphenyl diselenide-promoted electrophilic closure. It turns out that the calix[4]arenes give completely different bridging products than those described for the non-macrocyclic starting compounds. This can be demonstrated not only by the isolation of products with a six-membered ring (*6-exo-dig*), but mainly by the smooth formation of the *5-endo-dig* cyclization, which has never been observed in the aliphatic series. An attempt at electrocyclization led to a high yield of the 1,2-diketone (oxidation of the starting alkyne), again in contrast to the reaction described for the acyclic derivatives. The structures of the unexpected products were unequivocally established by X-ray analysis and clearly demonstrate how the preorganized macrocyclic skeleton favors a completely different regioselectivity of cyclization reactions compared to common aliphatic compounds.

Keywords: calixarene; Sonogashira coupling; *meta*-substitution; alkyne cyclization; diphenyl diselenide; *6-exo-dig* and *5-endo-dig* cyclization; regioselectivity; electrochemistry; X-ray analysis

Citation: Surina, A.; Salvadori, K.; Poupě, M.; Čejka, J.; Šimková, L.; Lhoták, P. Upper Rim-Bridged Calix[4]arenes via Cyclization of *meta* Alkynyl Intermediates with Diphenyl Diselenide. *Molecules* **2024**, *29*, 1237. https://doi.org/10.3390/molecules29061237

Academic Editor: Ming Wang

Received: 19 February 2024
Revised: 5 March 2024
Accepted: 7 March 2024
Published: 11 March 2024

1. Introduction

Calixarenes [1,2] have become an integral part of modern supramolecular chemistry, where they have found wide application due to their unique properties. Their use as building blocks or molecular scaffolds with the possibility of purposefully changing the 3D structure of the basic skeleton (e.g., immobilization of four different conformations (atropisomers) in the case of calix[4]arene) makes them an ideal choice for the design and synthesis of new receptors and similar functional molecular systems [3,4].

Calix[4]arenes show an almost unlimited possibility of derivatization of the basic skeleton, enabling the introduction of desired functional groups at virtually any place in the macrocycle. Currently, many strategies are known leading to regio- or stereo-selective derivatization of basic calix[4]arene, usually based on a combination of alkylation reactions on the lower rim (OH groups) and electrophilic substitution of the upper rim (aromatic part). In this context, however, we come across a limitation given by the directing effects of the substituents present in the macrocycle (OH, OR), where electrophilic substitution takes place preferentially in the *para* position to these groups. For this reason, in calixarene chemistry, *meta* substitution [5] represents a synthetic challenge, and the *meta* substituted derivatives appear only rarely.

Systematic studies on the calix[4]arene **I** reactivity surprisingly revealed that the reaction with mercury trifluoroacetate ($Hg(TFA)_2$) leads to a direct substitution [6] of the hitherto practically inaccessible *meta* position (Scheme 1). The formation of an organomercury compound **II** then enabled the preparation of new types of calixarenes with a *meta-meta* bridged upper rim [7]. These compounds are not only interesting because of their highly increased rigidity (see e.g., **III** and **IV**), but often also represent inherently chiral systems potentially suitable for the design of chiral receptors (**V–VII**) [8,9].

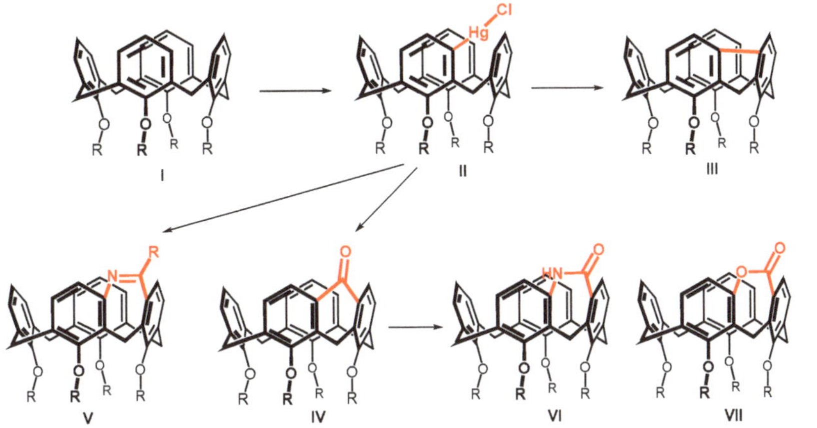

Scheme 1. The upper rim-bridged calix[4]arenes bearing five-membered, six-membered and seven-membered rings within the molecules.

The bridging of two adjacent aromatic nuclei in the *meta* position leads to the formation of condensed tricyclic systems having a five-, six-, or seven-membered ring (Scheme 1). While a number of systems having a five- or six-membered ring have been prepared, only three seven-membered systems (**V–VII**) have been described so far [8,9]. Thus, compounds of type **V** were obtained by cyclization of the corresponding *meta*-amides, and compounds **VI** and **VII** were obtained from ketone **IV** using a ring extension via Baeyer-Villiger or Beckmann rearrangements.

Since different cycle sizes introduce different internal strains into the macrocyclic system, we thought it interesting to apply a bridging reaction that could provide different sizes of cyclic moieties. Such a reaction could be the cyclization of alkynes proceeding either by the *exo-dig* or *endo-dig* mechanism (Baldwin's rules), where both processes are allowed (Scheme 2) [10]. This paper describes an effort to prepare the above systems, which show a surprising preference of calix[4]arene skeleton compared to acyclic molecules.

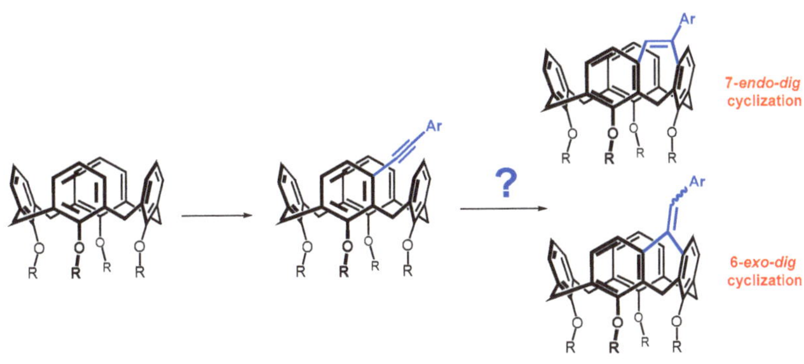

Scheme 2. *meta*-Alkynylcalix[4]arenes and possible regioisomers formed by their cyclization.

2. Results and Discussion

Starting calix[4]arene **1** immobilized in the *cone* conformation was successively reacted with mercury(II) trifluoroacetate ($Hg(TFA)_2$) and sodium chloride using a literature procedure [6]. The corresponding *meta*-substituted chloromercurio derivative **2** was obtained regioselectively in good yield (70%). A subsequent reaction with iodine provided the *meta*-iodo derivative **3** in 58% yield (Scheme 3) [8]. The introduction of triple bond into the calixarene skeleton was carried out by Sonogashira coupling [11,12], which is well known for terminal alkynes. This reaction has already been successfully used several times in calixarene chemistry, but it involved either the *para* position of the aromatic subunit [13–15] or the transformation of the lower rim of the macrocycle [16,17]. The reactivity of the halogen in the *meta* position was therefore unknown. To study this cross-coupling of compound **3**, a series of substituted aromatic acetylenes **4a–e** completed with a silyl- derivative **4f** was selected.

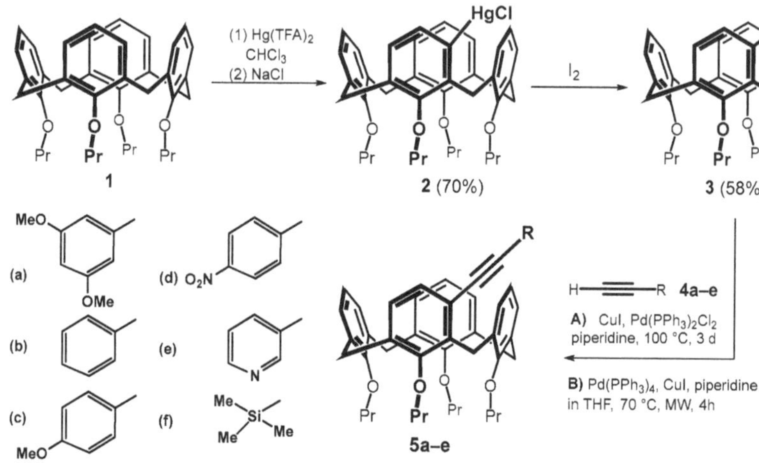

Scheme 3. Sonogashira coupling of *meta*-substituted calixarenes.

Based on the literature search [11], we applied two different implementations of the Sonogashira reaction: (i) classical heating on a magnetic stirrer and (ii) application of microwave reactor. Thus, heating compound **3** with **4a** and $Pd(PPh_3)_2Cl_2/CuI$ in piperidine at 100 °C for 3 days gave the expected product **5a** in 89% yield. The reaction carried out in a microwave reactor ($Pd(PPh_3)_4/CuI$, piperidine in THF) at 70 °C for 4 h provided the same product in 74% yield. A similar trend, i.e., a higher yield upon classical heating, was observed for all electron-donating substituents (**5a–c**) and the silyl-substituted derivative **5f**, see Table 1. In contrast, both derivatives bearing electron-withdrawing substituents **5d** and **5e** were obtained in a higher yield upon microwave reaction (compare 33% vs. 47% for **5e**).

Table 1. The yields of the Sonogashira cross-coupling of calixarene **3** with acetylenes **4a–4f**.

Product	Method A [(a)]	Method B [(b)]
5a	89%	74%
5b	96%	62%
5c	64%	58%
5d	27%	39%
5e	33%	47%
5f	83%	75%

[(a)] Method A: $Pd(PPh_3)_2Cl_2/CuI$ in piperidine at 100 °C, stirring for 3 days. [(b)] Method B: $Pd(PPh_3)_4/CuI$, piperidine in THF at 70 °C for 4 h, microwave oven.

Alkynes thus obtained were subjected to the recently described iron(III) chloride and diorganyl diselenide-promoted electrophilic closure [18–20], giving selectively the seven-membered cyclization products [18]. The reaction of **5b** carried out under the conditions described in the literature (alkyne + Ph_2Se_2 (0.55 equiv.) + $FeCl_3$ (1.0 equiv.) in CH_2Cl_2 at RT) gave practically unreacted starting compounds. In order to achieve higher conversion of the calixarene derivatives, we increased the number of equivalents of both reagents, and the reaction was carried out at a higher temperature (Scheme 4). Thus, heating of **5b** with 2.5 equivs. of iron(III) chloride and 2 equivs. of Ph_2Se_2 in dichloromethane at 30 °C for 4 h gave a crude reaction mixture with 42% of unreacted starting compound. The column chromatography on silica gel provided an expected cyclized product (20% yield) as indicated by the HRMS (ESI+) peak at m/z = 871.3240 well corresponding to the theoretical value for $C_{54}H_{56}O_4SeNa^+$ ion (m/z = 871.3245). Other fractions contained derivatives that were apparently bridged but lacked varying numbers of propyl groups (^{1}H NMR), indicating a concurrent dealkylation reaction. The ^{1}H NMR spectrum of the product ($CDCl_3$, 400 MHz) showed eight doublets with typical geminal coupling constants (13–15 Hz) corresponding to the signals of the methylene bridges (3.00 to 4.7 ppm) in the *cone* conformation. Unfortunately, despite our efforts, it was not possible to decide between the two possible structures **6b** and **7b** using NMR spectroscopy.

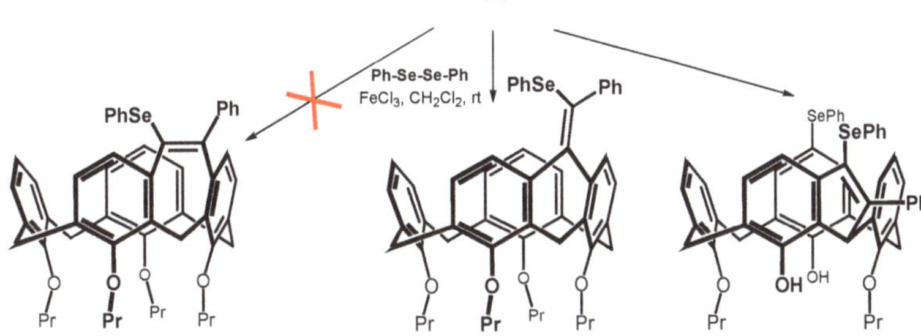

Scheme 4. An attempted cyclization of **5b** with $Ph_2Se_2/FeCl_3$ system.

The final proof of the product structure was performed using single crystal X-ray diffraction analysis. The product is in the form of a six-membered ring structure **7b** crystallized in a triclinic crystal system, space group *P*-1, and adopted the *cone* conformation (Figure 1a). The alkylidene bridge results not only in the rigidification of the calixarene skeleton, but also in the partial flattening of the originally square shape of the cavity. This can be demonstrated on different lengths of the main diagonals (the distance between two opposite Ar-CH_2-Ar carbons, compare 7.754 vs. 6.421 Å, Figure 1b).

The presence of two different substituents on the double bond makes the molecule inherently chiral, with both enantiomers included in the unit cell as the independent molecules (racemic mixture). The opposite enantiomers are held together throughout the crystal packing of **7b** (Figure 1c) by the CH-π interactions between the O-CH_2- group and the phenyl selenyl moiety (CH$\cdots C_{Ar}$ distance = 2.880 Å). At the same time, the close contacts between PhSe groups can be found, based on the T-shaped π-π interactions [21,22] between phenyl moieties (Figure 1c).

Compound **5a** also afforded the cyclization product **7a** in low yield, which, however, could not be sufficiently purified for analysis. Nevertheless, the structure was determined by comparing its ^{1}H NMR spectrum with that of **7b**. At the same time, the HRMS (ESI+) showed a molecular peak at m/z = 931.3453 (compared with the theoretical value 931.3447 for the $[M + Na]^+$ ion). Under the same conditions, nitro compound **5d** gave only a trace amount of product (TLC, HRMS), indicating a negative influence of the electron-withdrawing group on the course of the reaction (conversion ca 1% (^{1}H NMR)). Similar conclusions can be drawn for compound **5e**.

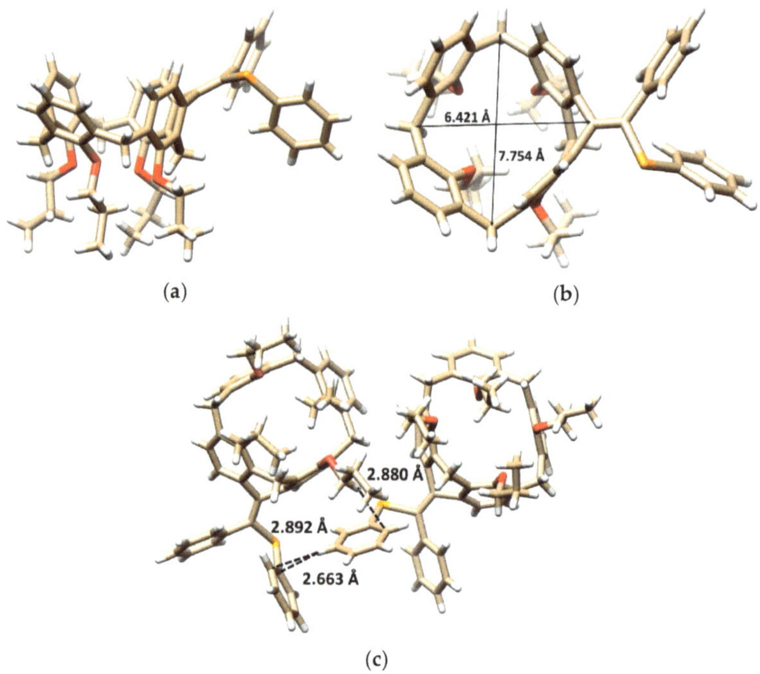

Figure 1. Single crystal X-ray structure of **7b**: (**a**) the side view, (**b**) top view showing the main diagonals, (**c**) dimeric motif within the crystal packing based on the CH$\cdots\pi$ and the π-π (T-shaped) interactions (red = O, orange = Se, white = H, beige = C atoms).

Since all cyclization attempts exhibited low conversion of the starting compound, we tried to increase the reaction temperature by using DCE as a solvent. Reaction of **5b** with Ph_2Se_2 and $FeCl_3$ (both 2.5 equiv) under reflux (85 °C) gave a very complex reaction mixture. As it was proved by means of model reactions, $FeCl_3$ is a strong dealkylating agent under these conditions, so that a partial non-selective dealkylation of both starting substances and products took place. The isolation by preparative TLC gave the bridged product **8b** in low yield (Scheme 4). The ^{1}H NMR spectra (400 MHz, $CDCl_3$) measured at room temperature showed two singlets at 8.21 and 8.31 ppm, indicating the presence of free phenolic groups. The aliphatic part of the spectrum contains signals of only two propyl groups, which proves the partial dealkylation of the starting macrocycle. Interestingly, instead of 8 doublets with a geminal coupling constants, only two sets of three doublets (J = 12.5–13.7 Hz) were seen for the methylene linkages (3.20, 3.28 and 3.37 ppm for the equatorial and 4.06, 4.36 and 4.44 ppm for axial CH bonds). The last bridge showed only singlet at 5.42 ppm corresponding to one hydrogen. This suggests that one methylene group was transformed during the reaction, but the actual structure remained unknown.

This problem was unequivocally solved by single crystal X-ray diffraction analysis. Compound **8b** crystallizes in a monoclinic crystal system, space group $P2_1/n$, and forms a CH_2Cl_2 solvate with 1:1 stoichiometry. As shown in Figure 2a,b, calixarene adopts the *cone* conformation containing the five-membered ring formed by the addition of the methylene bridge to the alkyne moiety. The unexpected cyclization is accompanied by the introduction of a second PhSe- group into the *para* position of the opposite aromatic subunit. Both phenols bearing the substitution are dealkylated at the same time. The *cone* conformation is held by hydrogen bonds between the free OH groups and the oxygens of the neighbor propoxy functions (OH$\cdots$O distance = 1.955 and 1.969 Å). The presence of the indene moiety within the backbone has surprisingly little effect on the shape of the cavity, which can be demonstrated by its approximately square shape (the length of the main diagonals is 6.952 vs. 7.426 Å), see Figure 2a.

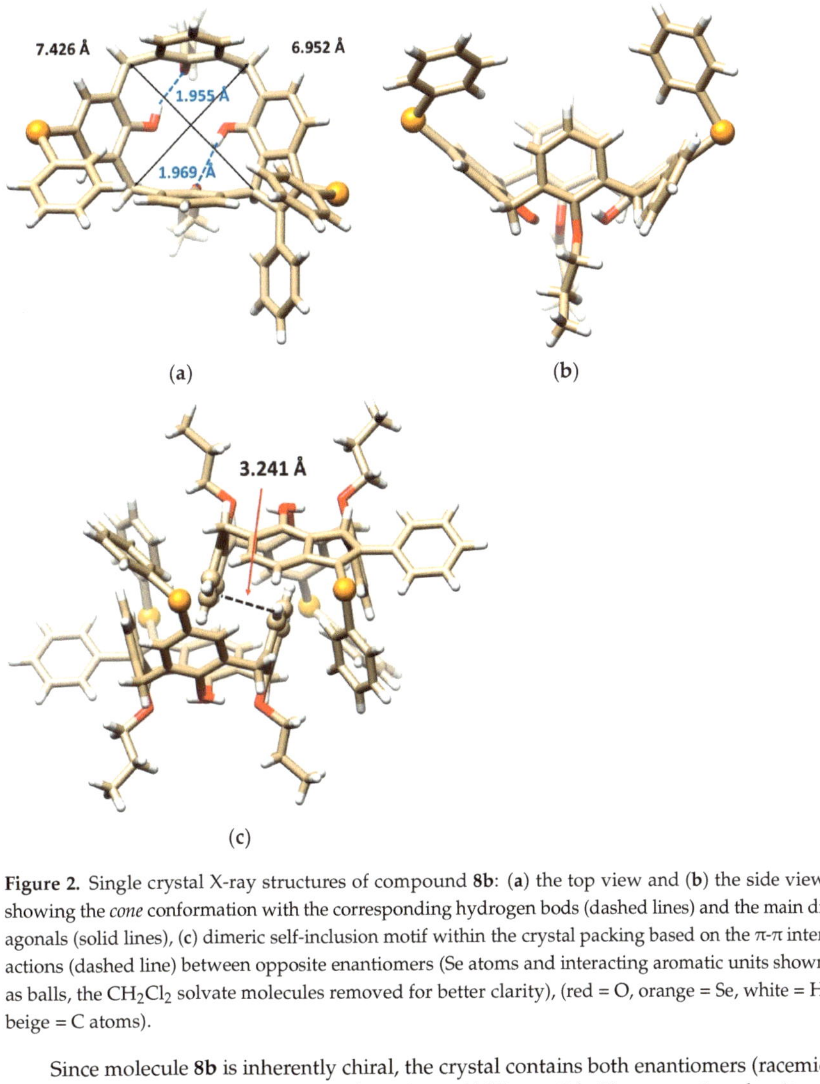

Figure 2. Single crystal X-ray structures of compound **8b**: (**a**) the top view and (**b**) the side view, showing the *cone* conformation with the corresponding hydrogen bods (dashed lines) and the main diagonals (solid lines), (**c**) dimeric self-inclusion motif within the crystal packing based on the π-π interactions (dashed line) between opposite enantiomers (Se atoms and interacting aromatic units shown as balls, the CH_2Cl_2 solvate molecules removed for better clarity), (red = O, orange = Se, white = H, beige = C atoms).

Since molecule **8b** is inherently chiral, the crystal contains both enantiomers (racemic mixture), which form an interesting dimeric motif (Figure 2c). The aromatic subunits of opposite enantiomers bearing the propoxy groups are nested into each other's adjacent cavity to form self-included dimers. The aromatic subunits are ideally coplanar interacting via the π-π interactions with each other in the so-called parallel stacked arrangement. The interplanar spacing of these aromatic subunits is exceptionally short (only 3.241 Å, while the usual distance is around 3.5 Å), indicating unusually strong π-π interactions.

The formation of product **8b** is completely unexpected and to the best of our knowledge, there is no example of similar structures in the literature. The selenium-mediated electrophilic cyclization reaction is well known in the art and has been used to construct a variety of structures [19,23–25]. On the other hand, although the $FeCl_3/Ph_2Se_2$ promoted cyclization of *o*-alkynyl substituted diaryl methanes (a structural fragment within a calixarene moiety) has been investigated in detail, [18] the formation of a five-membered product of type **8** has never been observed (Scheme 5). The formation of indene structures [26] from *o*-substituted aryl alkynes is known, but it is based on direct C(sp^3)-H activation using Au(I) [27] or Pt(II) [28] catalysts. A similar reaction was also achieved by the intramolecular addition of triarylmethanes [29] to the alkynyl group in the *ortho* position induced by the presence of a base (*t*-BuOK), but it was never reported for selenium-induced cyclization.

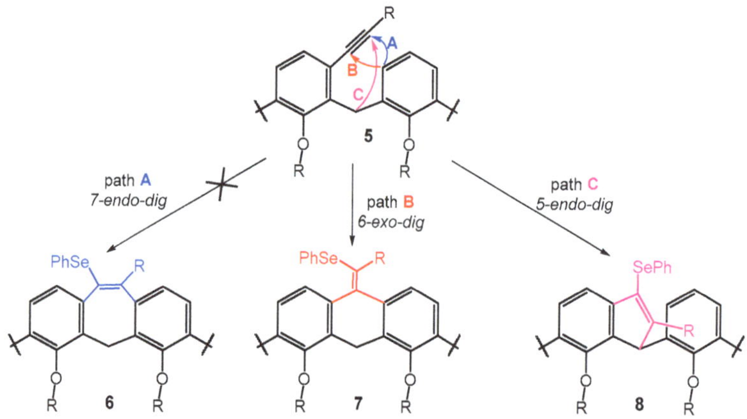

Scheme 5. The cyclization of *meta*-alkynylcalix[4]arenes the three different arrow colors show the three possible reaction pathways leading to the corresponding products.

Furthermore, it is noteworthy that the aforementioned $FeCl_3/Ph_2Se_2$ cyclization studies describe *endo-dig* cyclization leading to products of type **6** as the sole exclusive reaction pathway (path A, Scheme 5) [18,19]. In this respect, the calix[4]arene skeleton represents a unique system, since we did not observe the formation of a seven-membered cycle in any case. A possible explanation lies in the fact that the relative position of the neighboring aromatic nuclei imposed by the macrocyclic structure favors *6-exo-dig* (type **7**) or *5-endo-dig* (type **8**) over *7-endo-dig* (type **6**) cyclization, i.e., reaction pathways that are unknown for common acyclic derivatives (Scheme 5).

From this point of view, the reaction of the methoxy derivative **5c** was very surprising. The cyclization of the alkyne proceeded smoothly already at 30 °C (100% conversion), whereby two products were isolated from the reaction mixture (Scheme 6). The HRMS (ESI+) of the main product **9c** (37% yield) showed a molecular peak at m/z = 901.3345 which corresponded to the expected mass of the cyclized product m/z = 901.3342 ([M + Na]$^+$ ion). The ^{1}H NMR ($CDCl_3$, 298 K, 400 MHz) comprised two sets of three doublets for the methylene bridges ($J \approx 13.3$ Hz), together with a singlet at 5.70 ppm (1H). This splitting pattern is very similar to that of compound **8b** and refers to cyclization via a single methylene bridge. The second isolated product was assigned as the same type of structure, but lacking one propyl unit. Based on two-dimensional NMR experiments (HSQC, HMBC) it appears most likely that this derivative has the structure **10c** with the OH group in the indene fragment (Scheme 6). Thus, the result again reflects the remarkable regioselectivity of the reaction, where the cyclization proceeds only through the adjacent methylene bridge, while the expected bridging of the adjacent aromatic subunit was not observed at all.

The $FeCl_3$-mediated electrophilic selanylation/annulation of *meta*-substituted calixarenes showed very unexpected results; however, the use of a Lewis acid led to relatively low yields due to dealkylation of the lower rim of the calixarene. To avoid this, we decided to apply the recently described method based on the action of diphenyl diselenide under electrochemical conditions [30,31]. Thus, the reaction of **5b** with 0.6 equiv. of Ph_2Se_2 in MeCN in the presence of $LiClO_4$ as an electrolyte was carried out in an undivided cell under constant current electrolysis (CCE) at 10 mA for 3 h under aerobic atmosphere.

Although the original publications promise a virtually quantitative bridging yield, we were unable to demonstrate any cyclization. The only resulting product **11** was isolated in 73% yield (Scheme 7), and its 1,2-diketone [32] structure was confirmed by a combination of HRMS and NMR spectroscopy (see typical carbonyl signals at 197.92 and 195.89 ppm in ^{13}C NMR spectrum). Altering the reaction parameters, such as using a different electrode material (Pt vs. glassy carbon), did not lead to any changes. We attempted to eliminate the unwanted oxidation by performing electrolysis at a constant potential (+1.5 V), which was chosen from appropriate cyclic voltammogram (Figure S78) as the first irreversible step of Ph_2Se_2. Unfortunately, compound **11** was still the only isolated product. Finally, carrying out the reaction in a degassed solvent (purged with a stream of argon) prevented oxidation but gave no reaction. It indicates that the mentioned methodology is not suitable for the electrocyclization of *meta*-alkynyl calixarenes.

5c

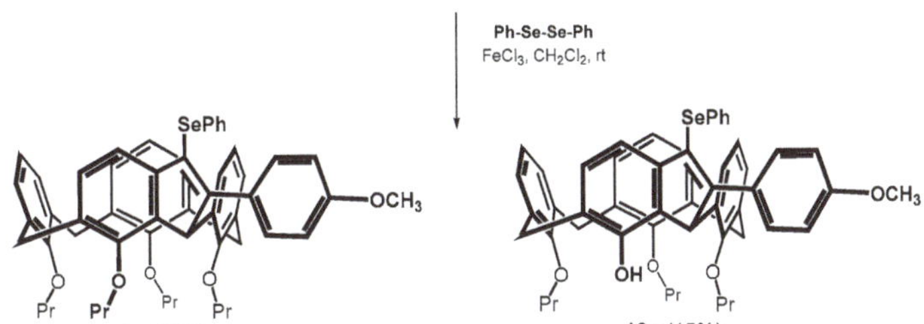

9c (37%) **10c** (15%)

Scheme 6. Cyclization of methoxy derivative **5c** (*5-endo dig*).

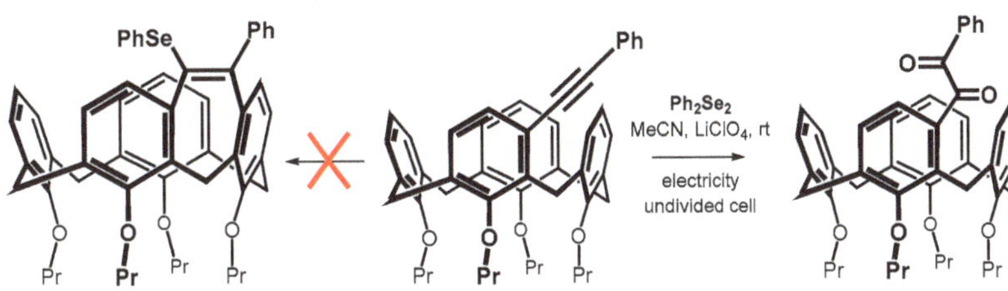

5b **11** (73%)

Scheme 7. Attempted electrocyclization of phenyl derivative **5b** under aerobic conditions.

3. Materials and Methods

3.1. General Experimental Procedures

All commercially obtained chemicals were purchased from commercial sources and used as received without further purification. Solvents were dried and distilled using conventional methods. NMR spectra were performed on Agilent 400-MR DDR2, JEOL-ECZL400G (^{1}H: 400 MHz, ^{13}C: 100 MHz) and Bruker Avance DRX 500 (^{1}H: 500 MHz, ^{13}C: 125 MHz) spectrometers at 298 K. Deuterated solvents used are indicated in each case. Chemical shifts (δ) are reported in parts per million (ppm) and were referenced to residual peak of the solvent or TMS as an internal standard; coupling constants (J) are expressed in Hz. NMR data were processed and displayed using MestReNova (Version: 14.3.3) and TopSpin (Version 3.6.5) softwares. The IR spectra were measured on an FT–IR spectrometer Nicolet iS50 (Thermo-Nicolet, Waltham, MA, USA) with a heatable Golden Gate Diamante ATR–Unit GladiATR (Specac, Orpington, UK). 64 scans for one spectrum were co-added at a spectral resolution of 4 cm^{-1}. Electrospray ionization mass spectra (ESI-MS) were recorded using a LTQ Orbitrap Velos—hybrid ion-trap-orbitrap (Thermo Scientific, Waltham, MA, USA). The purity of the substances and courses of the reactions were monitored by thin layer chromatography (TLC) using silica gel 60 F_{254} on aluminum-backed sheets (Merck,

Lowe, NJ, USA) and analyzed at 254 nm. The column chromatography was conducted on silica gel 60 with particle size 0.063–0.200 mm (Merck). For melting point studies, Heiztisch Mikroskop—Polytherm A (Wagner & Munz, Munich, Germany) was used.

3.2. *Synthetic Procedures*

3.2.1. Preparation of Alkynylcalixarenes **5a–f**

General Procedure for the Synthesis of Alkynylcalixarenes **5a–f** (Method A)

Starting calix[4]arene **3** (1 equiv), CuI (0.1 equiv) and $Pd(PPh_3)_2Cl_2$ (0.05 equiv) were dissolved in 3–4 mL of dry piperidine under an Ar atmosphere. Then, appropriate alkyne **4a–f** (1.1 equiv) was added, and the reaction mixture was stirred and heated at 100 °C for 3 days. The mixture was cooled to RT, and the solvent was evaporated under reduced pressure. The crude product was extracted (3 × 30 mL) with dichloromethane (DCM), the organic phase was washed with water, dried over $MgSO_4$, filtered and evaporated to dryness. The residue was purified by column chromatography on silica gel (eluent: DCM: cyclohexane) to afford the title compounds.

General Procedure for the Synthesis of Alkynylcalixarenes **5a–f** (Method B)

A mixture of starting calix[4]arene **3** (1 equiv), alkyne **4a–f** (4 equiv), piperidine (4 equiv), CuI (0.2 equiv), $Pd(PPh_3)_4$ (0.1 equiv) and dry THF was stirred under argon atmosphere at 70 °C for 4 h in the microwave reactor. The mixture was then evaporated to dryness, the residue was extracted with DCM (3 × 30 mL), washed with water and dried over $MgSO_4$. The purification of the crude product was done in the same way as described in Method A.

Alkynylcalixarene **5a**

Method A: Calixarene **3** (0.209 mmol, 150 mg), 3,5-dimethoxyphenylacetylene (0.230 mmol, 37 mg), CuI (0.0209 mmol, 4 mg) and $Pd(PPh_3)_2Cl_2$ (0.0104 mmol, 7 mg) were used under standard conditions; the product was purified by column chromatography on silica gel (DCM/cyclohexane 1:1) to afford **5a** in 89% yield (140 mg) as a white solid.

Method B: Calixarene **3** (0.139 mmol, 104 mg), 3,5-dimethoxyphenylacetylene (0.556 mmol, 90 mg), CuI (0.028 mmol, 5 mg), piperidine (0.417 mmol, 35 μL) and $Pd(PPh_3)_4$ (0.0139 mmol, 16 mg) were used under standard conditions. The product **5a** (77 mg, 74%) was obtained by column chromatography on silica gel (DCM/cyclohexane 60:40) as a white solid.

Data for compound **5a**: M.p. = 105–107 °C. ^1H NMR ($CDCl_3$, 500 MHz, 298 K) δ 7.19 (d, J = 7.7 Hz, 1H, Ar-H), 7.08 (d, J = 7.7 Hz, 2H, Ar-H), 7.06 (d, J = 7.7 Hz, 1H, Ar-H), 6.89 (t, J = 7.5 Hz, 1H, Ar-H), 6.67 (d, J = 2.3 Hz, 2H, Ar-H), 6.43 (t, J = 2.3 Hz, 1H, Ar-H), 6.33 (dd, J = 7.5 Hz, J = 1.1 Hz, 1H, Ar-H), 6.26 (t, J = 7.5 Hz, 1H, Ar-H), 6.24 (t, J = 7.5 Hz, 1H, Ar-H), 6.15 (d, J = 7.5 Hz, 2H, Ar-H), 6.13 (d, J = 7.5 Hz, 1H, Ar-H), 4.47 (d, J = 13.2 Hz, 1H, Ar-CH_2-Ar), 4.46 (d, J = 2.8 Hz, 1H, Ar-CH_2-Ar), 4.44 (d, J = 2.8 Hz, 1H, Ar-CH_2-Ar), 4.36 (d, J = 13.2 Hz, 1H, Ar-CH_2-Ar), 4.01 (m, 4H, -O-CH_2-), 3.95 (d, J = 13.2 Hz, 1H, Ar-CH_2-Ar), 3.79 (s, 6H, -O-CH_3), 3.73 (m, 2H, -O-CH_2-), 3.69 (m, 2H, -O-CH_2-), 3.16 (d, J = 13.2 Hz, 1H, Ar-CH_2-Ar), 3.15 (d, J = 13.2 Hz, 2H, Ar-CH_2-Ar), 1.98 (m, overlap, 4H, -CH_2-CH_3), 1.88 (m, 4H, -CH_2-CH_3), 1.10 (t, J = 7.3 Hz, 6H, -O-$(CH_2)_2$-CH_3), 0.91 (t, J = 7.4 Hz, 3H, -CH_3), 0.90 (t, J = 7.4 Hz, 3H, -CH_3). ^{13}C NMR ($CDCl_3$, 126 MHz, 298 K) δ 160.47, 157.98, 157.88, 155.17, 155.16, 139.31, 137.99, 137.01, 136.89, 133.47, 133.22, 132.92, 132.67, 128.84, 128.78, 128.51, 127.57, 127.41, 127.33, 127.24, 126.21, 125.01, 122.10 (2C), 121.82, 121.78, 109.33, 101.42, 91.1, 88.83, 76.94, 76.90, 76.58, 76.44, 55.43 (2C), 31.04, 30.97, 30.96, 28.11, 23.50 (2C), 22.99, 22.97, 10.79, 10.78, 9.86, 9.84 ppm. HRMS (ESI$^+$) calcd for $C_{50}H_{56}O_6$ 775.3969 [M + Na]$^+$, found m/z 775.3971 [M + Na]$^+$. IR (KBr) ν 2959, 2932, 2873, 1587, 1453, 1418, 1205, 1193, 1154, 1060, 965, 756 cm^{-1}.

Alkynylcalixarene **5b**

Method A: Calixarene **3** (0.315 mmol, 226 mg), phenylacetylene (0.3465 mmol, 40 μL), CuI (0.0315 mmol, 6 mg) and $Pd(PPh_3)_2Cl_2$ (0.0157 mmol, 11 mg) were used under standard conditions; the product was purified by column chromatography on silica gel (DCM/cyclohexane 1:1) to afford **5b** in 96% yield (209 mg) as a yellow oil.

Method B: Calixarene **3** (0.139 mmol, 103 mg), phenylacetylene (0.556 mmol, 60 μL), CuI (0.028 mmol, 5 mg), piperidine (0.417 mmol, 40 μL) and $Pd(PPh_3)_4$ (0.0139 mmol, 16 mg) were used under standard conditions. The product **5b** (59 mg, 62%) was obtained by column chromatography on silica gel (DCM/cyclohexane 70:30) as a yellow oil.

Data for compound **5b**: ^1H NMR ($CDCl_3$, 500 MHz, 298 K) δ 7.51 (dd, J = 7.7 Hz, J = 1.6 Hz, 2H, Ar-H), 7.27–7.35 (m, 3H, Ar-H) 7.19 (d, J = 7.7 Hz, 1H, Ar-H), 7.08 (d, J = 7.4 Hz, 2H, Ar-H), 7.06 (d, J = 7.7 Hz, 1H, Ar-H), 6.90 (t, J = 7.7 Hz, 1H, Ar-H), 6.34 (d, J = 7.7 Hz, 1H, Ar-H), 6.26 (d, J = 7.7 Hz, 1H, Ar-H), 6.22 (d, J = 7.7 Hz, 1H, Ar-H), 6.14 (d, J = 8.2 Hz, 2H, Ar-H), 6.12 (d, J = 7.7 Hz, 1H, Ar-H), 4.46 (d, J = 13.4 Hz, 1H, Ar-CH_2-Ar), 4.45 (d, J = 13.4 Hz, 1H, Ar-CH_2-Ar), 4.44 (d, J = 13.4 Hz, 1H, Ar-CH_2-Ar), 4.35 (d, J = 13.4 Hz, 1H, Ar-CH_2-Ar), 3.95–4.07 (m, 1H (Ar-CH_2-Ar) + 4H (-OCH_2-)), 3.63–3.76 (m, 4H, -OCH_2-), 3.16 (d, J = 13.4 Hz, 1H, Ar-CH_2-Ar), 3.15 (d, J = 13.4 Hz, 2H, Ar-CH_2-Ar), 1.93–2.03 (m, 4H, -CH_2-CH_3), 1.82–1.92 (m, 4H, -CH_2-CH_3), 1.10 (t, J = 7.3 Hz, 6H, -CH_3), 0.91 (t, J = 7.3 Hz, 3H, -CH_3), 0.89 (t, J = 7.3 Hz, 6H, -CH_3). ^{13}C NMR ($CDCl_3$, 125 MHz, 298 K) δ 160.47, 157.98, 157.88, 155.17, 155.16, 139.31, 137.99, 137.01, 136.89, 133.47, 133.22, 132.92, 132.67, 128.84, 128.78, 128.51, 127.57, 127.41, 127.33, 127.24, 126.21, 125.01, 122.10 (2C), 121.82, 121.78, 109.33, 101.42, 91.1, 88.83, 76.94, 76.90, 76.58, 76.44, 55.43 (2C), 31.04, 30.97, 30.96, 28.11, 23.50 (2C), 22.99, 22.97, 10.79, 10.78, 9.86, 9.84 ppm. HRMS (ESI$^+$) calcd for $C_{48}H_{52}O_4$ 715.3757 $[M + Na]^+$, found m/z 715.3764 $[M + Na]^+$. IR (KBr) ν 2960, 2931, 2873, 1493, 1454, 1382, 1208, 1192, 1087, 1044, 965, 754, 689 cm^{-1}.

Alkynylcalixarene **5c**

Method A: Calixarene **3** (0.209 mmol, 150 mg), 4-ethynylanisole (0.230 mmol, 30 μL), CuI (0.0209 mmol, 4 mg) and $Pd(PPh_3)_2Cl_2$ (0.011 mmol, 7 mg) were used under standard conditions; the product was purified by column chromatography on silica gel (DCM/cyclohexane 1:1) to afford **5c** in 64% yield (97 mg) as a white solid.

Method B: Calixarene **3** (0.139 mmol, 106 mg), 4-ethynylanisole (0.556 mmol, 70 μL), CuI (0.028 mmol, 5 mg), piperidine (0.417 mmol, 40 μL) and $Pd(PPh_3)_4$ (0.0139 mmol, 16 mg) were used under standard conditions. The product **5c** (58 mg, 58%) was obtained by column chromatography on silica gel (DCM/cyclohexane 50:50) as a white solid.

Data for compound **5c**: M.p. = 146–149 °C. ^1H NMR ($CDCl_3$, 400 MHz, 298 K) δ 7.49 (d, 2H, J = 8.4 Hz, Ar-H) 7.22 (d, J = 7.7 Hz, 1H, Ar-H) 7.13 (d, J = 7.4 Hz, 2H, Ar-H) 7.10 (d, J = 7.8 Hz, 1H, Ar-H) 6.94 (t, J = 7.6 Hz, 1H, Ar-H) 6.88 (d, J = 8.4 Hz, 2H, Ar-H) 6.39 (d, J = 7.8 Hz, 1H, Ar-H), 6.29 (t, J = 7.6 Hz, 1H, Ar-H), 6.27 (t, J = 7.6 Hz, 1H, Ar-H), 6.18 (br d, J = 7.4 Hz, 2H, Ar-H), 6.17 (br d, J = 7.4 Hz, 1H, Ar-H), 4.50 (d, J = 13.4 Hz, 1H, Ar-CH_2-Ar), 4.49 (d, J = 13.4 Hz, 1H, Ar-CH_2-Ar), 4.48 (d, J = 13.4 Hz, 1H, Ar-CH_2-Ar), 4.38 (d, J = 13.4 Hz, 1H, Ar-CH_2-Ar), 4.12–4.02 (m, 4H, -OCH_2-), 4.02 (d, J = 13.4 Hz, 1H, Ar-CH_2-Ar), 3.83 (s, 3H, -OCH_3), 3.79–3.64 (m, 4H, -OCH_2-), 3.19 (d, J = 13.4 Hz, 3H, Ar-CH_2-Ar), 2.09–1.96 (m, 4H, -CH_2-CH_3), 1.96–1.84 (m, 4H, -CH_2-CH_3), 1.13 (t, J = 7.4 Hz, 6H, -CH_3), 0.93 (t, J = 7.4 Hz, 3H, -CH_3), 0.92 (t, J = 7.4 Hz, 3H, -CH_3). ^{13}C NMR ($CDCl_3$, 125 MHz, 298 K) δ 160.35, 158.97, 158.99, 156.10 (2C), 140.02, 138.43, 137.93, 137.80, 135.25, 134.01, 133.89, 133.78 (2C), 133.58, 129.69, 129.62, 129.29, 128.35, 128.19, 128.11, 128.08, 126.76, 123.12, 122.92, 122.89, 122.58, 116.60, 114.64 (2C), 91.78, 88.45, 77.49, 77.46, 77.12, 76.99, 55.70, 31.34, 31.27 (2C), 28.30, 23.79 (2C), 23.27, 23.23, 11.03, 11.01, 10.06, 10.04 ppm. HRMS (ESI$^+$) calcd for $C_{49}H_{54}O_5$ 745.3863 $[M + Na]^+$, found m/z 775.3871 $[M + Na]^+$. IR (KBr) ν 2960, 2932, 2920, 2874, 1511, 1454, 1249, 1208, 1192, 1105, 964, 830, 794, 775, 756 cm^{-1}.

Alkynylcalixarene **5d**

Method A: Calixarene **3** (0.209 mmol, 150 mg), 1-ethynyl-4-nitrobenzene (0.230 mmol, 34 mg), CuI (0.0209 mmol, 4 mg) and $Pd(PPh_3)_2Cl_2$ (0.011 mmol, 7 mg) were used under standard conditions; the product was purified by column chromatography on silica gel (DCM/cyclohexane 1:1) to afford **5d** in 27% yield (42 mg) as a yellow oil.

Method B: Calixarene **3** (0.097 mmol, 70 mg), 1-ethynyl-4-nitrobenzene (0.388 mmol, 62 mg), CuI (0.019 mmol, 4 mg), piperidine (0.291 mmol, 30 μL) and $Pd(PPh_3)_4$ (0.0097 mmol, 11 mg) were used under standard conditions. The product **5d** (28 mg, 39%) was obtained by column chromatography on silica gel (DCM/cyclohexane 50:50) as a yellow oil.

Data for compound **5d**: ^{1}H NMR ($CDCl_3$, 500 MHz, 298 K) δ 8.19 (d, J = 8.9 Hz, 2H, Ar-H), 7.63 (d, J = 8.9 Hz, 2H, Ar-H), 7.21 (d, J = 7.7 Hz, 1H, Ar-H), 7.08 (d, J = 7.7 Hz, 1H, Ar-H), 7.07 (d, J = 7.5 Hz, 1H, Ar-H), 6.89 (t, J = 7.5 Hz, 1H, Ar-H), 6.31 (dd, J = 7.7 Hz, J = 1.1 Hz, 1H, Ar-H), 6.26 (td, J = 7.5 Hz, J = 3.2 Hz, 2H, Ar-H), 6.19 (t, J = 6.2 Hz, 2H, Ar-H), 6.15 (d, J = 7.5 Hz, 1H, Ar-H), 4.49 (d, J = 13.3 Hz, 2H, Ar-CH_2-Ar) 4.46 (dd, J = 13.3 Hz, J = 3.3 Hz 2H, Ar-CH_2-Ar), 4.41 (d, J =13.3 Hz, 1H, Ar-CH_2-Ar), 4.08–3.96 (m, 4H, -O-CH_2-), 3.90 (d, 1H, J = 13.3 Hz, Ar-CH_2-Ar), 3.78–3.65 (m, 4H, -O-CH_2-), 3.19 (d, J = 13.2 Hz, 1H, Ar-CH_2-Ar), 3.16 (d, J = 13.2 Hz, 1H, Ar-CH_2-Ar), 3.73 (m, 2H, -O-CH_2-), 1.99 (sextet, J = 7.7 Hz, 4H, -CH_2-CH_3), 1.89 (sextet, J = 7.7 Hz, 4H, -CH_2-CH_3), 1.11 (t, J = 7.5 Hz, 3H, -CH_3), 1.10 (t, J = 7.5 Hz, 3H, -CH_3), 0.92 (t, J = 7.5 Hz, 3H, -CH_3), 0.91 (t, J = 7.5 Hz, 3H, -CH_3).^{13}C NMR ($CDCl_3$, 125 MHz, 298 K) δ 158.02, 157.78, 155.24 (2C), 146.70, 139.56, 139.05, 136.87, 136.78, 133.62, 133.54, 132.59, 132.47, 132.10 (2C), 130.73, 128.80 (2C), 128.72, 127.71, 127.67, 127.21, 127.16, 126.46, 123.60 (2C), 122.11, 122.10, 121.85, 120.80, 95.05, 89.52, 76.97, 76.92, 76.68, 76.46, 31.10, 30.97 (2C), 28.20, 23.48, 23.47, 23.02, 22.98, 10.77, 10.76, 9.86, 9.85 ppm. HRMS (ESI$^+$) calcd for $C_{48}H_{51}NO_6$ 760.3609 [M + Na]$^+$, found m/z 760.3607 [M + Na]$^+$. IR (KBr) ν 2960, 2929, 2874, 1589, 1454, 1382, 1340, 1208, 1192, 1087, 1004, 966, 756 cm^{-1}.

Alkynylcalixarene **5e**

Method A: Calixarene **3** (0.280 mmol, 205 mg), 3-ethynylpyridine (0.421 mmol, 40 μL), CuI (0.028 mmol, 5 mg) and $Pd(PPh_3)_2Cl_2$ (0.014 mmol, 10 mg) were used under standard conditions; the product was purified by column chromatography on silica gel (DCM/cyclohexane 2:1) to afford **5e** in 33% yield (64 mg) as a yellow oil.

Method B: Calixarene **3** (0.280 mmol, 204 mg), 3-ethynylpyridine (0.88 mmol, 90 μL), CuI (0.056 mmol, 11 mg), piperidine (0.841 mmol, 80 μL) and $Pd(PPh_3)_4$ (0.028 mmol, 32 mg) were used under standard conditions. The product **5e** (91 mg, 47%) was obtained by column chromatography on silica gel (DCM/cyclohexane 2:1) as a yellow oil.

Data for compound **5e**: ^{1}H NMR ($CDCl_3$, 500 MHz, 298 K) δ 8.61 (br s, 1H, Ar-H) 7.64 (td, J = 7.7 Hz, J = 1.5 Hz, 1H, Ar-H) 7.49 (d, J = 7.8 Hz, 1H, Ar-H) 7.27 (d, J = 7.8 Hz, 1H, Ar-H) 7.20 (dd, J = 7.8 Hz, J = 4.8 Hz, 1H, Ar-H) 7.08 (d, J = 7.4 Hz, 3H, Ar-H) 6.89 (t, J = 7.4 Hz, 1H, Ar-H), 6.33 (d, J = 7.3 Hz, 1H, Ar-H), 6.21–6.28 (m, 2H, Ar-H), 6.10–6.19 (m, 3H, Ar-H), 4.48 (d, J = 13.0 Hz, 1H, Ar-CH_2-Ar), 4.46 (d, J = 13.0 Hz, 1H, Ar-CH_2-Ar), 4.45 (d, J = 13.0 Hz, 1H, Ar-CH_2-Ar), 4.38 (d, J = 13.0 Hz, 1H, Ar-CH_2-Ar), 3.94–4.10 (m, 1H (Ar-CH_2-Ar) + 4H (-OCH_2-)), 3.62–3.79 (m, 4H, -OCH_2-), 3.18 (d, 1H, J = 13.0 Hz, 1H, Ar-CH_2-Ar), 3.15 (d, 1H, J = 13.0 Hz, 2H, Ar-CH_2-Ar), 1.93–2.03 (m, 4H, -CH_2-CH_3), 1.84–1.92 (m, 4H, -CH_2-CH_3), 1.10 (t, J = 7.3 Hz, 6H, -CH_3), 0.91 (t, J = 7.3 Hz, 6H, -CH_3), 0.90 (t, J = 7.3 Hz, 6H, -CH_3).^{13}C NMR ($CDCl_3$, 125 MHz, 298 K) δ 157.98, 157.89, 155.19, 150.02, 143.88, 139.80, 138.70, 137.02, 136.90, 136.04, 133.50, 133.24, 132.79, 132.57, 128.86, 128.80, 128.61, 127.62, 127.45, 127.38, 127.26, 126.70, 122.46, 122.14, 122.12, 121.80, 121.04, 90.56, 89.46, 76.97, 76.90, 76.64, 76.47, 31.10, 31.00, 30.98, 23.51 (2C), 23.01, 23.99, 10.81, 10.79, 9.87, 9.86 ppm. HRMS (ESI$^+$) calcd for $C_{47}H_{51}NO_4$ 716.3710 [M + Na]$^+$, found m/z 716.3710 [M + Na]$^+$. IR (KBr) ν 2960, 2931, 2873, 1580, 1455, 1383, 1209, 1192, 1087, 1066, 1005, 956, 756 cm^{-1}.

Alkynylcalixarene **5f**

Method A: Calixarene **3** (0.349 mmol, 250 mg), trimethylsilylacetylene (0.418 mmol, 60 µL), CuI (0.035 mmol, 7 mg) and $Pd(PPh_3)_2Cl_2$ (0.017 mmol, 12 mg) were used under standard conditions; the product was purified by column chromatography on silica gel (DCM/cyclohexane 1:1) to afford **5f** in 83% yield (199 mg) as a yellow oil.

Method B: Calixarene **3** (0.138 mmol, 104 mg), trimethylsilylacetylene (0.552 mmol, 78 µL), CuI (0.028 mmol, 5 mg), piperidine (0.417 mmol, 40 µL) and $Pd(PPh_3)_4$ (0.014 mmol, 16 mg) were used under standard conditions. The product **5f** (71 mg, 75%) was obtained by column chromatography on silica gel (DCM/cyclohexane 1:1) as a yellow oil.

Data for compound **5f**: ^{1}H NMR ($CDCl_3$, 500 MHz, 298 K) δ 7.14 (d, 1H, J = 7.8 Hz, Ar-H) 7.10 (m, 2H, Ar-H) 7.03 (d, J = 7.8 Hz, 2H, Ar-H) 6.91 (t, J = 7.4 Hz, 1H, Ar-H) 6.27 (dd, J = 7.7 Hz, J = 1.7 Hz, 1H, Ar-H) 6.25 (t, J = 7.4 Hz, 1H, Ar-H) 6.20 (t, J = 7.4 Hz, 1H, Ar-H), 6.15–6.10 (m, 2H, Ar-H), 6.07 (dd, J = 7.4 Hz, J = 1.7 Hz, 1H, Ar-H), 4.43 (d, J = 13.5 Hz, 1H, Ar-CH_2-Ar), 4.44 (d, J = 13.5 Hz, 2H, Ar-CH_2-Ar), 4.28 (d, J = 13.4 Hz, 1H, Ar-CH_2-Ar), 4.04–3.95 (m, 4H, -OCH_2), 4.48 (d, J = 13.4 Hz, 1H, Ar-CH_2-Ar), 4.38 (d, J = 13.4 Hz, 1H, Ar-CH_2-Ar), 4.12–4.02 (m, 4H, -OCH_2-), 3.90 (d, J = 13.4 Hz, 1H, Ar-CH_2-Ar), 3.76–3.60 (m, 4H, -OCH_2-), 3.14 (d, J = 13.4 Hz, 2H, Ar-CH_2-Ar), 3.13 (d, J = 13.4 Hz, 1H, Ar-CH_2-Ar), 2.02–1.90 (m, 4H, -CH_2-CH_3), 1.91–1.80 (m, 4H, -CH_2-CH_3), 1.09 (t, J = 7.4 Hz, 3H, -CH_3), 1.08 (t, J = 7.4 Hz, 3H, -CH_3), 0.87 (t, J = 7.4 Hz, 6H, -CH_3), 0.21 (s, 9H, -CH_3) ^{13}C NMR ($CDCl_3$, 100 MHz, 298 K) δ 158.40, 158.36, 155.55, 155.50, 140.07, 138.56, 137.43, 137.35, 133.70, 133.47, 133.28, 132.93, 129.22, 129.12, 128.71, 127.84, 127.65, 127.54, 127.50, 126.79, 122.41, 122.36, 122.20, 122.06, 105.20, 96.13, 76.98 (2C), 76.57, 76.49, 30.82, 30.77, 30.74, 27.64, 23.28 (2C), 22.71 (2C), 10.52 (2C), 9.51 (2C), 0.28 (3C).ppm. HRMS (ESI$^+$) calcd for $C_{45}H_{56}SiO_4$ 711.3840 $[M + Na]^+$, found m/z 711.3847 $[M + Na]^+$. IR (KBr) ν 2960, 2932, 2874, 1455, 1383, 1247, 1207, 1192, 1088, 1006, 966, 841, 756 cm^{-1}.

3.2.2. Preparation of Bridged Calixarene **7b**

Diphenyl diselenide (0.609 mmol, 190 mg) and $FeCl_3$ (0.752 mmol, 122 mg) were dissolved in 4 mL of dry dichloromethane in a Schlenk flask under an argon atmosphere. The resulting solution was stirred for 15 min at room temperature. After this time, the solution changed color to brown-red. Calixarene **5b** (0.301 mmol, 209 mg) dissolved in 2 mL of dry dichloromethane was added to the above solution, and the resulting mixture was heated at 30 °C for 4 h. Dichloromethane (20 mL) was then added to the reaction mixture, the organic layer was washed with a saturated solution of NH_4Cl, dried with $MgSO_4$ and concentrated under vacuum. The residue was purified by column chromatography on silica gel (DCM/cyclohexane 1:1) to provide the respective product **7b** in 20% yield (51 mg) as a yellow solid.

Data for compound **7b**: M.p. = 117–120 °C. ^{1}H NMR ($CDCl_3$, 400 MHz, 298 K) δ 7.24 (dd, J = 8.0 Hz, J = 1.8 Hz, 2H, Ar-H), 7.19 (dd, J = 8.0 Hz, J = 1.8 Hz, 2H, Ar-H), 7.03–6.95 (m, 7H, Ar-H) 6.93 (d, J = 7.3 Hz, 1H, Ar-H), 6.89 (d, J = 8.0 Hz, 1H, Ar-H), 6.87 (dd, J = 8.0 Hz, J = 1.5 Hz, 1H, Ar-H), 6.82 (dd, J = 7.8 Hz, J = 1.5 Hz, 1H, Ar-H), 6.75 (d, J = 9.0 Hz, 1H, Ar-H), 6.73 (t, J = 7.6 Hz, 1H, Ar-H), 6.61 (t, J = 7.4 Hz, 1H, Ar-H), 6.43 (d, J = 7.8 Hz, 1H, Ar-H), 5.98 (d, J = 7.8 Hz, 1H, Ar-H), 4.70 (d, J = 13.4 Hz, 1H, Ar-CH_2-Ar), 4.50 (d, J = 12.5 Hz, 1H, Ar-CH_2-Ar), 4.41 (d, J = 12.0 Hz, 1H, Ar-CH_2-Ar), 4.33 (d, J = 12.5 Hz, 1H, Ar-CH_2-Ar), 4.04–3.79 (m, 6H, -OCH_2-), 3.64–3.54 (m, 2H, -OCH_2-), 3.28 (d, J = 13.2 Hz, 1H, Ar-CH_2-Ar), 3.27 (d, J = 12.5 Hz, 1H, Ar-CH_2-Ar), 3.11 (d, J = 12.5 Hz, 1H, Ar-CH_2-Ar), 3.07 (d, J = 12.5 Hz, 1H, Ar-CH_2-Ar), 2.21 (sextet, J = 7.7 Hz, 2H, -CH_2-CH_3), 2.13–2.01 (m, 2H, -CH_2-CH_3), 2.01–1.83 (m, 4H, -CH_2-CH_3), 1.20 (t, J = 7.4 Hz, 3H, -CH_3), 1.19 (t, J = 7.4 Hz, 3H, -CH_3), 1.00 (t, J = 7.4 Hz, 3H, -CH_3), 0.99 (t, J = 7.4 Hz, 3H, -CH_3). ^{13}C NMR ($CDCl_3$, 100 MHz, 298 K) δ 156.32, 155.80, 154.80, 154.77, 140.82, 140.37, 140.36, 140.08, 136.54, 136.23, 135.04, 134.42, 134.18, 132.98, 132.71, 132.70, 132.21, 131.16, 130.97 (2C), 130.39, 129.27, 128.86, 128.41 (2C), 128.09, 127.40, 127.37 (2C), 126.89, 126.58, 126.48, 126.44, 123.27, 123.05, 123.01, 122.94, 77.48, 77.45, 76.99, 76.70, 31.79, 31.56, 29.19, 24.82, 23.77, 23.73, 23.12, 22.99, 11.10 (2C), 10.29, 10.15 ppm. HRMS (ESI$^+$) calcd for $C_{54}H_{56}SeO_4$

871.3236 [M + Na]$^+$, found m/z 871.3246 [M + Na]$^+$. IR (KBr) ν 2959, 2924, 2872, 1455, 1250, 1208, 1066, 1007, 960, 734, 690 cm^{-1}.

3.2.3. Preparation of Bridged Calixarene **8b**

Diphenyl diselenide (0.28 mmol, 90 mg) and $FeCl_3$ (0.351 mmol, 57 mg) were dissolved in 3 mL of dry DCE in a Schlenk flask under an argon atmosphere, and the resulting solution was stirred for 15 min at room temperature. After this time, the color of the solution turned to brown-red. Calixarene **5b** (0.144 mmol, 100 mg) dissolved in 1 mL of dry DCE was added, and the reaction mixture was heated to 85 °C for 4 h. The eluent was then evaporated under reduced pressure, dichloromethane (20 mL) was added to the crude product, washed with a saturated solution of NH_4Cl, dried with $MgSO_4$ and then concentrated under vacuum. The residue was purified by preparative thin layer chromatography on silica gel (DCM/cyclohexane 1:1) to provide the methylene bridged product **8b** in 10% yield (12 mg) as a white solid.

Data for compound **8b**: M.p. = 101–104 °C. ^{1}H NMR ($CDCl_3$, 400 MHz, 298 K) δ 8.38 (s, 1H, -OH), 8.25 (s, 1H, -OH), 7.45 (d, J = 7.9 Hz, 2H, Ar-H), 7.39 (d, J = 1.4 Hz, 1H, Ar-H), 7.34 (t, J = 7.0 Hz, 3H, Ar-H), 7.30–7.11 (m, 10H, Ar-H) 7.07 (d, J = 7.3 Hz, 1H, Ar-H), 6.87 (d, J = 7.3 Hz, 1H, Ar-H), 6.83 (d, J = 7.3 Hz, 2H, Ar-H), 6.74 (d, J = 7.5 Hz, 1H, Ar-H), 6.73 (t, J = 7.4 Hz, 1H, Ar-H), 6.59 (t, J = 7.7 Hz, 1H, Ar-H), 6.50 (d, J = 7.2 Hz, 1H, Ar-H), 5.49 (s, 1H, Ar-CH(R)-Ar), 4.51 (d, J = 12.4 Hz, 1H, Ar-CH_2-Ar), 4.43 (d, J = 12.5 Hz, 1H, Ar-CH_2-Ar), 4.18–4.04 (m, 2H, -OCH_2-), 4.03–3.90 (m, 2H, -OCH_2-), 3.43 (d, J = 13.0 Hz, 1H, Ar-CH_2-Ar), 3.35 (d, J = 13.0 Hz, 1H, Ar-CH_2-Ar), 3.26 (d, J = 12.4 Hz, 1H, Ar-CH_2-Ar), 2.19–1.96 (m, 4H, -CH_2-CH_3), 1.37 (t, J = 7.3 Hz, 3H, -CH_3), 1.30 (t, J = 7.4 Hz, 3H, -CH_3). ^{13}C NMR ($CDCl_3$, 125 MHz, 298 K) δ 154.14, 153.59, 153.13, 152.32, 151.24, 144.49, 135.99, 134.72, 134.00, 133.82, 132.04, 131.70, 130.81, 130.35, 130.24 (2C), 130.24, 130.06, 130.02, 129.72 (2C), 129.46 (2C), 129.32 (2C), 129.21, 129.17, 128.68, 128.56, 128.23 (2C), 128.01, 127.57, 126.25, 125.94, 125.76, 125.50, 125.22, 116.72, 113.61, 78.29, 78.21, 50.51, 32.11, 31.45, 30.09, 23.88, 23.69, 11.18, 11.06 ppm. HRMS (ESI$^+$) calcd for $C_{54}H_{48}Se_2O_4$ 943.1775 [M + Na]$^+$, found m/z 943.1785 [M + Na]$^+$.

3.2.4. Preparation of Bridged Calixarenes **9c** and **10c**

Diphenyl diselenide (0.170 mmol, 53 mg) and $FeCl_3$ (0.213 mmol, 34 mg) were dissolved in 3 mL of dry dichloromethane in a Schlenk flask under an argon atmosphere. The resulting solution was stirred for 15 min at room temperature. After this time, the solution changed color to brown-red. Calixarene **5c** (0.065 mmol, 85 mg) dissolved in 1 mL of dry dichloromethane was added, and the reaction mixture was heated at 30 °C for 4 h. Dichloromethane (20 mL) was added to the mixture, the organic layer was washed with a saturated solution of NH_4Cl, dried over $MgSO_4$, and concentrated under vacuum. The residue was purified by column chromatography on a silica gel (DCM/cyclohexane 1:1) to provide the respective products **9c** in 37% yield (27 mg) and **10c** in 15% yield (11 mg) as yellow oils.

Data for compound **9c**: ^{1}H NMR ($CDCl_3$, 400 MHz, 298 K) δ 7.41 (dd, J = 7.4 Hz, J = 1.5 Hz, 2H, Ar-H), 7.31 (d, J = 8.4 Hz, 2H, Ar-H), 7.23–7.15 (m, 4H, Ar-H) 7.13 (dd, J = 7.5 Hz, J = 1.4 Hz, 1H, Ar-H), 7.07 (d, J = 7.5 Hz, 1H, Ar-H), 6.95 (t, J = 7.3 Hz, 1H, Ar-H), 6.87 (d, J = 7.4 Hz, 1H, Ar-H), 6.77 (d, J = 8.9 Hz, 2H, Ar-H), 6.20 (t, J = 7.4 Hz, 1H, Ar-H), 6.14–6.07 (m, 3H, Ar-H), 6.02 (dd, J = 7.5 Hz, J = 1.3 Hz, 1H, Ar-H), 5.80 (dd, J = 7.5 Hz, J = 1.5 Hz, 1H, Ar-H), 5.70 (s, 1H, Ar-CH(R)-Ar), 4.49 (d, J = 13.3 Hz, 1H, Ar-CH_2-Ar), 4.44 (d, J = 13.3 Hz, 2H, Ar-CH_2-Ar), 4.34 (td, J = 11.0 Hz, J = 5.3 Hz, 1H, -OCH_2-), 4.23–4.10 (m, 2H, -OCH_2-), 4.04–3.94 (m, 1H, -OCH_2-), 3.74 (s, 3H, -OCH_3), 3.71–3.64 (m, 3H, -OCH_2-), 3.54–3.46 (m, 1H, -OCH_2-), 3.19 (d, J = 13.3 Hz, 1H, Ar-CH_2-Ar), 3.15 (d, J = 13.3 Hz, 1H, Ar-CH_2-Ar), 3.13 (d, J = 13.3 Hz, 1H, Ar-CH_2-Ar), 2.04–1.81 (m, 8H, -CH_2-CH_3), 1.14 (t, J = 7.5 Hz, 3H, -CH_3), 1.10 (t, J = 7.4 Hz, 3H, -CH_3), 0.91 (t, J = 7.5 Hz, 3H, -CH_3), 0.89 (t, J = 7.4 Hz, 3H, -CH_3). We cannot acquire reliable ^{13}C NMR spectra due to degradation of compound. HRMS (ESI$^+$) calcd for $C_{55}H_{58}SeO_5$ 901.3341 [M + Na]$^+$, found *m/z* 901.3345 [M + Na]$^+$.

Data for compound **10c**: ^{1}H NMR ($CDCl_3$, 400 MHz, 298 K) δ 7.40 (dd, *J* = 7.4 Hz, *J* = 1.5 Hz, 2H, Ar-H), 7.37 (d, *J* = 8.6 Hz, 2H, Ar-H), 7.22–7.14 (m, 3H, Ar-H), 7.10 (d, *J* = 7.6 Hz, 3H, Ar-H), 6.93 (d, *J* = 7.6 Hz, 1H, Ar-H), 6.81 (d, *J* = 7.6 Hz, 1H, Ar-H), 6.77 (d, *J* = 8.6 Hz, 2H, Ar-H), 6.46–6.34 (m, 3H, Ar-H), 6.27 (~dd, *J* = 7.6 Hz, *J* = 1.5 Hz, 1H, Ar-H), 6.22 (t, *J* = 7.6 Hz, 1H, Ar-H), 5.98 (dd, *J* = 7.6 Hz, *J* = 1.5 Hz, 1H, Ar-H), 5.74 (s, 1H, Ar-CH(R)-Ar), 4.42 (d, *J* = 13.3 Hz, 1H, Ar-CH_2-Ar), 4.40 (d, *J* = 13.3 Hz, 1H, Ar-CH_2-Ar), 4.37 (d, *J* = 13.3 Hz, 1H, Ar-CH_2-Ar), 4.28–4.19 (m, 1H, -OCH_2-), 4.00–3.92 (m, 1H, -OCH_2-), 3.77–3.72 (m, 3H + 3H, -OCH_2-, -OCH_3), 3.65–3.59 (m, 1H, -OCH_2-), 3.33 (d, *J* = 13.3 Hz, 1H, Ar-CH_2-Ar), 3.23 (d, *J* = 13.3 Hz, 1H, Ar-CH_2-Ar), 3.22 (d, *J* = 13.3 Hz, 1H, Ar-CH_2-Ar), 2.26–2.20 (m, 1H, -CH_2-CH_3), 2.15–2.07 (m, 1H, -CH_2-CH_3), 1.97–1.84 (m, 4H, -CH_2-CH_3), 1.15 (t, *J* = 7.3 Hz, 3H, -CH_3), 1.10 (t, *J* = 7.3 Hz, 3H, -CH_3), 0.93 (t, *J* = 7.4 Hz, 3H, -CH_3). We cannot acquire ^{13}C NMR spectra due to degradation of compound. HRMS (ESI$^+$) calcd for $C_{52}H_{52}SeO_5$ 859.2881 [M + Na]$^+$, found *m/z* 859.2881 [M + Na]$^+$.

3.2.5. Electrochemical Experiments

Batch Electrolysis Setup

The electrolysis was performed in a 20 mL undivided cell filled with 15 mL of MeCN (HPLC gradient grade ≥99.9%, Merck), containing 0.02 M $LiClO_4 \cdot H_2O$ (BDH Chemicals Ltd., Poole, England) used as a supporting electrolyte. Additionally, the solution contained compound **5b** with 0.6 equiv. of Ph_2Se_2. As working electrodes, two different materials were tested: Pt sheet (5 mm × 30 mm), or glassy carbon (GC, diameter 3 mm). In the case of using GC as the working electrode, GC was conducted by Motorcontroller for Rotator, and 100 RPM rotation speed was used. As the reference electrode, a saturated calomel electrode (SCE) separated from the investigated sample by a salt-bridge filled by the blank (MeCN electrolyte solution) was used, and as the counter (auxiliary) electrode, a Pt sheet (5 mm × 30 mm) or GC (diameter 3 mm) were applied. The reaction mixture was stirred and electrolyzed at room temperature during (a) constant current conditions (10 mA was applied) (b) controlled-potential electrolysis (potential was set to +1.5 V). In the cases when electrolysis was performed under inert atmosphere, oxygen was removed from the solution by passing a stream of argon (99.998%, Messer, Bad Soden, Germany). Measurements were carried out using the computer-driven digital potentiostat PGSTAT204 (Autolab-Metrohm, Herisau, Switzerland) controlled by software NOVA 1.11.

Attempted Electrolytic Cyclization of 5b

Calixarene **5b** (25 mg, 0.036 mmol) with 0.6 equiv. of diphenyl diselenide (6.76 mg, 0.022 mmol) was dissolved in 15 mL of solution of a supporting electrolyte (0.02 M $LiClO_4 \cdot H_2O$). The solution was poured into an undivided cell, and then the Pt plate (used as working electrode), SCE (reference electrode), and Pt plate (auxiliary electrode) were placed in solution (Figure S79). The reaction mixture was stirred by a magnetic bar at room temperature during constant current (set to 10 mA) under aerobic condition. The progress of the reaction was monitored by TLC until the full conversion of **5b** was detected (3.5 h).

After completion, the crude reaction mixture was collected and evaporated to dryness. The solid was partitioned between DCM and water; the organic layer was washed with water and dried over $MgSO_4$. Crude evaporated product was purified using preparative TLC on silica gel (DCM/cyclohexane 1:1) to provide product 11 (19 mg, 73%) as a white solid.

Data for compound **11**: M.p. = 137–141 °C. ^{1}H NMR ($CDCl_3$, 400 MHz, 298 K) δ 7.92 (d, *J* = 7.4 Hz, 2H, Ar-H), 7.65 (t, *J* = 7.8 Hz, 1H, Ar-H), 7.51 (t, *J* = 7.1 Hz, 2H, Ar-H), 7.15 (d, *J* = 8.2 Hz, 1H, Ar-H), 6.97 (d, *J* = 7.5 Hz, 1H, Ar-H), 6.91 (t, *J* = 7.1 Hz, 2H, Ar-H), 6.70 (t, *J* = 7.0 Hz, 1H, Ar-H), 6.39–6.27 (m, 5H, Ar-H), 6.23 (d, *J* = 7.4 Hz, 1H, Ar-H), 4.55 (d, *J* = 13.4 Hz, 1H, Ar-CH_2-Ar), 4.52 (d, *J* = 13.4 Hz, 1H, Ar-CH_2-Ar), 4.45 (d, *J* = 13.4 Hz, 1H,

Ar-CH_2-Ar), 4.44 (d, J = 13.4 Hz, 1H, Ar-CH_2-Ar), 4.41 (d, J = 13.4 Hz, 1H, Ar-CH_2-Ar), 4.04–3.91 (m, 4H, -OCH_2-), 3.82–3.65 (m, 4H, -OCH_2-), 3.20 (d, J = 13.4 Hz, 1H, Ar-CH_2-Ar), 3.17 (d, J = 13.4 Hz, 1H, Ar-CH_2-Ar), 3.15 (d, J = 13.4 Hz, 1H, Ar-CH_2-Ar), 2.06–1.82 (m, 8H, -CH_2-CH_3), 1.06 (t, J = 7.5 Hz, 3H, -CH_3), 1.05 (t, J = 7.4 Hz, 3H, -CH_3), 0.95 (t, J = 7.5 Hz, 3H, -CH_3), 0.93 (t, J = 7.4 Hz, 3H, -CH_3). ^{13}C NMR ($CDCl_3$, 100 MHz, 298 K) δ 197.92, 195.89, 160.23, 158.43, 156.69, 156.58, 144.17, 140.08, 137.42, 137.02, 135.35, 135.17, 134.68, 134.09, 134.01, 132.94, 132.51, 130.79 (2C), 129.71 (2C), 129.56, 129.49, 129.42, 129.28, 129.07, 129.02, 128.92, 128.83, 128.72, 128.39, 128.22, 128.12, 122.97, 122.76, 122.59, 77.51, 77.49, 77.09, 77.05, 31.47, 31.37, 31.29, 25.94, 23.64, 23.48, 23.33, 23.30, 10.81, 10.76, 10.18, 10.12 ppm. HRMS (ESI^+) calcd for $C_{48}H_{52}O_6$ 747.3656 $[M + Na]^+$, found m/z 747.3662 $[M + Na]^+$. IR (KBr) ν 2961, 2924, 2872, 1674, 1665, 1586, 1454, 1384, 1290, 1245, 1228, 1105, 964, 785, 756, 719 cm^{-1}.

Cyclic Voltammetry

Cyclic voltammetry measurements (with scan rates 100, 200, and 500 $mV{\cdot}s^{-1}$) were performed in MeCN (HPLC gradient grade ≥99.9%, Merck) using 0.02 M $LiClO_4{\cdot}H_2O$ (BDH Chemicals Ltd., Poole, England) as the supporting electrolyte (Figures S77 and S78). Due to low conductivity, the three-electrode systems were applied for all measurements. As the working electrode, a glassy carbon electrode (diameter 3 mm) or Pt disk electrode (diameter 1 mm) was utilized. As the reference electrode, a saturated calomel electrode (SCE) separated from the investigated sample by a bridge filled by the blank (MeCN-electrolyte solution) was applied, and as the auxiliary electrode, a Pt sheet was used. All those experiments were standardly carried out in an undivided 20 mL cell (using 10 mL of the corresponding solution), and before measurements, the solutions were de-aerated by argon (99.998%, Messer). Measurements were carried out using the computer-driven digital potentiostat PGSTAT101 (Autolab-Metrohm) controlled by software NOVA 1.11.

3.3. X-ray Measurements

3.3.1. Crystallographic Data for **7b**

M = 848 $g{\cdot}mol^{-1}$, triclinic system, space group P-1, a = 11.6678 (4) Å, b = 17.7287(5) Å, c = 21.7145(6) Å, α = 82.4177(12)°, β = 86.9096(13)°, γ = 88.6006(13)°, Z = 4, V = 4445.3(2) Å^3, D_c = 1.267 $g{\cdot}cm^{-3}$, μ(Cu-Kα) = 1.494 mm^{-1}, crystal dimensions of 0.077 × 0.077 × 0.158 mm. Data were collected at 180 (2) K on a Bruker D8 Venture Photon II 7 diffractometer with Incoatec microfocus sealed tube Cu-Kα radiation. The data were integrated, scaled and corrected for absorption using Apex4 [33]. The structure was solved by SIR92 [34] and anisotropically refined by full matrix least squares on F squared using the CRYSTALS [35] to final value R = 0.048 and wR = 0.124 using 1486 independent reflections (θ_{max} = 72.2°), 1082 parameters and 42 restrains. The hydrogen atoms bonded to carbon atoms were placed in calculated positions and refined with riding constraints, while hydrogen atoms bonded to oxygen were refined using soft restraints. The disordered functional groups positions were found in difference electron density maps and refined with restrained geometry. MCE [36] was used for the visualization of electron density maps. The occupancy of disordered functional groups was constrained to full. The structure was deposited into Cambridge Structural Database under number CCDC 2332078. The full numbering scheme can be found in Supporting information Figures S74 and S75.

3.3.2. Crystallographic Data for **8b**

M = 1003.82 $g{\cdot}mol^{-1}$, monoclinic system, space group $P2_1/n$, a = 13.3272(4) Å, b = 13.0362(4) Å, c = 27.1526(7) Å, β = 94.7510(10)°, Z = 4, V = 4701.2(2) Å^3, D_c = 1.418 $g{\cdot}cm^{-3}$, μ(Cu-Kα) = 3.38 mm^{-1}, crystal dimensions of 0.048 × 0.126 × 0.426 mm. Data were collected at 180 (2) K on a Bruker D8 Venture Photon II 7 diffractometer with Incoatec microfocus sealed tube Cu-Kα radiation. The data were integrated, scaled and corrected for absorption using Apex4 [33]. The structure was solved by SIR92 [34] and anisotropically refined by full matrix least squares on F squared using the CRYSTALS [35] to final value R = 0.060 and wR = 0.178 using 9241 independent reflections (θ_{max} = 72.1°),

577 parameters and 0 restrains. The hydrogen atoms bonded to carbon atoms were placed in calculated positions and refined with riding constraints, while hydrogen atoms bonded to oxygen were first refined using soft restraints, then refined with riding constraints. The disordered solvent positions were found in difference electron density maps. MCE [36] was used for the visualization of electron density maps. The occupancy of disordered functional groups was first constrained to full, later fixed at values of 2/3 and 1/3, respectively. The structure was deposited into Cambridge Structural Database under number CCDC 2332079. The full numbering scheme can be found in Supporting information Figure S76.

4. Conclusions

A Sonogashira coupling of *meta*-iodocalix[4]arene with various terminal acetylenes confirmed that the *meta* position of calixarene is well addressable, and that both thermal and microwave protocols lead to good yields of the corresponding alkynylcalixarenes. Alkynes thus obtained were subjected to the $FeCl_3/Ph_2Se_2$-promoted electrophilic closure. It turns out that the calix[4]arenes give completely different bridging products than those described for the non-macrocyclic starting compounds. This can be demonstrated not only by the isolation of products with a six-membered ring (*6-exo-dig*), but mainly by the smooth formation of the *5-endo-dig* cyclization, which has never been observed in the aliphatic series. An attempt at electrocyclization led to a high yield of the 1,2-diketone, formed by oxidation of the starting alkyne. The structures of the unexpected products were unequivocally established by X-ray analysis, and clearly demonstrate how the preorganized macrocyclic skeleton favors a completely different regioselectivity of cyclization reactions compared to common aliphatic derivatives.

Supplementary Materials: The following supporting information can be downloaded at: https://www.mdpi.com/article/10.3390/molecules29061237/s1, Spectral characterization of all new compounds (^{1}H NMR, ^{13}C NMR, HRMS, IR), VT NMR experiments, X-ray structures and electrochemical experiments.

Author Contributions: Conceptualization, writing—review and editing, P.L.; experimental work, M.P.; synthesis, spectra analysis, editing, A.S.; X-ray measurement, J.Č.; electrochemical experiments, K.S. and L.Š. All authors have read and agreed to the published version of the manuscript.

Funding: This research was funded by the Czech Science Foundation, grant number 23-07154S (P.L., A.S.), 21-05926X (J.C., A.S.) and by Specific university research (A2_FCHT_2022_073) (A.S.).

Institutional Review Board Statement: Not applicable.

Informed Consent Statement: Not applicable.

Data Availability Statement: All experimental data are provided in the Supplementary Materials.

Conflicts of Interest: The authors declare no conflicts of interest.

References

1. Gutsche, C.D. *Calixarenes: An Introduction*; Royal Society of Chemistry, RSC Publishing: Cambridge, UK, 2008.
2. Asfari, Z.; Böhmer, V.; Harrowfield, J.; Vicens, J. *Calixarenes 2001*; Kluwer Academic Publishers: Dordrecht, The Netherlands, 2001.
3. Neri, P.; Sessler, J.L.; Wang, M.X. *Calixarenes and Beyond*; Springer: Cham, Switzerland, 2016. [CrossRef]
4. Mandolini, L.; Ungaro, R. *Calixarenes in Action*; World Scientific Publishing Company: Singapore, 2000.
5. Lhoták, P. Direct meta substitution of calix[4]arenes. *Org. Biomol. Chem.* **2022**, *20*, 7377–7390. [CrossRef] [PubMed]
6. Slavik, P.; Dudic, M.; Flidrova, K.; Sykora, J.; Cisarova, I.; Bohm, S.; Lhotak, P. Unprecedented Meta-Substitution of Calixarenes: Direct Way to Inherently Chiral Derivatives. *Org. Lett.* **2012**, *14*, 3628–3631. [CrossRef] [PubMed]
7. Flidrova, K.; Slavik, P.; Eigner, V.; Dvorakova, H.; Lhotak, P. meta-Bridged calix[4]arenes: A straightforward synthesis via organomercurial chemistry. *Chem. Commun.* **2013**, *49*, 6749–6751. [CrossRef] [PubMed]
8. Tlusty, M.; Spalovska, D.; Kohout, M.; Eigner, V.; Lhotak, P. Ketone transformation as a pathway to inherently chiral rigidified calix[4]arenes. *Chem. Commun.* **2020**, *56*, 12773–12776. [CrossRef] [PubMed]
9. Tlusty, M.; Slavik, P.; Kohout, M.; Eigner, V.; Lhotak, P. Inherently Chiral Upper-Rim-Bridged Calix[4]arenes Possessing a Seven Membered Ring. *Org. Lett.* **2017**, *19*, 2933–2936. [CrossRef] [PubMed]

10. Gilmore, K.; Alabugin, I.V. Cyclizations of Alkynes: Revisiting Baldwin's Rules for Ring Closure. *Chem. Rev.* **2011**, *111*, 6513–6556. [CrossRef] [PubMed]
11. Thomas, A.M.; Sujatha, A.; Anilkumar, G. Recent advances and perspectives in copper-catalyzed Sonogashira coupling reactions. *RSC Adv.* **2014**, *4*, 21688–21698. [CrossRef]
12. Karak, M.; Barbosa, L.C.; Hargaden, G.C. Recent mechanistic developments and next generation catalysts for the Sonogashira coupling reaction. *RSC Adv.* **2014**, *4*, 53442–53466. [CrossRef]
13. Dyker, G.; Mastalerz, M.; Müller, I.M. Electron-Rich Cavitands via Fourfold Sonogashira Cross-Coupling Reaction of Calix[4]arenes and Bromopyridines—Effect of the Nitrogen Position on Complexation Abilities. *Eur. J. Org. Chem.* **2005**, *2005*, 3801–3812. [CrossRef]
14. Böhmer, V.; Brusko, V.; Rissanen, K. Extended Calix[8]arenes by Sonogashira Cross-Coupling with Ethynylarenes. *Synthesis* **2002**, *2002*, 1898–1902. [CrossRef]
15. Boonkitpatarakul, K.; Yodta, Y.; Niamnont, N.; Sukwattanasinitt, M. Fluorescent phenylethynylene calix[4]arenes for sensing TNT in aqueous media and vapor phase. *RSC Adv.* **2015**, *5*, 33306–33311. [CrossRef]
16. Hennrich, G.; Murillo, M.T.; Prados, P.; Al-Saraierh, H.; El-Dali, A.; Thompson, D.W.; Collins, J.; Georghiou, P.E.; Teshome, A.; Asselberghs, I.; et al. Alkynyl Expanded Donor–Acceptor Calixarenes: Geometry and Second-Order Nonlinear Optical Properties. *Chem. Eur. J.* **2007**, *13*, 7753–7761. [CrossRef] [PubMed]
17. Al-Saraierh, H.; Miller, D.O.; Georghiou, P.E. Narrow-rim functionalization of calix[4]arenes via Sonogashira coupling reactions. *J. Org. Chem.* **2005**, *70*, 8273–8280. [CrossRef] [PubMed]
18. Prochnow, T.; Back, D.F.; Zeni, G. Iron(III) Chloride and Diorganyl Diselenide-Promoted Nucleophilic Closures of 1-Benzyl-2-alkynylbenzenes in the Preparation of 9-(Organoselanyl)-5H-benzo[7]annulenes. *Adv. Synth. Catal.* **2016**, *358*, 1119–1129. [CrossRef]
19. Goulart, T.A.C.; Kazmirski, J.A.G.; Back, D.F.; Zeni, G. Iron(III)-Promoted Synthesis of 3-(Organoselanyl)-1,2-Dihydroquinolines from Diorganyl Diselenides and N-Arylpropargylamines by Sequential Carbon-Carbon and Carbon-Selenium Bond Formation. *Adv. Synth. Catal.* **2019**, *361*, 96–104. [CrossRef]
20. Sperança, A.; Godoi, B.; Pinton, S.; Back, D.F.; Menezes, P.H.; Zeni, G. Regioselective Synthesis of Isochromenones by Iron(III)/PhSeSePh-Mediated Cyclization of 2-Alkynylaryl Esters. *J. Org. Chem.* **2011**, *76*, 6789–6797. [CrossRef]
21. Martinez, C.R.; Iverson, B.L. Rethinking the term "pi-stacking". *Chem. Sci.* **2012**, *3*, 2191–2201. [CrossRef]
22. Thakuria, R.; Nath, N.K.; Saha, B.K. The Nature and Applications of π–π Interactions: A Perspective. *Cryst. Growth Des.* **2019**, *19*, 523–528. [CrossRef]
23. Mukherjee, N.; Satyanarayana, A.N.; Singh, P.; Dixit, M.; Chatterjee, T. Recyclable iodine-catalyzed radical selenylative annulation of 2-alkynyl biaryls with diselenides in water: A green approach to selanyl polycyclic aromatic hydrocarbons and polycyclic heteroaromatics. *Green Chem.* **2022**, *24*, 7029–7038. [CrossRef]
24. An, S.; Zhang, Z.; Li, P. Metal-Free Synthesis of Selenodihydronaphthalenes by Selenoxide—Mediated Electrophilic Cyclization of Alkynes. *Eur. J. Org. Chem.* **2021**, *2021*, 3059–3070. [CrossRef]
25. Zhang, X.; Campo, M.A.; Yao, T.; Larock, R.C. Synthesis of Substituted Quinolines by Electrophilic Cyclization of N-(2-Alkynyl)anilines. *Org. Lett.* **2005**, *7*, 763–766. [CrossRef] [PubMed]
26. Gabriele, B.; Mancuso, R.; Veltri, L. Recent Advances in the Synthesis of Indanes and Indenes. *Chem. Eur. J.* **2016**, *22*, 5056–5094. [CrossRef]
27. Nahide, P.D.; Jiménez-Halla, J.O.C.; Wrobel, K.; Solorio-Alvarado, C.R.; Ortiz Alvarado, R.; Yahuaca-Juárez, B. Gold(i)-catalysed high-yielding synthesis of indenes by direct Csp3–H bond activation. *Org. Biomol. Chem.* **2018**, *16*, 7330–7335. [CrossRef]
28. Yang, S.; Li, Z.; Jian, X.; He, C. Platinum(II)-Catalyzed Intramolecular Cyclization of o-Substituted Aryl Alkynes through sp3 C-H Activation. *Angew. Chem. Int. Ed.* **2009**, *48*, 3999–4001. [CrossRef] [PubMed]
29. Chen, Y.-Y.; Chen, Z.-Y.; Zhang, N.-N.; Chen, J.-H.; Zhang, X.-J.; Yan, M. Intramolecular Addition of Triarylmethanes to Alkynes Promoted by KOtBu/DMF: A Synthetic Approach to Indene Derivatives. *Eur. J. Org. Chem.* **2016**, *2016*, 599–606. [CrossRef]
30. Sun, L.; Wang, L.; Alhumade, H.; Yi, H.; Cai, H.; Lei, A. Electrochemical Radical Selenylation of Alkenes and Arenes via Se–Se Bond Activation. *Org. Lett.* **2021**, *23*, 7724–7729. [CrossRef]
31. Mukherjee, N.; Chatterjee, T. Highly Atom-Economic and Efficient Electrochemical Selenylative Annulation of 2-Alkynyl Biaryls. *Adv. Synth. Catal.* **2023**, *365*, 2255–2263. [CrossRef]
32. Zhou, J.; Tao, X.-Z.; Dai, J.-J.; Li, C.-G.; Xu, J.; Xu, H.-M.; Xu, H.-J. Electrochemical synthesis of 1,2-diketones from alkynes under transition-metal-catalyst-free conditions. *Chem. Commun.* **2019**, *55*, 9208–9211. [CrossRef]
33. Bruker. *APEX4, SAINT and SADABS*; Bruker AXS Inc.: Madison, WI, USA, 2021.
34. Altomare, A.; Cascarano, G.; Giacovazzo, C.; Guagliardi, A.; Burla, M.C.; Polidori, G.; Camalli, M. SIRPOW.92—A program for automatic solution of crystal structures by direct methods optimized for powder data. *J. Appl. Crystallogr.* **1994**, *27*, 435–436. [CrossRef]

35. Betteridge, P.; Carruthers, J.; Cooper, R.; Prout, K.; Watkin, D. CRYSTALS version 12: Software for guided crystal structure analysis. *J. Appl. Crystallogr.* **2003**, *36*, 1487. [CrossRef]
36. Rohlicek, J.; Husak, M. MCE2005—A new version of a program for fast interactive visualization of electron and similar density maps optimized for small molecules. *J. Appl. Crystallogr.* **2007**, *40*, 600–601. [CrossRef]

molecules

MDPI

Article

Bismuth(III)-Catalyzed Regioselective Selenation of Indoles with Diaryl Diselenides: Synthesis of 3-Selanylindoles

Mio Matsumura, Airi Umeda, Yuika Sumi, Naoki Aiba, Yuki Murata and Shuji Yasuike *

School of Pharmaceutical Sciences, Aichi Gakuin University, 1-100 Kusumoto-cho, Chikusa-ku, Nagoya 464-8650, Japan; m-matsu@dpc.agu.ac.jp (M.M.); y-murata@dpc.agu.ac.jp (Y.M.)
* Correspondence: s-yasuik@dpc.agu.ac.jp

Abstract: Heterocyclic aryl selenides have recently attracted considerable research interest owing to their applications in biological and pharmaceutical fields. Herein, we describe a simple and general synthesis of 3-selanylindoles via a novel regioselective C–H selenation of indoles using a bismuth reagent as a catalyst. The reactions of indoles with diselenides in the presence of 10 mol% BiI_3 at 100 °C in DMF afforded the corresponding 3-selanylindoles in moderate-to-excellent yields. The reaction proceeded efficiently under aerobic conditions by adding only a catalytic amount of BiI_3, which was non-hygroscopic and less toxic, and both selanyl groups of the diselenide were transferred to the desired products.

Keywords: regioselective selenation; bismuth catalyst; bismuth(III) iodide; indole; diaryl diselenide

check for updates

Citation: Matsumura, M.; Umeda, A.; Sumi, Y.; Aiba, N.; Murata, Y.; Yasuike, S. Bismuth(III)-Catalyzed Regioselective Selenation of Indoles with Diaryl Diselenides: Synthesis of 3-Selanylindoles. *Molecules* **2024**, *29*, 3227. https://doi.org/10.3390/molecules29133227

Academic Editor: Ming Wang

Received: 19 June 2024
Revised: 3 July 2024
Accepted: 3 July 2024
Published: 8 July 2024

1. Introduction

Organoselenium compounds have received considerable attention in organic chemistry, as well as in biological and pharmaceutical sciences [1–14], and there is growing interest in biologically active unsymmetrical diaryl selenides containing heterocyclic rings (i.e., aryl heteroaryl selenides). For example, 3-selanylindoles, compounds with a selenium side chain substituted at the 3-position of indoles, which are widely used as a basic skeleton in natural products and medicines, have been reported to have biological activities, such as the inhibition of tubulin polymerization, antiproliferative activity, anti-inflammatory properties, and antioxidant activity, and are expected to be used as drug discovery resources (Figure 1) [15–20]. Therefore, the development of synthetic methods for these compounds has attracted attention. Direct selenation into indoles has been reported since the 2010s and is a powerful and commonly used method involving the reaction of available indole derivatives with stable and easy-to-handle diselenides as selenium sources. These reactions can be broadly classified into those involving the addition of oxidants [21–23] or bases [24–26], radical reactions using photoreactors [27–34] or electrolytic devices [35,36], and those using transition metal catalysts containing Pd, Cu, Ag, and Fe [37–42]. However, these reactions use excessive reagents, additives, and transition metal catalysts of toxicological concern even in catalytic reactions, and require special equipment and expensive photocatalysts or supporting electrolytes for the photoreactions and electrolytic reactions, respectively. Recently, four transition metal-free catalytic reactions were reported (Scheme 1). Braga et al. developed a catalytic reaction using DMSO as the oxidant in the presence of a catalytic quantity of I_2; however, the reaction required microwave irradiation [43]. The researchers also used KIO_3 as a catalyst, but this reaction required an excess (4 equiv.) of glycerol [44]. Roehrs et al. reported an I_2-catalyzed reaction that required the addition of stoichiometric amounts of urea hydrogen peroxide as an oxidant [45]. Jana et al. developed a reaction using Cs_2CO_3 as a catalyst, albeit in an oxygen atmosphere [46]. As mentioned above, catalytic reactions require additives; otherwise, the reaction conditions are restrictive.

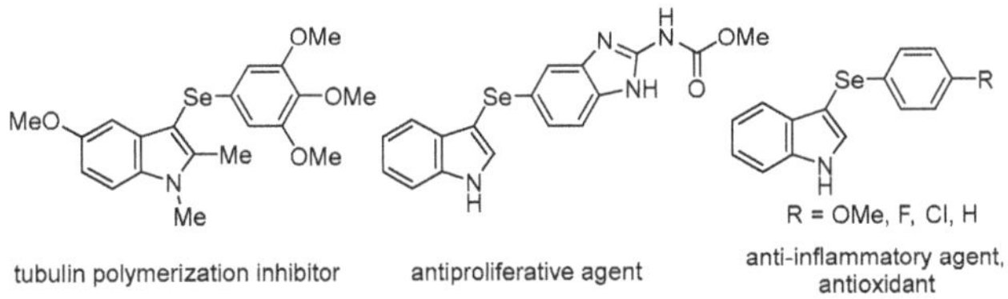

Figure 1. Biologically active 3-selanylindoles.

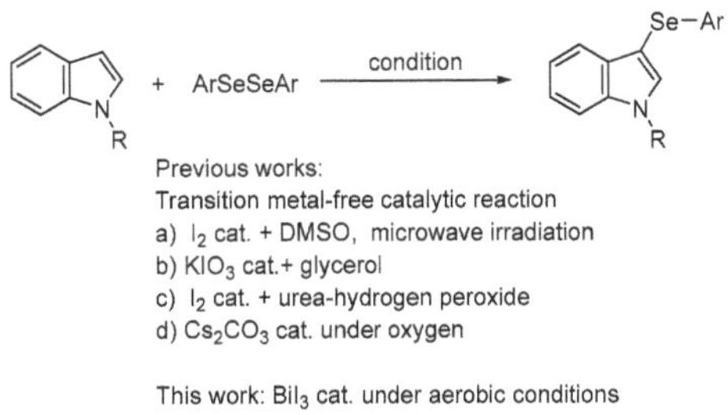

Scheme 1. Selenation of indoles with diaryl diselenides.

Inorganic bismuth compounds have attracted attention in the field of organic synthesis since the 1980s because of their excellent reactivity as mild Lewis acids, nontoxicity, and environmental friendliness [47–52]. For example, $BiCl_3$, a trivalent bismuth halide, has been reported to act as a catalyst for the following reactions: the Mukaiyama aldol reaction [53,54], the nucleophilic opening of epoxide [55], deoxygenative allylation [56], the Diels–Alder reaction [57,58], the three-component reaction of aldehydes, amines, and ketones or trimethylsilyl cyanide [59,60], the Friedel–Crafts reaction [61], the oxy-Michael addition [62], the aminooxygenation of propargyl amidine [63], and the tandem cyclization of tryptamine-ynamide [64]. More recently, $BiCl_3$ has been utilized in the catalytic coupling reactions of aryl iodides or aminobenzimidazoles with arylboronic acids for C(Ar)–C(Ar) and C(Ar)–N bond formation [65,66]. By contrast, bismuth iodide (BiI_3) is widely used in semiconductors and solar cell devices [67,68]. However, its chemical reactivity in organic reactions is largely unknown, and its use in catalytic reactions has been limited to the deprotection of acetals, guanylation with desulfurization using thioureas and amines, and S,S-acetalization of benzaldehyde [69–71]. Inspired by these reports, we present a facile Bi(III)-catalyzed regioselective C(Ar)–Se bond formation reaction of indoles with diaryl diselenides using BiI_3 as the catalyst for the synthesis of 3-selanylindoles under mild conditions. The system was simple, containing only substrates and a Bi catalyst.

2. Results and Discussion

We initially focused on determining the optimal experimental conditions, including screening for suitable catalysts and solvents, for the synthesis of 3-selanylindole **3aa** using *N*-methylindole **1a** and diphenyl diselenide **2a** as model substrates, the results of which, are summarized in Table 1. *N*-methylindole **1a** (0.5 mmol) was reacted with **2a** (0.25 mmol) in the presence of several Bi catalysts (0.05 mmol) in DMF at 100 °C under aerobic conditions (entries 1–7). $BiCl_3$, $BiBr_3$, BiI_3, and $Bi(OTf)_3$, which function as Lewis acids, afforded the corresponding 3-selanylindole **3aa** in good-to-excellent yields (77–97%). BiI_3 displayed the best yield and reaction time, and both selanyl groups were efficiently transferred from the diselenide to product **3aa** (entry 3). Furthermore, although bismuth halides such as $BiCl_3$ and $BiBr_3$ are hygroscopic, BiI_3 can be easily handled in air without such concerns. By contrast, antimony catalysts with the same group of atoms as bismuth and other Lewis acid catalysts were less effective than BiI_3 (entries 8–12). A comparison to iodine (I_2) was also attempted; however, the reaction barely progressed (entry 13). Solvent screening indicated that the reaction proceeded efficiently in DMF (97%), DMSO (89%), and THF (60%), whereas CH_3CN, MeOH, dioxane, 1,2-DCE, and toluene were inefficient (entries 3 and 14–20). When the reaction was performed at 60 °C, the reaction time increased markedly to 8 h (entry 21). The reaction performed under oxygen produced **3aa** in a high yield (94%), which was almost identical to that obtained under aerobic conditions (entries 3 and 22). However, the yield was notably suppressed (9%) under an argon atmosphere (entry 23). Decreasing the BiI_3 loading from 10 to 5 and 1 mol% markedly prolonged the reaction time, although the reaction afforded the desired product (entries 24 and 25). The best result was obtained under aerobic conditions at 100 °C when **1a** was treated with 0.5 equivalents of diselenide **2a** in the presence of BiI_3 (10 mol%) in DMF (entry 3). This selenation could also be scaled up to 10 mmol. The desired product **3aa** was obtained in an excellent yield (99%), generating up to 2.84 g of the product. Furthermore, the reaction of **1a** and **2a** with 1 equivalent of TEMPO [(2,2,6,6-tetramethylpiperidin-1-yl)oxyl] or 1,1-diphenylethylene as radical scavengers afforded **3aa** in yields of 94% and 96%, respectively (entries 26 and 27). These results indicate that the reaction system does not follow a radical mechanism. The regiochemistry of 3-selanylindole **3aa** was elucidated using ^{1}H-NMR and single-crystal X-ray analyses (Figure 2). The ^{1}H-NMR spectrum of **3aa** was consistent with that of the standard sample [41].

Table 1. Optimization of the reaction conditions [a].

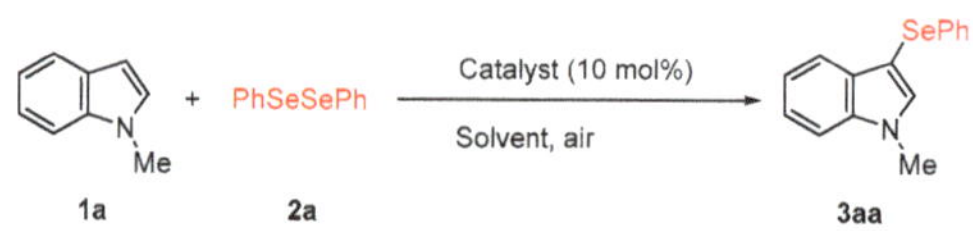

Entry	Catalyst	Solvent	Temp. [°C]	Time [h]	Yield (%) [b]
1	$BiCl_3$	DMF	100	24	85
2	$BiBr_3$	DMF	100	6	84
3	BiI_3	DMF	100	1	97 (91) [c]
4	BiF_3	DMF	100	24	14
5	$Bi(OTf)_3$	DMF	100	24	77
6	$Bi(ONO_2)_3$	DMF	100	24	49
7	Ph_3Bi	DMF	100	24	2
8	$SbBr_3$	DMF	100	24	77
9	SbI_3	DMF	100	24	74
10	$AlCl_3$	DMF	100	24	11
11	$InCl_3$	DMF	100	24	25
12	$FeCl_3$	DMF	100	24	41
13	I_2	DMF	100	24	20
14	BiI_3	DMSO	100	2	89
15	BiI_3	CH_3CN	80	24	45
16	BiI_3	THF	60	24	60
17	BiI_3	MeOH	60	24	51
18	BiI_3	1,2-DCE	80	2	12
19	BiI_3	Dioxane	100	24	19
20	BiI_3	Toluene	100	24	12
21	BiI_3	DMF	60	8	89
22 [d]	BiI_3	DMF	100	1	94
23 [e]	BiI_3	DMF	100	24	9
24 [f]	BiI_3	DMF	100	8	94
25 [g]	BiI_3	DMF	100	24	92
26 [h]	BiI_3	DMF	100	1	94
27 [i]	BiI_3	DMF	100	1	96

[a] Conditions: **1a** (0.5 mmol), **2a** (0.25 mmol), catalyst (10 mol%), and solvent (2 mL). [b] GC yield using biphenyl as the internal standard. [c] Isolated yield. [d] Under O_2. [e] Under argon. [f] BiI_3 (5 mol%). [g] BiI_3 (1 mol%). [h] TEMPO (0.5 mmol). [i] Diphenylethylene (0.5 mmol).

Figure 2. ORTEP drawing of **3aa** with 50% probability (CCDC 2291058).

To understand the scope and limitations of the developed regioselective selenation reaction, various indoles **1** (0.5 mmol) were reacted with diselenides **2** (0.25 mmol) under the optimized conditions (Figure 3). The reaction of *N*-methylindole **1a** with diaryl diselenides **2b–i** afforded the corresponding products, i.e., **3ab–ai**, in good-to-excellent yields, except for **3ah**. For **3ab–ae**, the presence of an electron-donating or electron-withdrawing group at the 4-position of the benzene ring of diselenides **3b–e** did not affect the reaction progression, although the reaction time was slightly prolonged when electron-donating groups were substituted. Sterically hindered *ortho*-substituted diselenides **2f** and **2g** reacted to give selenides **3af** and **3ag**, respectively. By contrast, for **2h**, which comprises a benzylamino group, the reaction did not proceed, and the starting materials were recovered. For the reaction using diaryl diselenide **2i**, which bears a heterocyclic ring, **3ai** was afforded in a good yield. Dibenzyl diselenide **2j**, which contains a benzyl moiety as the alkyl group, also afforded **3aj** in a good yield (82%). Next, the reaction of diphenyl diselenide **2a** with various *N*-methylindoles, i.e., **1b–i**, bearing electron-donating or electron-withdrawing groups on the benzene ring afforded the desired products **3ba–3ia** in satisfactory yields (75–99%). The reaction proceeded smoothly from the unsubstituted indoles **1j–l** to obtain the parent 3-selanylindoles **3ja–la** (79–93%). Furthermore, the reaction of the *N*-substituted indoles **1m** and **1n** with benzyl or phenyl groups on the nitrogen also gave the corresponding products **3ma** and **3na**; however, *N*-acetylindole **1o** with an electron-withdrawing group did not give **3oa**, and the starting materials were recovered. These results suggest that the reaction is electrically influenced by the substituents on the indole nitrogen. 2-Phenyl- and 2-methylindoles **1p** and **1q** were treated with **2a** to afford the 3-selanyl-2-substituted indoles **3pa** and **3qa**, respectively. The attempted double selenation of **1a** using two equivalents of diphenyl diselenide **2a** did not yield the corresponding 2,3-diselanylindole **3ra**; instead, 3-selanylindole **3aa** was isolated in a yield of 98%. These results suggest that this reaction proceeds only at the 3-position of the indole. Finally, the reaction of **1a** with dichalcogenides containing sulfur and tellurium was attempted. The reaction with diphenyl disulfide afforded the desired 3-sulfanylindole **4** in an excellent yield (99%), although the reaction time (24 h) was longer than that with diselenide, which is a selenium reagent. By contrast, the reaction proceeded to a certain extent with diphenyl ditelluride, and indole **5** was obtained in a yield of 26%.

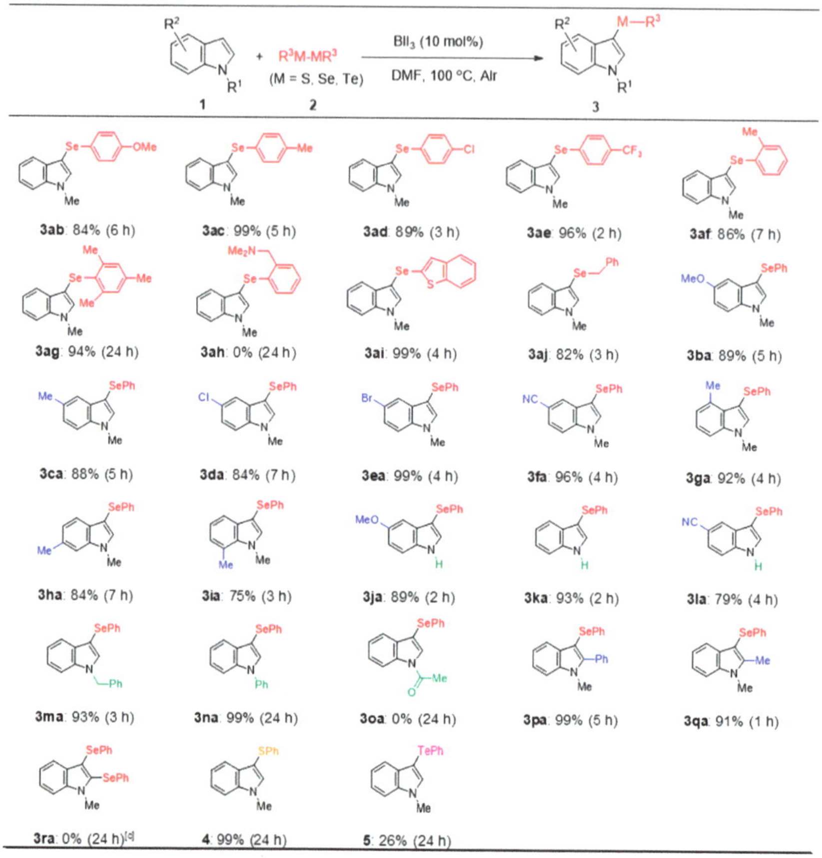

Figure 3. Substrate scope: reaction of indoles with dichalcogenides [a,b]. [a] **1** (0.5 mmol), **2** (0.25 mmol), BiI_3 (0.05 mmol), and DMF (2 mL). [b] Yield of isolated products. [c] **2a** (0.5 mmol); **3aa** was isolated in a yield of 98%.

However, the reaction mechanism for this selenation remains unclear. Circumstantial evidence indicates that the reaction was affected by the gaseous atmosphere and proceeded smoothly in the presence of a molecular oxygen atmosphere while being notably suppressed in an inert gas atmosphere (Table 1: entries 3, 22, and 23). BiI_3 forms a pentacoordinated complex with bismuth, the central atom, and the oxygen atoms of reagents and solvents such as Mo_8O_{26} and THF [72,73]. Therefore, a possible mechanism for this reaction is illustrated in Scheme 2. The initial step was the generation of the pentacoordinated Bi–peroxo complex **A** from BiI_3 and oxygen. While the selenium atom of the diselenide coordinates with complex **A**, the 3-position of the indole nucleophilically attacks another selenium atom, forming complex **B** and intermediate **C**. The aryl selenide anion formed during the interconversion between complexes **B** and **A** attacks intermediate **C** to form 3-selanylindole **3** and selenol **D**. Selenol **D** is converted to diselenide **2** via oxidation in air. Therefore, the reaction proceeds with 0.5 equivalents of diselenide, and both selanyl groups are used for the reaction. Bismuth complexes **A** and **B**, which are expected to form during this process, have not yet been confirmed or isolated.

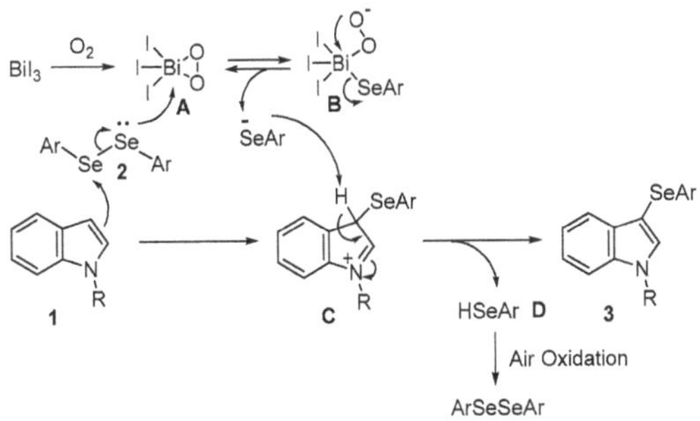

Scheme 2. Possible mechanism.

3. Conclusions

Herein, we report a simple Bi-catalyzed regioselective selenation protocol for the synthesis of 3-selanylindoles under mild reaction conditions. The reaction is atom-economical, with the participation of both selanyl groups of the diaryl diselenide. Indoles and diselenides bearing different functional groups afforded the corresponding products in satisfactory yields. This reaction is the first example of the Bi-catalyzed C–H selenation of aromatic heterocycles. Detailed studies on the exact mechanism of this reaction and the synthesis of asymmetric selenides containing other heterocyclic rings using this protocol are currently underway.

4. Materials and Methods

4.1. General Information

All the chemicals, including organic solvents, were obtained from commercial vendors and used as received without further purification. All chromatographic separations were accomplished with Silica Gel 60N (Kanto Chemical Co., Inc., Tokyo, Japan). Thin-layer chromatography (TLC) was performed using Macherey–Nagel Pre-coated TLC plates Sil G25 UV_{254}. Melting point measurements were conducted on a Yanagimoto micro-melting point hot-stage apparatus (MP-S3) and reported as uncorrected values. In addition, ^{1}H NMR (TMS: δ = 0.00 ppm as an internal standard), ^{13}C NMR ($CDCl_3$: δ = 77.00 ppm as an internal standard), ^{19}F NMR (376 MHz, benzotrifluoride; δ = −64.0 ppm as an external standard), and ^{77}Se NMR (76 MHz, diphenyldiselenide; δ = 463.15 ppm as an external standard) spectra were recorded on JEOL ECZ-400S (400, 100, 376, and 76 MHz for ^{1}H-, ^{13}C-, ^{19}F-, and ^{77}Se NMR, respectively) spectrometers (JEOL Ltd., Tokyo, Japan). GC-MS (EI) spectra were recorded on Agilent 5977 E Diff-SST MSD-230 V spectrometer. HRMS (ESI) spectra were recorded on Agilent 6230 (Agilent Technologies Japan, Ltd., Tokyo, Japan). X-ray measurements were recorded on Rigaku XtaLAB Synergy with a HyPix3000 diffractometer (Rigaku, Corp., Tokyo, Japan). IR spectra were recorded on an FTIR-8400S or IRAffinity-1S system from a Shimadzu spectrometer (SHIMADZU Corp, Kyoto, Japan) and are reported as the frequencies of absorption (cm^{-1}). Only selected IR absorbencies are reported. The spectroscopic data of the calcogenated indoles **3aa–ad**, **3ag**, **3ka**, **3ma**, **3qa**, **4** [41], **3aj** [74], **3ba** [42], **3ea**, **3ja**, **3na** [35], **3la** [27], **3pa** [75], and **5** [76] are in accordance with those in the literature, and their characterization data are in Supplementary Materials.

4.2. General Procedure for the Synthesis of Calcogenated Indoles

The indole derivative (**1**) (0.5 mmol) was added to a solution of dichalcogenide (**2**) (0.25 mmol, 0.5 eq.) and bismuth(III) iodide (30 mg, 0.05 mmol, and 10 mol%) in anhydrous dimethylformamide (2 mL). After stirring at 100 °C in an oil bath, the mixture was cooled to room temperature and evaporated to dryness under reduced pressure. The crude product was purified on a silica gel column chromatography to give the desired product **3**.

4.3. Characterization Data of Novel Compounds

4.3.1. 3-(4-Trifluoromethylphenyl)selanyl-1-methyl-1H-indole (**3ae**)

Yield: 171 mg (96%); Colorless prism (from CH_2Cl_2-Hexane); m.p. 133.0–135.0 °C; R_f = 0.54 (CH_2Cl_2-Hexane, 1:2). ^{1}H NMR (400 MHz, $CDCl_3$): δ = 7.57 (d, J = 7.8 Hz, 1H; Ar-H), 7.41 (d, J = 8.2 Hz, 1H; Ar-H), 7.36 (s, 1H; Ar-H), 7.35–7.30 (m, 3H; Ar-H), 7.26 (d, J = 8.2 Hz, 2H; Ar-H), 7.19 (td, J = 8.2, 0.9 Hz, 1H; Ar-H), 3.88 ppm (s, 3H; N-CH_3). ^{13}C NMR (100 MHz, $CDCl_3$): δ = 140.0 (C), 137.5 (C), 135.9 (CH), 130.4 (C), 128.0 (CH), 127.5 (q, J = 32 Hz, C), 125.5 (q, J = 3.9 Hz, CH), 124.2 (q, J = 272 Hz, C), 122.7 (CH), 120.7 (CH), 120.2 (CH), 109.7 (CH), 94.6 (C), 33.2 ppm (CH_3). ^{19}F NMR (376 MHz, $CDCl_3$): δ = −63.7 ppm. ^{77}Se NMR (76 MHz, $CDCl_3$): δ = 223.4 ppm. IR (ATR): $\tilde{\nu}$ = 739, 822, 1072, 1105, 1321 cm^{-1}. MS (EI, 70 eV): m/z (%) = 355 (21) [M]$^+$, 275 (100), 130 (14). HRMS (ESI): m/z calcd for $C_{16}H_{12}F_3NSe$: 355.0087 [M]$^+$; found: 355.0088.

4.3.2. 1-Methyl-3-(2-methylphenyl)selanyl-1H-indole (**3af**)

Yield: 129 mg (86%); Colorless plate (from CH_2Cl_2-Hexane); m.p. 129.0–132.0 °C; R_f = 0.20 (CH_2Cl_2-Hexane, 1:5). ^{1}H NMR (400 MHz, $CDCl_3$): δ = 7.59 (d, J = 7.8 Hz, 1H; Ar-H), 7.39 (d, J = 8.2 Hz, 1H; Ar-H), 7.31 (s, 1H; Ar-H), 7.30 (td, J = 8.2, 0.9 Hz, 1H; Ar-H), 7.17 (td, J = 7.3, 0.9 Hz, 1H; Ar-H), 7.10 (d, J = 7.3 Hz, 1H; Ar-H), 7.00 (td, J = 7.8, 2.3 Hz, 1H; Ar-H), 6.86–6.80 (m, 2H; Ar-H), 3.85 (s, 3H; N-CH_3), 2.46 ppm (s, 3H; CH_3). ^{13}C NMR (100 MHz, $CDCl_3$): δ = 137.6 (C), 135.92 (C), 135.87 (CH), 134.8 (C), 130.8 (C), 129.7 (CH), 127.8 (CH), 126.4 (CH), 125.2 (CH), 122.4 (CH), 120.5 (CH), 120.4 (CH), 109.5 (CH), 94.9 (C), 33.1 (CH_3), 21.2 ppm (CH_3). ^{77}Se NMR (76 MHz, $CDCl_3$): δ = 178.3 ppm. IR (ATR): $\tilde{\nu}$ = 411, 426, 729, 746, 1456 cm^{-1}. MS (EI, 70 eV): m/z (%) = 301 (40) [M]$^+$, 221 (60), 131 (100), 91 (30). HRMS (ESI): m/z calcd for $C_{16}H_{15}NSe$: 301.0370 [M]$^+$; found: 301.0370.

4.3.3. 3-(2-Benzothienyl)selanyl-1-methyl-1H-indole (**3ai**)

Yield: 171 mg (99%); Yellow needle (from CH_2Cl_2-Hexane); m.p. 144.0–147.0 °C; R_f = 0.21 (CH_2Cl_2-Hexane, 1:5). ^{1}H NMR (400 MHz, $CDCl_3$): δ = 7.78 (d, J = 7.8 Hz, 1H; Ar-H), 7.62 (d, J = 6.9 Hz, 1H; Ar-H), 7.60 (d, J = 7.3 Hz, 1H; Ar-H), 7.39 (s, 1H; Ar-H), 7.36 (d, J = 7.8 Hz, 1H; Ar-H), 7.32–7.16 (m, 5H; Ar-H), 3.83 ppm (s, 3H; N-CH_3). ^{13}C NMR (100 MHz, $CDCl_3$): δ = 142.2 (C), 140.5 (C), 137.2 (C), 135.1 (CH), 132.1 (C), 130.3 (C), 126.3 (CH), 124.1 (CH), 123.5 (CH), 122.51 (CH), 122.47 (CH), 121.5 (CH), 120.5 (CH), 120.3 (CH), 109.6 (CH), 97.1 (C), 33.1 ppm (CH_3). ^{77}Se NMR (76 MHz, $CDCl_3$): δ = 179.9 ppm. IR (ATR): $\tilde{\nu}$ = 426, 486, 556, 723, 735, 1236 cm^{-1}. MS (EI, 70 eV): m/z (%) = 343 (14) [M]$^+$, 263 (100), 207 (21), 131 (43), 89 (21), 44 (29). HRMS (ESI): m/z calcd for $C_{17}H_{13}NSSe$: 342.9934 [M]$^+$; found: 342.9932.

4.3.4. 1,5-Dimethyl-3-phenylselanyl-1H-indole (**3ca**)

Yield: 133 mg (88%); Colorless plate (from CH_2Cl_2-Hexane); m.p. 104.0–105.0 °C; R_f = 0.58 (CH_2Cl_2-Hexane, 1:5). ^{1}H NMR (400 MHz, $CDCl_3$): δ = 7.41 (s, 1H; Ar-H), 7.27–7.20 (m, 4H; Ar-H), 7.14–7.05 (m, 4H; Ar-H), 3.81 (s, 3H; N-CH_3), 2.43 ppm (s, 3H; CH_3). ^{13}C NMR (100 MHz, $CDCl_3$): δ = 135.83 (C), 135.76 (CH), 134.5 (C), 130.9 (C), 129.8 (C), 128.9 (CH), 128.3 (CH), 125.4 (CH), 124.1 (CH), 120.0 (CH), 109.2 (CH), 95.0 (C), 33.1 (CH_3), 21.4 ppm (CH_3). ^{77}Se NMR (76 MHz, $CDCl_3$): δ = 206.4 ppm. IR (ATR): $\tilde{\nu}$ = 424, 457, 731, 793, 1474, 1506 cm^{-1}. MS (EI, 70 eV): m/z (%) = 301 (20) [M]$^+$, 221 (100), 144 (13). HRMS (ESI): m/z calcd for $C_{16}H_{15}NSe$: 301.0370 [M]$^+$; found: 301.0369.

4.3.5. 5-Chloro-1-methyl-3-phenylselanyl-1H-indole (**3da**)

Yield: 135 mg (84%); Colorless plate (from CH_2Cl_2-Hexane); m.p. 132.0–133.5 °C; R_f = 0.54 (CH_2Cl_2-Hexane, 1:5). ^{1}H NMR (400 MHz, $CDCl_3$): δ = 7.60 (s, 1H; Ar-H), 7.33 (s, 1H; Ar-H), 7.28–7.19 (m, 4H; Ar-H), 7.15–7.08 (m, 3H; Ar-H), 3.82 ppm (s, 3H; N-CH_3). ^{13}C NMR (100 MHz, $CDCl_3$): δ = 136.9 (CH), 135.9 (C), 133.7 (C), 131.9 (C), 129.0 (CH), 128.6 (CH), 126.5 (C), 125.7 (CH), 122.8 (CH), 119.9 (CH), 110.7 (CH), 95.7 (C), 33.3 ppm (CH_3). ^{77}Se NMR (76 MHz, $CDCl_3$): δ = 209.6 ppm. IR (ATR): $\tilde{\nu}$ = 422, 457, 689, 733, 795, 1422,

1474 cm^{-1}. MS (EI, 70 eV): m/z (%) = 321 (25) $[M]^+$, 241 (100), 164 (15). HRMS (ESI): m/z calcd for $C_{15}H_{12}ClNSe$: 320.9823 $[M]^+$; found: 320.9825.

4.3.6. 1-Methyl-3-phenylselanyl-1H-indole-5-carbonitrile (**3fa**)

Yield: 149 mg (96%); Colorless prism (from CH_2Cl_2-Hexane); m.p. 198.0–199.5 °C; R_f = 0.34 (CH_2Cl_2-Hexane, 1:1). ^1H NMR (400 MHz, $CDCl_3$): δ = 7.95 (d, J = 1.4 Hz, 1H; Ar-H), 7.49 (dd, J = 8.7, 1.4 Hz, 1H; Ar-H), 7.45 (s, 1H; Ar-H), 7.41 (d, J = 8.7 Hz, 1H; Ar-H), 7.23–7.20 (m, 2H; Ar-H), 7.17–7.12 (m, 3H; Ar-H), 3.88 ppm (s, 3H; *N*-CH_3). ^{13}C NMR (100 MHz, $CDCl_3$): δ = 139.0 (C), 137.6 (CH), 132.8 (C), 130.5 (C), 129.11 (CH), 129.07 (CH), 126.14 (CH), 126.08 (CH), 125.4 (CH), 120.4 (C), 110.5 (CH), 103.6 (C), 97.9 (C), 33.3 ppm (CH_3). ^{77}Se NMR (76 MHz, $CDCl_3$): δ = 212.5 ppm. IR (ATR): $\tilde{\nu}$ = 461, 474, 631, 692, 741 cm^{-1}. MS (EI, 70 eV): m/z (%) = 312 (17) $[M]^+$, 232 (100), 155 (16). HRMS (ESI): m/z calcd for $C_{16}H_{12}N_2Se$: 312.0166 $[M]^+$; found: 312.0166.

4.3.7. 1,4-Dimethyl-3-phenylselanyl-1H-indole (**3ga**)

Yield: 138 mg (92%); Colorless plate (from CH_2Cl_2-Hexane); m.p. 84.0–85.0 °C; R_f = 0.25 (CH_2Cl_2-Hexane, 1:5). ^1H NMR (400 MHz, $CDCl_3$): δ = 7.29 (s, 1H; Ar-H), 7.24–7.18 (m, 4H; Ar-H), 7.15–7.11 (m, 2H; Ar-H), 7.07 (tt, J = 6.9, 1.4 Hz, 1H; Ar-H), 6.89 (d, J = 7.3 Hz, 1H; Ar-H), 3.80 (s, 3H; *N*-CH_3), 2.69 ppm (s, 3H; CH_3). ^{13}C NMR (100 MHz, $CDCl_3$): δ = 137.9 (C), 136.8 (CH), 136.4 (C), 132.3 (C), 129.0 (CH), 127.9 (CH), 125.2 (CH), 122.4 (CH), 122.0 (CH), 107.5 (CH), 94.6 (C), 33.1 (CH_3), 18.7 ppm (CH_3). ^{77}Se NMR (76 MHz, $CDCl_3$): δ = 251.5 ppm. IR (ATR): $\tilde{\nu}$ = 457, 667, 689, 727, 739, 1474 cm^{-1}. MS (EI, 70 eV): m/z (%) = 301 (33) $[M]^+$, 221 (100), 144 (44). HRMS (ESI): m/z calcd for $C_{16}H_{15}NSe$: 301.0370 $[M]^+$; found: 301.0369.

4.3.8. 1,6-Dimethyl-3-phenylselanyl-1H-indole (**3ha**)

Yield: 126 mg (84%); Colorless plate (from CH_2Cl_2-Hexane); m.p. 86.0–87.5 °C; R_f = 0.45 (CH_2Cl_2-Hexane, 1:5). ^1H NMR (400 MHz, $CDCl_3$): δ = 7.50 (d, J = 8.2 Hz, 1H; Ar-H), 7.26–7.22 (m, 3H; Ar-H), 7.18 (s, 1H; Ar-H), 7.14–7.06 (m, 3H; Ar-H), 7.01 (d, J = 7.8 Hz, 1H; Ar-H), 3.80 (s, 3H; *N*-CH_3), 2.52 ppm (s, 3H; CH_3). ^{13}C NMR (100 MHz, $CDCl_3$): δ = 137.8 (C), 135.1 (CH), 134.3 (C), 132.4 (C), 128.9 (CH), 128.5 (C), 128.4 (CH), 125.4 (CH), 122.1 (CH), 120.1 (CH), 109.5 (CH), 95.6 (C), 32.9 (CH_3), 21.8 ppm (CH_3). ^{77}Se NMR (76 MHz, $CDCl_3$): δ = 209.3 ppm. IR (ATR): $\tilde{\nu}$ = 430, 598, 689, 729, 797 cm^{-1}. MS (EI, 70 eV): m/z (%) = 301 (19) $[M]^+$, 221 (100), 144 (17). HRMS (ESI): m/z calcd for $C_{16}H_{15}NSe$: 301.0370 $[M]^+$; found: 301.0371.

4.3.9. 1,7-Dimethyl-3-phenylselanyl-1H-indole (**3ia**)

Yield: 113 mg (75%); Colorless plate (from CH_2Cl_2-Hexane); m.p. 110.0–111.0 °C; R_f = 0.34 (CH_2Cl_2-Hexane, 1:5). ^1H NMR (400 MHz, $CDCl_3$): δ = 7.46 (d, J = 7.8 Hz, 1H; Ar-H), 7.24–7.20 (m, 3H; Ar-H), 7.13–7.05 (m, 3H; Ar-H), 7.01 (t, J = 7.3 Hz, 1H; Ar-H), 6.96 (d, J = 6.9 Hz, 1H; Ar-H), 4.08 (s, 3H; *N*-CH_3), 2.79 ppm (s, 3H; CH_3). ^{13}C NMR (100 MHz, $CDCl_3$): δ = 137.2 (CH), 136.1 (C), 134.2 (C), 131.8 (C), 128.9 (CH), 128.5 (CH), 125.4 (CH), 125.1 (CH), 121.5 (C), 120.6 (CH), 118.7 (CH), 95.7 (C), 37.0 (CH_3), 19.6 ppm (CH_3). ^{77}Se NMR (76 MHz, $CDCl_3$): δ = 208.8 ppm. IR (ATR): $\tilde{\nu}$ = 689, 733, 748, 781, 1450 cm^{-1}. MS (EI, 70 eV): m/z (%) = 301 (18) $[M]^+$, 221 (100), 144 (17). HRMS (ESI): m/z calcd for $C_{16}H_{15}NSe$: 301.0370 $[M]^+$; found: 301.0370.

4.4. *Single-Crystal X-ray Diffraction Experiment of* **3aa**

A suitable crystal was selected and measured on an XtaLAB Synergy, Single source at home/near, HyPix3000 diffractometer. The crystal was kept at 103 K in an N_2 cold stream during data collection. Using Olex2 [77], the structure was solved with the SHELXT [78] structure solution program using Intrinsic Phasing and refined with the SHELXL [79] refinement package using Least Squares minimization. Crystal Data for **3aa**: $C_{15}H_{13}NSe$ (M = 286.22 g/mol), monoclinic, space group $P2_1/n$ (no. 14), a = 7.73810(10) Å, b = 9.03610(10) Å,

$c = 18.1620(2)$ Å, $\beta = 101.9160(10)°$, $V = 1242.56(3)$ Å^3, $Z = 4$, $T = 103$ K, μ(Cu Kα) = 3.873 mm^{-1}, D_{calc} = 1.530 g/cm^3, 6253 reflections measured (9.954° $\leq 2\Theta \leq$ 136.378°), and 2263 unique reflections (R_{int} = 0.0255 and R_{sigma} = 0.0218), which were used in all calculations. The final R_1 was 0.0243 (I > 2σ(I)), and wR_2 was 0.0652 (all data).

Supplementary Materials: The following supporting information can be downloaded at: https://www.mdpi.com/article/10.3390/molecules29133227/s1. The characterization data of known compounds and ^{1}H- and ^{13}C-NMR spectra are available online. The crystal structures have been deposited to the CCDC with the number 2291058, and the CIF files are also provided.

Author Contributions: All authors contributed to the writing and gave approval for the final version of the manuscript. M.M. and S.Y. designed chemical synthesis, analyzed results, and wrote the manuscript. M.M., A.U. and Y.S. performed chemical synthesis experiments and analyzed the results. Y.M. and N.A. analyzed the results and wrote the manuscript. M.M. performed single-crystal X-ray diffraction analysis and wrote the manuscript. All authors have read and agreed to the published version of the manuscript.

Funding: This research was supported by a research grant from the Institute of Pharmaceutical Life Sciences, Aichi Gakuin University.

Institutional Review Board Statement: Not applicable.

Informed Consent Statement: Not applicable.

Data Availability Statement: The data that support the findings of this study are available in the manuscript and Supplementary Materials of this article.

Conflicts of Interest: The authors declare no conflicts of interest.

References

1. Gandeepan, P.; Müller, T.; Zell, D.; Cera, G.; Warratz, S.; Ackermann, L. 3d Transition Metals for C–H Activation. *Chem. Rev.* **2019**, *119*, 2192–2452. [CrossRef] [PubMed]
2. Rampon, D.S.; Luz, E.Q.; Lima, D.B.; Balaguez, R.A.; Schneider, P.H.; Alves, D. Transition Metal Catalysed Direct Selanylation of Arenes and Heteroarenes. *Dalton Trans.* **2019**, *48*, 9851–9905. [CrossRef]
3. Hellwig, P.S.; Peglow, T.J.; Penteado, F.; Bagnoli, L.; Perin, G.; Lenardão, E.J. Recent Advances in the Synthesis of Selenophenes and Their Derivatives. *Molecules* **2020**, *25*, 5907. [CrossRef] [PubMed]
4. Jose, D.E.; Kanchana, U.S.; Mathew, T.V.; Anilkumar, G. Recent Developments and Perspectives in the C-Se Cross Coupling Reactions. *Curr. Org. Chem.* **2020**, *24*, 1230–1262. [CrossRef]
5. Sonawane, A.D.; Sonawane, R.A.; Ninomiya, M.; Koketsu, M. Diorganyl Diselenides: A Powerful Tool for the Construction of Selenium Containing Scaffolds. *Dalton Trans.* **2021**, *50*, 12764–12790. [CrossRef] [PubMed]
6. Guo, T.; Li, Z.; Bi, L.; Fan, L.; Zhang, P. Recent Advances in Organic Synthesis Applying Elemental Selenium. *Tetrahedron* **2022**, *112*, 132752. [CrossRef]
7. Beletskaya, I.P.; Ananikov, V.P. Transition-Metal-Catalyzed C–S, C–Se, and C–Te Bond Formations via Cross-Coupling and Atom-Economic Addition Reactions. Achievements and Challenges. *Chem. Rev.* **2022**, *122*, 16110–16293. [CrossRef] [PubMed]
8. Ranu, B.C.; Adak, L.; Mukherjee, N.; Ghosh, T. Benign-Metal-Catalyzed Carbon–Carbon and Carbon–Heteroatom Bond Formation. *Synlett* **2023**, *34*, 601–621. [CrossRef]
9. Mugesh, G.; du Mont, W.-W.; Sies, H. Chemistry of Biologically Important Synthetic Organoselenium Compounds. *Chem. Rev.* **2001**, *101*, 2125–2179. [CrossRef] [PubMed]
10. Nogueira, C.W.; Zeni, G.; Rocha, J.B.T. Organoselenium and Organotellurium Compounds: Toxicology and Pharmacology. *Chem. Rev.* **2004**, *104*, 6255–6285. [CrossRef]
11. Sarma, B.K.; Mugesh, G. Thiol Cofactors for Selenoenzymes and Their Synthetic Mimics. *Org. Biomol. Chem.* **2008**, *6*, 965–974. [CrossRef]
12. Nogueira, C.W.; Rocha, J.B.T. Toxicology and Pharmacology of Selenium: Emphasis on Synthetic Organoselenium Compounds. *Arch. Toxicol.* **2011**, *85*, 1313–1359. [CrossRef] [PubMed]
13. Álvarez-Pérez, M.; Ali, W.; Marć, M.A.; Handzlik, J.; Domínguez-Álvarez, E. Selenides and Diselenides: A Review of Their Anticancer and Chemopreventive Activity. *Molecules* **2018**, *23*, 628. [CrossRef] [PubMed]
14. Chuai, H.; Zhang, S.-Q.; Bai, H.; Li, J.; Wang, Y.; Sun, J.; Wen, E.; Zhang, J.; Xin, M. Small Molecule Selenium-Containing Compounds: Recent Development and Therapeutic Applications. *Eur. J. Med. Chem.* **2021**, *223*, 113621. [CrossRef]
15. Guan, Q.; Han, C.; Zuo, D.; Zhai, M.; Li, Z.; Zhang, Q.; Zhai, Y.; Jiang, X.; Bao, K.; Wu, Y.; et al. Synthesis and Evaluation of Benzimidazole Carbamates Bearing Indole Moieties for Antiproliferative and Antitubulin Activities. *Eur. J. Med. Chem.* **2014**, *87*, 306–315. [CrossRef] [PubMed]

16. Wen, Z.; Xu, J.; Wang, Z.; Qi, H.; Xu, Q.; Bai, Z.; Zhang, Q.; Bao, K.; Wu, Y.; Zhang, W. 3-(3,4,5-Trimethoxyphenylselenyl)-1*H*-indoles and Their Selenoxides as Combretastatin A-4 Analogs: Microwave-Assisted Synthesis and Biological Evaluation. *Eur. J. Med. Chem.* **2015**, *90*, 184–194. [CrossRef]
17. Wen, Z.; Li, X.; Zuo, D.; Lang, B.; Wu, Y.; Jiang, M.; Ma, H.; Bao, K.; Wu, Y.; Zhang, W. Ultrasound-Promoted Two-Step Synthesis of 3-Arylselenylindoles and 3-Arylthioindoles as Novel Combretastatin A-4 Analogues. *Sci. Rep.* **2016**, *6*, 23986. [CrossRef] [PubMed]
18. Casaril, A.M.; Ignasiak, M.T.; Chuang, C.Y.; Vieira, B.; Padilha, N.B.; Carroll, L.; Lenardão, E.J.; Savegnago, L.; Davies, M.J. Selenium-Containing Indolyl Compounds: Kinetics of Reaction with Inflammation-Associated Oxidants and Protective Effect against Oxidation of Extracellular Matrix Proteins. *Free Radic. Biol. Med.* **2017**, *113*, 395–405. [CrossRef] [PubMed]
19. Vieira, B.M.; Thurow, S.; da Costa, M.; Casaril, A.M.; Domingues, M.; Schumacher, R.F.; Perin, G.; Alves, D.; Savegnago, L.; Lenardão, E.J. Ultrasound-Assisted Synthesis and Antioxidant Activity of 3-Selanyl-1*H*-indole and 3-Selanylimidazo [1,2-*a*]pyridine Derivatives. *Asian J. Org. Chem.* **2017**, *6*, 1635–1646. [CrossRef]
20. Pedroso, G.J.; Costa, D.M.S.; Felipe Kokuszi, L.T.; da Silva, E.B.V.; Cavalcante, M.F.O.; Junca, E.; Moraes, C.A.O.; Pich, C.T.; de Lima, V.R.; Saba, S.; et al. Selenylated Indoles: Synthesis, Effects on Lipid Membrane Properties and DNA Cleavage. *New J. Chem.* **2023**, *47*, 2719–2726. [CrossRef]
21. Silveira, C.C.; Mendes, S.R.; Wolf, L.; Martins, G.M.; von Mühlen, L. Efficient Synthesis of 3-Selanyl- and 3-Sulfanylindoles Employing Trichloroisocyanuric Acid and Dichalcogenides. *Tetrahedron* **2012**, *68*, 10464–10469. [CrossRef]
22. Li, H.; Wang, X.; Yan, J. Selective Synthesis of 3-Selanylindoles from Indoles and Diselenides Using IK/*m*CPBA System. *Appl. Organomet. Chem.* **2017**, *31*, e3864. [CrossRef]
23. Wang, Y.-H.; Zhang, Y.-Q.; Zhou, C.-F.; Jiang, Y.-Q.; Xu, Y.; Zeng, X.; Liu, G.-Q. Iodine Pentoxide-Mediated Oxidative Selenation and Seleno/Thiocyanation of Electron-Rich Arenes. *Org. Biomol. Chem.* **2022**, *20*, 5463–5469. [CrossRef] [PubMed]
24. Ferreira, N.L.; Azeredo, J.B.; Fiorentin, B.L.; Braga, A.L. Synthesis of 3-Selenylindoles under Ecofriendly Conditions. *Eur. J. Org. Chem.* **2015**, *2015*, 5070–5074. [CrossRef]
25. Yu, Y.; Zhou, Y.; Song, Z.; Liang, G. An Efficient *t*-BuOK Promoted C3-Chalcogenylation of Indoles with Dichalcogenides. *Org. Biomol. Chem.* **2018**, *16*, 4958–4962. [CrossRef] [PubMed]
26. Xu, S.; Yi, R.; Zeng, C.; Cui, Y.; Xu, X.; Wang, X.-Q.; Li, N. CsOH-Promoted Regiospecific Sulfenylation, Selenylation, and Telluration of Indoles in H_2O. *Synlett* **2023**, *34*, 124–132.
27. Zhang, Q.-B.; Ban, Y.-L.; Yuan, P.-F.; Peng, S.-J.; Fang, J.-G.; Wu, L.-Z.; Liu, Q. Visible-Light-Mediated Aerobic Selenation of (Hetero)Arenes with Diselenides. *Green Chem.* **2017**, *19*, 5559–5563. [CrossRef]
28. Kumaraswamy, G.; Ramesh, V.; Gangadhar, M.; Vijaykumar, S. Catalyst and Sensitizer-Free Visible-Light-Induced C(sp^2)–H Chalcogenation of Arenes/Heteroarenes with Dichalcogenides. *Asian J. Org. Chem.* **2018**, *7*, 1689–1697. [CrossRef]
29. Saba, S.; Rafique, J.; Franco, M.S.; Schneider, A.R.; Espíndola, L.; Silva, D.O.; Braga, A.L. Rose Bengal Catalysed Photo-Induced Selenylation of Indoles, Imidazoles and Arenes: A Metal Free Approach. *Org. Biomol. Chem.* **2018**, *16*, 880–885. [CrossRef] [PubMed]
30. Rathore, V.; Kumar, S. Visible-Light-Induced Metal and Reagent-Free Oxidative Coupling of sp^2 C–H Bonds with Organo-Dichalcogenides: Synthesis of 3-Organochalcogenyl Indoles. *Green Chem.* **2019**, *21*, 2670–2676. [CrossRef]
31. Lemir, I.D.; Castro-Godoy, W.D.; Heredia, A.A.; Schmidt, L.C.; Argüello, J.E. Metal- and Photocatalyst-Free Synthesis of 3-Selenylindoles and Asymmetric Diarylselenides Promoted by Visible Light. *RSC Adv.* **2019**, *9*, 22685–22694. [CrossRef] [PubMed]
32. Heredia, A.A.; Soria-Castro, S.M.; Castro-Godoy, W.D.; Lemir, I.D.; López-Vidal, M.; Bisogno, F.R.; Argüello, J.E.; Oksdath-Mansilla, G. Multistep Synthesis of Organic Selenides under Visible Light Irradiation: A Continuous-Flow Approach. *Org. Process Res. Dev.* **2020**, *24*, 540–545. [CrossRef]
33. Huang, Q.; Peng, X.; Li, H.; He, H.; Liu, L. Visible-Light-Induced, Graphene Oxide-Promoted C3-Chalcogenylation of Indoles Strategy under Transition-Metal-Free Conditions. *Molecules* **2022**, *27*, 772. [CrossRef] [PubMed]
34. Quadros, G.T.; de Medeiros, S.P.; de Oliveira, C.A.; Rambo, M.W.; Abenante, L.; Lenardão, E.J.; Penteado, F. Benzeneseleninic Acids (BSA) and Photocatalysis: An Alternative Duo for the Synthesis of 3-Selanylindoles. *Asian J. Org. Chem.* **2023**, *12*, e202300517. [CrossRef]
35. Zhang, X.; Wang, C.; Jiang, H.; Sun, L. Convenient Synthesis of Selenyl-Indoles via Iodide Ion-Catalyzed Electrochemical C–H Selenation. *Chem. Commun.* **2018**, *54*, 8781–8784. [CrossRef] [PubMed]
36. Meirinho, A.G.; Pereira, V.F.; Martins, G.M.; Saba, S.; Rafique, J.; Braga, A.L.; Mendes, S.R. Electrochemical Oxidative C(sp^2)–H Bond Selenylation of Activated Arenes. *Eur. J. Org. Chem.* **2019**, *2019*, 6465–6469. [CrossRef]
37. Fang, X.-L.; Tang, R.-Y.; Zhong, P.; Li, J.-H. Iron-Catalyzed Sulfenylation of Indoles with Disulfides Promoted by a Catalytic Amount of Iodine. *Synthesis* **2009**, *24*, 4183–4189.
38. Vieira, B.M.; Thurow, S.; Brito, J.S.; Perin, G.; Alves, D.; Jacob, R.G.; Santi, C.; Lenardão, E.J. Sonochemistry: An Efficient Alternative to the Synthesis of 3-Selanylindoles Using CuI as Catalyst. *Ultrason. Sonochem.* **2015**, *27*, 192–199. [CrossRef] [PubMed]
39. Vásquez-Céspedes, S.; Ferry, A.; Candish, L.; Glorius, F. Heterogeneously Catalyzed Direct C-H Thiolation of Heteroarenes. *Angew. Chem. Int. Ed.* **2015**, *54*, 5772–5776. [CrossRef] [PubMed]
40. Luz, E.Q.; Seckler, D.; Araújo, J.S.; Angst, L.; Lima, D.B.; Rios, E.A.M.; Ribeiro, R.R.; Rampon, D.S. Fe(III)-Catalyzed Direct C3 Chalcogenylation of Indole: The Effect of Iodide Ions. *Tetrahedron* **2019**, *75*, 1258–1266. [CrossRef]

41. Rios, E.A.M.; Gomes, C.M.B.; Silvério, G.L.; Luz, E.Q.; Ali, S.; D'Oca, C.d.R.M.; Albach, B.; Campos, R.B.; Rampon, D.S. Silver-Catalyzed Direct Selanylation of Indoles: Synthesis and Mechanistic Insights. *RSC Adv.* **2023**, *13*, 914–925. [CrossRef] [PubMed]
42. Benchawan, T.; Maneewong, J.; Saeeng, R. Selective Synthesis of 3-Chalcogenylindoles via Silver-Catalyzed Direct Chalcogenation of Indoles with Dichalcogenides. *ChemistrySelect* **2023**, *8*, e202301988. [CrossRef]
43. Azeredo, J.B.; Godoi, M.; Martins, G.M.; Silveira, C.C.; Braga, A.L. A Solvent- and Metal-Free Synthesis of 3-Chacogenyl-Indoles Employing DMSO/I_2 as an Eco-Friendly Catalytic Oxidation System. *J. Org. Chem.* **2014**, *79*, 4125–4130. [CrossRef] [PubMed]
44. Rafique, J.; Saba, S.; Franco, M.S.; Bettanin, L.; Schneider, A.R.; Silva, L.T.; Braga, A.L. Direct, Metal-free C(sp^2)−H Chalcogenation of Indoles and Imidazopyridines with Dichalcogenides Catalysed by KIO_3. *Chem. Eur. J.* **2018**, *24*, 4173–4180. [CrossRef] [PubMed]
45. Menezes, J.R.; Gularte, M.M.; dos Santos, F.C.; Roehrs, J.A.; Azeredo, J.B. Synthesis of 3-Chalcogenyl-Indoles Mediated by the Safer Reagent Urea-Hydrogen Peroxide (UHP). *Tetrahedron Lett.* **2023**, *120*, 154446. [CrossRef]
46. Bhunia, S.K.; Das, P.; Jana, R. Atom-Economical Selenation of Electron-Rich Arenes and Phosphonates with Molecular Oxygen at Room Temperature. *Org. Biomol. Chem.* **2018**, *16*, 9243–9250. [CrossRef] [PubMed]
47. Leonard, N.M.; Wieland, L.C.; Mohan, R.S. Applications of Bismuth(III) Compounds in Organic Synthesis. *Tetrahedron* **2002**, *58*, 8373–8397. [CrossRef]
48. Gaspard-Iloughmane, H.; Le Roux, C. Bismuth(III) Triflate in Organic Synthesis. *Eur. J. Org. Chem.* **2004**, *2004*, 2517–2532. [CrossRef]
49. Bothwell, J.M.; Krabbe, S.W.; Mohan, R.S. Applications of Bismuth(III) Compounds in Organic Synthesis. *Chem. Soc. Rev.* **2011**, *40*, 4649–4707. [CrossRef] [PubMed]
50. Ondet, P.; Lemière, G.; Duñach, E. Cyclisations Catalysed by Bismuth(III) Triflate. *Eur. J. Org. Chem.* **2017**, *2017*, 761–780. [CrossRef]
51. Raţ, C.I.; Soran, A.; Varga, R.A.; Silvestru, C. C–H Bond Activation Mediated by Inorganic and Organometallic Compounds of Main Group Metals. *Adv. Organomet. Chem.* **2018**, *70*, 233–311.
52. Takasawa, R.; Jona, A.; Inoue, M.; Azuma, M.; Akahane, H.; Ueno, Y.; Nakagawa, Y.; Chimori, R.; Mano, Y.; Murata, Y.; et al. Triphenylbismuth Dichloride Inhibits Human Glyoxalase I and Induces Cytotoxicity in Cultured Cancer Cell Lines. *J. Toxicol. Sci.* **2022**, *47*, 539–546. [CrossRef] [PubMed]
53. Ohki, H.; Wada, M.; Akiba, K.Y. Bismuth Trichloride as a New Efficient Catalyst in the Aldol Reaction. *Tetrahedron Lett.* **1988**, *29*, 4719–4722. [CrossRef]
54. Wada, M.; Takeichi, E.; Matsumoto, T. Bismuth Trichloride as a New Efficient Catalyst in the Aldol Reaction and the Michael Reaction. *Bull. Chem. Soc. Jpn.* **1991**, *64*, 990–994. [CrossRef]
55. Ollevier, T.; Lavie-Compin, G. An Efficient Method for the Ring Opening of Epoxides with Aromatic Amines Catalyzed by Bismuth Trichloride. *Tetrahedron Lett.* **2002**, *43*, 7891–7893. [CrossRef]
56. De, S.K.; Gibbs, R.A. Bismuth(III) Chloride-Catalyzed Direct Deoxygenative Allylation of Substituted Benzylic Alcohols with Allyltrimethylsilane. *Tetrahedron Lett.* **2005**, *46*, 8345–8350. [CrossRef]
57. Sabitha, G.; Reddy, E.V.; Maruthi, C.; Yadav, J.S. Bismuth(III) Chloride-Catalyzed Intramolecular Hetero-Diels–Alder Reactions: A Novel Synthesis of Hexahydrodibenzo[*b*,*h*][1,6]Naphthyridines. *Tetrahedron Lett.* **2002**, *43*, 1573–1575. [CrossRef]
58. Sabitha, G.; Reddy, E.V.; Yadav, J.S.; Rama Krishna, K.V.S.; Ravi Sankar, A. Stereoselective Synthesis of Octahydro-3b*H*-[1,3]dioxolo [4″,5″:4′,5′]furo [2′,3′:5,6]pyrano [4,3-*b*]quinolines via Intramolecular Hetero-Diels–Alder Reactions Catalyzed by Bismuth(III) Chloride. *Tetrahedron Lett.* **2002**, *43*, 4029–4032. [CrossRef]
59. Li, Z.; Wei, C.; Chen, L.; Varma, R.S.; Li, C.-J. Three-Component Coupling of Aldehyde, Alkyne, and Amine Catalyzed by Silver in Ionic Liquid. *Tetrahedron Lett.* **2004**, *45*, 2443–2446. [CrossRef]
60. Li, H.; Zeng, H.-Y.; Shao, H.-W. Bismuth(III) Chloride-Catalyzed One-Pot Mannich Reaction: Three-Component Synthesis of β-Amino Carbonyl Compounds. *Tetrahedron Lett.* **2009**, *50*, 6858–6860. [CrossRef]
61. Wu, F.; Huang, W.; Yiliqi; Yang, J.; Gu, Y. Relay Catalysis of Bismuth Trichloride and Byproduct Hydrogen Bromide Enables the Synthesis of Carbazole and Benzo[α]carbazoles from Indoles and α-Bromoacetaldehyde Acetals. *Adv. Synth. Catal.* **2018**, *360*, 3318–3330. [CrossRef]
62. Wu, Z.; Feng, X.-X.; Wang, Q.-D.; Yun, J.-J.; Rao, W.; Yang, J.-M.; Shen, Z.-L. Bismuth Trichloride-Catalyzed Oxy-Michael Addition of Water and Alcohol to α,β-Unsaturated Ketones. *Chin. Chem. Lett.* **2020**, *31*, 1297–1300. [CrossRef]
63. Li, S.; Li, Y.; Feng, B.; Liang, J.; You, G.; Liu, X.; Xian, L. Bi(III)-Catalyzed Aminooxygenation of Propargyl Amidines to Synthesize 2-Fluoroalkyl Imidazole-5-carbaldehydes and Their Decarbonylations. *Chem. Commun.* **2020**, *56*, 6400–6403. [CrossRef] [PubMed]
64. Lin, X.-T.; Zhao, C.; Wang, D.-R.; Wu, G.-C.; Chen, G.-S.; Chen, S.-J.; Ren, H.; Deng, D.-S.; Xu, Y.-B.; Hu, X.-W.; et al. $BiCl_3$-Mediated Tandem Cyclization of Tryptamine-Derived Ynamide: Concise Synthesis of Pentacyclic Spiroindolines and Tricyclic Indole Derivatives. *Adv. Synth. Catal.* **2022**, *364*, 890–896. [CrossRef]
65. Malik, P.; Joseph, D.; Chakraborty, D. $BiCl_3$-catalyzed Carbon–Carbon Cross-Coupling of Organoboronic Acids with Aryl Iodides. *Appl. Organometal. Chem.* **2013**, *27*, 519–522. [CrossRef]
66. Riyaz, M.A.B.; Swu, T. Bismuth-catalyzed *N*-Arylation of 2-Aminobenzimidazole and Phosphorylation of Substituted Coumarins via C-H Functionalization. *ChemistrySelect* **2022**, *7*, e202203281. [CrossRef]

67. Zhang, J.Z. Interfacial Charge Carrier Dynamics of Colloidal Semiconductor Nanoparticles. *J. Phys. Chem. B* **2000**, *104*, 7239–7253. [CrossRef]
68. Ünlü, F.; Deo, M.; Mathur, S.; Kirchartz, T.; Kulkarni, A. Bismuth-Based Halide Perovskite and Perovskite-Inspired Light Absorbing Materials for Photovoltaics. *J. Phys. D Appl. Phys.* **2022**, *55*, 113002. [CrossRef]
69. Komatsu, N.; Uda, M.; Suzuki, H. Bismuth(III) Halides and Sulfate as Highly Efficient Catalyst for the Sulfenylation of Carbonyl and Related Compounds[1]. *Synlett* **1995**, *9*, 984–986. [CrossRef]
70. Cunha, S.; Rodrigues, M.T., Jr. The First Bismuth(III)-Catalyzed Guanylation of Thioureas. *Tetrahedron Lett.* **2006**, *47*, 6955–6956. [CrossRef]
71. Bailey, A.D.; Baru, A.R.; Tasche, K.K.; Mohan, R.S. Environmentally Friendly Organic Synthesis Using Bismuth Compounds: Bismuth(III) Iodide Catalyzed Deprotection of Acetals in Water. *Tetrahedron Lett.* **2008**, *49*, 691–694. [CrossRef]
72. Adonin, S.A.; Peresypkina, E.V.; Sokolov, M.N.; Korolkov, I.V.; Fedin, V.P. Polyoxomolybdate-Supported Bismuth Trihalides $[Mo_8O_{26}(BiX_3)_2]^{4-}$ (X = Cl, Br, I): Syntheses and Study of Polymorphism. *Inorg. Chem.* **2014**, *53*, 6886–6892. [CrossRef] [PubMed]
73. Wedal, J.C.; Ziller, J.W.; Evans, W.J. Expanding Bismuth Trihalide Coordination Chemistry with Trimethyltriazacyclohexane and Trimethyltriazacyclononane. *Inorg. Chem.* **2022**, *61*, 11766–11774. [CrossRef] [PubMed]
74. Guo, T.; Dong, Z.; Zhang, P.; Xing, W.; Li, L. Direct Selenation of Imidazoheterocycles and Indoles with Selenium Powder in a Copper-Catalyzed Three-Component One-Pot System. *Tetrahedron Lett.* **2018**, *59*, 2554–2558. [CrossRef]
75. Lin, M.; Kang, L.; Gu, J.; Dai, L.; Tang, S.; Zhang, T.; Wang, Y.; Li, L.; Zheng, X.; Zhu, W.; et al. Heterogeneous Synergistic Catalysis by Ru-RuO_x Nanoparticles for Se–Se Bond Activation. *Nano Res.* **2017**, *10*, 922–932. [CrossRef]
76. Chen, J.; Hu, L.; Wang, H.; Tan, H. Iodine-Catalyzed Telluration of Indole Derivatives with Diarylditellurides for Synthesis of 3-Aryltellurylindoles. *Chin. J. Org. Chem.* **2019**, *39*, 2048–2052. [CrossRef]
77. Dolomanov, O.V.; Bourhis, L.J.; Gildea, R.J.; Howard, J.A.K.; Puschmann, H. OLEX2: A Complete Structure Solution, Refinement and Analysis Program. *J. Appl. Cryst.* **2009**, *42*, 339–341. [CrossRef]
78. Sheldrick, G.M. SHELXT–Integrated Space-Group and Crystal-Structure Determination. *Acta Cryst.* **2015**, *A71*, 3–8. [CrossRef] [PubMed]
79. Sheldrick, G.M. Crystal Structure Refinement with SHELXL. *Acta Cryst.* **2015**, *C71*, 3–8.

molecules

MDPI

Communication

Electrochemical Nickel-Catalyzed Synthesis of Unsymmetrical Diorganyl Selanes from Diaryl Diselanes and Aryl and Alkyl Iodides

Jona Queder and Gerhard Hilt *

Institute of Chemistry, Oldenburg University, Carl-von-Ossietzky-Str. 9-11, 26129 Oldenburg, Germany; jona.queder@uni-oldenburg.de
* Correspondence: gerhard.hilt@uni-oldenburg.de

Abstract: The synthesis of unsymmetrical diorganyl selanes was accomplished under electrochemical conditions in an undivided cell utilizing a magnesium cathode and a carbon anode made out of aryl and alkyl iodides and diselanes. This electrochemical cross-electrophile coupling (eXEC) was accomplished using a simple nickel catalyst formed in situ out of $Ni(acac)_2$ and 2,2′-bipyridine in DMF at ambient temperatures. The reaction showed good functional group compatibility, and heteroaryl iodides, such as thiophene or pyridine derivatives, were well accepted.

Keywords: catalysis; diselanes; selanes; nickel; electrochemistry; aryl iodides; alkyl iodides

check for updates

Citation: Queder, J.; Hilt, G. Electrochemical Nickel-Catalyzed Synthesis of Unsymmetrical Diorganyl Selanes from Diaryl Diselanes and Aryl and Alkyl Iodides. *Molecules* **2024**, *29*, 4669. https://doi.org/10.3390/molecules29194669

Academic Editor: Ming Wang

Received: 27 August 2024
Revised: 10 September 2024
Accepted: 26 September 2024
Published: 1 October 2024

1. Introduction

A number of unsymmetrical diorganyl selane compounds exhibit useful (= non-lethal) biological applicability [1,2], such as phenylselenyl-substituted naphthol derivative **1**, which is applied as a 5-LOX inhibitor (Figure 1). Phenylselanyl-substituted quinoline derivative **2** is an antifungal agent, whereas multiple-substituted 3-indolyl selane **3** is used as an efficient cell growth inhibitor. Besides the well-known α-amino acids selenocysteine and selenomethionine, aryl-alkyl-selanes, such as α-amino acid **4**, exhibit useful biological activity; **4** acts as a chemopreventive agent.

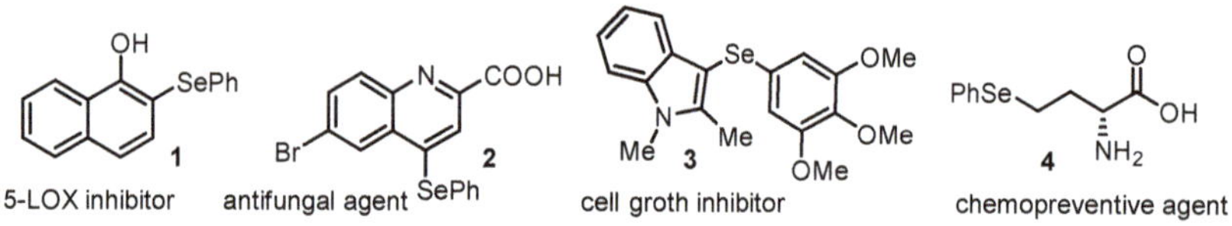

Figure 1. Biologically active unsymmetrical diaryl selanes and a biologically active alkyl–aryl-selane.

Also, chiral unsymmetrical diaryl selanes and alkyl–aryl-selanes are of synthetic value—for example, in the selenium-catalyzed oxidative cyclization of β,γ-unsaturated carboxylic acid, as reported by Maruoka in 2016 [3]. Recently, an excellent review concerning the application of selenium compounds to asymmetric synthesis was published by Back [4].

For the synthesis of such unsymmetrical selanes, the recent focus has been on the investigation of cross-coupling methodologies [5,6]. Transition metal-catalyzed coupling reactions, in most cases, rely on high temperatures, as shown by Engman [7]; the application of strong bases, as exemplified by Rueping [8]; photochemical activation [9,10]; or more sophisticated approaches, like the use of catalytically active nanoparticles, as described by Rao [11] (Scheme 1).

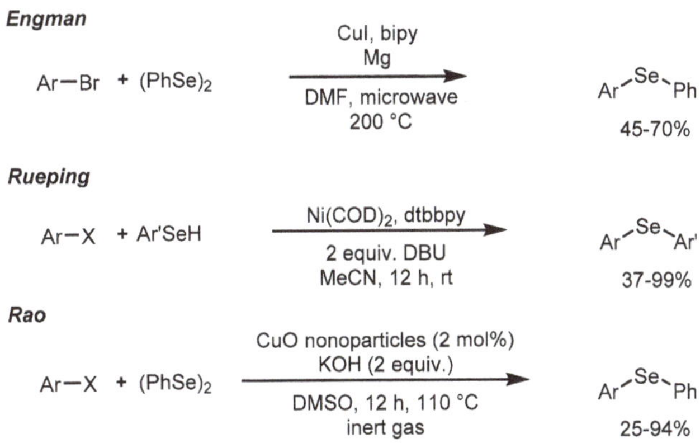

Scheme 1. Recent access routes for unsymmetrical diorganyl selanes.

To overcome the challenges faced in classical cross-coupling chemistry, electrochemically driven cross-electrophile coupling (eXEC) has emerged as a new methodology [12–14]. The electrochemical coupling of two electrophiles, formally accompanied by a two-electron reduction process, facilitates the desired cross-coupling via radical pathways [15], making the disadvantages of the classical pathways mentioned above obsolete (Scheme 2) [16]. A crucial point in this strategy is that the redox potentials of the two electrophiles are different so that selective reduction can occur. Naturally, electrochemical activation of one of the starting materials can either be accomplished by potentiostatic electrolysis with a fixed redox potential with respect to a reference electrode or by galvanostatic electrolysis at low current densities. Therefore, the larger the difference in the reduction potential, the more selective the activation process.

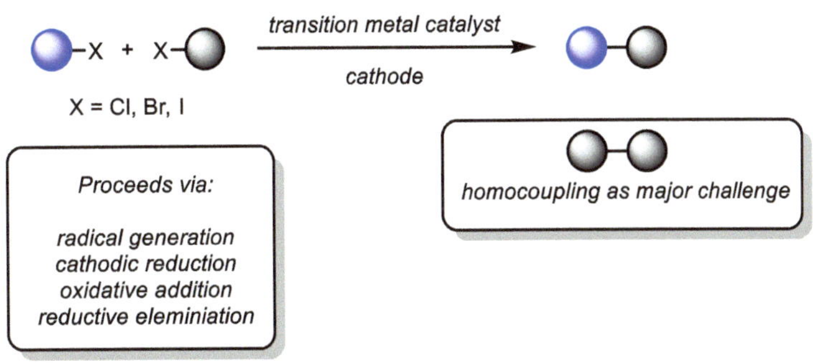

Scheme 2. General concept of electrochemical cross-electrophile coupling (eXEC).

The idea of an eXEC reaction pathway for mild synthesis of unsymmetrical diorganyl selanes inspired us to initiate an investigation of an eXEC process where electrochemically generated selanyl radical species could react with aryl or alkyl iodides under nickel catalysis. The focus of the proposed reaction was set to realize the synthesis of the diorganyl selane under mild reaction conditions with broad functional group tolerance from easily accessible starting materials.

2. Results and Discussion

As easily accessible test substrates, we chose 4-iodotoluene (**5a**) and diphenyl diselane (**6**) for the formation of the unsymmetrical phenyl-4-tolylselane (**7a**, Table 1).

Table 1. Optimization results for the nickel-catalyzed formation of unsymmetrical diaryl selane **7a**.

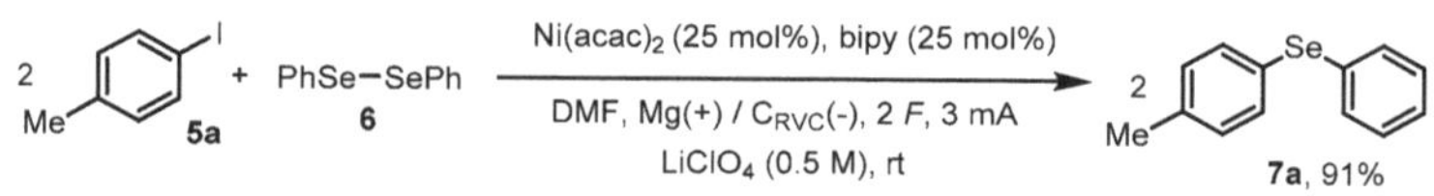

Entry	Change to the Conditions	Yield
1	Bu_4NPF_6 as supporting electrolyte	58%
2	$LiClO_4$ (0.5 M) as supporting electrolyte, glassy carbon cathode	66%
3	$LiClO_4$ (0.24 M) as supporting electrolyte, glassy carbon cathode	48%
4	$NiCl_2{\cdot}6H_2O$ as nickel catalyst precursor	62%
5	$NiBr_2$ as nickel catalyst precursor	52%
6	NiI_2 as nickel catalyst precursor	57%
7	$Ni(acac)_2$ (20 mol%)	43%
8	$Ni(acac)_2$ (30 mol%)	44%
9	1,10-phananthroline as ligand	46%
10	1,2-(diphenylphosphino)ethane (dppe) as ligand	2%
11	Dimethylaminoacetamide (DMA) as solvent	50%
12	Acetonitrile (MeCN) as solvent	36%
13	2,2,2-trifluoroethanol as solvent	0%
14	Divided cell (P4 frit) (2 *F*)	12%
15	Pseudo-divided cell (Pt wire anode)	0%
16	Glassy carbon cathode	57%
17	Magnesium cathode	56%
18	Platinum cathode	0%
19	Zinc anode	20%
20	Platinum anode	0%
21	1.5 *F* of electricity	42%
22	2.5 *F* of electricity	27%
23	None	91%

Conditions: $Ni(acac)_2$ (25.7 mg, 0.1 mmol), bipy (15.6 mg, 0.1 mmol), 4-iodotoluene (87.2 mg, 0.4 mmol), $LiClO_4$ (106 mg), and $(PhSe)_2$ (62.4 mg, 0.2 mmol) were weighed in an undivided cell, anhydrous DMF (3 mL) was added, and the solution was electrolyzed (Mg (+)/C_{RVC} (-), 2 *F*, 3 mA) under an argon atmosphere. The yield was determined via GC (FID) using mesitylene as the internal standard.

Optimization of the nickel-catalyzed formation of **7a** was performed with respect to the type and amount of the supporting electrolyte and the nickel precursor, the additive ligand, the solvent, and the typical electrochemical parameters, such as the cell type, the anode/cathode material, and the current required to reach the highest conversion and yield of product **7a**. Selected examples of the optimization process are given in Table 1 (for additional test reactions, see the Supporting Information). The individual optimized parameters were then combined, and the optimized process consisted of $Ni(acac)_2$/bipy-mediated electrochemical transformation with a magnesium sacrificial anode and an RVC (Reticulated Vitreous Carbon) cathode in an undivided cell, utilizing DMF as the solvent and $LiClO_4$ as the supporting electrolyte at room temperature and under a low current (3 mA), utilizing 2 *F* of electricity in an inert atmosphere to afford the desired product at a 91% yield.

During the optimization process, we checked the above-mentioned individual parameters, while some parameters were not yet optimized (see the Supporting Information). However, typical trends could be observed, and the most important set of information is given in Table 1.

As supporting electrolytes, tetrabutylammonium salts facilitated the reaction but were inferior to lithium perchlorate, which gave the highest yield of **7a** when a 0.5 M solution of $LiClO_4$ was used (entries 1–3). Several nickel salts could be applied as the catalyst precursor (entries 4–6,) among which $Ni(acac)_2$ proved to give the highest conversion when 25 mol% (Table 1) was used.

But for no obvious reason, the yield dropped significantly when the catalyst loading was only slightly reduced or increased (entries 7/8). Interestingly, several bipyridine derivatives (see the SI) and phenanthroline (entry 9) resulted in the formation of active nickel catalysts, whereas phosphine ligands, such as dppe (entry 10), inhibited the desired reaction completely.

Solvents with an amide subunit, such as DMF (Scheme 3) and DMA (entry 11), gave good results, while acetonitrile (entry 12) was less suited as the solvent. Slightly acidic solvents, such as trifluoroethanol, inhibited the reaction completely (entry 13). The reaction is best performed in an undivided cell (with a 5 mm electrode distance), while in a divided cell (P4 frit), only traces of the product are found, and no product is formed in a quasi-divided cell (entries 14/15). These results are of interest because they show that both electrode reactions are needed for the overall process to be successful. As cathode material, graphite, glassy carbon, or a Mg plate can be used (entries 16–18). The anode reaction is best performed using a Mg anode or with much less efficiency using a Zn anode (20%), but the reaction fails when a Pt anode is used (entries 19−20). The electrolysis should be stopped after the consumption of 2.0 F, resulting in the best results, while less or more electricity results in lower yields (entries 21/22).

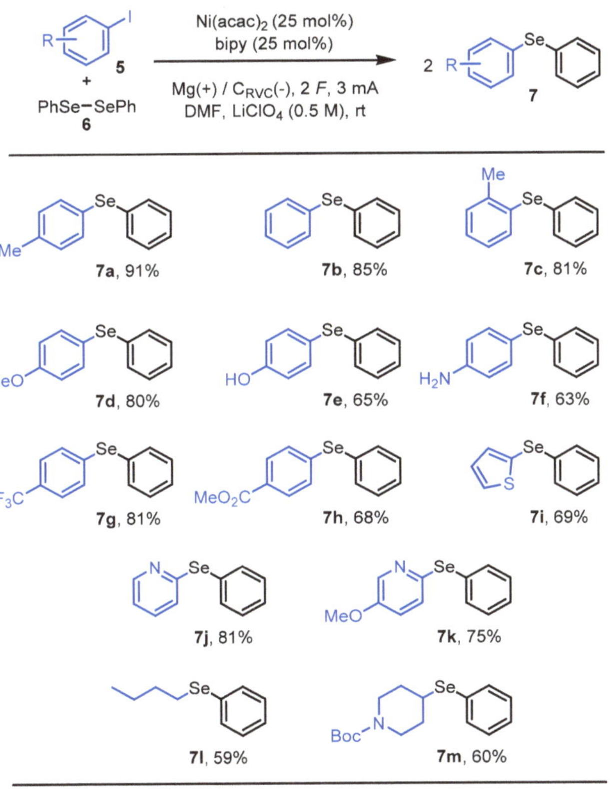

Scheme 3. Scope of the nickel-catalyzed electrochemical synthesis of unsymmetrical diaryl selanes.

With the optimized reaction conditions hitherto at hand, the application of several selected functionalized iodobenzene derivatives (**5**) with diphenyl diselane (**6**) was investigated for the synthesis of unsymmetrical selanes **7**. The results of this investigation are summarized in Scheme 3.

Under the optimized conditions, the use of iodobenzene and alkyl-substituted iodobenzenes led to the formation of **7a**–**7c** at good isolated yields ranging from 80 to 91%. Of considerably more interest were heteroatom-functionalized iodobenzene derivatives of type

5 in terms of determining the limiting factors based on their electronic nature. Therefore, we investigated alkoxy-functionalized iodobenzene derivatives, such as 4-iodanisole and 4-iodophenole, as typical electron-rich arenes. The desired products **7d** and **7e** were formed at 80% and 65% yields, respectively. With a similar efficiency, 4-iodoaniline was converted into product **7f**, which was isolated at a 63% yield. Substrates with electron-withdrawing substituents on their iodoarene moiety were also reactive under the reaction conditions, and the trifluoromethyl (**7g**) and methyl benzoate derivatives (**7h**) were produced at moderate to good yields of 81% and 68%, respectively. Heterocyclic iodoarenes were then investigated, and the 2-thienyl, 2-pyridinyl, and 4-methoxy-2-pyridinyl moieties were well accepted in this process. The desired products **7i–7k** could be isolated again at moderate to good yields, ranging from 69 to 81%. Finally, alkyl iodides were also investigated, and as the test substrate, a simple primary alkyl iodide, such as butyl iodide, was reacted with diphenyl diselane. The desired product **7l** was isolated at a moderate yield (59%), but this result indicated that the reaction was not strictly limited to aryl iodides. Last but not least, the secondary Boc-protected piperidinyl iodide was converted into **7m**, and the product was isolated at a 60% yield. In this reaction, a small portion (5–10%) of a Boc-deprotected piperidine selane was also observed but could not be isolated in its pure form.

According to these results, electron-deficient, electron-neutral, and electron-rich aryl iodides could be transformed into the desired selanes at moderate to good yields. Also, heteroaryl iodides were well accepted as starting materials, as were alkyl iodides, showing few restrictions concerning the application to forming selanes of primary and secondary alkyl derivatives.

The modification of diaryl diselanes is more complex since very few diaryl diselanes are commercially available. Diselanes **9** must be prepared from suitable aryl bromides **8** in multi-step synthesis, as is outlined in Scheme 4. The synthesis of diselanes was described by us in a preliminary study [17] and therefore should only be discussed briefly. Aryl bromide **8** is treated with excess *t*BuLi at a low temperature before elemental grey selenium is added and the mixture is stirred for 1 h. The solution is then acidified, and the selenole intermediate is oxidized by air at room temperature to yield the corresponding diselane **9**, which is then purified by column chromatography.

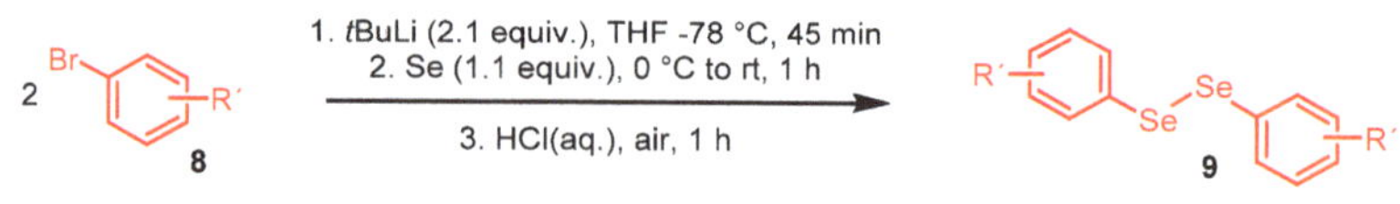

Scheme 4. Synthesis of bis-functionalized diaryl diselanes.

Thereby, we obtained two diselane candidates with different electronic natures, with diselane **9a** bearing a methoxy group in the 4-position and diselane **9b** having an electron-abstracting 4-trifluoromethyl group. These diselanes were then subjected to the optimized reaction conditions and reacted with 4-iodotoluene **5a** as an electron-neutral test substrate. The results of these two reactions are given in Scheme 5.

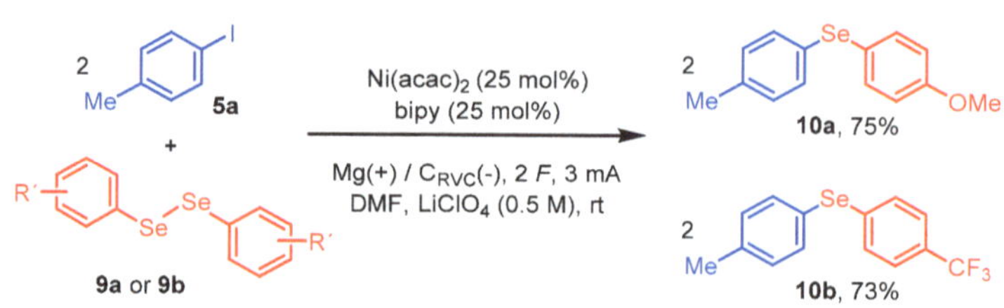

Scheme 5. Nickel-catalyzed synthesis of unsymmetrical bis-substituted and functionalized diaryl selanes.

Both products (**10a** and **10b**) were obtained at a good yield (>73%) and showed that the nickel-catalyzed synthesis of unsymmetrical diaryl selanes is capable of accepting all sorts of functionalized iodo(hetero)arenes and electron-rich and electron-deficient diselanes as starting materials.

The observations made during our examination of the reaction led us to a mechanistic interpretation consistent with one of the most prominent mechanisms for XEC reactions [18,19]. It is proposed that the modified radical chain mechanism, which depicts the net coupling of two electrophiles, proceeds via anodic radical formation and electrochemical manipulation of the oxidation state of the Ni (II) intermediates. As demonstrated in recent research [20], Ni (I) (bipy)halogen complexes are of noteworthy stability, whereas the corresponding Ni (0) species and Ni (II) species tend to comproportionate with Ni (I), leading us to believe that the latter is the resting state of the catalyst [21]. Once the Ni (II) precursor has been reduced, complex **A** is formed by trapping the anodically generated selanyl radical (Scheme 6). Following cathodic reduction, intermediary Ni (III) complex **C** is formed through following the oxidative addition of iodobenzene to Ni (I)SePh complex **B**.

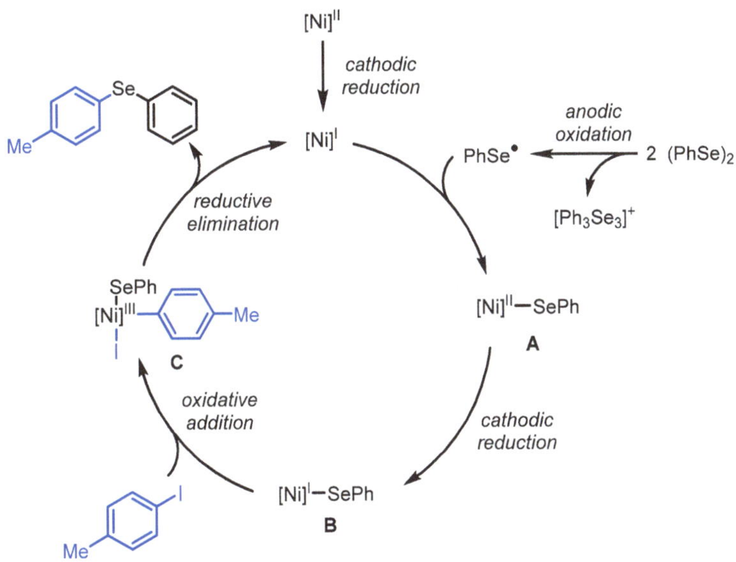

Scheme 6. Proposed mechanism of the nickel-catalyzed synthesis of bis-functionalized diaryl selanes. The ligands are omitted for clarity.

In a classical radical chain mechanism, Ni (I) intermediate **B** would react with one of the electrophiles to form a reactive radical [22] and a Ni (II) species. We surmised that conducting the reaction in a divided cell, where the generation of the selanyl radical is anodically precluded, should therefore still result in the formation of the desired product. Subsequent experimental investigations indicate a markedly reduced yield (12%), suggesting a relatively slow rate of radical formation and the absence of a requirement for Mg ions in the reaction process. When the rate of radical formation was controlled electrochemically [23] in an undivided cell, the yield was increased to 91%, presumably by bypassing the previously described mode of radical formation through anodic oxidation of diphenyl diselane. To rule out the premature oxidation of diselane by non-electrochemical oxidation on the surface of the Mg electrode, a control experiment was carried out using Mg turnings. Even after two days of reaction and additional heating, no formation of the desired product was observed.

3. Materials and Methods

3.1. General Information

To work under an inert gas atmosphere, the solvents used were purchased water- and oxygen-free from Thermo Fischer Scientific and used in an Acroseal system. The solvents used for column chromatography were purified by distillation prior to use. All chemicals and reagents were purchased from commercial suppliers without further purification, if not stated otherwise, or were prepared according to the known procedures from the literature. Where water- or air-sensitive compounds were used, the experiments were carried out in oven-dried glassware using conventional Schlenk techniques under an argon atmosphere. NMR spectroscopy: The multiplicities of individual detected signals were given as they could be observed, not as theoretically expected. s (singlet), d (doublet), t (triplet), and m (multiplet) signals are indicated, as are their combinations. ^{1}H NMR spectra were measured with an Avance 500 (500 MHz) from Bruker at 305.0 K. The samples were measured in deuterated chloroform ($CDCl_3$). A pre-set pulse program was used for all the measurements. ^{1}H NMR: ($CHCl_3$: δ = 7.26 ppm); ^{13}C NMR: ($CDCl_3$: δ = 77.00 ppm); ^{19}F NMR: (CCl_3F: δ = 0.00 ppm); ^{77}Se NMR: (MeSe-SeMe: δ = 275.00 ppm). Measurements of the ^{77}Se and ^{19}F spectra were taken using external standards. IR spectra: All the IR spectra were obtained on an IRSpirit from Shimadzu with the QATR-S Single Reflection ATR Accessory. The position of the absorption bands is indicated in wavenumbers (cm^{-1}). Column chromatography: Column chromatography, used for purification of the products, was carried out with silica gel 60 (grain size: 40–63 μm) from Macherey Nagel. Thin-layer chromatography was carried out on Merck TLC plates coated with silica gel 60 F_{254} with a fluorescence indicator. For the detection of the ultraviolet light signals (λ = 254 nm), GC analysis or heating the plate after it had been dipped in a $KMnO_4$ solution was used. Melting point determination: The melting point was determined in glass capillaries on a Mel-Temp from Laboratory Devices, Cambridge.

3.2. The General Procedure

An oven-dried undivided cell was charged with diphenyldiselane (0.5 equiv.), the organo halogen compound (1.0 equiv.), bipyridine (0.25 equiv.), nickel bis-acetylacetonate (0.25 equiv.), and lithium perchlorate (3.75 equiv.). The solids were purged with inert argon gas thrice and left in vacuo for 30 min. Afterwards, anhydrous DMF (7 mL) was added, and the programmed electrolysis was performed (3 mA, 2.0 F, Mg (+)/C_{RVC} (-), 1500 rpm) under an argon atmosphere. After complete consumption of the educt was confirmed via GC/MS, the resulting blackish mixture was filtered over silica gel (eluent Et_2O) and washed with water (50 mL) once. The aqueous phase was extracted with Et_2O (50 mL) once, and the combined organic phase was dried over $MgSO_4$. The solvents were removed under reduced pressure, and the crude product was purified by flash column chromatography.

3.3. Characterization Data of Novel Compounds

5-Methoxy-2-(phenylselanyl)pyridine (**7k**)

Yield: 75%, colorless oil. R_f = 0.33 (*n*-pentane:Et_2O = 3:1) ^{1}H NMR (500 MHz, $CDCl_3$): δ = 8.36 (d, J = 2.32 Hz, 1H), 7.74 (dd, J = 8.54, 2.35 Hz, 1H), 7.36–7.34 (m, 2H), 7.25–7.21 (m, 2H), 6.71 (d, J = 8.60 Hz, 2H), 3.95 (s, 3H) ppm. ^{13}C{^{1}H} NMR (125 MHz, $CDCl_3$): δ = 164.1, 152.4, 145.3, 132.2, 131.3, 129.4, 127.0, 118.0, 112.3, 53.8 ppm. ^{77}Se{^{1}H} NMR (95.4 MHz, $CDCl_3$): δ = 364.7 ppm. IR (ATR): λ^{-1} = 3057, 2981, 2941, 2868, 2840, 1577, 1557, 1474, 1427, 1358, 1300, 1278, 1244, 1178, 1157, 1122, 1092, 1067, 1018, 1004, 998, 925, 825, 731, 688, 667, 647, 628, 531, 497 cm^{-1}.

All the other compounds are known in the literature, and the analytical data are in accordance with the published data. For details, see the Supporting Information.

In conclusion, the electrochemical eXEC reaction was designed to utilize readily available aryl iodides and diselanes as electrophiles, where electron transfer generates selanyl radicals, and in the presence of a nickel catalyst, the radicals are trapped, and further reduction of the nickel–selanyl complex facilitates the oxidative addition of the aryl iodide. Thereby, several unsymmetrical diaryl selanes and alkyl–aryl-selanes were generated at moderate to good yields. Moreover, the process accepted electron-rich as well as electron-deficient diaryl-diselanes as the starting materials, which resulted in the formation of more complex selanes at good yields.

Supplementary Materials: The following supporting information can be downloaded at https://www.mdpi.com/article/10.3390/molecules29194669/s1. The characterization data on the known compounds and the ^{1}H and ^{13}C NMR spectra are available online.

Author Contributions: Conceptualization, optimization of the reactions, experiments, data acquisition: J.Q. Writing and supervision: G.H. All authors have read and agreed to the published version of the manuscript.

Funding: This research received no external funding.

Data Availability Statement: All the relevant data regarding this investigation can be found in the Supporting Information.

Conflicts of Interest: The authors declare no conflicts of interest.

References

1. Nogueira, C.W.; Zeni, G.; Rocha, J.B.T. Organoselenium and organotellurium compounds: Toxicology and pharmacology. *Chem. Rev.* **2004**, *104*, 6255–6285. [CrossRef] [PubMed]
2. Sonego, J.M.; de Diego, S.I.; Szajnman, S.H.; Gallo-Rodriguez, C.; Rodriguez, J.B. Organoselenium compounds: Chemistry and applications in organic synthesis. *Chem. Eur. J.* **2023**, *52*, e202300030. [CrossRef] [PubMed]
3. Kawamata, Y.; Hashimotot, T.; Maruoka, K. A chiral electrophilic selenium catalyst for highly enantioselective oxidative cyclization. *J. Am. Chem. Soc.* **2016**, *138*, 5206–5209. [CrossRef] [PubMed]
4. Stadel, J.T.; Back, T.G. Asymmetric synthesis with organoselenium compounds—The past twelve years. *Chem. Eur. J.* **2024**, *31*, e202304074. [CrossRef] [PubMed]
5. Taniguchi, N. Convenient synthesis of unsymmetrical organochalcogenides using organoboronic acids with dichalcogenides via cleavage of the S-S, Se-Se, or Te-Te bond by a copper catalyst. *J. Org. Chem.* **2007**, *72*, 1241–1245. [CrossRef]
6. Kumar, S.; Engman, L. Microwave-assisted copper-catalyzed preparation of diaryl chalcogenides. *J. Org. Chem.* **2006**, *71*, 5400–5403. [CrossRef]
7. Taniguchi, N.; Onami, T. Copper-catalyzed synthesis of diaryl selenide from aryl iodide and diphenyl diselenide using magnesium metal. *Synlett* **2003**, *2003*, 829–832. [CrossRef]
8. Zhumagazy, S.; Zhu, C.; Yue, H.F.; Rueping, M. Nickel-catalyzed carbon–selenium bond formations under mild conditions. *Synlett* **2023**, *34*, 1381–1384. [CrossRef]
9. Bonciolini, S.; Pulcinella, A.; Leone, M.; Schiroli, D.; Ruiz, A.L.; Sorato, A.; Dubois, M.A.J.; Gopalakrishnan, R.; Masson, G.; Ca', N.D.; et al. Metal-free photocatalytic cross-electrophile coupling enables C1 homologation and alkylation of carboxylic acids with aldehydes. *Nat. Commun.* **2024**, *15*, 1509. [CrossRef]
10. Wang, R.; Wang, X.; Mao, S.; Zhao, Y.; Yuan, B.; Yang, X.-Y.; Li, J. Metal-free photochemical C-Se cross-coupling of aryl halides with diselenides. *Adv. Synth. Catal.* **2022**, *364*, 1607–1612. [CrossRef]
11. Reddy, V.P.; Kumar, A.V.; Swapna, K.; Rao, K.R. Copper oxide nanoparticle-catalyzed coupling of diaryl diselenide with aryl halides under ligand-free conditions. *Org. Lett.* **2009**, *11*, 951–953. [CrossRef] [PubMed]
12. Liu, Y.; Li, P.; Wang, Y.; Qui, Y. Electroreductive cross-electrophile coupling (eXEC) reactions. *Angew. Chem. Int. Ed.* **2023**, *62*, e202306679. [CrossRef] [PubMed]
13. Zhang, W.; Lu, L.; Zhang, W.; Wang, Y.; Ware, S.D.; Mondragon, J.; Rein, J.; Strotman, N.; Lehnherr, D.; See, K.A.; et al. Electrochemically driven cross-electrophile coupling of alkyl halides. *Nature* **2022**, *604*, 292–297. [CrossRef] [PubMed]
14. Ding, L.; Zhao, Y.; Lu, H.; Shi, Z.; Wang, M. Nickel-catalyzed asymmetric propargyl-aryl cross-electrophile coupling. *Angew. Chem. Int. Ed.* **2023**, *63*, e202313655. [CrossRef] [PubMed]
15. Pang, X.; Su, P.-F.; Shu, X.-Z. Reductive cross-coupling of unreactive electrophiles. *Acc. Chem. Res.* **2022**, *55*, 2491–2509. [CrossRef]
16. Cai, Y.-M.; Lui, X.-T.; Xu, L.-L.; Shang, M. Electrochemical Ni-catalyzed decarboxylative C(sp^{3})-N cross-electrophile coupling. *Angew. Chem. Int. Ed.* **2024**, *63*, e202315222. [CrossRef]
17. Queder, J.; Hilt, G. The electrochemical *trans*-chloroformyloxylation of unactivated alkenes. *Synlett* **2024**, *34*, *in press*. [CrossRef]
18. Lucas, L.E.; Jarvo, E.R. Stereospecific and stereoconvergent cross-couplings between alkyl electrophiles. *Nat. Chem. Rev.* **2017**, *1*, 65. [CrossRef]
19. Biswas, S.; Weix, D.J. Mechanism and Selectivity in Nickel-Catalyzed Cross-Electrophile Coupling of Aryl Halides with Alkyl Halides. *J. Am. Chem. Soc.* **2013**, *135*, 16192–16197. [CrossRef]
20. Day, C.S.; Rentaria-Gomez, A.; Ton, S.J.; Gogoi, A.R.; Gutierrez, O.; Martin, R. Elucidating electron-transfer events in polypyridine nickel complexes for reductive coupling reactions. *Nat. Catal.* **2023**, *6*, 244–253.

21. Beromi, M.M.; Banerjee, G.; Brudvig, G.W.; Hazari, N.; Mercado, B.Q. Nickel (I) Aryl species: Synthesis, properties, and catalytic activity. *ACS Catal.* **2018**, *8*, 2526–2533. [CrossRef] [PubMed]
22. Gong, Y.; Hu, J.; Qui, C.; Gong, H. Insights into recent nickel-catalyzed reductive and redox C-C coupling of electrophiles, C(sp^3)-H bonds and alkenes. *Acc. Chem. Res.* **2024**, *57*, 1149–1162. [CrossRef] [PubMed]
23. Zhan, L.; Wang, Q.; Tang, H.-T.; Mo, Z.-Y.; Pan, Y.-M. Recent advances in electrochemically mediated reactions of diselenides. *SynOpen* **2023**, *7*, 521–534.

molecules

MDPI

Article

A Computational Study of Heteroatom Analogues of Selenoxide and Selenone *syn* Eliminations

Adrian I. Doig †, **Jessica T. Stadel** † **and Thomas G. Back** *

Department of Chemistry, University of Calgary, Calgary, AB T2N 1N4, Canada
* Correspondence: tgback@ucalgary.ca; Tel.: +1-403-220-6256
† These authors contributed equally to this work.

Abstract: Selenoxide *syn* elimination is a widely used method for the synthesis of alkenes because it proceeds under exceptionally mild conditions, typically with excellent regio- and stereoselectivity. Surprisingly, hetero-selenoxide eliminations, where one or both olefinic carbon atoms are replaced with heteroatoms, have been little investigated, and their selenonyl counterparts even less so. A variety of such reactions, where the heteroatoms included combinations of O, N and S, as well as C, were investigated computationally. Selenoxides typically have lower activation energies and are slightly endothermic, while the corresponding selenones display higher activation energies and are exothermic in the gas state. The results are consistent with concerted, five-centre processes, leading to the formation of dioxygen, aldehydes, diazenes and imines from seleninyl or selenonyl peroxides, esters, hydrazines and amines, respectively. The more acidic selenenyl hydrodisulfide analogue undergoes proton transfer to the basic selenoxide oxygen atom instead of concerted elimination, resulting in the formation of a zwitterion. However, the formation of the corresponding selenonyl zwitterion is disfavoured compared to concerted *syn* elimination. The effects of solvents were also computed along with changes in enthalpy, entropy and free energy. Solvent effects were variable, while free energy calculations indicated overall ΔG values ranging between 3.60 and −32.12 kcal mol^{-1} for the syn eliminations of methyl methanethioseleninate and methaneperoxyselenonic acid, respectively. These computations suggest that the olefin-forming selenoxide *syn* elimination may be more general than currently understood and that replacement of the two carbon atoms with heteroatoms can lead to viable processes.

Keywords: organoselenium chemistry; DFT computations; selenoxides; selenones; *syn* eliminations; heteroatom *syn* eliminations; zwitterions

Citation: Doig, A.I.; Stadel, J.T.; Back, T.G. A Computational Study of Heteroatom Analogues of Selenoxide and Selenone *syn* Eliminations. *Molecules* **2024**, *29*, 4915. https://doi.org/10.3390/molecules29204915

Academic Editors: Ming Wang and Hajime Hirao

Received: 30 July 2024
Revised: 28 September 2024
Accepted: 7 October 2024
Published: 17 October 2024

1. Introduction

The 1,2 eliminations that afford alkenes are among the most fundamental and widely used processes in synthetic organic chemistry. *Syn* eliminations are a subset that typically proceed by concerted, unimolecular processes [1] involving five- or six-centred transition states and afford complementary stereochemistry to that of E_2 eliminations. Early examples of *syn* processes include ester pyrolyses, as well as Chugaev eliminations of xanthates and Cope eliminations of amine oxides. Unfortunately, these reactions typically require high temperatures that make them unsuitable for many applications. The related sulfoxide elimination has also been employed in alkene synthesis, but it is similarly hampered by the requirement for elevated temperatures. In contrast, selenoxide *syn* eliminations (for selected reviews, see [2–5]; for seminal papers, see [6–10]) often proceed readily at room temperature and are typically highly regio- and stereoselective. Moreover, the introduction of the required selenium residue into an organic substrate can be easily achieved via either electrophilic, nucleophilic or radical processes, thus further enhancing the popularity of this useful reaction.

The widely employed alkyl selenoxide elimination has been studied previously by computational methods and is included here for comparison with the heteroatom analogues that are the focus of the present investigation. For example, early work by Kwart et al. [11] demonstrated that tunnelling plays an important role in lowering the activation energies of these processes compared to those of other *syn* eliminations. A computational investigation of the regioselectivity of selenoxide eliminations by Fujimoto et al. [12] indicated that oxygen and nitrogen substituents resulted in the preferential formation of allylic ether, alcohol and amine products, while delocalizing groups such as cyano substituents afforded vinylic products exclusively. They also reported that the transition states were asynchronous, with earlier hydrogen transfer compared with retarded C-Se bond cleavage. Schiesser and coworkers [13] performed computations that revealed that substituents on α-arylseleno ketones had little effect on elimination rates and that the oxidation of selenium was generally rate-determining in the overall process. Selenoxide eliminations of selenocysteine derivatives were modelled by Bayse and Allison [14], who reported that increased chain lengths in RSeCys (vs. R = Me or Ph) decreased activation energies, as did *ortho* Lewis base substituents (when R = aryl) that stabilized the transition state by electron donation to the positive selenium centre. More recently, Orian and coworkers [15] reported a rigorous computational study, including a comparison with analogous sulfoxides, telluroxides and eliminations via their higher oxidation states. These researchers found that the activation energies decreased in the order S > Se > Te in chalcogenoxide eliminations and were higher in the selenonyl analogues compared to their selenoxides. The facile hydration of telluroxides impeded their ability to eliminate. Furthermore, activation energies were correlated with the basicities of the chalcogenoxide oxygen atoms and the chalcogen–carbon bond strengths. Their research included studies of biologically relevant selenocysteine derivatives.

In contrast with the well-known and synthetically useful selenoxide *syn* elimination for the preparation of alkenes, to our knowledge, there has been no systematic computational study of related processes where either or both of the two alkene-generating carbon atoms are replaced by heteroatoms (X and Y in **1** in Scheme 1). Several such reactions, in addition to the alkene-forming process of selenoxide **2**, are illustrated in Scheme 2 for selenoxide analogues **3–12**, which are of interest for a variety of reasons. For example, during recent studies of selenium-catalyzed epoxidations of alkenes, it was discovered that peroxyseleninic acids **3** decompose to the corresponding selenonium selenonate salts **13**, with the concomitant evolution of dioxygen [16] (Scheme 3, path A). The vigorous liberation of oxygen was also reported during the preparation of selenonic acids **15** in the presence of hydrogen peroxide [17] (Scheme 3, path B). The salt **13** (R = Ph), containing mixed oxidation states of selenium, had been previously misidentified as the selenonic acid **14** [18], but its structure was unequivocally established by X-ray crystallography [16]. It was therefore of interest to ascertain whether or not selenoxide eliminations of peroxy acids **3** or their selenonyl analogues **16** could be responsible for the observed formation of oxygen, by analogy with the formation of alkenes from alkyl selenoxides **2**. The oxidations of alcohols to ketones with benzeneseleninic anhydride and its congeners have also been reported, most likely via the elimination of seleninate ester intermediates **4** [19–21].

1

Scheme 1. Hetero-selenoxide *syn* eliminations.

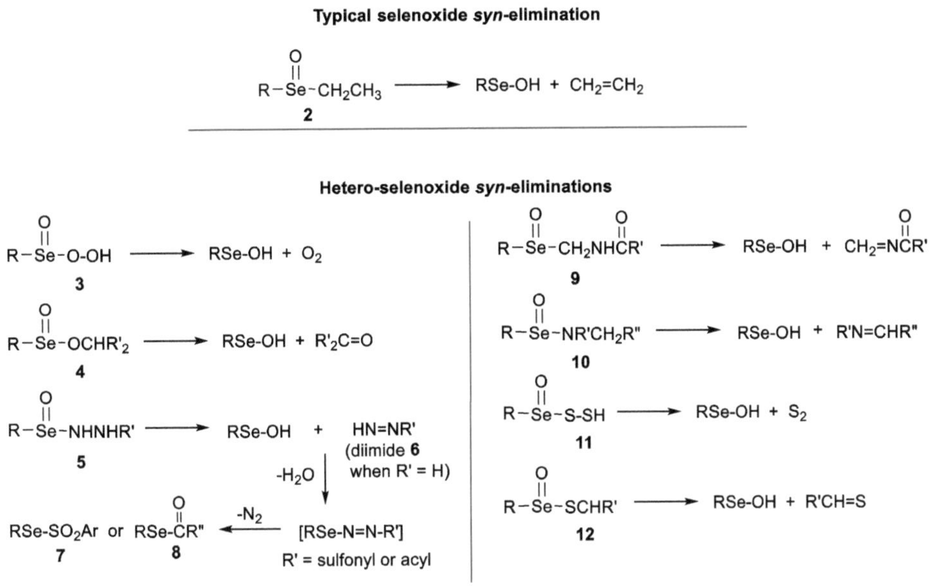

Scheme 2. Examples of known and potential hetero-selenoxide *syn* eliminations.

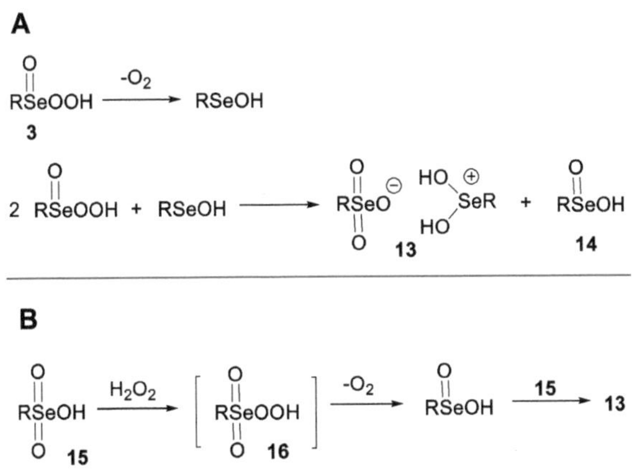

Scheme 3. (**A**) Formation of dioxygen by elimination of seleninic peroxy acids. (**B**) Formation of dioxygen during the reaction of selenonic acids with hydrogen peroxide.

Similarly, the oxidation of hydrazines with seleninic acids or anhydrides presumably involves the formation of *N*-(seleninyl)hydrazines **5**, which afford diimide **6** when R′ = H [22] (Scheme 2). Alternatively, in substituted hydrazines, further reaction of the intermediate diazenes with the byproduct selenenic acid, followed by the loss of dinitrogen, affords selenosulfonates **7** [23] or selenoesters **8** [24]. The conversion of amines and amides to the corresponding imines via selenoxides **9** [25] or by the treatment of amines with benzeneseleninic anhydride and related oxidants via postulated seleninamides **10** has also been reported [26–28], as was the similar oxidation of primary amines to nitriles [26].

Furthermore, the generation of diatomic sulfur (S_2) from silyl or germanyl trisulfides was reported by Steliou, Gareau and Harpp et al. [29], who trapped it by its Diels–Alder cycloaddition with dienes. Although the multiplicity of S_2 was not established, the authors noted that the ground state of this species is a triplet, while the energy of the corresponding singlet state is 13 kcal mol^{-1} higher. The possibility that diatomic sulfur could be produced from seleninyl hydrodisulfides **11** by selenoxide elimination remains to be explored. Finally, the similar oxidation and elimination of thioseleninates **12** (Scheme 2) was first reported by Reich and Jasperse [30], and later by Glass et al. [31] and by us [32] as a step in the mechanism by which the antioxidant drug ebselen catalyzes the reduction of peroxides with sacrificial thiols, accompanied by the formation of the corresponding thioaldehyde. In each case in Scheme 2, the unstable byproduct selenenic acid (RSeOH) undergoes disproportionation, reduction, oxidation or reaction with nucleophiles, depending on the specific conditions. Since the mechanisms reported for most of the reactions shown in Scheme 2 are speculative, a computational study of these postulated hetero-selenoxide eliminations appeared warranted. The corresponding selenonyl analogues of selenoxides **1** and their possible role in such processes have been much less studied and so were included in this investigation. Thus, in order to gain insight into the viability of such reactions and their kinetic and thermodynamic properties, we now report geometry optimization and transition-state computations of a variety of hetero-*syn* eliminations of species of general structure **1** and their Se(VI) analogues in both the gaseous state and in the presence of solvents.

2. Results

In principle, if X and Y in selenoxide structure **1** are confined to C, O, N and S, then sixteen structures are possible, along with sixteen others based on the corresponding selenonyl analogues that can undergo known or hypothetical *syn* elimination processes. Table 1 includes the computed values of activation energies (ΔE‡) and overall energy changes (ΔE) for sixteen of these processes that correspond to synthetically useful or theoretically interesting transformations, as summarized in Scheme 2. For consistency, ease of comparison and simplicity of computation, all examples are based on methyseleno derivatives.

Table 1. Electronic energies in hetero-selenoxide *syn* eliminations [a,b].

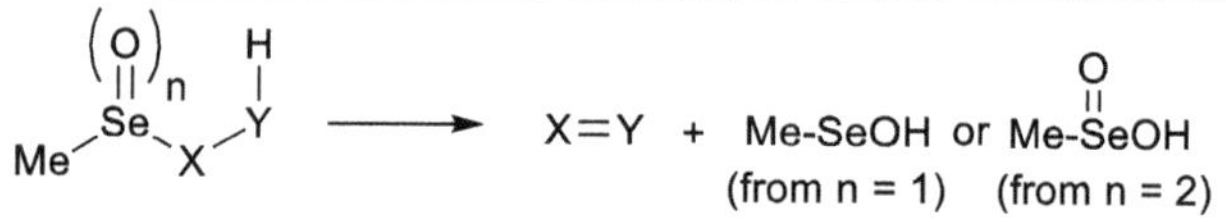

Entry	n	X	Y	ΔE‡ Gas Phase	ΔEGas Phase	ΔE‡ H_2O	ΔE H_2O	ΔE‡ CH_2Cl_2	ΔE CH_2Cl_2	ΔE‡ MeOH	ΔE MeOH
1	1	CH_2	CH_2	21.4	4.9	24.3	7.4	23.8	7.0	24.2	7.3
2	2			30.7	−8.5	31.4	−6.0	31.3	−6.4	31.4	−6.1
3 [c,d]	1	O	O	36.4	4.9	32.5	7.5	33.2	7.1	32.6	7.4
4 [c]	2			29.1	−13.7	28.7	−12.9	28.8	−12.9	28.7	−12.9
5	1	O	CH_2	26.6	−1.6	28.1	−2.5	27.9	−2.3	28.1	−2.4
6	2			42.5	−16.7	42.7	−19.2	42.7	−18.7	42.7	−19.1
7 [e]	1	NH	NH	10.1	5.1	12.3	6.0	11.9	5.8	12.2	6.0
8 [e]	2			13.9	−11.4	9.9	−11.9	11.5	−11.8	10.1	−11.8
9	1	CH_2	NH	8.2	3.8	10.1	5.0	9.8	4.8	10.1	5.0
10	2			9.6	−10.7	5.0	−9.2	6.2	−9.4	5.3	−9.2
11	1	NH	CH_2	23.8	0.6	26.4	0.6	26.0	0.6	26.3	0.6
12	2			37.0	−13.8	37.8	−15.3	37.7	−15.0	37.8	−15.2
13 [f]	1	S	S	4.3	1.1	3.7	−1.6	3.8	−1.1	3.7	−1.5
14	2			13.1	−14.5	12.3	−13.9	12.5	−14.0	12.4	−13.9
15	1	S	CH_2	18.3	17.1	19.7	17.0	19.4	17.0	19.6	17.0
16	2			27.0	−2.4	28.3	−3.2	28.1	−3.1	28.3	−3.2

Legend for Table 1. [a] ΔE‡ = Activation energy. [b] ΔE = Overall change in energy between the starting material and products in the reaction. [c] Based on the formation of triplet O_2. [d] These energies may be overestimated, as rotation of the methyl group was frozen to avoid additional imaginary frequencies. [e] Based on the formation of *trans*-diimide (**6**). [f] Based on the formation of zwitterion **17**.

With respect to computations in the gaseous state, entries 1 and 2 in Table 1 correspond to an alkyl selenoxide and selenone elimination, respectively, and are included for comparison. These computations indicate that the selenoxide reacts slightly endothermically, while the selenone reacts exothermically, but with a higher activation energy. The analogous eliminations of the peroxyseleninic and peroxyselenonic acids are shown in entries 3 and 4. The mechanism for dioxygen generation in the above processes was unclear, but the possibility of a selenoxide *syn* elimination was adumbrated [16]. The salts **13** were the products of protonation of the amphoteric seleninic acids **14** by the stronger selenonic acids **15**, which were in turn formed by the redox equilibration of selenium species of different oxidation state (Scheme 3). Furthermore, while the peroxyseleninic acid in entry 3 of Table 1 reacts with a relatively high activation energy in an endothermic process, the corresponding peroxyselenonic acid in entry 4 has a similar activation energy (29.1 kcal mol^{-1}) to that of alkene formation in entry 2 (30.7 kcal mol^{-1}) and is exothermic overall. The lower activation energy for peroxyselenonic acid (29.1 vs 36.4 kcal mol^{-1}) and its more exothermic reaction energy (−13.7 vs. 4.9 kcal mol^{-1}) compared to its seleninic counterpart indicate that the process in entry 4 is more facile than that in entry 3. These results cast doubt on the facile liberation of dioxygen from the peroxyseleninic acids **3** at or below room temperature, as shown in Scheme 3A, via a concerted *syn* elimination and suggest that further oxidation to the selenonic species **15** and **16** may be required for this process. The more facile *syn* elimination of oxygen from postulated peroxyacids **16**, produced from the oxidation of selenonic acids **15** with hydrogen peroxide, would again generate salts **13**, as shown in Scheme 3B [33].

The oxidations of alcohols with benzeneseleninic anhydride $[PhSe(=O)]_2O$ likely produce intermediate seleninate esters **4** (Scheme 2) that in turn fragment to afford aldehydes or ketones. The results shown in entry 5 of Table 1 reveal a slightly exothermic process with an activation energy of 26.6 kcal mol^{-1} for the transformation of the methyl seleninate to formaldehyde, indicating that a *syn* elimination is a viable mechanism for this type of oxidation. Interestingly, the reaction of the corresponding selenonate ester (entry 6) is more exothermic, but it is kinetically disfavoured with a higher activation energy of 42.5 kcal mol^{-1}.

Hydrazines are readily oxidized by seleninic acids or anhydrides via the formation of postulated seleninyl hydrazine intermediates **5**, as shown in Scheme 2, followed by their spontaneous *syn* elimination. Thus, the unsubstituted hydrazine (R′ = H) generated diimide **6**, which was confirmed by the in situ transfer hydrogenation of cinnamic acid to hydrocinnamic acid and azobenzene to hydrazobenzene [22]. When substituted hydrazines were employed, the resulting azenes reacted further with the selenenic acid byproduct, followed by the extrusion of dinitrogen to afford selenosulfonates **7** or selenoesters **8**, as shown in Scheme 2. The selenoxide elimination of intermediate **5** is characterized by a lower activation energy and moderate reaction energy (10.1 and 5.1 kcal mol^{-1}, respectively; Table 1, entry 7) for the formation of *trans*-diimide, while higher values of 11.4 and 7.0 kcal mol^{-1} were obtained for the formation of *cis*-diimide. The elimination of the selenonyl analogue of **5** was again characterized by a higher activation energy and a significantly more exothermic reaction (entry 8).

Syn eliminations are also plausible mechanisms for the conversion of α-aminoalkyl selenoxides to imines, as shown in Table 1, entry 9 (X = CH_2, Y = NH), and in the transposed seleninamide in entry 11 (X = NH, Y = CH_2). Like the hydrazine in entry 7, these systems also comprise slightly endothermic processes. However, the seleninamide reaction in entry 11 possesses a considerably higher activation energy than that in entry 9 (23.8 vs. 8.2 kcal mol^{-1}), attributed in part to the greater acidity of the amino moiety compared to that of the methyl group in the proton transfer step to the basic selenoxide oxygen atom. Again, the corresponding selenone systems in entries 10 and 12 display higher activation energies and more exothermic reactions than their selenoxide counterparts.

The reactions of seleninyl and selenonyl hydrodisulfides are summarized in Table 1, entries 13 and 14, respectively. The seleninyl compound revealed anomalous behaviour, where computations indicated an activation energy of 4.3 kcal mol^{-1}, which is actually lower than that of the sum of the presumed fragmentation products (MeSeOH and S_2) [34,35] (Scheme 4). However, IRC calculations revealed that proton transfer from the hydrodisulfide moiety to the selenoxide oxygen atom was more advanced than cleavage of the Se-S bond, suggesting the possible formation of zwitterion **17** instead of elimination products, as shown in Scheme 4. Indeed, geometry optimization of **17** indicated that it has an energy 3.2 kcal mol^{-1} lower than that of the transition state and is only 1.1 kcal mol^{-1} higher than the starting hydrodisulfide (Figure 1A). This can be attributed to the considerably higher acidity of the S-S-H moiety (pKa 6–7 [36]) compared to the β-hydrogens (C-H, O-H or N-H) of the other starting materials in Table 1, which facilitates proton transfer. Further evidence for the formation of **17** stems from a comparison of key interatomic distances in the transition state and products (Table 2; vide infra). Thus, it appears that in the case of the seleninyl hydrodisulfide, the proton transfer leads to a discrete zwitterion intermediate instead of to a merely asynchronous concerted elimination. This was confirmed by the corresponding IRC computation.

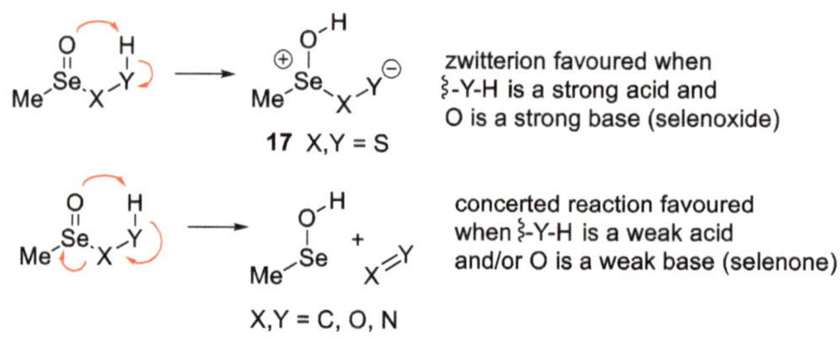

Scheme 4. Zwitterion formation and concerted elimination reactions of hetero-selenoxides.

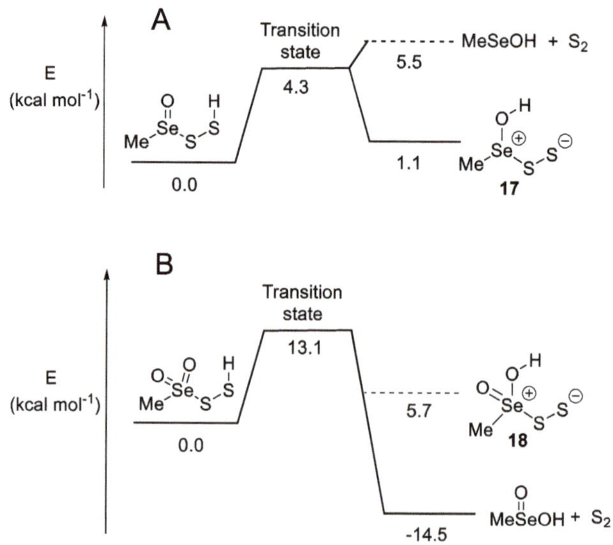

Figure 1. Activation and reaction energies (kcal mol^{-1}} for seleninyl (**A**) and selenonyl (**B**) hydrodisulfides in the gaseous state.

Table 2. Key interatomic distances (Å) in starting materials and transition states in the gaseous state [a].

				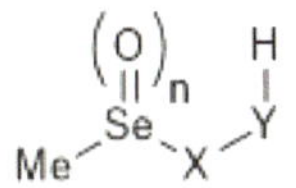					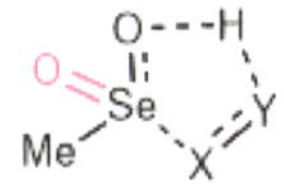		
		Starting Material					**Transition State**				
Entry	***n*, X, Y**	**X-Se**	**X-Y**	**H-OSe**	**H-Y**	**Se=O**	**X-Se**	**X-Y**	**H-OSe**	**H-Y**	**Se=O**
1	1, C, C	2.005	1.521	2.492	1.093	1.651	2.481	1.415	1.278	1.334	1.707
2	2, C, C	1.966	1.524	2.995 [b]	1.091	1.618, 1.618	2.513	1.415	1.384	1.278	1.689, 1.629
3	1, O, O	1.858	1.443	2.332	0.975	1.630	1.824	1.426	1.180	1.298	1.999
4	2, O, O	1.831	1.446	2.744	0.971	1.605, 1.600	2.055	1.337	1.103	1.376	1.707, 1.615
5	1, O, CH2	1.818	1.435	2.420	1.093	1.629	2.283	1.308	1.310	1.300	1.685
6	2, O, CH2	1.775	1.446	2.950 [b]	1.092	1.607, 1.607	2.237	1.317	1.370	1.278	1.621, 1.682
7	1, NH, NH	1.940	1.418	2.190	1.027	1.647	2.499	1.314	1.334	1.180	1.708
8	2, NH, NH	1.883	1.414	2.425 [c]	1.014	1.613, 1.617	2.581	1.292	1.700	1.059	1.677, 1.633
9	1, CH2, NH	2.074	1.421	2.214	1.023	1.660	2.530	1.335	1.341	1.181	1.715
10	2, CH2, NH	2.029	1.420	3.006	1.014	1.622, 1.622	2.641	1.320	1.632	1.072	1.682, 1.634
11	1, NH, CH2	1.885	1.475	2.554	1.096	1.642	2.370	1.370	1.310	1.303	1.693
12	2, NH, CH2	1.840	1.474	2.870 [d]	1.093	1.613, 1.612	2.370	1.371	1.372	1.281	1.684, 1.628
13	1, S, S	2.371	2.086	3.142	1.352	1.636	2.338	2.094	1.328	1.555	1.684
14	2, S, S	2.324	2.086	3.262	1.352	1.612, 1.612	2.290	2.105	1.243	1.634	1.677, 1.607
15	1, S, CH2	2.330	1.830	2.190	1.096	1.642	2.631	1.699	1.182	1.417	1.710
16	2, S, CH2	2.279	1.837	2.356	1.090	1.617, 1.610	2.576	1.705	1.247	1.375	1.697, 1.620

Legend for Table 2. [a] When the hydrogen source contains several hydrogen atoms, the H-Y distance is listed for the closest hydrogen to the Se=O oxygen. [b] The two respective Y-H---O=Se distances were equal. [c] The H-bonded (Y)NH---O=Se distance is indicated; the shortest (X)NH---O=Se distance was 2.683 Å. [d] The (X)NH---O=Se distance was 2.613 Å.

In contrast, the selenonyl hydrodisulfide provided a calculated exothermic reaction energy of −14.5 kcal mol^{-1} and a higher activation energy of 13.1 kcal mol^{-1}, in contrast to the seleninyl derivative (Table 1, entry 14). Moreover, the energy of the corresponding zwitterion product **18** was 20.2 kcal mol^{-1} higher in energy than that of its *syn* elimination products (Figure 1B). Since the selenonyl hydrodisulfide is expected to be even more acidic than the corresponding selenoxide hydrodisulfide, we attribute the difference in behaviour, at least in part, to the lower basicity of the selenone oxygen atom [37,38] and the weaker Se-S bond when the selenium atom is in a higher oxidation state. In entry 14, the S_2 product was assumed to be in the triplet ground state [35], but even if the less stable singlet state was formed, the products would still be lower in energy than the zwitterion. It therefore appears that in this case, *syn* elimination would be the preferred pathway.

Finally, the thioseleninate and thioselenonate in Table 1, entries 15 and 16, displayed activation energies of 18.3 and 27.0 kcal mol^{-1}, respectively, which are slightly lower than those in entries 1 and 2. The former compound also provided the most endothermic elimination in Table 1, with a reaction energy of 17.1 kcal mol^{-1}. This can be attributed to the instability of both the thioformaldehyde and selenenic acid products.

The results shown in Table 1 also include solvent effects from water, dichloromethane and methanol. In the case of the conventional selenoxide elimination ($n = 1$, X,Y = CH_2) and its selenone counterpart (entries 1 and 2), both the activation and overall energies increased in the presence of the solvents, indicating that solvation stabilized the starting materials more strongly than the corresponding transition states and products. The same was noted for entries 7 and 9. The solvents lowered the activation energies in entries 3, 4, 8, 10, 13 and 14, relative to their gaseous states, and raised them in the remaining entries 5, 6, 11, 12, 15 and 16. The overall changes in energy were rendered more exothermic or were insignificant in the presence of the solvents in the latter entries.

The overall results of the gas-state computations are summarized in Figures 2 and 3, illustrating the reaction energies (ΔE) and activation energies ($\Delta E^{\ddagger}$) for the processes in Table 1. It is striking that all of the selenonyl eliminations are exothermic, while their selenoxide counterparts are endothermic, except in the case of the methyl seleninate ($n = 1$; X = O, Y = CH_2). On the other hand, activation energies are higher for selenones compared to selenoxides, except for the peroxides (X,Y = O). Thus, in general, the concerted eliminations of selenoxides are kinetically favoured over the corresponding selenone eliminations, while the opposite is true for their overall thermodynamic outcomes.

Key bond lengths and interatomic distances in starting materials and transition states are provided in Table 2. As expected, the transition states typically show elongation of the X-Se and proton-accepting Se=O bonds, contraction of the X-Y bond and the partial transfer of H from Y to O. However, in entries 3, 13 and 14 of Table 2, the X-Se bond is actually slightly shorter in the transition state than in the starting material, while the S-S bond in entries 13 and 14 is slightly longer. Furthermore, the H---O=Se separation shows a dramatic decrease from 3.142 Å to 1.328 Å for the selenoxide in entry 13 and from 3.262 Å to 1.243 Å for the selenone in entry 14. Hydrogen migration in the transition state is indicated by the relative H---O=Se and H---Y distances and is most advanced (closer to O=Se than to Y) in entries 1, 3, 4 and 13–16. An asynchronous concerted reaction, as noted earlier for the alkene-forming selenoxide elimination in entry 1 [12], is also postulated for the selenonyl hydrodisulfide; however, the formation of a discrete zwitterion intermediate is ruled out on the basis of its thermodynamic instability.

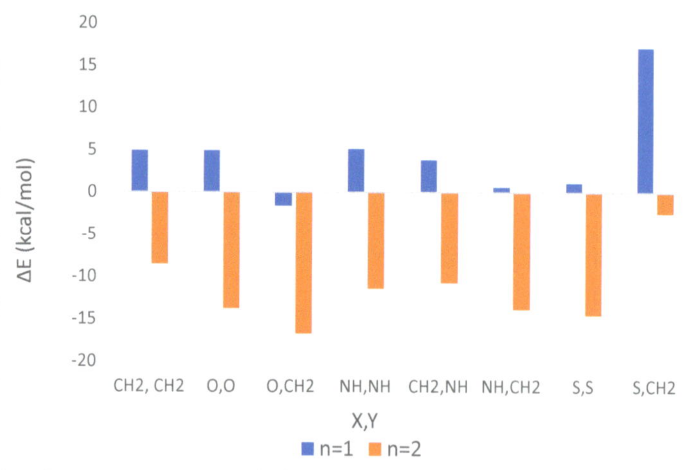

Figure 2. Reaction energies ΔE for heteroatom selenoxide and selenone *syn* eliminations in the gaseous state. All results represent concerted *syn* eliminations, except for the seleninyl hydrodisulfide ($n = 1$, X,Y = S), where the product is zwitterion **17**.

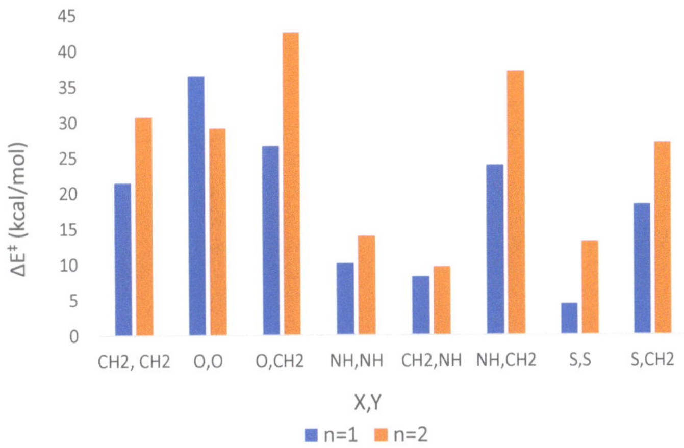

Figure 3. Activation energies $\Delta E^{\ddagger}$ for heteroatom selenoxide and selenone *syn* eliminations in the gaseous state. All results represent concerted *syn* eliminations, except for the seleninyl hydrodisulfide (n = 1, X,Y = S), where the product is zwitterion **17**.

The thermochemical values of ΔG, ΔH and ΔS for the conversion of starting materials to transition states and final products are provided in Table S1 on p. 17 of the Supporting Information. All reactions were exothermic, except for the formation of zwitterion **17** in entry 13 and the elimination of the thioseleninate in entry 15, which were slightly endothermic. Electron densities in transition states were calculated with Mulliken electron charges by summing and comparing the charges for the XY and seleninic or selenonic acid components. The results are shown in Table S2 on p. 17 of the Supporting Information, which revealed that electron density [39] resides preferentially on the XY component of the transition states, except in the *syn* elimination of ethyl methyl selenoxide in entry 1, where both components were essentially neutral.

3. Methods

Computations were performed using the DFT B3LYP platform in Gaussian 2016 [40]. A mixed basis set was employed, using 6-311G(d,p) for C, H, N, O and S, and the cc-pVTZ triple zeta variation of Dunning's correlation-consistent basis set for Se [41]. Transition-state energies were computed by means of the Berny algorithm. All transition states except for the hydroperoxide in Table 1, entry 3, resulted in single imaginary frequencies involving hydrogen migration and partial cleavage of the X-Se bond, while geometry optimizations of products and starting materials showed no imaginary frequencies. IRC computations were performed on all transition states (forward and backward pathways), resulting in agreement with the original geometry optimizations, except in the case of the hydroperoxide in entry 3 [42]. Enthalpies of the reactions were calculated as the difference in the sums of the electronic and thermal enthalpies of the products and reactants. Gibbs free energies of the reactions were calculated similarly, using the sums of electronic and thermal free energies. Computations were performed on the Advanced Research Computing (ARC) computer cluster at the University of Calgary.

4. Conclusions and Summary

Figures 2 and 3 clearly show that in all cases except for the peroxy species (entries 3 and 4 in Table 1), the selenoxide eliminations in the gas state produced significantly lower activation energies than their selenonyl analogues. Furthermore, all of the selenoxide reactions, except for the selenate ester in entry 5 in Table 1, were endothermic, while all of the selenone eliminations were exothermic. These observations are parallel to those made by Orian et al. [15] in the *syn* eliminations of oxidized selenocysteine and related compounds. The asynchronous nature of the alkyl selenoxide eliminations that was noted previously by Fujimoto et al. [12] applies similarly to the heteroatom-substituted systems in the present work, where proton transfer from Y to O in the transition states advances more rapidly than X-Se cleavage. In the extreme case of the seleninyl hydrodisulfide in entry 13 of Table 1, facile proton transfer produces a discrete intermediate in the form of zwitterion **17**, instead of the concerted fragmentation to MeSeOH and S_2. Solvent effects varied considerably in the examples in Table 1, increasing activation energies in some instances and decreasing them in others, attributed to the preferential solvation of starting materials and transition states, respectively. Enthalpies, entropies and free energies were also obtained, and ΔG values indicated exothermic reactions for all examples except entries 13 and 15, which were slightly endothermic. Most of the selenoxide eliminations in Table 1 have been little studied, except for entry 1, while their selenonyl counterparts remain virtually unexplored. In a more general sense, these results demonstrate that several extensions of the classical olefin-forming selenoxide elimination to systems containing heteroatoms instead of simple alkyl substituents are kinetically and thermodynamically viable, except where acidic hydrogens are transferred to strongly polarized Se=O bonds, resulting in the initial formation of discrete zwitterion intermediates.

Supplementary Materials: The following supporting information can be downloaded at https://www.mdpi.com/article/10.3390/molecules29204915/s1: Energies, Z-coordinates, numbers of imaginary frequencies and depictions of optimized structures of starting materials (pp. 2–8), transition states (pp. 9–15) and zwitterion products (p. 15), along with overall free energies and free energies of activation (Thermochemistry, Table S1, p. 17); electron densities in transition states (Table S2, p. 18).

Author Contributions: Conceptualization, A.I.D., J.T.S. and T.G.B.; Methodology, A.I.D., J.T.S. and T.G.B.; Formal Analysis, A.I.D., J.T.S. and T.G.B.; Investigation, A.I.D., J.T.S. and T.G.B.; Validation, A.I.D., J.T.S. and T.G.B.; Writing—Original Draft Preparation, T.G.B.; Writing—Review and Editing, A.I.D., J.T.S. and T.G.B.; Supervision, T.G.B.; Project Administration, T.G.B.; Funding Acquisition, T.G.B. All authors have read and agreed to the published version of the manuscript.

Funding: The authors thank the Natural Sciences and Engineering Research Council of Canada (NSERC) for financial support: Grant RGPIN-2019-04373 to T.G.B.

Institutional Review Board Statement: Not applicable.

Informed Consent Statement: Not applicable.

Data Availability Statement: The original contributions presented in the study are included in the article and Supplementary Materials. Further inquiries can be directed to the corresponding author.

Acknowledgments: J.T.S. thanks the University of Calgary and the Province of Alberta for an Alberta Graduate Excellence Scholarship.

Conflicts of Interest: The authors declare no conflicts of interest.

References and Notes

1. In principle, concerted reactions may be synchronous or asynchronous. Both types proceed without the formation of intermediates, but bonding changes in asynchronous processes do not occur simultaneously. For example, see Tantillo, D.J. Recent Excursions to the Lands between Concerted and Stepwise: From Natural Products Biosynthesis to Reaction Design. *J. Phys. Org. Chem.* **2008**, *21*, 561–570. To avoid these distinctions regarding the timing of events, some authors prefer to designate such reactions as "one-step" instead of "concerted".
2. Reich, H.J. Functional Group Manipulation Using Organoselenium Reagents. *Acc. Chem. Res.* **1979**, *12*, 22. [CrossRef]
3. Reich, H.J.; Wollowitz, S. Preparation of α,β-Unsaturated Carbonyl Compounds and Nitriles by Selenoxide Elimination. *Org. React.* **1993**, *44*, 1–296.
4. Back, T.G. Selenoxide Eliminations. In *Organoselenium Chemistry—A Practical Approach*; Chapter 2; Back, T.G., Ed.; Oxford University Press: Oxford, UK, 1999.
5. Nishibayashi, Y.; Uemura, S. Selenoxide Elimination and [2,3] Sigmatropic Rearrangement. In *Organoselenium Chemistry—Synthesis and Reactions*; Chapter 7; Wirth, T., Ed.; Wiley—VCH: Weinheim, Germany, 2012.

6. Walter, R.; Roy, J. Selenomethionine, a potential catalytic antioxidant in biological systems. *J. Org. Chem.* **1971**, *36*, 2561–2563. [CrossRef]
7. Jones, D.N.; Mundy, D.; Whitehouse, R.D. Steroidal Selenoxides Diastereoisomeric at Selenium; Syn-Elimination, Absolute Configuration, and Optical Rotatory Dispersion Characteristics. *J. Chem. Soc. D Chem. Commun.* **1970**, *2*, 86–87. [CrossRef]
8. Sharpless, K.B.; Lauer, R.F.; Teranishi, A.Y. Electrophilic and Nucleophilic Organoselenium Reagents. New Routes to α,β-Unsaturated Carbonyl Compounds. *J. Am. Chem. Soc.* **1973**, *95*, 6137–6139. [CrossRef]
9. Reich, H.J.; Reich, I.L.; Renga, J.M. Organoselenium Chemistry. Alpha.-Phenylseleno Carbonyl Compounds as Precursors for α,β-Unsaturated Ketones and Esters. *J. Am. Chem. Soc.* **1973**, *95*, 5813–5815. [CrossRef]
10. Clive, D.L.J. Fragmentation of selenoxides. New method for dehydrogenation of ketones. *J. Chem. Soc. Chem. Commun.* **1973**, *18*, 695–696. [CrossRef]
11. Kwart, L.D.; Horgan, A.G.; Kwart, H. Structure of the Reaction Barrier in the Selenoxide-Mediated Formation of Olefins. *J. Am. Chem. Soc.* **1981**, *103*, 1232–1234. [CrossRef]
12. Kondo, N.; Fueno, H.; Fujimoto, H.; Makino, M.; Nakaoka, H.; Aoki, I.; Uemura, S. Theoretical and Experimental Studies of Regioselectivity in Selenoxide Eliminations. *J. Org. Chem.* **1994**, *59*, 5254–5263. [CrossRef]
13. Macdougall, P.E.; Smith, N.A.; Schiesser, C.H. Substituent Effects in Selenoxide Elimination Chemistry. *Tetrahedron* **2008**, *64*, 2824–2831. [CrossRef]
14. Bayse, C.A.; Allison, B.D. Activation Energies of Selenoxide Elimination from Se-Substituted Selenocysteine. *Mol. Model.* **2007**, *13*, 47–53. [CrossRef] [PubMed]
15. Madabeni, A.; Zucchelli, S.; Nogara, P.A.; Rocha, J.B.T.; Orian, L. In the Chalcogenoxide Elimination Panorama: Systematic Insight into a Key Reaction. *J. Org. Chem.* **2022**, *87*, 11766–11775. [CrossRef] [PubMed]
16. Sands, K.N.; Mendoza Rengifo, E.; George, G.N.; Pickering, I.J.; Gelfand, B.S.; Back, T.G. The Unexpected Role of Se(VI) Species in Epoxidations with Benzeneseleninic Acid and Hydrogen Peroxide. *Angew. Chem. Int. Ed.* **2020**, *59*, 4283–4287. [CrossRef] [PubMed]
17. Sands, K.N.; Gelfand, B.S.; Back, T.G. One-Pot Synthesis of Aryl Selenonic Acids and Some Unexpected Byproducts. *J. Org. Chem.* **2021**, *86*, 9938–9944. [CrossRef]
18. Syper, L.; Młochowski, J. Benzeneperoxyseleninic Acids—Synthesis and Properties. *Tetrahedron* **1987**, *43*, 207–213. [CrossRef]
19. Barton, D.H.R.; Brewster, A.G.; Hui, R.A.H.F.; Lester, D.J.; Ley, S.V.; Back, T.G. Oxidation of Alcohols using Benzeneseleninic Anhydride. *J. Chem. Soc. Chem. Commun.* **1978**, *21*, 952–954. [CrossRef]
20. Back, T.G. Oxidation of Azasteroid Lactams and Alcohols with Benzeneseleninic Anhydride. *J. Org. Chem.* **1981**, *46*, 1442–1446. [CrossRef]
21. Kuwajima, I.; Shimizu, M.; Urabe, H. Oxidation of Alcohols with tert-Butyl Hydroperoxide and Diaryl Diselenide. *J. Org. Chem.* **1982**, *47*, 837–842. [CrossRef]
22. Back, T.G. Dehydrogenation of Hydrazines and of 4-Azacholestan-3-one with Benzeneseleninic Acid and Benzeneseleninic Anhydride. *J. Chem. Soc. Chem. Commun.* **1978**, *6*, 278–279. [CrossRef]
23. Back, T.G.; Collins, S.; Krishna, M.V. Reactions of Sulfonhydrazides with Benzeneseleninic Acid, Selenium Halides and Sulfur Halides. A Convenient Preparation of Selenosulfonates and Thiosulfonates. *Can. J. Chem.* **1987**, *65*, 38–42. [CrossRef]
24. Back, T.G.; Collins, S.; Kerr, R.G. Oxidation of Hydrazines with Benzeneseleninic Acid and Anhydride. *J. Org. Chem.* **1981**, *46*, 1564–1570. [CrossRef]
25. Branchaud, B.P.; Tsai, P. Relative Ease of Transient Imine Formation via Selenoxide, Sulfoxide and Sulfone β N-H Elimination. A Feasibility Study on the Preparation of Novel Peptide Analogues. *J. Org. Chem.* **1987**, *52*, 5475–5478. [CrossRef]
26. Czarny, M.R. Oxidation of amines with diphenylseleninic anhydride. *J. Chem. Soc. Chem. Commun.* **1976**, *3*, 81. [CrossRef]
27. Czarny, M.R. Oxidation of amines with benzeneseleninyl chloride. *Synth. Commun.* **1976**, *6*, 285–293. [CrossRef]
28. Barton, D.H.R.; Lusinchi, X.; Milliet, P. Studies on the reaction of primary and secondary amines with phenylseleninic anhydride and with phenylseleninic acid. *Tetrahedron* **1985**, *41*, 4727–4738. [CrossRef]
29. Steliou, K.; Gareau, Y.; Harpp, D.N. Molecular Sulfur (S2): Generation and Synthetic Application. *J. Am. Chem. Soc.* **1984**, *106*, 799–801. [CrossRef]
30. Reich, H.J.; Jasperse, C.P. Organoselenium Chemistry. Redox Chemistry of Selenocysteine Model Systems. *J. Am. Chem. Soc.* **1987**, *109*, 5549–5551. [CrossRef]
31. Glass, R.S.; Farooqui, F.; Sabahi, M.; Ehler, K.W. Formation of Thiocarbonyl Compounds in the Reaction of Ebselen Oxide with Thiols. *J. Org. Chem.* **1989**, *54*, 1092–1097. [CrossRef]
32. Sands, K.N.; Burman, A.L.; Ansah-Asamoah, E.; Back, T.G. Chemistry Related to the Catalytic Cycle of the Antioxidant Ebselen. *Molecules* **2023**, *28*, 3732. [CrossRef]
33. These results are based on the assumption that triplet oxygen is formed; singlet oxygen is reported to be 23 kcal mol^{-1} higher in energy, based on spectroscopic data. Lowry, T.H.; Schueller Richardson, K. *Mechanism and Theory on Organic Chemistry*, 2nd ed.; Harper and Row: New York, NY, USA, 1981; p. 960.
34. The sum of the calculated energies of MeSeOH and S_2 would be 1.2 kcal mol^{-1} higher than that of the transition state, if the triplet form $^3S_2$ was produced, while formation of the singlet $^1S_2$ would further increase the energy of products by another 13 kcal mol^{-1}

35. Steliou et al. indicated that the singlet sulfur species $^1S_2$ is 13 kcal. mol^{-1} higher in energy than the triplet ground state, but the singlet might be formed by their method because of spin conservation rules. Steliou, K.; Gareau, Y.; Harpp, D.N. Molecular Sulfur (S2): Generation and Synthetic Application. *J. Am. Chem. Soc.* **1984**, *106*, 799–801. [CrossRef]
36. Bailey, T.S.; Zakharov, L.N.; Pluth, M.D. Understanding Hydrogen Sulfide Storage: Probing Conditions for Sulfide Release from Hydrodisulfides. *J. Am. Chem. Soc.* **2014**, *136*, 10573–10576. [CrossRef] [PubMed]
37. Paetzold, R.; Bochmann, G. Selenium compounds. XLVI. Aliphatic selenium oxides and selenones. *Z. Anorg. Allg. Chem.* **1968**, *360*, 293–299.
38. Agenäs, L.-B. *Organic Selenium Compounds: Their Chemistry and Biology*; Klayman, D.L., Günther, W.H.H., Eds.; Wiley: New York, NY, USA, 1973; Chapter 5; p. 210.
39. For a lead reference on electron density transfer, see: Domingo, L.R.; Ríos-Gutiérez, M.; Pérez, P. How does the global electron density transfer diminish activation energies in polar cycloaddition reactions? A Molecular Electron Density Theory study. *Tetrahedron* **2017**, *73*, 1718–1724. [CrossRef]
40. Gaussian 16, Revision B.01, Frisch, M.J.; Trucks, G.W.; Schlegel, H.B.; Scuseria, G.E.; Robb, M.A.; Cheeseman, J.R.; Scalmani, G.; Barone, V.; Petersson, G.A.; et al. Gaussian, Inc.: Wallingford, CT, USA, 2016.
41. Kendall, R.A.; Dunning, T.H., Jr.; Harrison, R.J. Electron Affinities of the First-Row Atoms Revisited. Systematic Basis Sets and Wave Functions. *J. Chem. Phys.* **1992**, *96*, 6796–6806. [CrossRef]
42. The forward IRC computation for the hydroperoxide in entry 3 did not proceed to the fragmentation products, but instead indicated formation of the corresponding zwitterion. Nevertheless, fragmentation appears to be more likely since the optimized zwitterion energy was 29.4 kcal mol^{-1} higher than that of the sum of the fragmentation products (triplet O2 and MeSeOH). This discrepancy, along with the need to freeze the methyl group in the transition state calculation to avoid additional imaginary frequencies, suggest that these results should be regarded with caution.

molecules MDPI

Review

An Examination of Chemical Tools for Hydrogen Selenide Donation and Detection

Rynne A. Hankins and John C. Lukesh *

Department of Chemistry, Wake Forest University, Wake Downtown Campus, 455 Vine Street, Winston-Salem, NC 27101, USA
* Correspondence: lukeshjc@wfu.edu

Abstract: Hydrogen selenide (H_2Se) is an emerging biomolecule of interest with similar properties to that of other gaseous signaling molecules (i.e., gasotransmitters that include nitric oxide, carbon monoxide, and hydrogen sulfide). H_2Se is enzymatically generated in humans where it serves as a key metabolic intermediate in the production of selenoproteins and other selenium-containing biomolecules. However, beyond its participation in biosynthetic pathways, its involvement in cellular signaling or other biological mechanisms remains unclear. To uncover its true biological significance, H_2Se-specific chemical tools capable of functioning under physiological conditions are required but lacking in comparison to those that exist for other gasotransmitters. Recently, researchers have begun to fill this unmet need by developing new H_2Se-releasing compounds, along with pioneering methods for selenide detection and quantification. In combination, the chemical tools highlighted in this review have the potential to spark groundbreaking explorations into the chemical biology of H_2Se, which may lead to its branding as the fourth official gasotransmitter.

Keywords: gasotransmitters; signaling; donors; sensors; hydrogen selenide; H_2Se

check for updates

Citation: Hankins, R.A.; Lukesh, J.C. An Examination of Chemical Tools for Hydrogen Selenide Donation and Detection. *Molecules* **2024**, *29*, 3863. https://doi.org/10.3390/molecules29163863

Academic Editor: Ming Wang

Received: 26 July 2024
Revised: 12 August 2024
Accepted: 13 August 2024
Published: 15 August 2024

1. Introduction

The discovery of elemental selenium can be traced back to a Swedish sulfuric acid plant in the early 19th Century where chemist Jöns Jakob Berzelius first observed a reddish-brown sediment in the acid being produced there [1]. Berzelius initially mistook this new substance for tellurium due to its odor and appearance. Famed for his advancement of modern chemical notation, the principle of stoichiometry, and the determination of atomic weights of most known elements at the time, Berzelius chemically compared the red-brown byproduct with a known sample of tellurium and determined that the two were, in fact, different elements with the new substance having the properties of a metal combined with that of sulfur. In the words of Berzelius, he had discovered "a new kind of sulfur".

Today, selenium is regarded as an essential micronutrient that is acquired through dietary means with an optimal daily dose of 55 μg for adults [2]. There are at least 25 selenoproteins [3–5], many of which play a central role in cellular redox homeostasis, including glutathione peroxidases (GPx) and thioredoxin reductases (TrxR), and require dietary selenium for their production. Low selenium levels in humans are associated with a myriad of illnesses [6–9]. Keshan Disease [10,11], a potentially fatal form of cardiomyopathy, is primarily observed in selenium-deficient regions in China. The same is true for Kashin–Beck Disease [12,13], a chronic joint disease predominantly found in parts of the world where selenium is scarce. Weakened immune function [14,15], cardiovascular diseases [16,17], and certain cancers [18–21] are also strongly associated with selenium deficiency. These correlative studies suggest that the role of selenium in human health and biology could extend beyond its incorporation into selenium-containing proteins and point towards its possible involvement in other cellular processes.

In terms of its chemistry, selenium, like sulfur, can exist in numerous oxidation states, including selenate (SeO_4^{2-}, +6), selenite (SeO_3^{2-}, +4), and selenide (Se^{2-}, −2) [22]. Selenide, the most reduced form of selenium, is the heavier chalcogen counterpart to sulfide (S^{2-}). Both sulfide and selenide exist in different protonation states, dependent on environmental pH, with the fully protonated forms of both—hydrogen sulfide (H_2S, pK_{a1}: 6.9) [23] and hydrogen selenide (H_2Se, pK_{a1}: 3.9) [24]—being formerly dismissed as highly toxic gases with little biological relevance [25,26]. However, in the case of hydrogen sulfide, this malodorous gas has recently experienced a rebirth as an important, endogenous signaling molecule (or gasotransmitter) in mammals [27–30].

H_2S is primarily produced by three principal enzymes—cystathionine β-synthase (CBS) [31], cystathionine γ-lyase (CSE) [32], and 3-mercaptopyruvate sulfur transferase (3-MST) [33]—providing exquisite control over its production. As such, endogenous H_2S is known to be involved in numerous signaling processes throughout the body, including the brain and central nervous system [34,35], and within specific cellular compartments (i.e., mitochondria) [36,37]. Much of what is known about H_2S pharmacology stems from the advent of donor compounds, synthetic small molecules designed to slowly liberate H_2S in a controlled fashion that mimics its natural biosynthesis, and the use of said compounds in various cellular and animal models of disease [38–41].

Alongside nitric oxide (NO) and carbon monoxide (CO), hydrogen sulfide is the most recent, widely recognized member of the gasotransmitter family [42–44]. Its inclusion was suggested in the early 2000s and was based on five key observations [45]. Namely, H_2S is (i) a small molecule gas that (ii) freely permeates cellular membranes. It is (iii) endogenously and enzymatically generated, with (iv) well-defined biological functions that stem from its (v) action at specific cellular targets.

It is interesting to note that H_2Se (Predominantly HSe^- at physiological pH) already checks several of these boxes. It exists as a lipophilic gas in its diprotic form and is expressed enzymatically in mammals where it serves as a key intermediate in the production of selenium-containing biomolecules (Figure 1) [46]. H_2Se was shown to act on at least some protein targets, including inhibition of cytochrome c oxidase, which modulates aerobic respiration [46,47]. Moreover, it is the reduced selenide form of selenium, and not the often administrated selenite (or other oxidized forms), that is believed to be responsible for the observed biological activity of selenium, including its anticancer effects and noted protection against myocardial ischemia reperfusion injury [48–50].

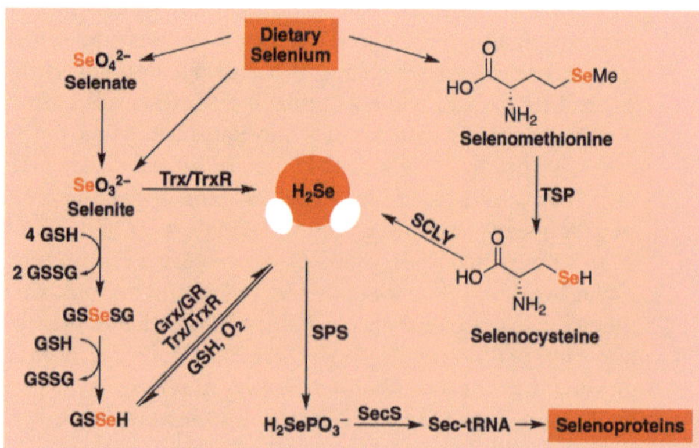

Figure 1. A simplified schematic of H_2Se production in mammalian systems. Glutathione (GSH), glutathione disulfide (GSSG), glutaredoxin (Grx), glutathione reductase (GR), thioredoxin (Trx), thioredoxin reductase (TrxR), transsulfuration pathway (TSP), selenocysteine lyase (SCLY), selenophosphate synthetase (SPS), and selenocysteine synthase (SecS).

Still, compared to H_2S, very little is known about the (patho)physiological effects of H_2Se. This is due, at least in part, to a lack of refined donor compounds that can increase the bioavailability of selenide and be used to effectively modulate cellular concentrations. We and others have begun to fill this unmet need by developing new H_2Se-releasing compounds, which we will highlight below, along with current methods for selenide detection and quantification. In combination, the chemical tools outlined in this review have the potential to serve as invaluable exploratory tools for uncovering H_2Se biology and its potential in medicine.

2. Chemical Tools for H_2Se Donation

2.1. Selenotrisulfides

A key intermediate in the biosynthesis of H_2Se is believed to be glutathione selenotrisulfide (GSSeSG), which forms in vitro from the reduction of selenite with four equivalents of glutathione (GSH) [51–53]. While the isolation of GSSeSG is difficult due to its chemical instability, the reduction of selenite is not specific to GSH, meaning the preparation and evaluation of alternative selenotrisulfides (RSSeSR) as potential H_2Se-donating motifs could be a viable option (Figure 2A).

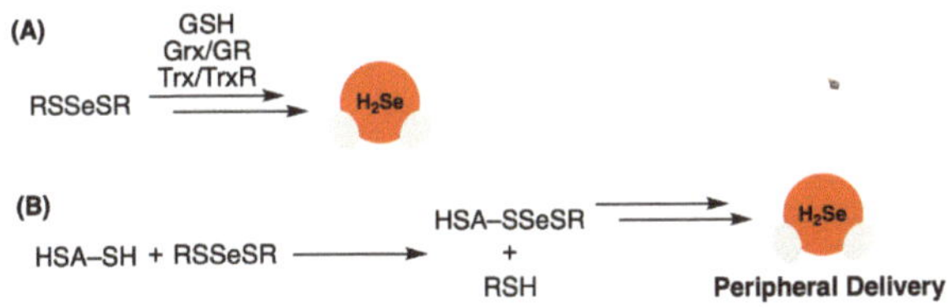

Figure 2. (**A**) Selenotrisulfides as H_2Se donors. (**B**) Peripheral delivery of H_2Se via human serum albumin. Glutathione (GSH), glutaredoxin (Grx), glutathione reductase (GR), thioredoxin (Trx), thioredoxin reductase (TrxR), and human serum albumin (HSA).

In a pioneering study by Nakayama and co-workers, the more robust penicillamine selenotrisulfide (PenSSeSPen) was prepared, which is isolable due to increased steric bulk near the S–Se–S motif, and its bioavailability was compared with that of selenite, an established dietary source of selenium, in Se-deficient mice [54]. Following oral administration, the selenium content in selected organs was quantified fluorometrically using 2,3-diaminonaphthalene. Similar to selenite-treated mice, the administration of PenSSeSPen led to a significant increase in selenium in the heart, liver and blood. Additionally, PenSSeSPen-fed mice exhibited similar GPx activity to that of selenite-fed mice, indicating that the selenium content from PenSSeSPen was available for selenoprotein production.

In a later study by Nakayama, it was proposed that human serum albumin (HSA), the most abundant plasma protein, serves as a selenium carrier via a selenotrisulfide linkage that enables it to distribute selenide to peripheral tissues and organs throughout the body [55]. The authors observed that treatment of red blood cells (RBCs) with selenite led to selenium efflux that was dependent on HSA concentration. Moreover, pretreatment of HSA with iodoacetamide, a thiol-blocking agent, appeared to inhibit selenium transfer from RBCs to HSA, confirming that the thiol functional group on HSA played a key role in the transfer event. When selenium-bound HSA was treated with penicillamine (Pen), selenotrisulfide PenSSeSPen was produced. Additionally, the same PenSSeSPen product was formed when the selenium efflux experiment was conducted in the presence of Pen rather than HSA. These observations led to the conclusion that selenium is likely to be exported from RBCs as a selenotrisulfide (RSSeSR), which forms from the reaction between selenite and an RBC thiol. The ensuing selenotrisulfide then binds to HSA via a thiol exchange reaction, enabling the transport of selenium throughout the body (Figure 2B).

In a very recent study by Wang, Xu, Xie, and co-workers, it was reported that a stable selenotrisulfide (AcidSSeSAcid) could be formed by treating 2-mercaptoacetic acid with selenium dioxide [56]. However, when the same reaction was run with 2-mercaptoethanol, the resultant selenotrisulfide (HydSSeSHyd) behaved more like GSSeSG and proved difficult to characterize. Thus, only the reactivity of the more robust AcidSSeSAcid was assessed further in the presence of glutathione. Using ESI-MS, the byproducts of this reaction were found to be GSSG, AcidSH, AcidSSeSG, AcidSSG, and AcidSSeH, providing some indirect evidence of glutathione-promoted H_2Se release. At the outset of this study, the authors hypothesized that H_2Se might function as an "H_2S disguiser" and thereby assist in overcoming H_2S-indcued antibiotic resistance. This theory was tested with AcidSSeSAcid in an H_2S-induced antibiotic-resistant MRSA model ($MRSA^{S+}$). While the antibiotic gentamicin alone proved ineffective against $MRSA^{S+}$ it, in combination with AcidSSeSAcid, displayed impressive bactericidal activity. The authors attributed this reduction in antibiotic resistance to the release of H_2Se, which increases bacterial membrane permeability and reactivates bacterial respiratory flux.

Although the direct liberation of H_2Se was not confirmed in any of these studies with trapping experiments, they do underscore the biological relevance of the RSSeSR motif and its likely ability to function as an endogenous selenide delivery agent. The general instability of this framework, however, may limit its overall utility as an exogenous source of H_2Se, as the ability to generate a large library of selenotrisulfides with tunable rates of release could prove difficult. Thus, the search continues for alternative frameworks with the potential to supply selenide in a controlled and sustained fashion.

2.2. Selenide Salts

Logically, many H_2Se donors were inspired by previously reported H_2S-releasing compounds. To this end, selenide salts, which serve as a convenient H_2Se equivalent in buffer, were examined for convenience [47,50,57], drawing parallels to sulfide salts being used in initial studies aimed at exploring H_2S chemical biology [58].

Using sodium hydroselenide (NaHSe), the Dyson group was among the first to explore in detail the pharmacology and therapeutic utility of H_2Se [47]. Given its instability and absence of a reliable commercial source, the group elected to generate NaHSe in-house by reducing elemental selenium with an aqueous solution of sodium borohydride [59]. Once in hand, they evaluated the metabolic effects of NaHSe ex vivo using dissected rat soleus muscle and homogenized liver tissue. In these models, NaHSe was shown to inhibit oxygen consumption in a concentration-dependent manner, albeit to a lesser extent than sodium hydrosulfide (NaHS) and potassium cyanide (KCN), which were used as positive controls. The authors also investigated the mechanism of inhibition of O_2 consumption and found its inhibition of cytochrome C oxidase to be a likely candidate, similar to NaHS.

The influence of NaHSe on selenoprotein expression in HepG2 (human hepatocyte) cells was also inspected [47]. Using a Western blot analysis, a significant increase in the production of glutathione peroxidase-1 (GPx-1) was observed upon treatment with NaHSe. Interestingly, the addition of selenite (SeO_3^{2-}), a common dietary source of selenium, had the opposite effect with a notable reduction in GPx-1 expression compared to nontreated controls. Moreover, the Dyson group also demonstrated that the addition of DL-propargylglycine (PAG), an established CSE inhibitor, led to significant reductions in both GPx-1 and thioredoxin reductase-1 (TrxR), presumably due to its inhibition of the H_2Se producing enzyme selenocysteine lyase (SCLY, Figure 1). However, the expression of both proteins was restored by the addition of exogenous NaHSe, suggesting that endogenous selenoprotein expression can be regulated by H_2Se administration.

The therapeutic value of H_2Se supplementation was also examined in HepG2 cells exposed to hydrogen peroxide (H_2O_2) insult [47]. Indeed, incubation with NaHSe (0.3 and 1 μM) for 1 h conferred cellular protection in a dose-dependent manner. The authors noted that multiple mechanisms could be in play to account for the improved ROS management by NaHSe-treated cells, including its direct ROS scavenging, its functioning as a metabolic modulator, and/or its serving as the catalytic component of antioxidant selenoproteins.

Finally, the in vivo pharmacological effects of NaHSe were also investigated in this study [47]. Anesthetized rats were given an escalated dose of 0.01 mg/kg to 10 mg/kg of selenide. Both blood pressure and heart rate decreased notably at the highest dose level but, overall, cardiac output remained relatively unaffected. The authors also noted significant hyperlactatemia (inhibition of oxidative phosphorylation) at the highest dose of NaHSe.

To date, this remains one of the most comprehensive studies, detailing the chemical biology and pharmacological effects of H_2Se [47]. Still, a severe limitation, even noted by the authors, was the use of NaHSe as an H_2Se source. As cited above, the use of selenide salts is analogous to early work on H_2S which employed sulfide salts as a convenient method for H_2S delivery. Their use, however, creates a bolus effect that poorly mimics the endogenic production of H_2S. The same is likely true for the use of selenide salts as a research tool for examining the pharmacology and medicinal value of H_2Se. Thus, refined selenium-containing compounds with suitable pharmacokinetics and exquisite control over their selenide release are highly desirable and early attempts to access such compounds are summarized below.

2.3. Selenoanhydrides

Selenoanhydrides were among the first small molecules assessed for their ability to deliver selenide in a controlled fashion and in response to biologically pertinent molecules [60]. Domínguez-Álvarez and co-workers reported on the impressive anticancer effects of selenoanhydride R-Se (Figure 3A) in earlier studies and suspected it might be due to its release of H_2Se [61–63]. To examine this, they monitored the fragmentation pattern of R-Se in a 50% methanol/water solution. Under electrospray ionization conditions (ESI), initial attack by methanol led to the observed fragmentation products with the loss of H_2Se (Figure 3B). Moreover, the addition of Na_2S appeared to amplify the decomposition of the donor with selenium-containing fragments being observed but with a significantly lower abundance. While experimental conditions were not biologically relevant, this simple ESI-MS analysis did provide some insight into the propensity of selenoanhydrides to expel selenide upon nucleophilic exposure.

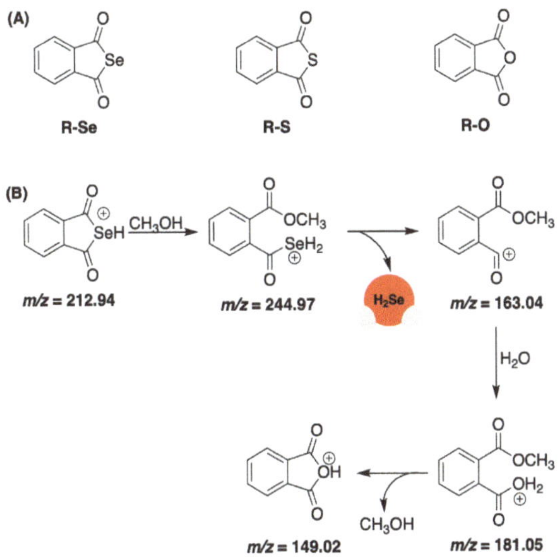

Figure 3. (**A**) Chalcogen anhydrides (R-Se, R-S, and R-O) under examination; (**B**) Proposed fragmentation pattern for R-Se in 50% methanol/water under ESI conditions.

The reducing capacity of R-Se and related chalcogens (R-S and R-O, Figure 3A) was assessed using cPTIO, a nitric oxide radical scavenger [60]. The authors noted that the addition of H_2S potentiated the radical scavenging ability of R-Se and R-S (albeit to a lesser extent) but not R-O. The same trend was observed with the antioxidant glutathione (GSH). While the addition of GSH alone did not effectively quench cPTIO, it in combination with R-Se was found to significantly reduce cPTIO radicals. Similar results were obtained with superoxide ($O_2^{\bullet-}$). Using BMPO as an EPR spin trap reagent, the authors found that H_2S/R-Se and H_2S/R-S (but not H_2S/R-O) could effectively scavenge BMPO-OOH/OH adducts and, perhaps, $O_2^{\bullet-}$ directly.

In total, these observations indicate that the reducing potency of thiols (i.e., GSH and H_2S) are significantly boosted upon their interaction with R-Se (and to a lesser extent R-S), which is likely to liberate reactive selenium (or sulfur) species, including hydrogen selenide. Although these studies never provided any direct evidence of H_2Se release from R-Se, the radical scavenging activity of this purported donor provides a strong indication.

2.4. *P=Se Motifs*

With inspiration from GYY4137, an early and extensively studied H_2S donor that gradually decomposes to release H_2S via P=S bond cleavage [64], Pluth and co-workers developed an analogous H_2Se donor, TDN1042, that delivered selenide (H_2S/H_2Se^-) via a similar hydrolytic pathway (Figure 4) [65]. Synthetically, TD1042 was accessed upon treatment of Woollins' reagent with excess morpholine, akin to the preparation of GYY4137.

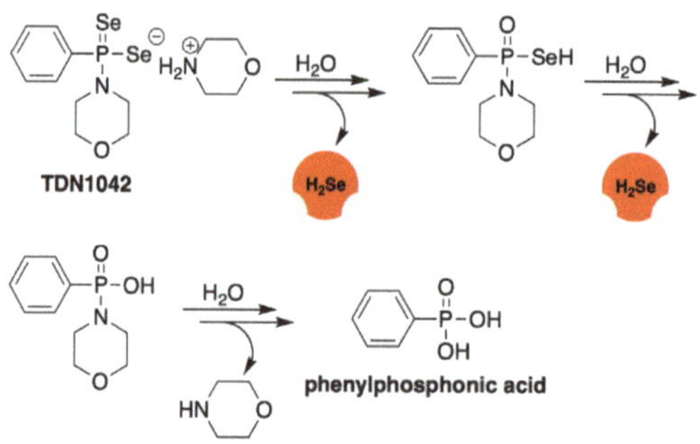

Figure 4. Proposed hydrolytic pathway of TDN1042, which results in the release of two equivalents of H_2Se and the formation of phenylphosphonic acid.

Once prepared and structurally characterized by NMR and single crystal X-ray diffraction, H_2Se release from TDN1042 in wet DMSO was evaluated using ^{31}P NMR spectroscopy. These studies revealed the clean conversation of TD1042 to phenylphosphonic acid, as expected (Figure 4). This clean transformation to phenylphosphonic acid was also observed in citrate buffer (pH 3.0 to 6.0), with higher rates of release occurring at more acidic pH values, which is consistent with a hydrolysis-based mechanism.

The authors also confirmed the direct liberation of H_2Se from TD1042, a key experiment that was omitted from previous reports of H_2Se donating motifs. To accomplish this, an aqueous solution of TD1042 was acidified with HCl and sparged with N_2 to assist in volatilizing any released H_2Se into the vial headspace where it was then bubbled through a separate trapping solution of dinitrofluorobenzene (DNFB). Using HPLC, Pluth and co-workers observed the formation of both di(2,4-dinitrophenyl) selenide ($(DNP)_2Se$) and the related diselenide ($(DNP)_2Se_2$) in the trapping solution.

In a later study, the Pluth group sought to augment the rate of H_2Se donation from this platform through the introduction of cyclic-PSe donors [66]. To accomplish this, Woollins' reagent was treated with various *ortho*-substituted phenols to generate a small library of donors with a single P=Se motif (Cat-PSe, 2AP-PSe, and NMe2AP-PSe, Figure 5A).

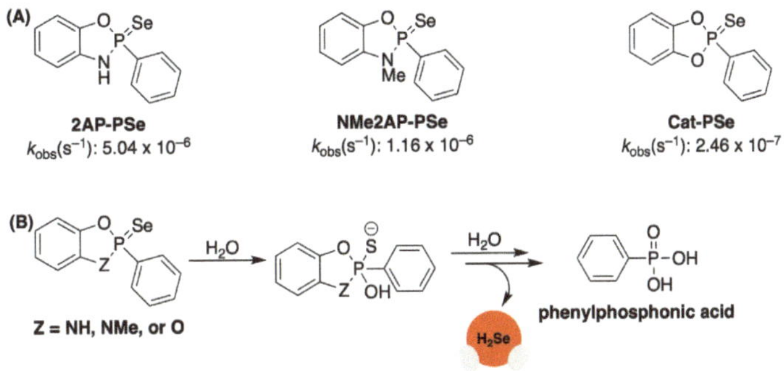

Figure 5. (**A**) Cyclic-PSe donors arranged in order of decreasing rates of hydrolysis; (**B**) Proposed hydrolytic pathway of cyclic-PSe donors, which results in the release of one equivalent of H_2Se and the formation of phenylphosphonic acid.

By introducing a second electronegative heteroatom (i.e., oxygen or nitrogen), it was suspected that the rate of hydrolysis would increase due to the enhanced electrophilic character of the phosphorous center. This was confirmed by ^{31}P NMR, which was used to monitor donor hydrolysis in PIPES buffer (pH 7.4). Like TD1042, cyclic-PSe compounds were found to release selenide while cleanly forming phenylphosphonic acid as a byproduct (Figure 5B). However, unlike TD1042, these donors were shown to operate at neutral pH due to their enhanced hydrolytic lability. The rates of hydrolysis among cyclic-PSe donors varied significantly with Cat-PSe displaying the slowest rate of hydrolysis and 2AP-PSe hydrolyzing the quickest. This result was somewhat unexpected and suggests that factors other than electronic effects influence the rates of hydrolysis.

Cell permeability studies were also conducted using time-of-flight secondary ion mass spectrometry (TOF-SIMS). A dose-dependent increase in intracellular selenium levels was observed in HeLa cells with increasing concentrations of 2AP-PSe, the most efficient H_2Se donor identified in this study.

The in cellulo antioxidant activity of 2AP-PSe was also assessed. When live HeLa cells were treated with exogenous H_2O_2 (500 μM) and peroxide sensor DCFH-DA, a significant fluorescence response was observed due to the formation of 2′,7′-dichlorofluorescein (DCF) [67]. However, when cells were first pre-treated with 2AP-PSe (5–25 μM) a notable decrease in ROS-generated fluorescence was observed. Due to the instability of H_2Se, the authors stress that the observed effects are unlikely due to the buildup of H_2Se, but rather an increase in selenocompounds (likely antioxidant proteins) that effectively scavenge ROS.

2.5. Selenocarbonyls

In a comprehensive study by Yi and co-workers, both selenocyclopropenones and arylselenoamides were found to provide highly tunable rates of H_2Se delivery in the presence of supraphysiological concentrations of cysteine [68].

Initially, the authors generated a selenium analogue of Michler's ketone but found that it quickly hydrolyzed in a buffer, producing a red residue (Se^0), presumably due to its rapid discharge of selenide (Figure 6A). Searching for a selenocarbonyl with a more tractable H_2Se-releasing profile, selenocyclopropenones were then investigated. A small library was generated by treating the corresponding ketone with Woollins' reagent (Figure 6B). Compound 1 was chosen for initial studies, evaluating its reactivity and selectivity as a selenide donor. HPLC analysis confirmed that 1 was stable in a 50% PBS/CH_3CN (pH 7.4) mixture. The introduction of cysteine (2–10 mM), however, led to the expulsion of selenide and the subsequent formation of a red solid (Se^0). This was further corroborated by an H_2Se-selective gas detector. Donors 2 and 3, with electron-withdrawing substituents at the *para* position, were found to liberate even more H_2Se than 1 in the presence of cysteine, implying that the rates of donation from this platform can be easily tuned via simple structural modifications that alter the electrophilicity of the selenocarbonyl. The reaction between 1 and cysteine was further scrutinized by high-resolution mass spectrometry (HRMS) and identifiable byproducts were uncovered. Based on these observations, a mechanism for cys/thiol-triggered H_2Se release from selenocyclopropenones was put forward by the authors (Figure 6B).

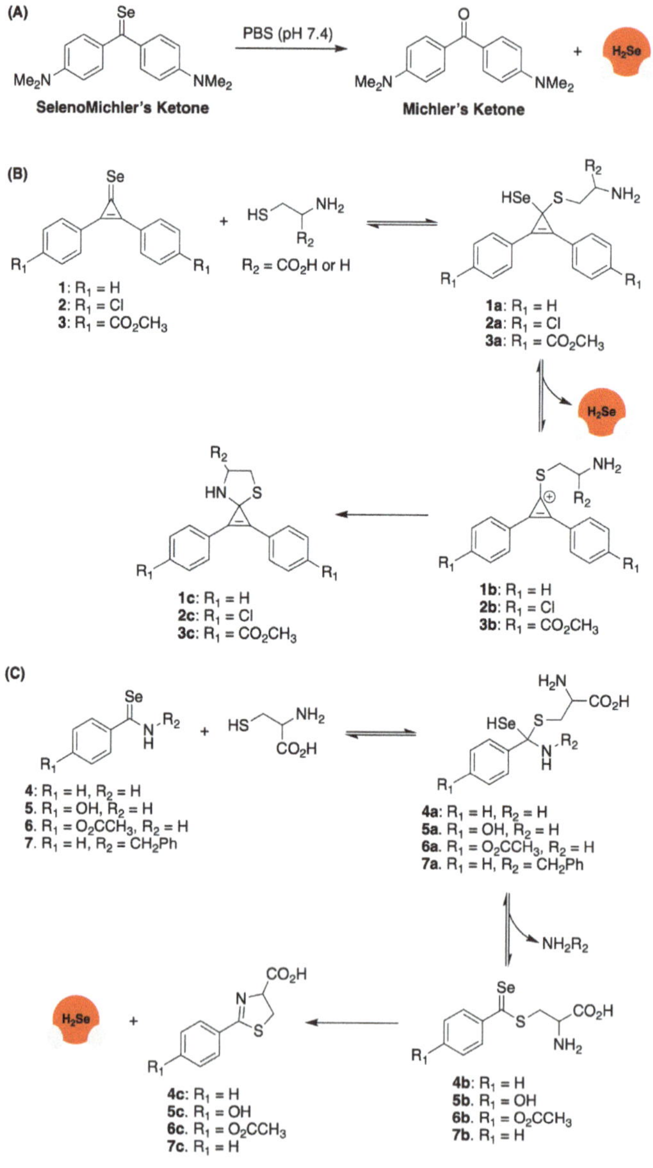

Figure 6. (**A**) Rapid hydrolysis of SelenoMichler's ketone generates H_2Se in PBS (pH 7.4). (**B**) Proposed mechanism for cysteine/thiol-triggered H_2Se donation from selenocyclopropenones. (**C**) Proposed mechanism for cysteine-triggered H_2Se donation from arylselenoamides.

Aryl thioamides were previously shown to function as an advantageous platform for the controlled delivery of H_2S under biologically relevant conditions [69–71]. Yi and co-workers suspected that selenoanalogues would provide an avenue for H_2Se donation under similar conditions (Figure 6C) [68]. To test their hypothesis, a small library of selenoamides was generated by treating benzamide derivatives with Woollins' reagent or by subjecting 4-hydroxybenzonitrile to a mixture of selenium powder and $NaBH_4$. Selenobenzamide (4) was shown to be stable in PBS (pH 7.4), but the addition of cysteine led to the formation of red Se^0. This was further substantiated with an H_2Se gas detector. As expected, the rate of H_2Se release from 4 was found to be significantly slower than 1. However, it was noted that the selenoamide platform could be structurally modified to alter reaction kinetics. For example, the introduction of an electron-donating hydroxyl group at the *para* position on the phenyl ring (5, Figure 6C) appeared to amplify the speed of selenide release. However, when the hydroxyl group was converted to a methyl ester (6, Figure 6C), its donating efficiency diminished somewhat, even relative to 4. On the other hand, amide *N*-alkylation (7, Figure 6C) was shown to suppress the rate of selenide delivery even further. HPLC analysis and DFT calculations were used to establish a mechanism for cysteine-triggered H_2Se donation (Figure 6C). This proposed pathway aligns with the observation that treatment of arylselenoamides with other thiols that lack a nucleophilic amine (i.e., *N*-acetylcysteine, glutathione, and β-mercaptoethanol) fail to generate H_2Se.

2.6. Selenocarbamates

Caged thiocarbamates have offered a reliable avenue for H_2S delivery (Figure 7A). They also provide an opportunity to selectivity tune the release of H_2S to a specific biological trigger (i.e., ROS [72–76], light [77,78], pH [79,80], and enzymes [81,82]). However, the production of H_2S from this donor class is multi-layered, with the triggering event causing the donor to undergo a self-immolating process that first generates carbonyl sulfide (COS) prior to its quick conversation to H_2S by the ubiquitous enzyme carbonic anhydrase (Figure 7A) [83].

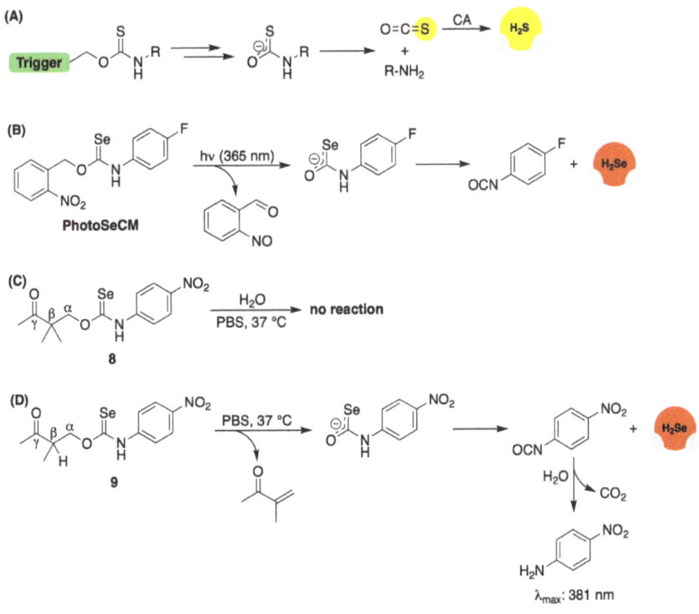

Figure 7. (**A**) General strategy for generating stimuli-responsive COS/H_2S donors. (**B**) Proposed mechanism for the direct release of H_2Se from PhotoSeCM upon irradiation at 365 nm. (**C**) Control compound used to highlight the stability of γ-ketoselenocarbamates in water. (**D**) Proposed mechanism for the direct release of H_2Se from γ-ketoselenocarbamates containing a deprotonatable hydrogen at the β position.

Continuing with the common theme of prior H_2S-releasing motifs laying the foundation for H_2Se donor development, Pluth and co-workers were curious whether this framework could be reengineered to release COSe, providing a new means for the controlled delivery of selenide [84]. To test their hypothesis, the authors treated *p*-fluorophenyl isoselenocyanate with 2-nitrobenzyl alcohol in the presence of NaH to generate light-activated PhotoSeCM (Figure 7B). To analyze product formation upon photoactivation, the authors used ^{19}F NMR to streamline the process. Using this method, consumption of PhotoSeCM (−117 ppm) was confirmed but the expected *p*-fluoroaniline product (due to COSe release) at −130 ppm was not observed. Instead, a signal matching *p*-fluoroisocyanate was detected, suggesting the direct liberation of selenide rather than COSe serving as an intermediary (Figure 7B).

Interested in assessing a second COSe/H_2Se-releasing system but without the requirement of photoactivation to simplify the overall mechanism, Pluth and team also examined γ-ketoselenocarbamates, which were reconstituted from their earlier work on γ-ketothiocarbamates [80]. This H_2S donor class undergoes an enol-mediated self-immolating process that delivers COS/H_2S alongside *p*-nitroaniline, which can be used to track reaction progress due to its UV–Vis signature at 381 nm. γ-Ketoselenocarbamates were constructed from *p*-nitrophenyl isoselenocyanate and their activation at different pH values was monitored by UV–Vis. While compound 8, which lacks deprotonatable hydrogens at the β position, failed to generate *p*-nitroaniline (Figure 7C), compound 9, as expected, showed a pH-dependent rate of release, increasing at higher pH values (Figure 7D). In buffer, the authors were unable to provide direct experimental evidence of selenide release from this donor system as the ensuing *p*-nitrophenyl isocyanate byproduct is unstable and likely to hydrolyze quickly to *p*-nitroaniline and CO_2. Nevertheless, computational studies corroborate the likelihood of direct selenide liberation as it was calculated to be by far the lowest energy decomposition pathway, underscoring the fundamental differences in thiocarbamate/selenocarbamate reactivity.

2.7. γ-Ketoselenides

An analogous base-mediated selenide delivery system was first reported by us in 2022 [85]. In this study, we prepared a library of γ-ketoselenides (10–13, Figure 8A) upon treatment of the corresponding γ-ketohalide/tosylate with a solution of sodium selenide that was generated in situ by reducing elemental selenium with $NaBH_4$.

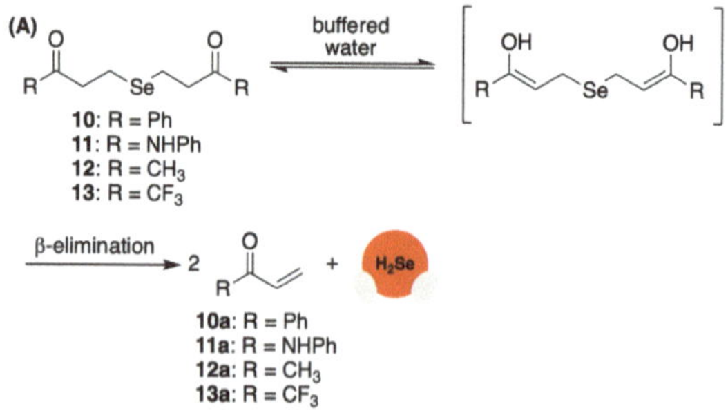

(B)

IC$_{50}$ (μM)

Cell Line	10	11	12	13	12EC
HeLa	3.7	>50	6.2	10.6	>50
HCT116	4.1	>50	5.1	10.1	>50

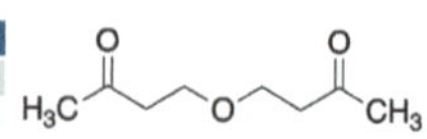

12EC

Figure 8. (**A**) A library of γ-ketoselenides that undergo base-promoted α-deprotonation/β-elimination to release H_2Se. (**B**) Cell growth inhibition of HeLa and HCT116 cells in culture. IC$_{50}$ values were determined after a 24 h incubation period with donor.

Once in hand, we first monitored the release of selenide from 10 (Figure 8A) in a 1:1 mixture of CD_3CN and deuterated phosphate buffer (50 mM, pD 7.4) by tracking the formation of byproduct 10a (Figure 8A) using ^1H NMR. During the experiment, we observed a decrease in the intensity of the protons alpha to selenium while the signal of the terminal enone protons simultaneously increased. The addition of 1,4-dioxane as an internal standard allowed us to track the concentration of both and confirm that the consumption of 10 correlated well with the production of 10a. During these studies, we also observed the formation of a red film in our NMR tube, which we attributed to the rapid oxidation of H_2Se in an aqueous buffer and its conversion to Se^0. This was confirmed by ^{31}P NMR and the formation of triphenylphosphine selenide upon treatment with triphenylphosphine.

Appearing to proceed through an α-deprotonation/β-elimination sequence, we suspected that an enhanced rate of H_2Se donation from γ-ketoselenides would be observed at both higher pH values and with donors that possess more acidic alpha protons. To test our hypotheses, the same kinetic studies were repeated at pH 6 and 8.5. As anticipated, the rate of selenide donation was found to be intensified under more basic conditions (pH 8.5) and hindered under more acidic conditions (pH 6). Similarly, alpha proton acidity appeared to play a significant role in donor reactivity with 13 undergoing rapid selenide liberation (confirmed with trapping experiments using benzyl bromide), while 11 was found to be considerably more stable than both 10 and 12. In total, these experiments not only verified the release of selenide from γ-ketoselenides but also the mechanism through which it occurs.

The anticancer activity of these compounds was also assessed in HeLa (human cervical cancer) and HCT116 (human colon cancer) cells in culture (Figure 8B). Predictably, 11, which was found to be stable in buffer for months at a time, liberating only trace amounts of selenide, was found to be completely inactive in both cell lines. Conversely, 10, 12, and 13 exhibited low micromolar activity. As a key control, compound 12EC, an oxygen conger of 12, was also tested and found to be completely inactive, confirming that the activity of donors is due to their release of selenide and not the other components of the reaction. It is also worth mentioning that 11, which liberates selenide at an order of magnitude faster than both 10 and 12, was found to be two-fold less potent in both cell lines. This implies that greater antiproliferative activity may be achieved through continuous exposure to low levels of H_2Se for a prolonged period as opposed to a rapid surge in selenide concentration that is likely afforded by 11.

2.8. 5′-O-Selenophosphate Nucleosides

It had been previously shown by Kaczmarek and colleagues that 5′-*O*-thiophosphate nucleosides function as an H_2S source in the presence of histidine triad nucleotide-binding protein 1 (HINT1) [86,87]. Consistent with the common theme throughout this review, Kaczmarek speculated that a selenium congener might function as a HINT1 substrate [88]. If so, then perhaps H_2Se, like H_2S, would be afforded as a byproduct of the enzyme's hydrolase activity.

To test this premise, 2′-deoxyguanosine-5′-*O*-selenophosphate (dGMPSe) was produced using a known method for generating phosphoroselenoates [89]. After confirming the stability of dGMPSe in the buffer alone, its HINT1-catalyzed hydrolysis was assessed. This selenophosphate derivative was found to be a substrate for HINT1 with both deoxyguanosine monophosphate (dGMP), confirmed by HPLC, and H_2Se, detected by fluorescence spectroscopy and the use of the reaction-based probe SF7 [90], were observed as products of the enzyme-catalyzed hydrolysis (Figure 9).

After confirming the HINT1-promoted release of H_2Se in vitro, the cytotoxicity of dGMPSe was evaluated in HeLa cells. Compared to dGMP, which was found to be nontoxic at all concentrations tested, dGMPSe exhibited dose-dependent cytotoxicity with an IC_{50} value of 8 μM after a 24 h incubation period. Furthermore, dead cells were shown to exhibit higher fluorescence in the presence of SF7, further supporting the notion that released H_2Se from dGMPSe is responsible for cell death.

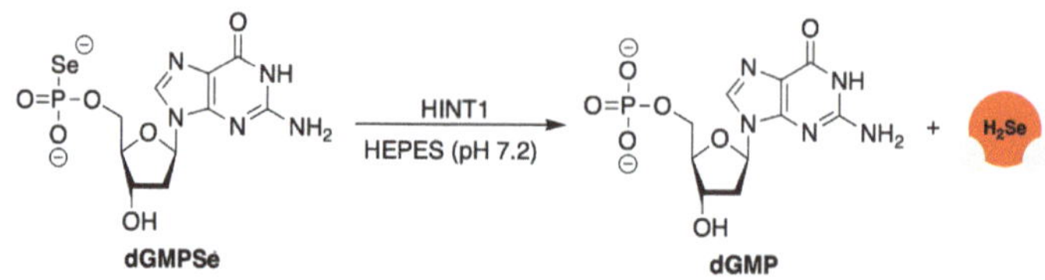

Figure 9. 5′-*O*-Selenophosphates (**dGMPSe**) undergo HINT1-catalyzed hydrolysis to generate H_2Se.

3. Chemical Tools for H_2Se Detection

3.1. Nonspecific Electrophilic Traps: Dinitrofluorobenzene, Benzyl Bromide, and Iodoacetamide

While selective microsensors and a plethora of reaction-based fluorescent probes offer reliable methods for sulfide detection, analogous chemical tools with selenide specificity are lacking. Furthermore, the methylene blue assay, a ubiquitous method for sulfide quantification in buffer, is unlikely to translate to accurate selenide detection due to its rapid oxidation and overall chemical instability compared to that of sulfide. As such, many researchers have turned to the use of nonspecific electrophilic traps as a quick and dirty method for selenide sensing and validation of H_2Se donor activation.

As mentioned earlier, Pluth and co-workers relied on dinitrofluorobenzene (DNFB) for confirmation of H_2Se donation from TDN1042 (Figure 10A) [65]. However, the high electrophilicity of DNFB, coupled with the augmented nucleophilic character of selenium, made this a challenging endeavor. The authors had previously observed the P=Se moiety of the donor reacting directly with other electrophiles, such as benzyl bromide, making it impossible to distinguish donor alkylation from the alkylation of H_2Se. Thus, to unequivocally establish selenide release from TDN1042, the authors placed the DNFB trapping solution in a vial separate from the donor. Under this setup, when TDN1042 was acidified with HCl and sparged with N_2, the liberated H_2Se was volatilized into the headspace and bubbled through the separate trapping solution containing an excess of DNFB. The contents of the trapping solution were then analyzed by HPLC with the expected mono and diselenide products being clearly visible (Figure 10A).

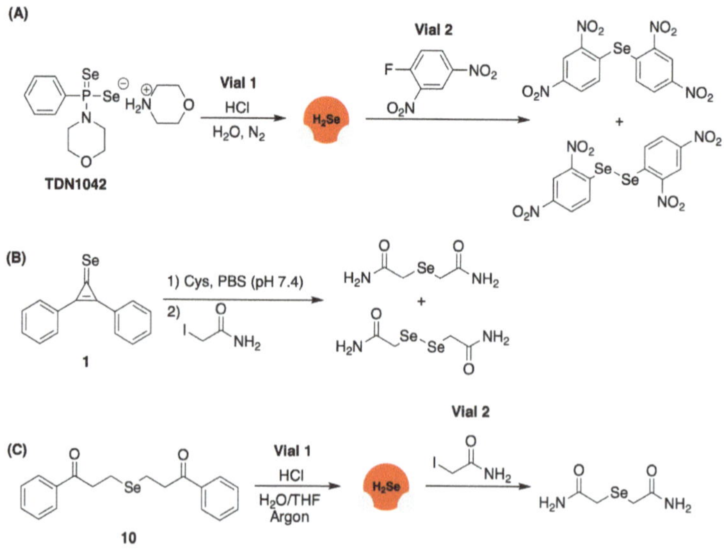

Figure 10. Use of nonspecific electrophilic traps to confirm H_2Se release from donors ((**A**) TDN1042, (**B**) 1, (**C**) 10) by forming stable selenide/diselenide products that can be easily identified by spectroscopic methods.

Likewise, Yi and colleagues first utilized iodoacetamide for quick confirmation of selenide donation from their selenocyclopropenone-based donors (Figure 10B) [68]. This analysis was further complicated by the fact that cysteine, which can be consumed by the added iodoacetamide, was used to trigger H_2Se release from this donor class. Nevertheless, using HRMS the authors observed the corresponding mono and diselenide products, which is consistent with the release of selenide and the autooxidation process.

We, too, relied on nonspecific electrophiles, both benzyl bromide and iodoacetamide, for confirmation of H_2Se liberation from our γ-ketoselenide-based system [85]. Like the Pluth group, we also sought to unambiguously confirm H_2Se donation from 10 by trapping the gas in a separate vial (Figure 10C). We elected to use iodoacetamide for these experiments as the ensuing product from its trapping of volatilized H_2Se would be easily recognizable by HRMS. Indeed, when 10 was acidified and sparged with argon, released H_2Se was transferred through a cannula needle and into a separate trapping solution where the expected selenide product was clearly visible by HRMS.

3.2. Fluorescent Sensors Based on Benzoselenadiazole Se–N Bond Cleavage

A small molecule fluorescent probe with high selectivity towards H_2Se was first reported by Tang and colleagues in 2016 (NIR-H_2Se, Figure 11A) [91]. By fabricating a benzoselenadiazole moiety onto a mercaptan dye, the authors found that Se–N bond cleavage occurred quickly in the presence of H_2Se, but not when exposed to other selenols or thiols, to afford to a fluorescent diamine reporter (λ_{ex}: 688 nm, λ_{em}: 735 nm).

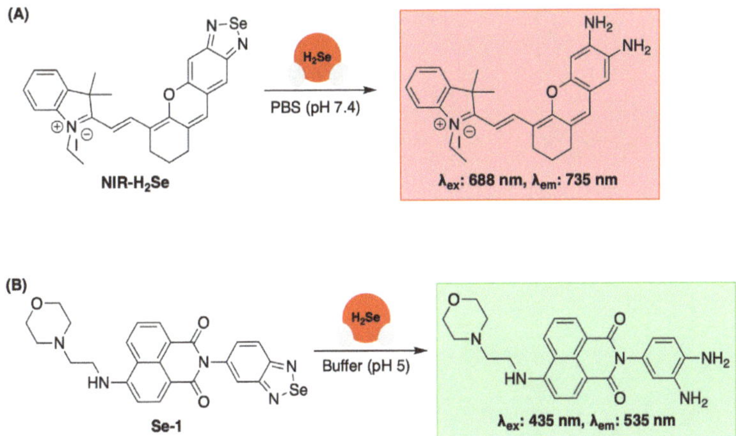

Figure 11. (**A**) Reaction-based fluorescent sensor with selectivity towards H_2Se based on benzoselenadiazole Se–N bond cleavage. (**B**) A lysosomal-targeting fluorescent sensor for H_2Se bioimaging.

With an H_2Se-selective sensor in hand, the authors then used NIR-H_2Se to monitor cellular H_2Se levels in HepG2 cells using Na_2SeO_3 as a metabolic precursor. In this experiment, a dose and time-dependent increase in fluorescence was observed, as expected. Moreover, these investigations were conducted under both hypoxic (1% pO_2) and normoxic (20% O_2) conditions, with a significant reduction in fluorescence being observed during the latter. This was attributed to rapid H_2Se oxidation and the production of superoxide and other reactive oxygen species while exposed to higher O_2 levels. Based on these observations, it was concluded that the anticancer activity of Na_2SeO_3 in HepG2 cells under normoxic conditions can be attributed to ROS-induced cell death. Conversely, under hypoxic conditions, which is a hallmark of solid tumors, a non-oxidative stress mechanism is likely in play due to a notable buildup in cellular H_2Se. This hypothesis was further validated in a solid tumor mouse model using NIR-H_2Se.

A lysosomal-specific H_2Se sensor based on this same framework was reported by Zhang and Jing in 2019 (Se-1, Figure 11B) [92]. In addition to a benzoselenadiazole moiety for selective H_2Se detection, a morpholino group was appended to the sensor for lysosomal-targeting [93]. Under simulated lysosomal conditions (acetate buffer, pH 5), a notable increase in fluorescence at 535 nm (λ_{ex}: 435 nm) was observed when a solution of Se-1 was exposed to H_2Se (but not other biologically relevant analytes). Moreover, the fluorescent enhancement of Se-1 in the presence of H_2Se was found to be consistent within a pH range of 4.5–7, confirming its compatibility with lysosomal conditions.

After establishing its reactivity and photophysical properties in buffer, lysosomal-targeting of Se-1 was confirmed in HepG2 cells using LysoTracker Blue (Pearson's correlation coefficient of 0.91). Additionally, using Se-1, the authors observed elevated levels of lysosomal H_2Se in hypoxic HepG2 cells, whereas little lysosomal fluorescence was observed under normoxic conditions.

To date, benzoselenadiazoles have not been used as a recognition subunit to validate selenide delivery from novel donor scaffolds. However, its reported high reactivity towards H_2Se could prove useful in future studies as the transient nature of H_2Se requires rapid detection for accurate monitoring.

3.3. Fluorescent Sensors Based on Disulfide Bond Cleavage

A second-generation sensor from Tang and co-workers employed disulfide bond reduction as a recognition mechanism for H_2Se-initiated turn-on fluorescence (Hcy-H_2Se, Figure 12) [94]. Thiol-activated prodrugs and chemosensors that utilize disulfide bond reduction as their initiation step have been widely reported [95]. Based on these accounts, it was suspected that a more stable cyclic disulfide (i.e., a six-membered ring) would respond much more quickly to H_2Se, given its heightened nucleophilicity, thereby imparting selectivity over other selenols and thiols.

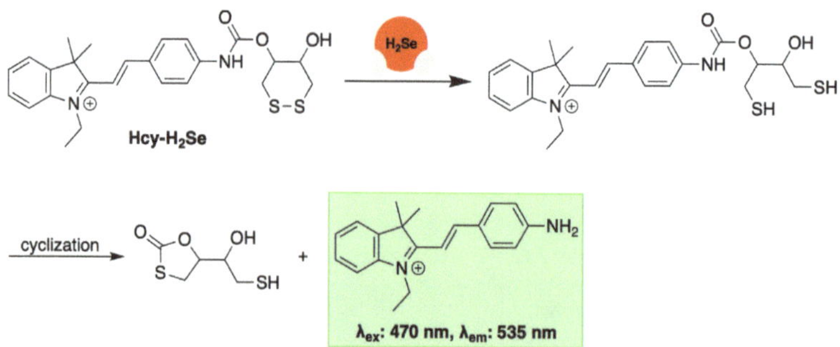

Figure 12. Proposed mechanism for Hcy-H_2Se turn-on fluorescence initiated by H_2Se-promoted disulfide reduction.

Hcy-H_2Se was generated by combining oxidized dithiothreitol [96] with a masked hemicyanine dye via a carbamate linker. When exposed to H_2Se, a maximum fluorescence intensity (λ_{ex}: 470 nm, λ_{em}: 535 nm) was reached almost immediately, indicating rapid H_2Se-initiated disulfide bond reduction and cyclization (Figure 12). Selectivity studies indicated that the addition of glutathione, H_2S, selenocysteine, dithiothreitol, bovine serum albumin, and thioredoxin reductase yielded little fluorescence compared to H_2Se. Moreover, the emission intensity at 535 nm correlated well with increasing concentrations of H_2Se, affording a good linear relationship between the two.

Bioimaging of H_2Se in live human cells was accomplished with Hcy-H_2Se [94]. HepG2 cells exposed to Na_2SeO_3 and Hcy-H_2Se under hypoxic conditions exhibited fluorescence that was both dose- and time-dependent. The authors also demonstrated that hypoxic tumor regions in mice injected with sodium selenite could be imaged with Hcy-H_2Se. This further supports the anticancer effects of selenium in hypoxic solid tumors being due to an increase in reductive rather an oxidative stress.

3.4. Spectroscopic Sensors Based on Nucleophilic Substitution

While examining selenocyclopropenones and selenoamides as potential cysteine-activated donors (Figure 6), Yi and collaborators developed a quantitative assay for H_2Se based on its nucleophilic substitution with a carefully chosen electrophilic species [68]. Upon testing several potential candidates, the authors discovered that commercially available Cy7-Cl reacted much more quickly with selenide than other selenols and thiols, to form Cy7-SeH (Figure 13A). As the reaction progressed, it was noted that the starting absorbance of 780 nm (Cy7-Cl) was shifted to 710 nm (Cy7-SeH), offering a convenient colorimetric method for monitoring reaction progress. The reaction between Cy7-Cl (500 μM) and various amounts of Na_2Se (25–350 μM) was also monitored HPLC, with a plot of Cy7-SeH peak area vs. selenide concentration yielding a straight line. This calibration curve was then used to effectively determine the H_2Se-releasing efficiency from selenocyclopropenones in the presence of cysteine.

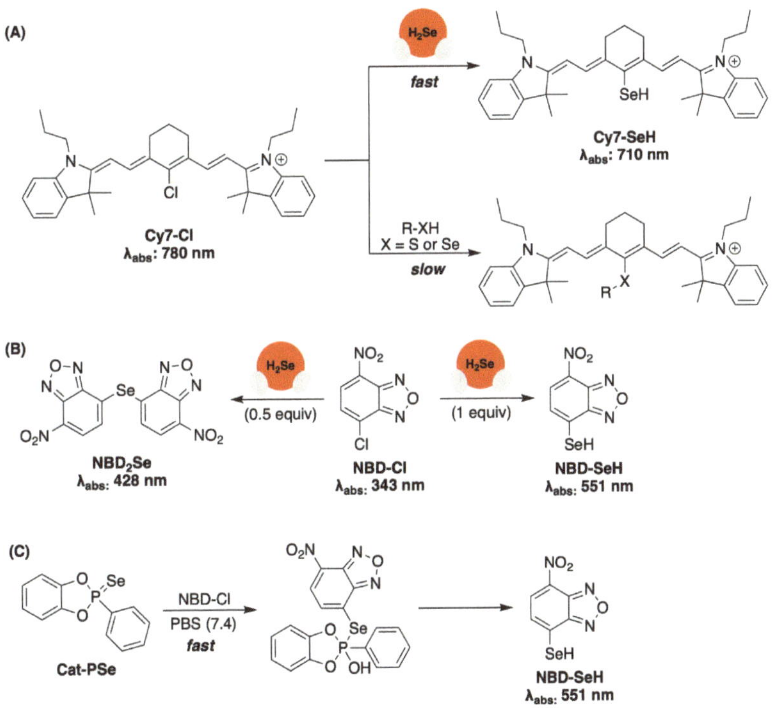

Figure 13. (**A**,**B**) Commercially available electrophilic traps that provide a colorimetric readout for monitoring H_2Se donor progress. (**C**) Cautioning researchers that complementary H_2Se measurements should be employed to avoid confusing donor alkylation with the alkylation of released H_2Se in solution.

While evaluating the H_2Se-releasing efficiencies of cyclic-PSe donors (Figure 5), the Pluth group found that commercially available 4-chloro-7-nitrobenzofurazan (NBD-Cl), which had been previously used to detect H_2S [97], could also be used to trap selenide via a nucleophilic aromatic substitution reaction (Figure 13B) [66]. Initially, the authors analyzed the reaction between NBD-Cl and tetrabutylammonium hydroselenide (NBu_4SeH) in PBS (pH 7.4). They found that substoichiometric amounts of selenide provided NBD_2Se (λ_{abs}: 428 nm), while stoichiometric HSe^- yielded NBD-SeH as the primary product (λ_{abs}:

551 nm). With confirmation of a colorimetric response, NBD-Cl was then used to monitor the hydrolysis of Cat-PSe in PBS (Figure 13C). While the formation of NBD-SeH was clearly visible (λ_{abs}: 551 nm), the rate of apparent H_2Se liberation appeared to occur much faster when compared to earlier ^{31}P NMR hydrolysis experiments. The authors attributed this notable rate enhancement to donor alkylation, rather than hydrolysis, while in the presence of a strong electrophile, such as NBD-Cl. Therefore, while the use of NBD-Cl appears to provide a convenient colorimetric method for monitoring donor progress, additional trapping experiments are likely necessary to untangle donor alkylation from the alkylation of released H_2Se in solution.

3.5. Fluorescent Sensors Based on Azide Reduction

Like H_2S-responsive NBD-Cl being reintroduced for H_2Se detection, H_2S-sensitive fluorescent sensors that rely on aryl azide reduction were reexamined for their capacity to detect H_2Se [89]. Probes SF4 and SF7 (Figure 14) were previously engineered by Chang and co-workers for monitoring H_2S levels under physiological conditions [90,98]. While stable towards other sulfur-containing molecules, H_2S initiates aryl azide reduction on the masked rhodamine dye, resulting in a turn-fluorescent response (Figure 14A).

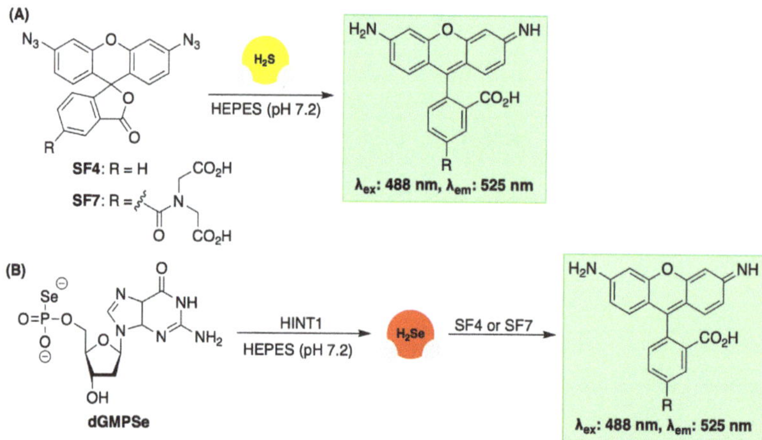

Figure 14. (**A**) Aryl azide reduction for H_2S-initiated turn-on fluorescence. (**B**) The use of aryl azides as fluorescent probes for tracking H_2Se liberation from donor compounds.

Given the increased reactivity of H_2Se, Kaczmarek and co-workers suspected that these same sensors would be responsive towards H_2Se and used them to confirm selenide release from dGMPSe in both in a buffer and in live HeLa cells (Figure 14B) [88]. In fact, when using SF7 and SF4, higher fluorescence values were observed with dGMPSe compared to its sulfur congener, which likely stems from the higher reactivity and reductive properties of H_2Se compared to that of H_2S. Thus, while not selective for H_2Se, aryl azide reduction appears to be a potential method for the real-time monitoring of donor progress.

4. Conclusions and Outlook

The continued evolution of H_2Se-specific chemical tools is a prime objective for those interested in uncovering the (patho)physiological effects of hydrogen selenide. While donor compounds capable of providing the slow and sustained release of H_2Se have begun to emerge, the introduction of additional stimulus-responsive donors (i.e., enzyme or bioanalyte-triggered) with greater spatiotemporal control over their biological delivery of H_2Se will serve as a great advancement in the field. H_2Se detection and quantification, especially when liberated from a donor scaffold, remains a great challenge. Most current methods rely on electrophiles to trap the released selenide, which often leads to interference from the donor itself given the intensified nucleophilic character of selenium. New approaches for detection that avoid the introduction of strong electrophiles should continue to be explored, including additional reaction-based fluorescent sensors that can be used in cellular and in vivo imaging experiments. Together, these compounds will continue to unearth the unique aspects of H_2Se chemical biology, while providing new evidence that further supports (or refutes) the addition of H_2Se as the fourth gasotransmitter.

Funding: This research was funded by the National Science Foundation (Grant No. 2143826).

Institutional Review Board Statement: Not applicable.

Informed Consent Statement: Not applicable.

Data Availability Statement: Not applicable.

Conflicts of Interest: The authors declare no conflict of interest.

References

1. Weeks, M.E. The Discovery of the Elements. VI. Tellurium and Selenium. *J. Chem. Educ.* **1932**, *9*, 474. [CrossRef]
2. Levander, O.A. Scientific Rationale for the 1989 Recommended Dietary Allowance for Selenium. *J. Am. Diet. Assoc.* **1991**, *91*, 1572–1576. [CrossRef]
3. Labunskyy, V.M.; Hatfield, D.L.; Gladyshev, V.N. Selenoproteins: Molecular Pathways and Physiological Roles. *Physiol. Rev.* **2014**, *94*, 739–777. [CrossRef]
4. Atkins, J.F.; Gesteland, R.F. The Twenty-First Amino Acid. *Nature* **2000**, *407*, 463–464. [CrossRef]
5. Böck, A.; Forchhammer, K.; Heider, J.; Leinfelder, W.; Sawers, G.; Veprek, B.; Zinoni, F. Selenocysteine: The 21st Amino Acid. *Mol. Microbiol.* **1991**, *5*, 515–520. [CrossRef]
6. Brown, K.M.; Arthur, J.R. Selenium, Selenoproteins and Human Health: A Review. *Public Health Nutr.* **2001**, *4*, 593–599. [CrossRef]
7. Kieliszek, M.; Błażejak, S. Current Knowledge on the Importance of Selenium in Food for Living Organisms: A Review. *Molecules* **2016**, *21*, 609. [CrossRef]
8. Rayman, M.P. Selenium and Human Health. *Lancet* **2012**, *379*, 1256–1268. [CrossRef]
9. Schomburg, L. Dietary Selenium and Human Health. *Nutrients* **2016**, *9*, 22. [CrossRef] [PubMed]
10. Chen, J. An Original Discovery: Selenium Deficiency and Keshan Disease (an Endemic Heart Disease). *Asia Pac. J. Clin. Nutr.* **2012**, *21*, 320–326. [PubMed]
11. Shi, Y.; Yang, W.; Tang, X.; Yan, Q.; Cai, X.; Wu, F. Keshan Disease: A Potentially Fatal Endemic Cardiomyopathy in Remote Mountains of China. *Front. Pediatr.* **2021**, *9*, 576916. [CrossRef] [PubMed]
12. Wang, L.; Yin, J.; Yang, B.; Qu, C.; Lei, J.; Han, J.; Guo, X. Serious Selenium Deficiency in the Serum of Patients with Kashin–Beck Disease and the Effect of Nano-Selenium on Their Chondrocytes. *Biol. Trace Elem. Res.* **2020**, *194*, 96–104. [CrossRef]
13. Yao, Y.; Pei, F.; Kang, P. Selenium, Iodine, and the Relation with Kashin-Beck Disease. *Nutrition* **2011**, *27*, 1095–1100. [CrossRef] [PubMed]
14. Arthur, J.R.; McKenzie, R.C.; Beckett, G.J. Selenium in the Immune System. *J. Nutr.* **2003**, *133*, 1457S–1459S. [CrossRef]
15. Hoffmann, P.R.; Berry, M.J. The Influence of Selenium on Immune Responses. *Mol. Nutr. Food Res.* **2008**, *52*, 1273–1280. [CrossRef] [PubMed]
16. Al-Mubarak, A.A.; van der Meer, P.; Bomer, N. Selenium, Selenoproteins, and Heart Failure: Current Knowledge and Future Perspective. *Curr. Heart Fail. Rep.* **2021**, *18*, 122–131. [CrossRef] [PubMed]
17. Bomer, N.; Grote Beverborg, N.; Hoes, M.F.; Streng, K.W.; Vermeer, M.; Dokter, M.M.; IJmker, J.; Anker, S.D.; Cleland, J.G.F.; Hillege, H.L.; et al. Selenium and Outcome in Heart Failure. *Eur. J. Heart Fail.* **2020**, *22*, 1415–1423. [CrossRef]
18. Rataan, A.O.; Geary, S.M.; Zakharia, Y.; Rustum, Y.M.; Salem, A.K. Potential Role of Selenium in the Treatment of Cancer and Viral Infections. *Int. J. Mol. Sci.* **2022**, *23*, 2215. [CrossRef]
19. Rayman, M.P. Selenium in Cancer Prevention: A Review of the Evidence and Mechanism of Action. *Proc. Nutr. Soc.* **2005**, *64*, 527–542. [CrossRef]
20. Sayehmiri, K.; Azami, M.; Mohammadi, Y.; Soleymani, A.; Tardeh, Z. The Association between Selenium and Prostate Cancer: A Systematic Review and Meta-Analysis. *Asian Pac. J. Cancer Prev.* **2018**, *19*, 1431–1437. [CrossRef]
21. Vinceti, M.; Filippini, T.; Del Giovane, C.; Dennert, G.; Zwahlen, M.; Brinkman, M.; Zeegers, M.P.; Horneber, M.; D'Amico, R.; Crespi, C.M. Selenium for Preventing Cancer. *Cochrane Database Syst. Rev.* **2018**, *1*, CD005195. [CrossRef]
22. Barceloux, D.G.; Barceloux, D. Selenium. *J. Toxicol. Clin. Toxicol.* **1999**, *37*, 145–172. [CrossRef]
23. Hughes, M.N.; Centelles, M.N.; Moore, K.P. Making and Working with Hydrogen Sulfide. *Free Radic. Biol. Med.* **2009**, *47*, 1346–1353. [CrossRef]
24. Haynes, W.M.; Lide, D.R.; Bruno, T.J. (Eds.) *CRC Handbook of Chemistry and Physics 2016–2017: A Ready-Reference Book of Chemical and Physical Data*, 97th ed.; CRC Press: Boca Raton, FL, USA, 2017; ISBN 978-1-4987-5429-3.

25. Jiang, J.; Chan, A.; Ali, S.; Saha, A.; Haushalter, K.J.; Lam, W.-L.M.; Glasheen, M.; Parker, J.; Brenner, M.; Mahon, S.B.; et al. Hydrogen Sulfide—Mechanisms of Toxicity and Development of an Antidote. *Sci. Rep.* **2016**, *6*, 20831. [CrossRef] [PubMed]
26. National Research Council (U.S.) (Ed.) *Acute Exposure Guideline Levels for Selected Airborne Chemicals*; The Compass Series; National Academy Press: Washington, DC, USA, 2000; ISBN 978-0-309-07294-6.
27. Filipovic, M.R.; Zivanovic, J.; Alvarez, B.; Banerjee, R. Chemical Biology of H_2S Signaling through Persulfidation. *Chem. Rev.* **2018**, *118*, 1253–1337. [CrossRef]
28. Li, L.; Rose, P.; Moore, P.K. Hydrogen Sulfide and Cell Signaling. *Annu. Rev. Pharmacol. Toxicol.* **2011**, *51*, 169–187. [CrossRef]
29. Wang, R. Hydrogen Sulfide: The Third Gasotransmitter in Biology and Medicine. *Antioxid. Redox Signal.* **2010**, *12*, 1061–1064. [CrossRef]
30. Wang, R. The Gasotransmitter Role of Hydrogen Sulfide. *Antioxid. Redox Signal.* **2003**, *5*, 493–501. [CrossRef]
31. Miles, E.W.; Kraus, J.P. Cystathionine β-Synthase: Structure, Function, Regulation, and Location of Homocystinuria-Causing Mutations. *J. Biol. Chem.* **2004**, *279*, 29871–29874. [CrossRef]
32. Pan, L.L.; Liu, X.H.; Gong, Q.H.; Yang, H.B.; Zhu, Y.Z. Role of Cystathionine γ-Lyase/Hydrogen Sulfide Pathway in Cardiovascular Disease: A Novel Therapeutic Strategy? *Antioxid. Redox Signal.* **2012**, *17*, 106–118. [CrossRef]
33. Shibuya, N.; Tanaka, M.; Yoshida, M.; Ogasawara, Y.; Togawa, T.; Ishii, K.; Kimura, H. 3-Mercaptopyruvate Sulfurtransferase Produces Hydrogen Sulfide and Bound Sulfane Sulfur in the Brain. *Antioxid. Redox Signal.* **2009**, *11*, 703–714. [CrossRef] [PubMed]
34. Zhang, X.; Bian, J.-S. Hydrogen Sulfide: A Neuromodulator and Neuroprotectant in the Central Nervous System. *ACS Chem. Neurosci.* **2014**, *5*, 876–883. [CrossRef]
35. Abe, K.; Kimura, H. The Possible Role of Hydrogen Sulfide as an Endogenous Neuromodulator. *J. Neurosci.* **1996**, *16*, 1066–1071. [CrossRef]
36. Guo, W.; Kan, J.; Cheng, Z.; Chen, J.; Shen, Y.; Xu, J.; Wu, D.; Zhu, Y. Hydrogen Sulfide as an Endogenous Modulator in Mitochondria and Mitochondria Dysfunction. *Oxidative Med. Cell. Longev.* **2012**, *2012*, 878052. [CrossRef]
37. Murphy, B.; Bhattacharya, R.; Mukherjee, P. Hydrogen Sulfide Signaling in Mitochondria and Disease. *FASEB J.* **2019**, *33*, 13098–13125. [CrossRef]
38. Powell, C.R.; Dillon, K.M.; Matson, J.B. A Review of Hydrogen Sulfide (H_2S) Donors: Chemistry and Potential Therapeutic Applications. *Biochem. Pharmacol.* **2018**, *149*, 110–123. [CrossRef]
39. Xu, S.; Hamsath, A.; Neill, D.L.; Wang, Y.; Yang, C.; Xian, M. Strategies for the Design of Donors and Precursors of Reactive Sulfur Species. *Chem. A Eur. J.* **2019**, *25*, 4005–4016. [CrossRef] [PubMed]
40. Levinn, C.M.; Cerda, M.M.; Pluth, M.D. Activatable Small-Molecule Hydrogen Sulfide Donors. *Antioxid. Redox Signal.* **2020**, *32*, 96–109. [CrossRef]
41. Hu, Q.; Lukesh, J.C. H_2S Donors with Cytoprotective Effects in Models of MI/R Injury and Chemotherapy-Induced Cardiotoxicity. *Antioxidants* **2023**, *12*, 650. [CrossRef] [PubMed]
42. Papapetropoulos, A.; Foresti, R.; Ferdinandy, P. Pharmacology of the 'Gasotransmitters' NO, CO and H2S: Translational Opportunities. *Br. J. Pharmacol.* **2015**, *172*, 1395–1396. [CrossRef]
43. Wang, R. Shared Signaling Pathways among Gasotransmitters. *Proc. Natl. Acad. Sci. USA* **2012**, *109*, 8801–8802. [CrossRef]
44. Yang, G.; Sener, A.; Ji, Y.; Pei, Y.; Pluth, M.D. Gasotransmitters in Biology and Medicine: Molecular Mechanisms and Drug Targets. *Oxidative Med. Cell. Longev.* **2016**, *2016*, 4627308. [CrossRef] [PubMed]
45. Wang, R. Two's Company, Three's a Crowd: Can H_2S Be the Third Endogenous Gaseous Transmitter? *FASEB J.* **2002**, *16*, 1792–1798. [CrossRef]
46. Kuganesan, M.; Samra, K.; Evans, E.; Singer, M.; Dyson, A. Selenium and Hydrogen Selenide: Essential Micronutrient and the Fourth Gasotransmitter? *Intensive Care Med. Exp.* **2019**, *7*, 71. [CrossRef]
47. Samra, K.; Kuganesan, M.; Smith, W.; Kleyman, A.; Tidswell, R.; Arulkumaran, N.; Singer, M.; Dyson, A. The Pharmacology and Therapeutic Utility of Sodium Hydroselenide. *Int. J. Mol. Sci.* **2021**, *22*, 3258. [CrossRef]
48. Tarze, A.; Dauplais, M.; Grigoras, I.; Lazard, M.; Ha-Duong, N.-T.; Barbier, F.; Blanquet, S.; Plateau, P. Extracellular Production of Hydrogen Selenide Accounts for Thiol-Assisted Toxicity of Selenite against Saccharomyces Cerevisiae. *J. Biol. Chem.* **2007**, *282*, 8759–8767. [CrossRef] [PubMed]
49. Pan, X.; Song, X.; Wang, C.; Cheng, T.; Luan, D.; Xu, K.; Tang, B. H_2Se Induces Reductive Stress in HepG2 Cells and Activates Cell Autophagy by Regulating the Redox of HMGB1 Protein under Hypoxia. *Theranostics* **2019**, *9*, 1794–1808. [CrossRef]
50. Iwata, A.; Morrison, M.L.; Blackwood, J.E.; Roth, M.B. Selenide Targets to Reperfusing Tissue and Protects It from Injury. *Crit. Care Med.* **2015**, *43*, 1361–1367. [CrossRef]
51. Painter, E.P. The Chemistry and Toxicity of Selenium Compounds, with Special Reference to the Selenium Problem. *Chem. Rev.* **1941**, *28*, 179–213. [CrossRef]
52. Ganther, H.E. Selenotrisulfides. Formation by the Reaction of Thiols with Selenious Acid. *Biochemistry* **1968**, *7*, 2898–2905. [CrossRef] [PubMed]
53. Nakagawa, T.; Aoyama, E.; Kobayashi, N.; Tanaka, H.; Chikuma, M.; Sakurai, H.; Nakayama, M. Thiol Exchange Reactions Involving Selenotrisulfides. *Biochem. Biophys. Res. Commun.* **1988**, *150*, 1149–1154. [CrossRef]
54. Haratake, M.; Ono, M.; Nakayama, M. Penicillamine Selenotrisulfide as a Selenium-Source in Mice. *J. Health Sci.* **2004**, *50*, 366–371. [CrossRef]

55. Haratake, M.; Hongoh, M.; Miyauchi, M.; Hirakawa, R.; Ono, M.; Nakayama, M. Albumin-Mediated Selenium Transfer by a Selenotrisulfide Relay Mechanism. *Inorg. Chem.* **2008**, *47*, 6273–6280. [CrossRef]
56. Liu, M.; Bu, F.; Li, G.; Xie, W.; Xu, H.; Wang, X. S-Se-S Type Molecule: A Bactericidal Promoter against H_2S-Induced Antibiotic Resistance. *Innov. Life* **2024**, *2*, 100076. [CrossRef]
57. Asfar, P.; Radermacher, P. Drug-Induced "Suspended Animation": Can a Dream Become True? *Crit. Care Med.* **2015**, *43*, 1528–1530. [CrossRef] [PubMed]
58. Szabo, C. Gaseotransmitters: New Frontiers for Translational Science. *Sci. Transl. Med.* **2010**, *2*, 59ps54. [CrossRef]
59. Klayman, D.L.; Griffin, T.S. Reaction of Selenium with Sodium Borohydride in Protic Solvents. A Facile Method for the Introduction of Selenium into Organic Molecules. *J. Am. Chem. Soc.* **1973**, *95*, 197–199. [CrossRef]
60. Kharma, A.; Misak, A.; Grman, M.; Brezova, V.; Kurakova, L.; Baráth, P.; Jacob, C.; Chovanec, M.; Ondrias, K.; Domínguez-Álvarez, E. Release of Reactive Selenium Species from Phthalic Selenoanhydride in the Presence of Hydrogen Sulfide and Glutathione with Implications for Cancer Research. *New J. Chem.* **2019**, *43*, 11771–11783. [CrossRef]
61. Domínguez-Álvarez, E.; Gajdács, M.; Spengler, G.; Palop, J.A.; Marć, M.A.; Kieć-Kononowicz, K.; Amaral, L.; Molnár, J.; Jacob, C.; Handzlik, J.; et al. Identification of Selenocompounds with Promising Properties to Reverse Cancer Multidrug Resistance. *Bioorg. Med. Chem. Lett.* **2016**, *26*, 2821–2824. [CrossRef] [PubMed]
62. Domínguez-Álvarez, E.; Plano, D.; Font, M.; Calvo, A.; Prior, C.; Jacob, C.; Palop, J.A.; Sanmartín, C. Synthesis and Antiproliferative Activity of Novel Selenoester Derivatives. *Eur. J. Med. Chem.* **2014**, *73*, 153–166. [CrossRef]
63. Gajdács, M.; Spengler, G.; Sanmartín, C.; Marć, M.A.; Handzlik, J.; Domínguez-Álvarez, E. Selenoesters and Selenoanhydrides as Novel Multidrug Resistance Reversing Agents: A Confirmation Study in a Colon Cancer MDR Cell Line. *Bioorg. Med. Chem. Lett.* **2017**, *27*, 797–802. [CrossRef]
64. Li, L.; Whiteman, M.; Guan, Y.Y.; Neo, K.L.; Cheng, Y.; Lee, S.W.; Zhao, Y.; Baskar, R.; Tan, C.-H.; Moore, P.K. Characterization of a Novel, Water-Soluble Hydrogen Sulfide–Releasing Molecule (GYY4137): New Insights into the Biology of Hydrogen Sulfide. *Circulation* **2008**, *117*, 2351–2360. [CrossRef] [PubMed]
65. Newton, T.D.; Pluth, M.D. Development of a Hydrolysis-Based Small-Molecule Hydrogen Selenide (H_2Se) Donor. *Chem. Sci.* **2019**, *10*, 10723–10727. [CrossRef]
66. Newton, T.D.; Bolton, S.G.; Garcia, A.C.; Chouinard, J.E.; Golledge, S.L.; Zakharov, L.N.; Pluth, M.D. Hydrolysis-Based Small-Molecule Hydrogen Selenide (H_2Se) Donors for Intracellular H_2Se Delivery. *J. Am. Chem. Soc.* **2021**, *143*, 19542–19550. [CrossRef]
67. Myhre, O.; Andersen, J.M.; Aarnes, H.; Fonnum, F. Evaluation of the Probes 2′,7′-Dichlorofluorescin Diacetate, Luminol, and Lucigenin as Indicators of Reactive Species Formation. *Biochem. Pharmacol.* **2003**, *65*, 1575–1582. [CrossRef]
68. Kang, X.; Huang, H.; Jiang, C.; Cheng, L.; Sang, Y.; Cai, X.; Dong, Y.; Sun, L.; Wen, X.; Xi, Z.; et al. Cysteine-Activated Small-Molecule H_2Se Donors Inspired by Synthetic H_2S Donors. *J. Am. Chem. Soc.* **2022**, *144*, 3957–3967. [CrossRef]
69. Martelli, A.; Testai, L.; Citi, V.; Marino, A.; Pugliesi, I.; Barresi, E.; Nesi, G.; Rapposelli, S.; Taliani, S.; Da Settimo, F.; et al. Arylthioamides as H_2S Donors: L-Cysteine-Activated Releasing Properties and Vascular Effects In Vitro and In Vivo. *ACS Med. Chem. Lett.* **2013**, *4*, 904–908. [CrossRef]
70. Hu, Q.; Suarez, S.I.; Hankins, R.A.; Lukesh, J.C. Intramolecular Thiol- and Selenol-Assisted Delivery of Hydrogen Sulfide. *Angew. Chem. Int. Ed.* **2022**, *61*, e202210754. [CrossRef]
71. Hu, Q.; Zhu, C.; Hankins, R.A.; Murmello, A.R.; Marrs, G.S.; Lukesh, J.C. An ROS-Responsive Donor That Self-Reports Its H_2S Delivery by Forming a Benzoxazole-Based Fluorophore. *J. Am. Chem. Soc.* **2023**, *145*, 25486–25494. [CrossRef] [PubMed]
72. Zhao, Y.; Pluth, M.D. Hydrogen Sulfide Donors Activated by Reactive Oxygen Species. *Angew. Chem. Int. Ed.* **2016**, *55*, 14638–14642. [CrossRef]
73. Zhao, Y.; Henthorn, H.A.; Pluth, M.D. Kinetic Insights into Hydrogen Sulfide Delivery from Caged-Carbonyl Sulfide Isomeric Donor Platforms. *J. Am. Chem. Soc.* **2017**, *139*, 16365–16376. [CrossRef] [PubMed]
74. Chauhan, P.; Jos, S.; Chakrapani, H. Reactive Oxygen Species-Triggered Tunable Hydrogen Sulfide Release. *Org. Lett.* **2018**, *20*, 3766–3770. [CrossRef]
75. Hu, Y.; Li, X.; Fang, Y.; Shi, W.; Li, X.; Chen, W.; Xian, M.; Ma, H. Reactive Oxygen Species-Triggered off-on Fluorescence Donor for Imaging Hydrogen Sulfide Delivery in Living Cells. *Chem. Sci.* **2019**, *10*, 7690–7694. [CrossRef] [PubMed]
76. Zhu, C.; Suarez, S.I.; Lukesh, J.C. Illuminating and Alleviating Cellular Oxidative Stress with an ROS-Activated, H_2S-Donating Theranostic. *Tetrahedron Lett.* **2021**, *69*, 152944. [CrossRef]
77. Zhao, Y.; Bolton, S.G.; Pluth, M.D. Light-Activated COS/H_2S Donation from Photocaged Thiocarbamates. *Org. Lett.* **2017**, *19*, 2278–2281. [CrossRef] [PubMed]
78. Sharma, A.K.; Nair, M.; Chauhan, P.; Gupta, K.; Saini, D.K.; Chakrapani, H. Visible-Light-Triggered Uncaging of Carbonyl Sulfide for Hydrogen Sulfide (H_2S) Release. *Org. Lett.* **2017**, *19*, 4822–4825. [CrossRef] [PubMed]
79. Gilbert, A.K.; Zhao, Y.; Otteson, C.E.; Pluth, M.D. Development of Acid-Mediated H_2S/COS Donors That Respond to a Specific pH Window. *J. Org. Chem.* **2019**, *84*, 14469–14475. [CrossRef]
80. Zhao, Y.; Steiger, A.K.; Pluth, M.D. Colorimetric Carbonyl Sulfide (COS)/Hydrogen Sulfide (H_2S) Donation from γ-Ketothiocarbamate Donor Motifs. *Angew. Chem. Int. Ed.* **2018**, *57*, 13101–13105. [CrossRef] [PubMed]
81. Chauhan, P.; Bora, P.; Ravikumar, G.; Jos, S.; Chakrapani, H. Esterase Activated Carbonyl Sulfide/Hydrogen Sulfide (H_2S) Donors. *Org. Lett.* **2017**, *19*, 62–65. [CrossRef]

82. Steiger, A.K.; Marcatti, M.; Szabo, C.; Szczesny, B.; Pluth, M.D. Inhibition of Mitochondrial Bioenergetics by Esterase-Triggered COS/H_2S Donors. *ACS Chem. Biol.* **2017**, *12*, 2117–2123. [CrossRef]
83. Steiger, A.K.; Zhao, Y.; Pluth, M.D. Emerging Roles of Carbonyl Sulfide in Chemical Biology: Sulfide Transporter or Gasotransmitter? *Antioxid. Redox Signal.* **2018**, *28*, 1516–1532. [CrossRef] [PubMed]
84. Newton, T.D.; Li, K.; Sharma, J.; Champagne, P.A.; Pluth, M.D. Direct Hydrogen Selenide (H_2Se) Release from Activatable Selenocarbamates. *Chem. Sci.* **2023**, *14*, 7581–7588. [CrossRef] [PubMed]
85. Hankins, R.A.; Carter, M.E.; Zhu, C.; Chen, C.; Lukesh, J.C. Enol-Mediated Delivery of H_2Se from γ-Keto Selenides: Mechanistic Insight and Evaluation. *Chem. Sci.* **2022**, *13*, 13094–13099. [CrossRef] [PubMed]
86. Ozga, M.; Dolot, R.; Janicka, M.; Kaczmarek, R.; Krakowiak, A. Histidine Triad Nucleotide-Binding Protein 1 (HINT-1) Phosphoramidase Transforms Nucleoside 5′-O-Phosphorothioates to Nucleoside 5′-O-Phosphates. *J. Biol. Chem.* **2010**, *285*, 40809–40818. [CrossRef] [PubMed]
87. Krakowiak, A.; Kaczmarek, R.; Baraniak, J.; Wieczorek, M.; Stec, W.J. Stereochemistry of rHint1 Hydrolase Assisted Cleavage of P–N Bond in Nucleoside 5′-O-Phosphoramidothioates. *Chem. Commun.* **2007**, 2163–2165. [CrossRef] [PubMed]
88. Krakowiak, A.; Czernek, L.; Pichlak, M.; Kaczmarek, R. Intracellular HINT1-Assisted Hydrolysis of Nucleoside 5′-O-Selenophosphate Leads to the Release of Hydrogen Selenide That Exhibits Toxic Effects in Human Cervical Cancer Cells. *Int. J. Mol. Sci.* **2022**, *23*, 607. [CrossRef] [PubMed]
89. Guga, P.; Maciaszek, A.; Stec, W.J. Oxathiaphospholane Approach to the Synthesis of Oligodeoxyribonucleotides Containing Stereodefined Internucleotide Phosphoroselenoate Function. *Org. Lett.* **2005**, *7*, 3901–3904. [CrossRef]
90. Lin, V.S.; Lippert, A.R.; Chang, C.J. Cell-Trappable Fluorescent Probes for Endogenous Hydrogen Sulfide Signaling and Imaging H_2O_2-Dependent H_2S Production. *Proc. Natl. Acad. Sci. USA* **2013**, *110*, 7131–7135. [CrossRef] [PubMed]
91. Kong, F.; Ge, L.; Pan, X.; Xu, K.; Liu, X.; Tang, B. A Highly Selective Near-Infrared Fluorescent Probe for Imaging H_2Se in Living Cells and In Vivo. *Chem. Sci.* **2016**, *7*, 1051–1056. [CrossRef]
92. Tian, Y.; Xin, F.; Jing, J.; Zhang, X. Fluorescence Imaging of Lysosomal Hydrogen Selenide under Oxygen-Controlled Conditions. *J. Mater. Chem. B* **2019**, *7*, 2829–2834. [CrossRef]
93. Choi, N.-E.; Lee, J.-Y.; Park, E.-C.; Lee, J.-H.; Lee, J. Recent Advances in Organelle-Targeted Fluorescent Probes. *Molecules* **2021**, *26*, 217. [CrossRef]
94. Kong, F.; Zhao, Y.; Liang, Z.; Liu, X.; Pan, X.; Luan, D.; Xu, K.; Tang, B. Highly Selective Fluorescent Probe for Imaging H_2Se in Living Cells and In Vivo Based on the Disulfide Bond. *Anal. Chem.* **2017**, *89*, 688–693. [CrossRef]
95. Lee, M.H.; Yang, Z.; Lim, C.W.; Lee, Y.H.; Dongbang, S.; Kang, C.; Kim, J.S. Disulfide-Cleavage-Triggered Chemosensors and Their Biological Applications. *Chem. Rev.* **2013**, *113*, 5071–5109. [CrossRef] [PubMed]
96. Cleland, W.W. Dithiothreitol, a New Protective Reagent for SH Groups. *Biochemistry* **1964**, *3*, 480–482. [CrossRef] [PubMed]
97. Montoya, L.A.; Pearce, T.F.; Hansen, R.J.; Zakharov, L.N.; Pluth, M.D. Development of Selective Colorimetric Probes for Hydrogen Sulfide Based on Nucleophilic Aromatic Substitution. *J. Org. Chem.* **2013**, *78*, 6550–6557. [CrossRef] [PubMed]
98. Lin, V.S.; Lippert, A.R.; Chang, C.J. Azide-Based Fluorescent Probes. In *Methods in Enzymology*; Elsevier: Amsterdam, The Netherlands, 2015; Volume 554, pp. 63–80. ISBN 978-0-12-801512-4.

MDPI AG
Grosspeteranlage 5
4052 Basel
Switzerland
Tel.: +41 61 683 77 34

Molecules Editorial Office
E-mail: molecules@mdpi.com
www.mdpi.com/journal/molecules

www.ingramcontent.com/pod-product-compliance
Lightning Source LLC
Chambersburg PA
CBHW041412210926
51850CB00013B/3256